Viruses and Man: A History of Interactions

Milton W. Taylor

Viruses and Man: A History of Interactions

Springer

Milton W. Taylor
Indiana University
Bloomington, IN
USA

ISBN 978-3-319-07757-4 ISBN 978-3-319-07758-1 (eBook)
DOI 10.1007/978-3-319-07758-1
Springer Cham Heidelberg New York Dordrecht London

Library of Congress Control Number: 2014940749

Printed on acid-free paper

Springer is part of Springer Science+Business Media (www.springer.com)

I dedicate this book to my wife, Mimi,
without whom I could not have succeeded,
and to my sons: Yuval, Jonathan, and
grandchildren, Thalia, Jacob, Rina and Alex

Preface

The idea for this book came from a class that I taught at Indiana University in the years 2005–2008. The class, "Viruses and human history: the effect of viruses on human society," was intended for freshman non-biology majors, and fulfilled a science requirement. Michael Oldstone's *Viruses, Plagues and History* was the assigned text. Although I considered this an excellent text, some students (specifically those with no science background) found it difficult and too detailed. I initially introduced material on molecular biology, including nucleic acid structure and protein structure as simply as I could in class, but since a proportion of the students were from the business or music school, this, too, proved too difficult. In addition, many students had no chemistry background. I therefore taught the course with minimal references to Molecular biology and chemistry—a difficult task for someone who had taught both graduate and undergraduate courses to virology majors for many years. I have used the same approach in this book—a minimal amount of chemistry—but I do describe viral structure and viral cell interaction.

I had originally thought of dividing the book into three segments: past, present and future. The "future" would not be predictions of new epidemics, rather the use of viruses as tools to cure disease, viral vectors carrying genes to cure genetic diseases, viral treatment of tumors (viral oncolysis), and phage therapy against antibiotic-resistant bacteria. However the "past" and "present" are difficult to separate. What is past in the USA and the Western World is unfortunately present in the Third World. The development of vaccines has eliminated most common childhood diseases in "our world," but there are still millions dying from these diseases in Africa and parts of Asia. We have conquered smallpox, and almost wiped out polio, but millions are still dying from diseases such as measles, hepatitis A, hepatitis B and other viruses in the rest of the world.

In the course of my research for this book, I was struck by the effect of infectious disease on armies and military campaigns. This should not have come as a surprise, since in the military, the conditions are such that hundreds of thousands of men live together, in crowded barracks with little sanitation. Not only did smallpox interfere with military campaigns during the American War of

Independence and the American Civil War, but other diseases—measles, yellow fever, and hepatitis A (jaundice)—affected the military during World War II. One historian claims that jaundice forced the U.S. to alter its plans for D-Day. The great influenza epidemic of 1918 spread due to the proximity of soldiers to each other in the trenches, and, as pointed out by Sir Martin Gilbert, by the celebrations that accompanied the victory at the end of the war see http://virus.stanford.edu/uda/ and Martin Gilbert.

The introductory chapter covers the development of the germ theory and history of infectious disease until the early 1900s. Viruses were elusive, unseen and "imaginary" entities. Chapters on the history of virology follow it in more detail, as do the development of cell culture (without which modern virology would not exist), the discovery of bacteriophage, and the immense contribution such organisms made to molecular biology. There is also a brief history of immunology.

This last chapter was one of the hardest to write, since modern immunology is unbelievably complex, and in order to understand it one needs knowledge of cell biology, signaling mechanism, the functions of various types of cells, biochemistry, etc. I have concentrated on the early history of the events leading to clonal selection of antibodies by going back to the original literature. Discussed, albeit briefly, are the roles of each of the immune cell types, as they are known today. It is possible that within 2 or 3 years additional immune cells and cytokines will be discovered by immunologists. Although interferon is now considered to be one of many cytokines produced by the body as a component of the immune system, it has played a prominent role in the history of virology and the direction that viral research has taken. Interferon biology was an area of research that I have been active in for some 25 years; once predicted as a "cure" for virus infections, this expectation unfortunately has not been realized.

Likewise, viruses, for a considerable time during my own career, were thought to be the causative agent of most cancers. Extensive research was done connecting cancer and viruses, and many Nobel Prizes were awarded for this work. However, although most viruses are not directly connected, the oncogene theory and mutational basis of cancer arose from this research. For this reason, I have included a chapter on viruses and cancer, and of course there are viruses that *do* cause cancer, such as Epstein-Barr, hepatitis B and C, retroviruses and papilloma—all of which are discussed in separate chapters.

Continuing sequentially, I have described in some detail the history of smallpox epidemics and its impact on society; the story of yellow fever and its profound effect on American history; polio—a disease of the affluent as well as the poor; herpes viruses and hepatitis viruses, both of which are modern-day scourges; and the great influenza epidemic with the amazing story of how this virus was reconstructed after 75 years.

I have attempted to bring the HIV and AIDS story up to date and present the current situation. Much of it is troubling, with society biases, problems with blood supplies, and denial by scientists as well as government bodies of the connection between HIV and AIDS—and the personal arguments over the discovery of HIV.

However, there is a bright side to the AIDS story, in that thanks to modern-day medications, although expensive, AIDS is no longer a fatal disease.

Under emerging viruses, I discuss those viruses that are relatively new to man. They are not "new" viruses appearing out of nowhere; rather, they are viruses of other animals—usually bats—that accidentally infect man. They are only new in the sense that man has become an accidental victim and host by encroaching on new territory and, in some cases, altering the environment.

It is impossible in a text of this kind to cover all the classes of viruses. I have selected those viruses where there is an interesting story attached to their discovery or vaccine development, with all of them having caused immense suffering at one time. I have also confined myself to viruses that attack man, and not other vertebrates, fish, or plants. I have included bacterial viruses, since their discovery led to the molecular biology revolution.

One of the most successful chapters in human history has been the development of vaccines against most childhood illnesses. New vaccines are being introduced and licensed in the U.S. and subsequently in the rest of the world at a surprisingly high pace. In my own lifetime, vaccines have been introduced for all the childhood illnesses I was aware of while growing up: measles, chickenpox, mumps, whooping cough, and diphtheria, to list just a few. Within the last few years, vaccines have been developed against hepatitis A, hepatitis B, and papilloma viruses. With the advent of commercially available, cheap nucleic acid synthesis and sequencing and the technological know-how of synthesizing a virus in vitro, it should be possible to develop vaccines, either killed or attenuated against most viral infections.

As to the future, viruses are now being clinically tested in the U.S., U.K., and China as tools against deadly cancers. Brain tumors are among the most untreatable cancers, and a positive diagnosis of such is a death sentence. Today, adenovirus and herpesvirus vectors are being used as a mode of treatment in China and are being tested elsewhere. Viral treatment does not have the side effects found with radiation therapy or chemotherapy, and initial results look promising. I have delved into the background of viral oncolysis, an old idea that is now at the forefront of research. Another old idea being revisited is that of phage therapy, selecting bacteriophage to target specific antibiotic-resistant strains of bacteria, or using bacteriophage in the food industry to destroy salmonella contaminants. The history of d'Herelle and phage therapy is a fascinating one, and has been told in detail in two publications; Summers WC: *Felix d'Herelle and the Origins of Molecular Biology*, Yale University Press, 1999, and Hausler T: *Viruses versus Superbugs*, Macmillan, 2006. I cover this history only briefly here, and I recommend these texts for more details.

In the future we will see more and more viruses being constructed outside the cell, either being made in cell-free solutions or by adding DNA to cells in culture. A dangerous aspect of this research is the alteration of a wide range of viruses, which is a daring experiment and will lead to publication, but may be hazardous to the public.

Finally, I close with a chapter on biological weapons. Again, this is not a new topic, since man has always looked for the most efficient way to kill other humans. In the twentieth century this was developed to perfection with the atom bomb, gas chambers, and the development of tons of smallpox ready to be used in a global war. Luckily, this last never happened.

In most cases, I have gone to the available primary literature. This book would have been much more difficult to write had it not been for the Internet and access to information on Wikipedia, the Centers for Disease Control (CDC), and the World Health Organization (WHO). All three of these Internet sites have astonishing amounts of information; wherever possible, I checked their facts against primary sources.

This book is being written in a new era of information technology. In earlier days, I would have spent my time in the library looking for material among the stacks. In other times, I would have traveled to museums and repositories to search for archival material. With the availability of various search engines and access to downloading from the Indiana University Library, I have been able to write this book from home. The Indiana University Library has to be commended for easy access to most journals, and to the ease of obtaining books through interlibrary loans; even the rarest material is available.

I wish to thank those who spent time reading and reviewing a few chapters; they include Dr. Maurizio Mirolli, Professor Emeritus of the IU Medical Sciences program, and my colleague Dr. Pranav Danthi, friends Phyllis Guskin, Laura Kao, and Luca Peragallo. I would also like to thank all those who gave permission to use their illustrations, many of which are on the Internet. I also thank my wife, Mimi, for her patience in spending time at home while I was writing, for putting up with my occasional frustrations, and for her encouragement to complete the book.

I of course also thank Andrea Schlitzberger of Springer for handling the manuscript and for her useful suggestions.

I am grateful to those who gave me permission to use their excellent figures and photographs of viruses.

Contents

About the Author

Milton Taylor, Professor Emeritus, Department of Biology, Indiana University has taught Virology and a Hands-on Laboratory course in cell culture and virological techniques to senior and graduate students for 40 years. He also taught a course on Viruses and Human History to non-majors for 5 years before retiring in 2010.

Dr. Taylor has over 200 publications in the areas of molecular genetics, virology, viral oncolysis, interferon, and hepatitis C. He was the recipient of many awards and grants, both from the USPHS and private foundations and companies. He was awarded Fogarty International Fellowships from the USPHS to pursue research in the United Kingdom and Italy, and was a recipient of a fellowship from the Weizmann Institute of Science, Israel.

Dr. Taylor received his B.S. from Cornell University and Ph.D. from Stanford University in Biology.

Introduction: A Short History of Virology

1

Abstract

Viral infections have been recorded unknowingly from the beginning of recorded history. The ancient Greeks and Romans described plagues of unknown origin. In general, infections were blamed on sins and punishment, balances of "vital humors" or on "miasma," (rotten smells). In the sixteenth century, Girolamo Fracastoro suggested that infectious agents might spread disease, as did Agostino Bassi, studying diseases of silkworms. The development of the microscope by Hook and Van Leeuwenhoek in the seventeenth century led to the discovery of a new living world inhabited by very small creatures. Edward Jenner in England demonstrated that smallpox, a dreaded disease, could be prevented by inoculation with an organism that caused pockmarks on cows and dairymaids; this was the beginning of the concept of vaccination. Louis Pasteur proved that fermentation only occurred in the presence of air and was due to microorganisms. Pasteur and Pierre Roux, a colleague, as well as Jacob Henle and Robert Koch, proved that germs caused bacterial diseases such as anthrax and tuberculosis; Pasteur and Roux developed a vaccine against rabies by passaging the infectious material through rabbits. By the end of the nineteenth century it had been established that most infectious diseases were the result of germs. In parallel with this research, plant scientists had isolated material that passed through a low pore filter that was infectious to tobacco plants. This was called a "virus," from the Latin for poison. Viruses were also found to be associated with leukemia and other cancers of chickens. The twentieth century saw the discovery of bacteriophage, viruses that attack bacteria, and the use of such bacteriophage to launch studies of molecular biology, and DNA and RNA structure.

M. W. Taylor, *Viruses and Man: A History of Interactions*,
DOI: 10.1007/978-3-319-07758-1_1,

1.1 What Is a Virus?

Although historians identified pandemic diseases, which spread around the world, the agents responsible for these were not recognized until the early and mid-twentieth century. Some of these pandemics, such as the Black Death, were caused by bacteria, and others by viruses such as smallpox. The word virus comes from the Latin meaning poison or slimy matter, and entered the English language to signify infectious disease.

Even today there is discussion as to whether viruses are "living" entities. Viruses are so small that they are invisible even at high magnification with a light microscope. Viruses cannot be grown independently in culture as can bacteria and other free-living microorganisms; viruses require a living cell for reproduction, whether animal, bacterial or plant. A virus contains genetic information—DNA or RNA—yet the statement that a virus is alive is a controversial one. In the 1930s, Wendell Stanley was able to form crystals in solution of the tobacco mosaic virus, a feat that had been performed up until then only with pure chemical molecules infinitesimally smaller than a virus. Viruses form aggregates (clumping) in solution and bond chemically with each other to form a crystal. Many different viruses have been crystallized, and their basic shape and composition revealed. Viruses are at the edge of life, having attributes of inert objects, such as the ability to form crystals, yet also possessing characteristics associated with life, such as genetic material and reproduction. One virus can replicate and give rise to millions more, which is certainly an attribute of life. A virus exists in two stages, one as an inert chemical with a chemical formula; for example, the chemical formula for poliovirus is C332,652 H492,388 N98,245 O131,196 P7,501 S2,340 [1]; when the virus enters a living cell, it becomes an active replicating organism.

Viruses are genetic parasites—that is, they require the protein synthetic machinery of the host to reproduce. Viruses, as visualized by the electron microscope, are symmetrical and complicated structures. Modern-day computers and electron microscopy allow us to produce fantastic pictures of viruses (Figs. 1.1, 1.2).

1.2 Plagues Before the Germ Theory

The concept of "germs"—something invisible that can cause illness—is a quite recent one. The germ theory is in reality a product of the nineteenth century; until then, disease was thought of as either a divine intervention and punishment, or the result of "miasmas" (bad air, stench of rotting material), or on the balance of four vital humors: blood, phlegm, black bile, or yellow bile. The theory of "in balance in humors" (body fluids) dates back to the ancient Greeks or even earlier. Currently, some individuals, and even whole societies, believe that AIDS and other diseases are the result of divine punishment for specific lifestyles. The idea of miasmas being responsible for disease lasted until the early nineteenth century.

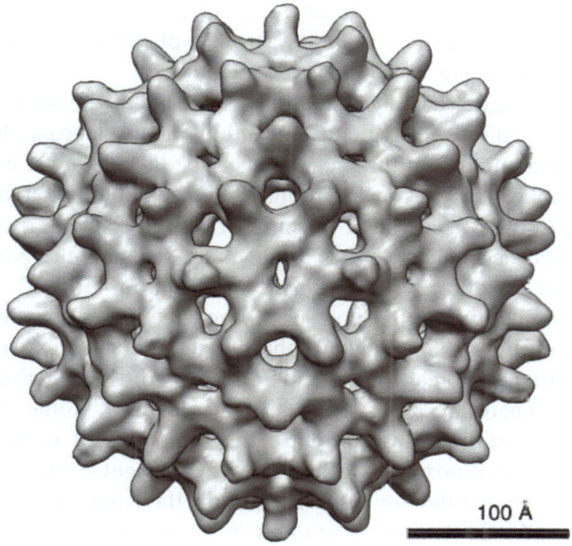

Fig. 1.1 Electron micrograph of hepatitis B virus capsid (courtesy of Adam Zlotnik)

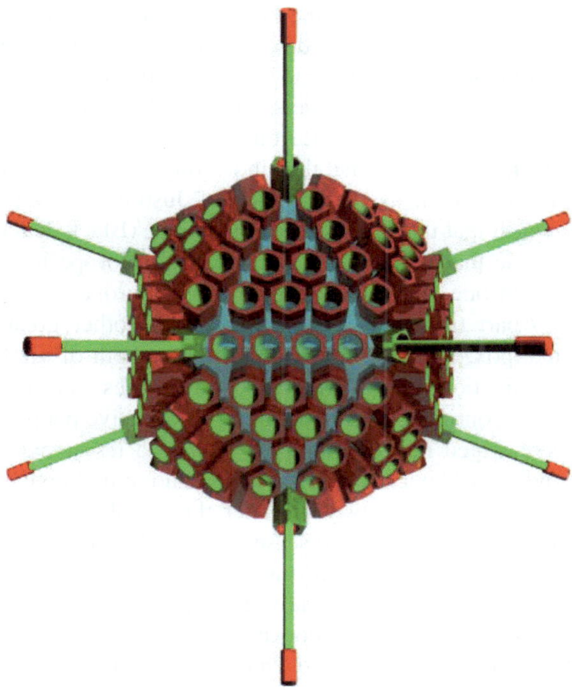

Fig. 1.2 Artist's rendition of adenovirus (Cronodon.com, with permission)

In the eighteenth century, the smell of rotting coffee beans was considered the causative agent by some, as an explanation for the Yellow Fever epidemic in Philadelphia.

Plagues have been recorded since the beginning of written history. Certainly the Bible discusses hygienic precautions against leprosy, the isolation of lepers (a kind of quarantine) and the precautions to be taken against infection by lepers. Two of the biblical plagues of Egypt may have been bubonic plague (boils), and rinderpest of cattle, still prevalent in parts of Africa today. The ancient Greeks and Romans recorded many plagues. Thucydides recounts the great plague that swept through Athens during the Peloponnesian Wars in 430 B.C.: “The dead lay unburied, the temples were full of corpses. Dying wretches were gathered at every fountain, seeking to relieve their unquenchable thirst.” Thucydides writes that the plague began in Ethiopia, spread throughout Egypt, over the Persian Empire, and then reached Athens.

There are reports of a similar plague in Rome at about the same time. From Thucydides's description of the symptoms, it is difficult to discern whether this was smallpox or measles in a very virulent form or, as has been suggested, some bacterial infection such as salmonella typhimurium, or Yesinia pestis, the black death. He describes the onset of fever, sneezing and coughing, with the discharge of blood and a fetid smell, followed by the appearance of ulcers and small pustules on the skin. This was followed by diarrhea and death. Thucydides claims that birds of prey that devoured the bodies also died, and that the disease could spread to domestic animals. His description of the effect on physicians and others who nursed the sick is similar to our descriptions today of outbreaks of dreaded diseases such as hemorrhagic fevers, Ebola and dengue.

Thucydides also writes that those who recovered from the disease were rarely affected a second time, thus indicating a recognition of immunity. In 542 A.D., the plague known as the “Plague of Justinian” hit Rome. The description would suggest that this was the bubonic plague (Black death—*Yersinia pestis*), a bacterial disease that periodically swept through Europe in the late Middle Ages and decimated one-third of the population of Europe.

Apart from the bubonic plague, the other great plagues affecting man were smallpox, and, later on, yellow fever—both of viral origin. With the Europeans' colonization of the Americas, viral plagues were important in shaping the history of the continent. Whole populations of native peoples of North and South America were wiped out by smallpox, and, to a lesser extent, by measles. When Cortes landed in Mexico in 1519, the population of Mexico was estimated to be between 25–50 million. Fifty years later the population was only 3 million. Most of the ravages were due to diseases to which Europeans were semi-immune, not the superior forces of the Spaniards.

The fact that viruses have been afflicting humans since the beginning of written history is indisputable. Pockmarks similar to those of smallpox have been found on the face of a mummified Pharaoh recovered from the pyramids of Egypt. Some drawings on the tombs from the same period indicate wasting of limbs that could be due to poliovirus. It has been suggested that viral infection of humans occurred as a result of the transition from a hunter-gatherer society to a settled agricultural society with the domestication of other mammalian species and the transfer of viruses from these animals to humans. The primitive hunter may have been a healthier person than the settled "gatherer," since he lived a more isolated life, without any crowding. The proximity to domesticated animals and groups of people living under one roof probably resulted in the introduction of new infections to the human population. Examples of this have occurred with human infections in the last few years, with the spread of avian flu, SARS, and HIV— all from other animals.

1.3 Development of the Microscope and Germ Theory

The development of a simple light microscope in the mid-seventeenth century by Robert Hooke (1635–1703) in England and Antonie van Leeuwenhoek (1632–1723) in Holland gave rise to the identification of "animalcules," microorganisms of various shapes, such as bacterial bacilli (rods) and cocci (spheres), and even spirochetes (twisted structures with flagella—tail-like appendages) reviewed in [2]. Hook described the basic cell, using slices of cork under his crude microscope. Leeuwenhoek explored the contents of many substances, including rainwater collected in barrels, flora from the mouth, semen, etc. He described tiny organisms using his microscopes that could magnify up to 300 times. These observations opened up a previously invisible world to science and allowed advances in many areas beyond diseases. An entire invisible world existed but it did not, however, include viruses, which are too small to be seen even with the best light microscope (Figs. 1.3, 1.4).

The first person to really describe diseases as caused by infectious agents was an Italian physician, Girolamo Fracastoro (1478–1553) (Fig. 1.5). He was an acute observer of the plague in Verona and he published many books on syphilis, a new disease of the time, as well as the basis of its contagion. Not only was he a physician, he was also a poet. The name "syphilis" is derived from Fracastoro's 1530 poem in three books, *Syphilis sive morbus gallicus* ("Syphilis or The French Disease"), about a shepherd named Syphilis, who angered the god Apollo and was thus cursed with this disease. (http://en.wikipedia.org/wiki/Girolamo_Fracastoro). The poem goes on to describe the symptoms of the disease.

> A shepherd once (distrust not ancient fame)
> Possest these Downs,
> and Syphilus his Name;
> Some destin'd Head t'attone the Crimes of all,
> On Syphilus the dreadful
> Lot did fall.
> Through what adventures this unknown Disease
> So lately did astonisht Europe seize,
> Through Asian coasts and
> Libyan Cities ran,
> And from what Seeds the Malady began,
> Took from France his name.
> Our Song shall tell: to Naples first it came
> From France, and justly named.

Fig. 1.3 The Leeuwenhoek microscope

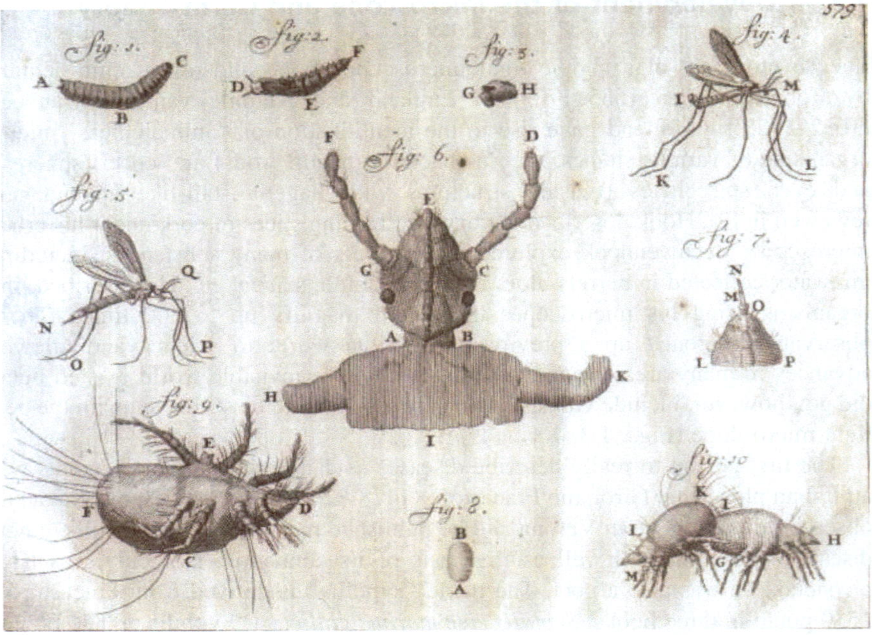

Fig. 1.4 Drawings of protozoa and other "animalcules" from Leuwenhoek (image from Clendening Library, U. Kansas Medical Center)

Fig. 1.5 Portrait of Girolamo Fracastoro by Titian, National Gallery, London, U.K. painted 1528

Fracastoro discussed the spread of disease that resulted from touching contaminated clothing (fomites) as well as from contact among individuals. He also discussed how various diseases infected and affected various age groups in a population and in various ways. Syphilis was later shown to be a bacterial disease and was called the "great pox," as opposed to smallpox, the virus disease, although smallpox was the more deadly of the two [3].

Another pioneer of the germ theory was Agostino Bassi (1773–1856), an Italian lawyer and later agriculturist. He devoted a large part of his life to studying muscardine, a disease of the silkworm. Muscardine was an important fungal pest of the economically important silk industry of Northern Italy. Many farms were abandoned because of this disease. Bassi concluded from his studies that an infectious agent that could be transmitted by contact between worms, and also could be transmitted by contaminated cages or utensils, caused the disease. He identified the infectious agent as a fungus, later named *Beauvaria bassiana*. He is also credited with rescuing the vitally important silk industry with recommendations such as the use of disinfectants; separating the rows of feeding caterpillars; isolating and destroying infected caterpillars; and keeping the silk farms clean. He used "semi-sterile" techniques to prevent infection. Pasteur and others who came after him were unaware of his research and made similar, independent discoveries. Bassi also proposed that human diseases such as syphilis and measles were transmitted from one person to another by contact. Both Fracastoro and Bassi have been overlooked in the history of medicine, whereas Pasteur has become a household name [2].

1.4 Vaccination in the Time of Jenner

Vaccine development began with attempts to prevent smallpox infections using variolation (extracting pus from smallpox lesions) and rubbing the pus onto the arm or leg of an uninfected person. Variolation was first performed in China and later in the Middle East in the fifteenth and sixteenth centuries. The practice of variolation was introduced into England by Lady Mary Montagu, wife of the British ambassador to Turkey. She had discovered the use of variolation during her sojourn in Turkey in 1716–1718; variolation was widely practiced, but often resulted in cases of smallpox, although not as severe as a natural infection.

Dr. Edward Jenner (1749–1823), a practicing physician in Berkeley, England, had noted, as had others, that milkmaids very rarely contracted smallpox. Instead, their hands often had sores later shown to be caused by the cowpox virus (vaccinia) (Fig. 1.6). Jenner had a small laboratory not far from Berkeley Castle in northern England, where he did experiments.

Jenner was not the first to recognize the connection between cowpox and immunity to smallpox. A number of farmers had previously made this observation. For example, Benjamin Jesty, a farmer in Dorset (southern England), had inoculated his wife and children with cowpox material during an outbreak in

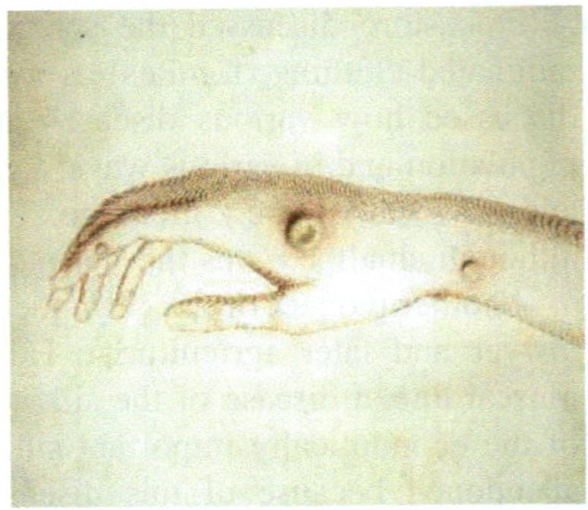

Fig. 1.6 Hand with cowpox (from Jenner, history of vaccines)

1774 and, I assume, induced immunity in the family. In 1796 Jenner tested his hypothesis that the pus from the wounds that occurred on the udder of milking cows contained material that could protect against smallpox by inoculating James Phipps, the son of his gardener, with pus from a wound on the hand of his milkmaid, Sarah Nelmes, who had been infected by a virus from a cow named Blossom. Blossom is probably one of the most famous cows in history, her hide immortalized in St George's Hospital Medical School in London. James was then variolated with the smallpox virus a number of times and had basically no reaction, showing complete immunity. Jenner then tested vaccination on 23 other people, none of whom came down with smallpox. He presented his data to the Royal Society and published his data [4]. His idea of preventing smallpox with a "vaccine" was not easily accepted on religious grounds ("interfering with the work of God"); also ridiculed was the use of material from a cow (see cartoon, Fig. 1.7) (see Chap. 19, Vaccines).

In this cartoon, the British satirist James Gillray drew a caricature of a scene at the Smallpox and Inoculation Hospital at St Pancras, showing Edward Jenner administering cowpox vaccine to frightened young women, and cows emerging from different parts of people's bodies. The cartoon was inspired by the controversy over inoculating against the dreaded disease—smallpox. The inoculation agent, cowpox vaccine, was rumored to have the ability to sprout cow-like appendages. A serene Edward Jenner stands amid the crowd. A boy next to him holds a container labeled "VACCINE POCK hot from ye COW"; papers in the boy's pocket are labeled "Benefits of the Vaccine." The tub on the desk next to Jenner is labeled "OPENING MIXTURE," and bottle next to the tub is labeled "VOMIT." The painting on the wall depicts worshippers of the Golden Calf. (from Wikipedia).

Around 1800, Dr. J. Clinch, a medical missionary in Newfoundland, began using Jenner's vaccine and by the end of 1801 he had vaccinated approximately 700 people. This was probably the first vaccination of an entire community in North America. Jenner died in 1823 as the result of a stroke. The British government passed the vaccination act in 1840, making variolation illegal and supplying smallpox vaccine (cowpox) free of charge. In 1853 the vaccination act

Fig. 1.7 1802 caricature of Jenner vaccinating patients who feared it would make them sprout cow-like appendages. The Cow-Pock—or—the wonderful effects of the new inoculation!—*vide*. The publications of the anti-vaccine society. Print (color engraving) published June 12, 1802, by H. Humphrey, St James's Street

was passed, requiring all children to receive smallpox vaccines by the age of 4 months. Later on, the act was changed to allow parents who objected to vaccination to appeal. In 1905 the U.S. Supreme Court ruled that states could impose mandatory vaccination on children.

Smallpox was declared eradicated from the world in 1980, after an intensive challenge to stamp it out. In fact, eradication really began in 1803 when the Spanish crown attempted to vaccinate the entire population of the Philippines. In 1813 the U.S. Congress passed the vaccination act to insure safe vaccines for the U.S. public. By 1832 the U.S. government had set up a program for the vaccination of American Indians; known as the Indian Vaccination Act. Unfortunately it did not eradicate the spread of the virus and was misused to remove Indians to reservations. Thus there was a piecemeal attempt to eradicate the virus in a few areas of the world, such as India and the East Indies. By 1897 smallpox had been eradicated from the US and by the beginning of the twentieth century from most of the industrialized countries of Europe. Despite this, 2,000,000 people died annually from the disease in the 1950s; therefore, a campaign was launched in 1959 to completely eradicate the virus from the hemisphere. Although it took longer than expected, the last cases of smallpox occurred in Somalia in 1977 [5, 6].

1.5 Louis Pasteur and Robert Koch

The two giants of microbiology who changed the history of medicine and led to "modern" clinical medicine were Louis Pasteur (1822–1895) in Paris, and Robert Koch (1843–1910) in Berlin. In 1857–1858 Pasteur made the discovery that put to rest the theory of spontaneous generation of life. He showed that liquids in sealed containers remained sterile, but that exposure to the air resulted in contaminations that gave rise to fermentations or growth of bacteria or fungi. He wrote, "*germs* of microscopic organisms abound in the surface of all objects, in the air, and in water," concluding that these organisms cause fermentation, such as occur in wine, beer, vinegar etc., and that under sterile conditions with no access to air these fermentations do not take place.

About the same time, working in Germany, Robert Koch, a country doctor, described the organism responsible for anthrax, a disease of cattle. He showed that the bacteria had a life cycle, including a stage of spore formation, a stage not easily destroyed by heat or other means, which could account for infections occurring at various periods, and allowed organisms to live in the soil for a long time. Among other achievements, Koch went on to isolate and describe the organism responsible for tuberculosis. Koch was influenced by his mentor, Jacob Henle, who in 1840 proposed that infectious diseases were caused by living organisms and reproduced outside the infected individual. Earlier, Aloys Pollender, Pierre-Francois Olive Rayer and Casamir-Joseph Davaine had discovered the anthrax bacterium. Koch himself set out to prove scientifically that this bacillus is, in fact, the cause of the disease. He inoculated mice with homemade slivers of wood containing anthrax bacilli taken from the spleens of farm animals that had died of the disease, and found that these mice were susceptible to the disease and died, whereas mice inoculated at the same time with blood from the spleens of healthy animals did not. This confirmed the work of others who had shown that the disease could be transmitted by means of the blood of infected animals. He went on to develop methods of growing and staining the bacteria, and with a colleague developed the Petri dish (named after his colleague) that is still used today [7].

> Koch also isolated the bacterium responsible for tuberculosis, and from this developed specific postulates. To prove the causative role of the bacillus
>
> ···the bacilli had to be isolated from the body and cultivated in pure culture until devoid of all adherent products of disease originating from the animal organism; and, finally, through transfer of the isolated bacilli to animals, the same clinical picture of tuberculosis as is obtained empirically by the injection of naturally developed tuberculous material had to be produced.

Koch or Emile Roux (1853–1933) proposed this simplification of his famous postulates:

> An organism must be isolated from the diseased animal.
> It must be possible to grow to a pure culture.
> It must be possible to re-infect the animal.
> It should be possible to re-isolate the organism.

These postulates are still used as the basis for characterizing an infectious agent, but have to be modified in that it is important to emphasize that the disease produced has to be similar to the one from which the material is taken. For example, in using an animal model system, such as the mouse, the disease produced may not be identical in that animal to the one produced in humans. It is also important that the bacteria or virus be found in *all* cases of the particular disease. As will be discussed later, in the case of work with yellow fever, bacteria were isolated from most yellow fever victims (not all, however) and was claimed to be the source of the disease. Only by showing that it was not present in ALL patients could it be proven that it was not the causative agent. It has proven ethically difficult to carry out Koch's postulates in the case of some viral diseases that only infect humans (e.g., human immunodeficiency virus [HIV] and hepatitis C). If no animal model is present, it is difficult to prove beyond a doubt—without human experimentation—that a specific agent is the cause of the disease. Smallpox and measles viruses also are exclusively human pathogens, and the early study of smallpox in particular involved unregulated research on human subjects. In more recent times, and particularly in the era of laboratory-driven investigations, we have had to create very specific ethical guidelines. Many of the discoveries, such as the role of mosquitoes in transmitting yellow fever, and vaccination against smallpox, involved the use of humans in a manner not acceptable today.

In 1878, in a paper read before the French Academy of Sciences, Pasteur presented evidence that disease was caused by discrete germs. He demonstrated this for the anthrax bacterium (*Bacillus anthrax*) isolated as described above by Koch. Furthermore, he showed that minimal numbers of the bacteria are required by serial dilution, growing the new cultures in test tubes and injecting material into animals (see Koch's postulates above). Anthrax today is better known as a possible agent of biological warfare than as a disease of cattle. In 2001, envelopes containing anthrax were sent through the U.S. mail to congressmen and caused quite a panic. It was initially assumed that this was a case of bio-terrorism conducted by foreign terrorists, although now it is suspected that this was the work of a disturbed American scientist.

Pasteur also successfully isolated organisms involved in sepsis, which results in septic shock due to the growth of bacteria and resulting toxins in the blood. During this study he demonstrated differences between anaerobic and aerobic organisms and identified bacterial spores that could give rise to septicemia (blood poisoning). Such spores survive for a long time outside the body. Koch had made a similar finding—that anthrax could exist in a spore form resistant to heat and desiccation. In a paper in 1880, Pasteur described the application of his findings and techniques to other conditions, particularly furuncles (boils). He showed that the pus that developed from such lesions contained infectious chains made of round cells (cocci) and that this organism could also be grown in culture. He showed that every furuncle contained cocci that caused the inflammation and pus formation. These organisms were injected into rabbits and formed abscesses containing pus, and the same organism could be isolated again (Koch's postulate). Of great interest were his observations on puerperal fever, associated with women who had recently given birth or had a miscarriage, and which was extremely fatal, and the isolation of the bacteria that caused this disease [8] (Figs. 1.8, 1.9).

Ignaz Semmelweis (1818–1865) (Fig. 1.10), working independently in Vienna, had shown that puerperal fever could be prevented by handwashing in a chlorine-lime solution and that it was often spread by physicians after autopsy, or following

Fig. 1.8 Pasteur working in his laboratory (painting by A. Edelfeldt)

Fig. 1.9 Robert Koch, circa 1900 (from Wikipedia)

treatment of sick women. Semmelweis was ridiculed for suggesting that doctors, midwives or nurses spread it. Handwashing was not a common practice, and Semmelweis insulted the egos of the medical profession by accusing them of spreading the bacteria. His ideas were accepted only after his death, which occurred in an asylum at the age of 47. Only a few weeks after being confined to the asylum, his guards beat him to death. [8, 9].

It could be shown that the organism causing puerperal fever could be grown in rabbits. In his paper of 1880, Pasteur also proposed the use of antiseptics to kill the infecting organisms. He proposed the use of boric acid rather than the carbolic acid then in use. Boric acid was used in the treatment of urinary tract infections by injection into the bladder. He also proposed placing a pad soaked in boric acid at the bedside of each woman during confinement to prevent puerperal fever. This infection was shown to be caused by streptococci and today is treatable with antibiotics (Fig. 1.10).

Fig. 1.10 Hungarian stamp in honor of Ignaz Semmelweis

1.6 How Can We Differentiate a Virus from a Bacterium?

The differentiation of viruses from other types of disease-causing organisms was initially based on size and filterability. Viruses are 100–1,000 times smaller than bacteria, and this could be determined only when one began to filter cultures from infected individuals, plants or animals. Various types of filters were invented in the nineteenth and early twentieth centuries to purify water and other liquids. These early filters were made of diatomaceous earth from organisms that left a glass shell after they died, and small matter would pass through the filters, but larger organisms, such as most bacteria, fungi, etc., would be retained by the filter. Thus viruses could be defined as "filterable agents." These porcelain filters were developed in the Pasteur group by one of Pasteur's colleagues, Charles Chamberland (1858–1901), and were called "Chamberland filters."

However, it was difficult to differentiate "viruses" from toxins. Work in the lab of Louis Pasteur, performed by his colleagues Emile Roux and Alexander Yersin (1863–1943), identified the toxin that caused diphtheria. Roux and Yersin developed an anti-serum to this toxin, which was used to save the lives of many young adults affected by diphtheria, a fatal disease. Filtration was an important tool in differentiating bacteria that had been retained by the filters from toxins and what were later called viruses. Diphtheria toxin is actually produced by a strain of bacillus containing a bacteriophage (virus). The toxin is produced by the virus.

Until very recently, the concept of a virus being a filterable agent, and much smaller than bacteria, was an accepted dogma of virology. However, within the last few years, viruses that parasitize protozoa and other lower forms of life have been isolated; they are as large as bacteria and contain considerable genetic material. Thus the idea that viruses are filterable, small and contain a small number of genes, is no longer tenable. There appears to be a continuum of size, from the smallest picornavirus to the large Mimivirus and Megavirus, larger than the smallest free-growing bacterium [10].

Emile Roux and Louis Pasteur also developed a vaccine against the rabies virus, a filterable agent. This vaccine was tested on a young boy, Joseph Meister, who was bitten by a rabid dog and was certain to die. He was given many shots of the vaccine into his belly and he survived the attack. There was much controversy at the time, since some considered the treatment unethical because the vaccine had not been properly tested [11]. However, the argument was that without this, the boy would have died. Although Pasteur had carried out vaccine experiments on dogs, the claim was made that this was not the vaccine used in the case of the boy. However, its success led to its adaption as a mode of treatment until very recently and enhanced Pasteur's reputation.

Yersin, a researcher at the Pasteur Institute, went on to discover the bacterium involved in the plague (Black Death) in 1894, initially called *Pasteurella pestis* but renamed *Yersinia pestis* in 1967. He isolated the bacterium from rats and showed that it was identical to the bacterium associated with the bubonic plague in humans. Thus the foundations of bacteriology and immunology were laid without seeing the organisms involved, and without differentiating virus from bacterium.

1.7 Plant Viruses

Pioneering work in microbiology was also carried out on the tobacco mosaic virus (TMV). This was a "germ" that caused areas of variegation (mosaic) and wilting of tobacco plants, a crop of economic importance. Many plant viruses, all species-specific, cause similar symptoms, the most famous of them the virus that causes the variegation of tulips. The pattern that developed as the result of the infection had great economic importance. In fact, it was the rarity of such tulips that led to the market concept of futures, and to the exchange of large sums of money in the seventeenth century in Holland (1634–1637), during a period known as tulipmania. Such tulips did not breed true, and the pattern died out quite suddenly. This caused a financial crash, and may be thought of as the first stock market crash or bursting of a financial bubble [12] (Fig. 1.11).

None of this related historical experience was understood at the time that TMV became a topic of scientific, and perhaps commercial, concern in the 1890s. Adolf Mayer (1843–1942), director of the Agricultural Experimental Station in Wageningen, Holland, first discovered that TMV was an infectious agent. He is credited with being the first person to extract juice (sap) from infected plants and used it as an inoculum to infect other plants. He tried to culture the organism but failed to grow it on media and it was not visible in the then crude microscopes. He concluded wrongly that the agent was a bacterium that lost activity upon filtration [13].

Dmitri Ivanovski (1864–1920), a Russian scientist, also reported that the infectious agent for tobacco mosaic disease was filterable (1892). However, like Mayer, he concluded that there must be something wrong with the filter, or that the agent could not be grown in vitro, or was some kind of spore-forming organism.

Fig. 1.11 A tulip, known as "the Viceroy," displayed in a 1637 Dutch catalog. Its bulb cost between 3,000–4,150 guilders (florins), depending on the size. A skilled craftsman at the time earned about 300 guilders a year. (http://en.wikipedia.org/wiki/Tulip_mania)

Fig. 1.12 Martinus Beijerinck in his laboratory. Delft University Archives, (http://en.wikipedia.org/wiki/File:Mwb_in_lab.JPG)

It was Martinus Beijerink (1851–1931) who made the seminal discovery that the agent was a "contagious living fluid." Since the filters should remove known bacteria, he used the term "virus" (the Latin word for "poison") to describe such fluids. TMV went on to be a very important tool in the study of virology and molecular biology. [13] Beijerink went on to make many major contributions to microbiology, including the discovery of nitrogen fixation, the process by which nitrogen in converted by bacterium (rhizobium) into ammonia, a form that plants can metabolize—a natural organic fertilizer! (Fig. 1.12).

1.8 Human and Animal Viruses

The first animal virus (filterable) shown to be infectious was foot and mouth disease virus (FMDV). This virus has great economic importance, particularly in the dairy industry of western Europe today. While investigating outbreaks of foot and mouth disease virus in cattle in 1897–1898, Friedrich Loeffler (1852–1915), (who had previously worked with Koch and helped form the "Koch Postulates"), together with Paul Frosch (1860–1928), described an agent from the pockmarks of diseased animals that could not be cultivated on the media devised for growing bacteria. This agent was filterable, and could be shown, by dilution, to be infectious. By sequential inoculation from animal to animal and serial dilution, these researchers proved that this was not a toxin but an infectious agent. Thus FMDV resembled TMV in character. By the turn of the twentieth century, it had been established that viruses were infectious agents associated with many diseases [14].

As we will see in a later chapter, among the first viral diseases to be characterized and prevented was yellow fever, identified as an infectious disease in 1900, although it took many years before the actual virus was isolated. This would have to wait until viruses could be grown in cell culture.

1.9 Viruses and Cancer

In parallel with the work described above, major advances were being made in research linking viruses to cancer. The relationship between cancer and viruses was first discovered in an avian species (domestic poultry). Unfortunately, the relevance of disease in non-humans, particularly in non-mammals, to humans was not realized until 50 or 60 years later. We now know, through examples such as "avian flu" or "swine flu," that other animals are stricken with the same vectors as man, can transmit these vectors to man, or act as a reservoir for these diseases. The first reported case of a possible virus related to cancer was the observation in 1908 by two Danish veterinary researchers, Vilhelm Ellerman and Oluf Bang, who found that leukemia of chickens could be transmitted by a cell-free extract. Peyton Rous (1879–1970) at the Rockefeller Institute in New York reported a few years later that not only leukemia but also solid tumors (sarcomas) were transmissible [15].

Peyton Rous came from a poor Texan family, settled in Baltimore, and received a BA from John Hopkins University in 1900. Due to an accident that affected his health, he worked at a Texas ranch. Returning to medical school, he received his MD degree in 1905 and then worked in the Department of Pathology at the University of Michigan, and later in Dresden, Germany. Subsequently he received a grant to do research at the Rockefeller Institute in New York. The director of the Institute, Simon Flexner, who headed the cancer laboratories, asked him to take his place since he, Flexner, wished to devote his time to poliovirus research, and Peyton Rous succeeded him. While working with spontaneous tumors of chickens, he demonstrated in 1910 that a cell-free extract made from the tumor was transmittable, and that different types of tumors arose from different cell-free extracts of different tumors. Thus the work of two groups (Ellerman/Bang and Rous) laid the groundwork for an analysis of the relationship between tumors and viruses.

Rous received the Nobel Prize in Medicine in 1966, many years after his discoveries. Not only did he discover tumor viruses (the Rous sarcoma virus [RSV], the first prototype retrovirus discovered), his research also contributed greatly to methods of storing blood, crucial during World War I. He gave up attempting to isolate mammalian tumor viruses a few years after identifying RSV. It took another 50 years or so before tumor-causing viruses were discovered in mammals.

1.10 Bacteriophage

The fact that all organisms can harbor viruses was confirmed by the discovery of bacterial viruses, known as bacteriophage. These were first discovered by Frederick Twort (1877–1950) [16] in England and by Felix d'Herelle (1873–1949), a Canadian working in France. In 1910 d'Herelle, working in Mexico, was alerted to the presence of dysentery in locusts [17]. He cultured the organism from the insects, and noted during the growth of the bacteria that there were areas of clearing in the culture in Petri dishes; however, he did not realize the importance of the discovery until World War I, when he was working with *shigella*, a bacterium that also caused dysentery in humans. D'Herelle realized then that the areas of clearing were plaques caused by a filterable agent, namely a virus-infecting bacteria [18]. He also noted that the culture of bacteria cleared during this infection so that no bacteria were left (or so it appeared). D'Herelle had hoped that such bacteriophage could be used in therapy against bacterial diseases [19]. This was an active area of research in the 1930s–1980s in Eastern Europe and Russia and today is an area of renewed research. The same idea was presented in Sinclair Lewis's novel *Arrowsmith*, in which a physician experiments with bacteriophage to cure common diseases [20].

While working with cultures of *Staphylococcus aureus* (the bacterium that causes the common boil) in 1915, Frederick Twort noticed that colonies of these bacteria were being destroyed, and, like d'Herelle, he found that clear plaques were formed. He isolated the substance that produced this effect and found that it was transmitted indefinitely to subsequent generations of the bacterium. He then suggested that the substance was a virus. Twort was unable to continue this work, and the importance of bacteriophage was not recognized until the 1950s, when it became a major tool of molecular biologists and revolutionized the whole field of biology. In chapters on bacteriophage (Chaps. 4 and 16), I describe the controversy surrounding the discovery of bacteriophage, its possible use in the treatment of bacterial infections, and the role bacteriophage plays in molecular biology.

1.11 Modern Molecular Biology

An understanding of the molecular organization of viruses began with the crystallization of the, tobacco mosaic virus by Wendell Stanley in 1934. Working with highly purified TMV, he was able to identify conditions where virus particles are arranged in a lattice, so that their molecular features can be elucidated by the diffraction of X-ray beams. Wendell Stanley received the Nobel Prize in Chemistry in 1946 for his seminal work. Even at that time it was still believed that proteins were the genetic material, although the virus did contain some nucleic acids. It was not until further work with bacteriophage, and simultaneously with bacteria, that it could be shown that the nucleic acid contained the genetic information. These were experiments involving DNA transformation in bacillus [21], and in later experiments with bacteriophage, it could be shown, using radio-labeled material, that the DNA of the bacteriophage entered the bacterial cells and replicated and produced progeny phage and not the protein [22] that remained attached to the outside of the cell (Fig. 1.13).

The advent of modern molecular biology came about with the analysis of bacteriophage growth and replication. Using bacteriophages such as T2 and lambda, it was possible to dissect each stage of virus replication, the role of nucleic acid in the process, quantify the appearance of and the mechanism of mutation, and develop a model from phage for tumor development, based on the fundamental observation that some phage genetic material was integrated into the bacterial genome. These early studies with bacteriophage, and the parallel studies

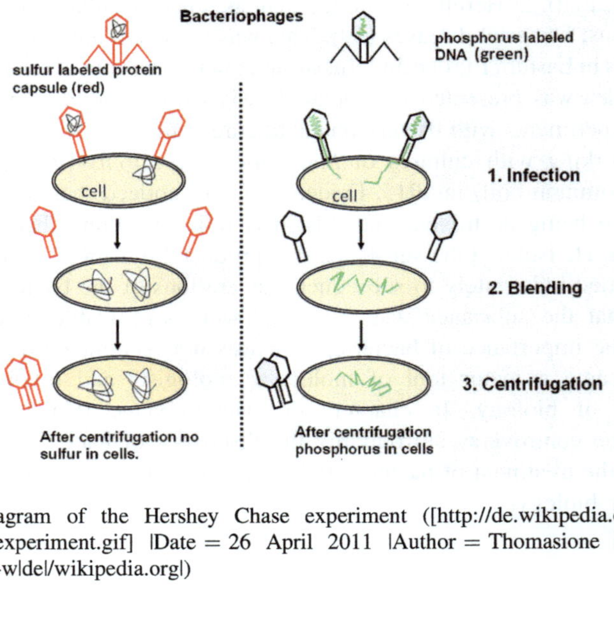

Fig. 1.13 Diagram of the Hershey Chase experiment ([http://de.wikipedia.org/wiki/Datei:Hersheychaseexperiment.gif] |Date = 26 April 2011 |Author = Thomasione |Permission = {{GFDL-user-w|de|/wikipedia.org|)

with cell culture and the growth of viruses, allowed for the quantization of virus growth. The number of scientists involved is too many to list, but all the principal players constituted the phage group, headed by Max Delbruck, that met at the Cold Spring Harbor Laboratories in New York. They were later joined by James Watson and Francis Crick of DNA double helix fame.

1.12 Viral Studies in the 1950s

The great virus scourge of the 1940s and 1950s was poliovirus (see Chaps. 3 and 12 on cell culture and poliovirus). A campaign in the United States, spearheaded by the March of Dimes, was determined to find a vaccine against polio. Based on the earlier work of John Franklin Enders and others, in 1953 Jonas Salk announced the achievement of a polio vaccine [23]. Salk became not only a celebrated scientist, but also a hero. His vaccine was based on a "dead" virus, which was followed two years later by a rival vaccine, the Sabin vaccine, based on a series of attenuated (weakened) viruses [24]. Both vaccines have their pros and cons, and both are still used today, although in the U.S. the Salk vaccine is considered safer since it does not involve viruses that can reproduce. However, the introduction of these vaccines was not without risk since the first batch of commercially produced Salk vaccine contained some improperly inactivated virus particles.

As a result of these vaccines, polio has almost been wiped out, although hopes to completely eradicate the virus have recently met with setbacks. Those carrying out the eradication campaign have been murdered in Pakistan, and in other countries the anti-polio inoculation campaign has been viewed as a political or cultural threat. Another oral vaccine was developed by Hilary Koprowski and was used in Africa, but not in the U.S. [25]. This later led to a controversy as to whether the AIDS virus (HIV) arose from the chimpanzee cells used to grow the polioviruses, and was thus introduced into the human population.

A major mystery surrounding viruses was (and to some extent still is) their involvement in cancer. Rous's sarcoma virus, and later other viruses associated with cancer in mice, were shown to be RNA viruses (at one time it was thought that most viruses were DNA viruses). In culture, these viruses were inhibited by Actinomycin D, a drug known to inhibit DNA- dependent RNA polymerase, the enzyme involved in mRNA synthesis from DNA, suggesting that these RNA viruses had a DNA stage. Although this was difficult to believe, eventually two groups in 1970 independently discovered an enzyme in infected cells (but not in uninfected cells) that copied RNA into DNA by a complicated mechanism. This was "reverse transcriptase," and the name "retroviruses" was given to this class of virus. The two labs involved were those of David Baltimore and Howard Temin, classmates from Swarthmore College in Pennsylvania [26, 27]. Each made independently parallel discoveries and shared the Nobel Prize for their work. The discovery of this enzyme opened the way to studies of cancer viruses, some of which could be linked to human leukemia (HTLV-1) and later on to HIV, the virus causing AIDS.

The 1970s to the present was an era of major discoveries and technological break-throughs, and most of the new advances were due to the emergence of molecular biology. The technological advancement is DNA sequencing that allowed the sequencing of the DNA and RNA of thousands of viruses. DNA sequencing can now be done very quickly and inexpensively. The development of the "polymerase chain reaction" (PCR) revolutionized the identification of specific genes, the measurement of gene expression, and virus identification, because nucleic acids that were previously in short supply can be amplified and studied carefully. Specific antibodies could be made in cultured cells (hybridomas) using cell culture techniques. New emerging viruses were discovered and linked to specific diseases or plagues (such as Ebola, HIV, Nypah, Hantan, all of which are known as "emerging viruses").

All of these topics will be covered in more detail in their respective chapters.
What is the situation at the time of this writing?

Smallpox. This is the only viral disease that has been wiped out worldwide. However, there is still the fear that this virus will be used in biological warfare, or by terrorists; therefore, stockpiles of the vaccine are necessary, and new methods of producing the vaccine in culture are necessary for stockpiling. Although vaccination has stopped, there are now calls for it to be reintroduced.

Measles. An effective vaccine has been available since 1963; this disease could be eliminated with a worldwide effort. However, false rumors based on poor science relating the vaccine to autism inhibited the vaccination program for some time. The linkage to autism was proven to be false.

Influenza. Strain-specific vaccines are available, but new variant strains emerge periodically and there are significant animal reservoirs worldwide (aquatic birds). Our battle with the flu, including pandemic flu, will be an ongoing one for the foreseeable future. New vaccines are produced each year based on predictions, and newer technologies, such as cloning of specific viral genes resulting in newer vaccines, are being clinically tested.

Polio. Effective vaccines are available. Scientists are optimistic that this will be the second viral disease to be completely wiped out. However, there are cultural problems and resistance to vaccination in certain parts of the world. Efforts are currently focused on replacing the Sabin vaccine with the Salk vaccine in hopes of removing attenuated poliovirus, although the oral Sabin vaccine has a "herd" effect and may thus spread in a population and be more useful in Third World countries. Vaccinating has been held up by civil wars and mistrust of the motives for vaccination.

HIV. A massive effort to develop vaccines is under way. There are effective drugs that can be costly and toxic, but they do allow for a normal life span and chronic condition. Worldwide spreading continues via sexual contact and use of recreational drugs. An estimated 250,000,000 people have been infected thus far, with over 30 million deaths. However, with changes in sexual behavior in some parts of the world, the use of condoms, and the encouragement of male circumcision, this disease is on the decline. With time, anti-retroviral drugs will become more available and cheaper. In the U.S., the disease is now often considered a chronic infection that gives a feeling of false security, that the disease is no longer dangerous of lethal.

Ebola. No vaccine currently exists. There is a need to identify the host reservoir, possibly bats. A vaccine is being developed.

Hepatitis C. This is now considered the number one challenge in the U.S. and other developed countries. Spread by the use of shared needles among drug users, it takes a long time, perhaps 10–20 years, before the symptoms of the disease are recognized; it can lead to liver cancer.

New viruses are being discovered and entering the human population continuously. As man moves into new environments, or changes social mores, new viruses are likely to exploit these situations and lead to new pandemics.

References

1. Wimmer, E., & Paul, A. V. (2011). Synthetic poliovirus and other designer viruses: What have we learned from them? *Annual Review of Microbiology, 65*, 583–609.
2. Gest, H. (2003). *Microbes: An invisible universe*. Washington, DC: ASM press.
3. Barquet, N., & Domingo, P. (1997). Smallpox: The triumph over the most terrible of the ministers of death. *Annals of Internal Medicine, 127*(8 Pt 1), 635–642.
4. Riedel, S. (2005). Edward Jenner and the history of smallpox and vaccination. *Proceedings, Baylor University Medical Center 18*(1), 21–25.
5. Fenner, F., Henderson, D. A, Arita, I, Jezek, Z., & Ladnyi, I. D. (1998). History of Smallpox and its spread around the world. *History of International Public Health* 1988, vol. 6, Chap. 5.
6. Tucker, J. B. (2001). *Scourge. the Once and Future Threat of Smallpox*. New York: Atlantic Monthly Press.
7. Brock, T. D. (1988). *Robert Koch: A life in medicine and bacteriology*. Washington, DC: ASM press.
8. Pasteur, L. (1880). On the extension of the germ theory to the etiology of certain common diseases. *Comptes Rendus de l'Academie des Sciences Paris, 60*, 1033–1044.
9. Semmelweis, I. (1983). *Excerpted from: The etiology, concept and prophylaxis of childbed fever*. vol. 1, Madison: University of Wisconsin Press.
10. Raoult, D., Audic, S., Robert, C., Abergel, C., Renesto, P., Ogata, H., et al. (2004). The 1.2-megabase genome sequence of Mimivirus. *Science, 306*(5700), 1344–1350.
11. Geison, G. L. (1990). Pasteur, Roux, and rabies: scientific versus clinical mentalities. *Journal of the History of Medicine and Allied Sciences, 45*(3), 341–365.
12. Goldgar, A. (2007). *Tulipmania: Money, honor and knowledge in the Dutch golden age*. Chicago: University of Chicago Press.
13. Lustig, A., & Levine, A. J. (1992). One hundred years of virology. *Journal of Virology 66*(8), 4629–4631.
14. Taylor, M. W. (2003). M430/L105 virology. http://courses.bio.indiana.edu/M430-Taylor/history.html.
15. Rous, P. (1910). A transmissible avian neoplasm. (Sarcoma of the common fowl). *Journal of Experimental Medicine, 12*(5), 696–705.
16. Twort, F. W., & Mellanby, E. (1912). On creatin-destroying bacilli in the intestine, and their isolation. *The Journal of Physiology, 44*(1–2), 43–49.

17. d'Herelle, F. (1917). An invisible microbe that is antagonistic to the dysentery bacillus. *Comptes Rendus de l'Academie des Sciences Paris, 165*, 373–375.
18. Fld, Hérelle. (1926). *The Bacteriophage and Its Behavior*. Baltimore, Md: The Williams & Wilkins Company.
19. d'Herelle, F. (1931). Bacteriophage as a treatment in acute medical and surgical infections. *Bulletin of the New York Academy of Medicine, 7*(5), 329–348.
20. Lewis, S. (1925) Arrowsmith. http://courses.bio.indiana.edu/M430-Taylor/history.html.
21. Avery, O. T., Macleod, C. M., & McCarty, M. (1944). Studies on the chemical nature of the substance inducing transformation of pneumococcal types: Induction of transformation by a deoxyribonucleic acid fraction isolated from pneumococcus type Iii. *Journal of Experimental Medicine, 79*(2), 137–158.
22. Hershey, A. D., & Chase, M. (1952). Independent functions of viral protein and nucleic acid in growth of bacteriophage. *Journal of General Physiology, 36*(1), 39–56.
23. Salk, J. E., Bazeley, P. L., Bennett, B. L., Krech, U., Lewis, L. J., Ward, E. N., et al. (1954). Studies in human subjects on active immunization against poliomyelitis. II. A practical means for inducing and maintaining antibody formation. *American Journal of Public Health and the Nation's Health, 44*(8), 994–1009.
24. Sabin, A. B. (1954). Current status of research on vaccination against poliomyelitis. *Journal of Michigan State Medical Society, 53*(9), 985. passim.
25. Koprowski, H. (2006). First decade (1950-1960) of studies and trials with the polio vaccine. *Biologicals: Journal of the International Association of Biological Standardization, 342*, 81–86.
26. Baltimore, D. (1970). RNA-dependent DNA polymerase in virions of RNA tumour viruses. *Nature, 226*(5252), 1209–1211.
27. Temin, H. M., & Mizutani, S. (1970). RNA-dependent DNA polymerase in virions of Rous sarcoma virus. *Nature, 226*(5252), 1211–1213.

What Is a Virus?

2

Abstract

Viruses are built from short sequences of nucleic acid, either DNA or RNA wrapped in a protein shell. Until the invention of the electron microscope, it was impossible to visualize a virus. The first viruses to be visualized were bacteriophage, which appeared to have a head and tail-like structure. Only the nucleic acid entered the bacterial cell through the tail. Animal viruses were described as spherical or rod-shaped; they were bound to receptors and were taken up by the cell. After the crystallization of the tobacco mosaic virus, there was much discussion as to whether viruses were "living" organisms; the controversy continues to this day. Although viruses were defined in part on the basis of size and filterability, viruses much larger than the traditional viruses have recently been isolated. Studies of viral replication indicate that most viruses self-assemble as a result of interactions between the viral proteins to form a viral capsid that interacts with the nucleic acid to form the whole. The viral replication cycle and synthesis is presented in this chapter. Viral classification into a Linnaean scheme has been proposed, but newer methods using nucleic acid homologies are changing classification. Viruses are spread in the human population by various means, including airborne particles, fecal-oral contact, clothing, insects, and contact with other animals (zoonosis).

2.1 Definition of a Virus

Although viruses tend to be diverse in terms of the diseases they cause and the organs they attack, all viruses have a unity of structure and consist of proteins and nucleic acid. Some viruses are also encased in a lipid membranous envelope (Fig. 2.1). Their mode of replication is not binary (one divides into two, two divide into four, etc.), as in most other organisms, but occurs as a burst of thousands of virus particles from a single virus over a short time. The number of viruses

M. W. Taylor, *Viruses and Man: A History of Interactions*,
DOI: 10.1007/978-3-319-07758-1_2,

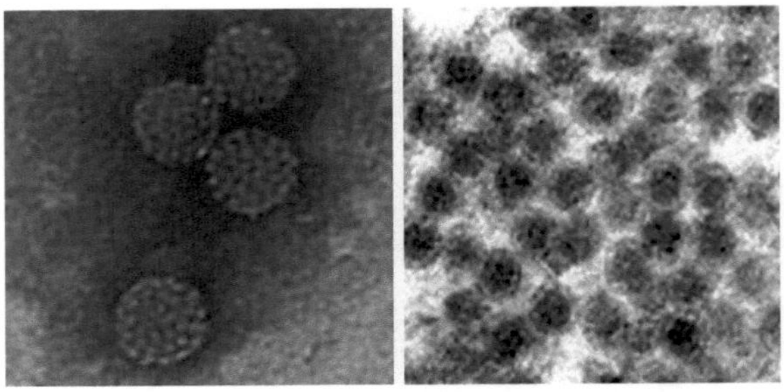

Fig. 2.1 Papilloma virus-16 and dengue virus (ICTVdB picture gallery with permission). Dengue virus is an enveloped virus

produced in cell culture or the blood is in the tens of millions per milliliter of media or blood. This mode of replication alone makes viruses unique. All other life forms contain DNA as genetic material and RNA as a message or intermediate for the formation of proteins or other structures. Viruses are also unique in that they contain either RNA or DNA as the genetic material. To date, no virus has been discovered that contains both types of nucleic acid as genetic material, although both types are used during virus replication in the cell. Viruses do not contain ribosomes, mitochondria or other cell-like organelles, and are thus completely parasitic. Since they cannot replicate without the metabolic processes of the host cell, they are genetic parasites. This differentiates viruses from bacteria or other single-cell microorganisms (e.g., protozoa), most of which can replicate in culture on their own, although they can also be parasitic. One other feature distinguishing viruses from other organisms in general is their small size and ability to pass through filters. This was certainly a major criterion in the early days of virology, although large viruses have recently been discovered. Viruses were not retained by filters made from diatomaceous earth (*kieselguhr*), extensively used at the beginning of the twentieth century. This characteristic was originally used as the major characteristic in differentiating a virus from bacteria, fungus or protozoa.

André Lwoff, a French microbiologist (who studied *lysogeny*, which is the insertion of bacteriophage DNA into the chromosome of a host), defined a virus in negative terms, using the following characteristics [1]:

- Possessing only one type of nucleic acid.
- Multiplying in the form of their genetic material, i.e., either RNA or DNA. This is not strictly true, since hepatitis B virus is a DNA virus but replicates as an RNA intermediate. Retroviruses are RNA viruses but replicate through a DNA.
- Unable to undergo binary fission and
- Lacking an energy system, including mitochondria and ribosomes.

This is a negative definition, and stresses the **non-cellular** nature of viruses. Fundamentally, then, a virus is a package of genetic information protected by a protein shell for delivery into a host cell to be expressed and replicated. The virus eventually takes over the cell completely for its own replication and in the process may kill (lyse) the cell or, in the case of tumor viruses, can permanently alter the cell [2].

Fig. 2.2 First electron microscope as invented by Ruska and Knoll (Wikipedia)

It was impossible to define a virus in terms of shape or physical characteristics until one could be seen; this had to await the development of the electron microscope. In 1931, two German scientists, Ernst Ruska and Max Knoll, invented the electron microscope (Fig. 2.2). This was a great improvement over the light microscope and allowed one to actually "see" viruses. One major disadvantage, however, is that the preparation for visualizing material by EM may distort the specimen because viable material cannot be used. The EM, as it became, uses electrons speeded up in a vacuum, aimed at the specimen, and gives an outline of the structure on an electron-sensitive photographic plate. Ernst Ruska later moved to the Siemens Company in Germany, where the first commercial electron microscope was developed. He received the Nobel Prize in Physics in 1986.

2.2 Are Viruses Alive?

Since the crystallization of viruses in 1935, there has been controversy as to whether viruses are alive or inanimate. The best analogy is to that of a seed. Is a bean seed alive? It has the properties of life under certain circumstances, but can be kept in an inert state for years. As soon as it is planted under appropriate conditions, it becomes "alive." A virus differs in that it cannot reproduce unless it is in a host cell. Thus, one may think of the virus as the bean and the host cell as the soil. Viruses use the metabolism of the host cell, although there are large viruses, such as smallpox, that carry genes similar to those of the host cell, which

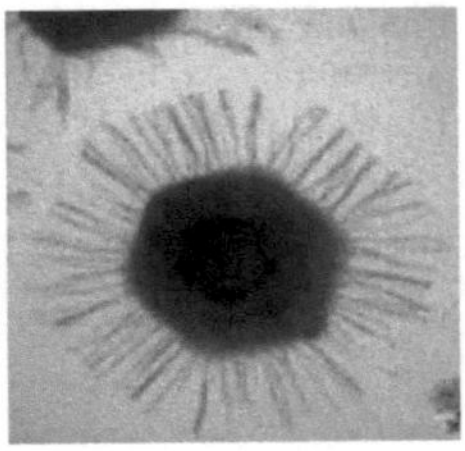

Fig. 2.3 Micrograph of Mimivirus (ICTVdB)

code for metabolic enzymes, and some genes mimic those of the host cell and thus disrupt cellular metabolism by acting as decoys. These viral products are termed virokines. However, even such viruses cannot multiply on their own since they need cellular components for their replication and protein synthetic apparatus. Where did those mimicking genes come from? Were these genes derived from the host cell since they mimic host functions and interfere with these functions, or were they acquired by the host eons ago from the virus? (This is a controversial and open question). Thus the virus is at the edge of life, and can exist in two states—an inert state and a living state.

In 2004, a large virus (Mimivirus) was discovered in a strain of amoeba, [3]. It is considered and called a virus since it cannot replicate outside its host, and although it carries genes for many metabolic functions, it does not code for ribosomal genes, making it dependent on the host cell for protein synthesis. It is as large as many bacteria, but despite this, it seems to be another example of an entity at the boundary between living and inert forms. One has to consider a continuum from the inert chemical (organic chemical) to more complex ones with the ability to reproduce, which then gives life. The definition of life or living is an arbitrary one (see Chap. 4 on bacteriophage), and is defined in different ways by different scientists and philosophers. The discovery of the Mimivirus has given rise to the discussion as to whether this is an ancestor of other viruses or is the result of genes breaking away from the host and becoming independent. An electron micrograph of Mimivirus is shown in Fig. 2.3, and a schematic in Fig. 2.4.

This virus has an icosahedral capsid of 400 nm in diameter and 125 nm long closely packed fibers projecting out from the capsid surface (750 nm total length). The capsid contains the internal core surrounded by an internal lipid layer. A fivefold axis displays a starfish-shaped structure.

2.3 Viral Structure

Viruses come in two main shapes as seen from electron micrographs: spherical, or rod-shaped. Some viruses are naked, while others have a lipid envelope around them, often derived from the host cell (see Fig. 2.1).

In some cases, viruses approach the shape of regular solids. In some of these viruses, 20 replicas of the same proteins make up the capsid. This can be a single protein, or three or many proteins, which interact spontaneously when in contact

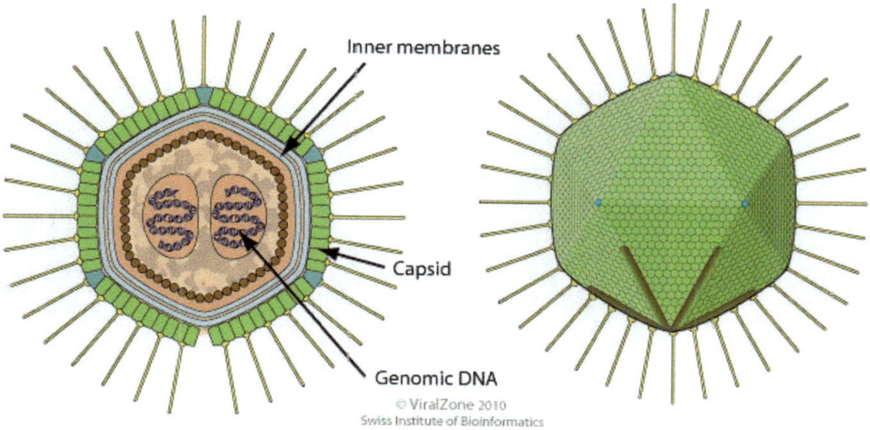

Fig. 2.4 Mimivirus structure (ViralZone, SIB Swiss Institute of Bioinformatics, with permission)

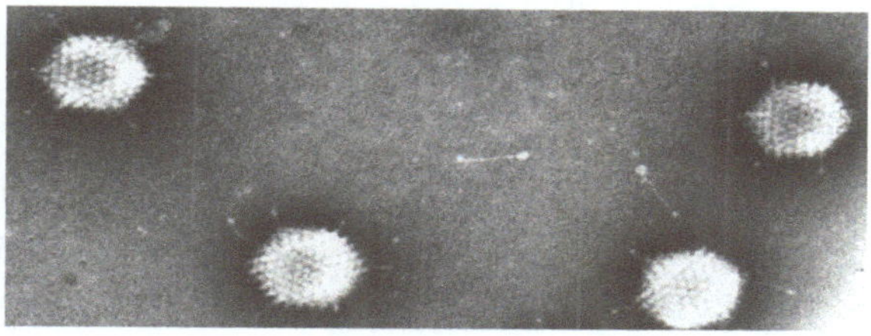

Fig. 2.5 Structure of adenovirus. Note peplomers (spikes) from vertices and free peplomer (ICTVdB picture gallery)

through template domains at their edges. The bonds between the capsid proteins are angled so that the complete structure acquires the form of a regular icosahedron.

Figures 2.1, 2.4, 2.5 and 2.6 illustrate the basic shape of icosahedral viruses.

Spherical viruses exist in two classes: non-enveloped but may have attachments to the vertices (corners) of the icosahedron (see figure of adenovirus); or enveloped.

All mammalian living cells are covered by a membrane—the viability of the cell depends on the integrity of this membrane. Enveloped viruses leaving the cell must, therefore, allow this membrane to remain intact if the cell is to survive, or even if the cell eventually dies. This is achieved by the budding of the viral nucleo-capsid through the membrane, during which the virus becomes coated in a lipid envelope derived from the host cell membrane or modified membrane (see Fig. 2.7, 2.8).

Some viruses are rod-shaped, as illustrated by the tobacco mosaic virus (TMV) and vesicular stomatitis virus (VSV). In the case of TMV this rod is made up of repeating units of a single coat protein enclosing the nucleic acid in Fig. 2.9.

Since viruses utilize the same protein repeatedly for their structure, they require very few genes. In order to replicate, they need genes to code for their coat protein and replication. Thus many viruses, such as poliovirus, may contain only enough nucleic acid for 5–6 genes. In some cases, the same nucleic acid codes for different

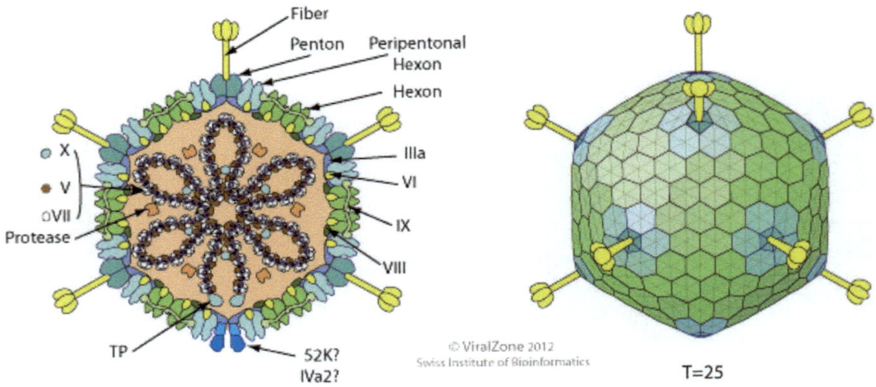

Fig. 2.6 Schematic of adenovirus structure showing icosahedral proteins and peplomers (spikes) (Courtesy of ViralZone, SIB Swiss Institute of Bioinformatics)

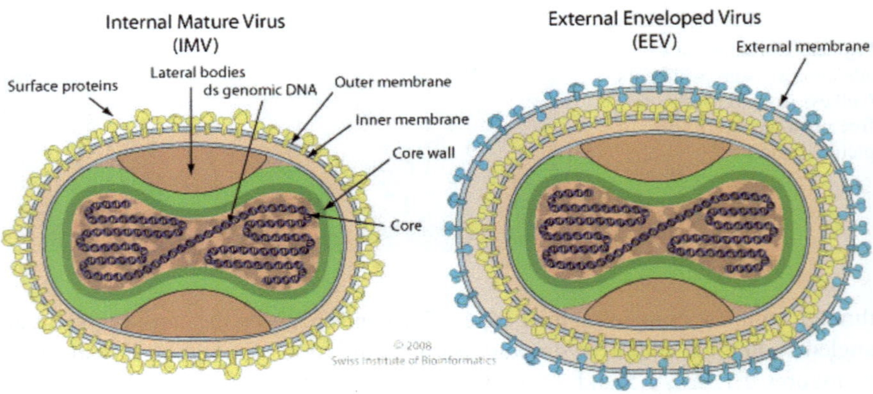

Fig. 2.7 Enveloped virus, such as poxvirus (vaccinia) (ViralZone, SIB Swiss Institute of Bioinformatics)

proteins by starting "read-out" at various locations on the nucleic strand, or using both strands of DNA.

In most cases, viruses undergo self-assembly. Apart from a few bacteriophages in which a scaffolding protein exists, viral proteins interact with each other or with the nucleic acid almost as if they have magnetic properties to spontaneously form a stable structure. The ability to undergo self-assembly is influenced by both pH (acidity or basic conditions) and salt concentrations. Poliovirus capsids will undergo self-assembly in the test tube from sub-viral particles. Classical experiments have shown that mixing RNA of Holmes' ribgrass virus and the tobacco mosaic virus protein will give rise to a viable virus. The species propagated will depend on the source of the RNA, in this case, RNA from Holmes' ribgrass virus [4].

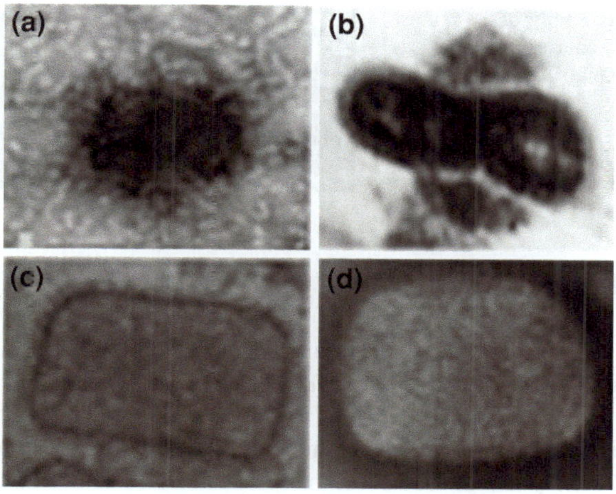

Fig. 2.8 Enveloped poxvirus (**a**) and removal of membrane by detergent and reducing agent. Note inner core remains (**b**) (ViralZone, SIB Swiss Institute of Bioinformatics)

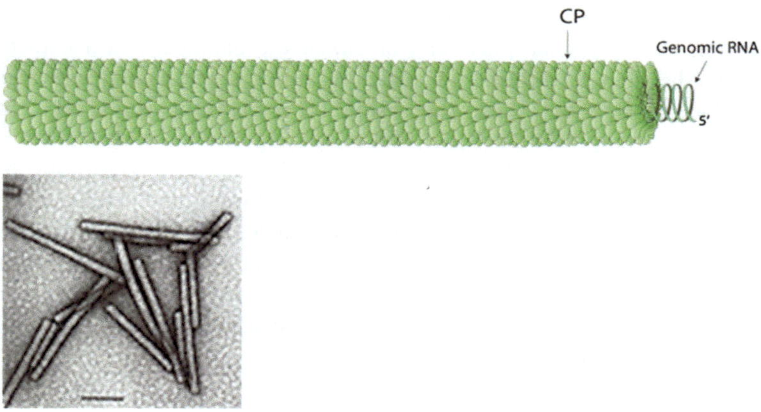

Fig. 2.9 Tobacco mosaic virus (ViralZone, SIB Swiss Institute of Bioinformatics. and ICTVdB). Non-enveloped, rigid helical rods with a helical symmetry. Virion is about 18 nm in diameter, and 300–310 nm in length

The ability to undergo self-assembly makes it easy to reconstruct viruses in the test tube, and has been done for polio and other viruses. Figure 2.10 illustrates the self-assembly of the hepatitis B capsid.

Thus viruses are small particles containing only a few genes that succeed in controlling the host's protein-synthesizing machinery to make more virus. Outside the living cell, they are completely inert and can only be seen at high resolution via electron microscopy.

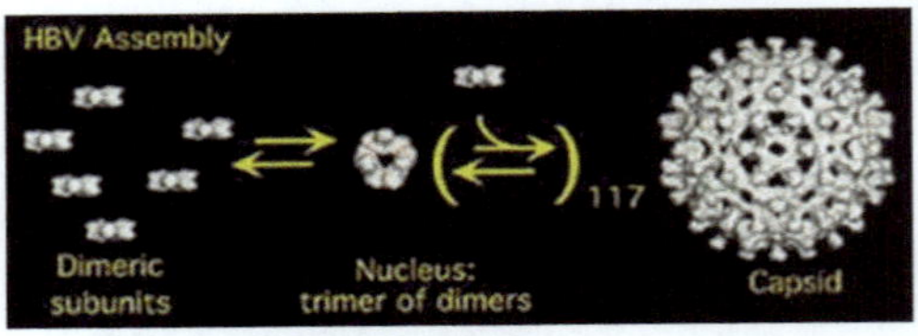

Fig. 2.10 Self-assembly of hepatitis B virus (courtesy of A. Zlotnick, Indiana University)

2.4 Nucleic Acid

Two types of nucleic acid exist in nature—DNA and RNA. In humans, as in all replicating organisms (*archaea*, *bacteria* and *eukaryotes*) the genes that make up the chromosomes are strands of DNA. Viruses contain either RNA or DNA as their genetic material. DNA in most organisms is a double-stranded structure, based on Watson and Crick pairing; however, in viruses, DNA and RNA can be either single-stranded or double-stranded. Since the viral genome is small, consisting in many cases of just a few genes—some of which code for enzymes involved in replication—there is a limited amount of genetic material.

The single virus is referred to as the "virion." It consists of an outer shell (protein capsid, or membrane), the function of which is to protect the genetic information from physical, chemical, or enzymatic damage, and a nucleocapsid containing the genetic information and any required replicating enzymes. The outer surface of the virus is also responsible for recognition of and attachment to the host cell. Initially, this takes the form of binding of a virus-attachment protein to a cellular receptor molecule. One can imagine this as a key and lock mechanism. Once the virus attaches to the cell, it is engulfed through the cellular membrane and the viral coat removed in small cellular vesicles, releasing the viral nucleic acid into the cytoplasm for replication (Fig. 2.11). In the case of bacteriophage, the viral coat is not taken into the cell, but the genetic material is injected into the bacterial cell as if through a syringe, with the bacteriophage proteins remaining attached through their tails.

2.5 Virus Classification

The common names of viruses are associated in many cases with the place the virus was first isolated (e.g., Ebola is the location in the Congo where the disease was first recognized); the organ from which the virus is isolated (e.g., adenovirus from the adenoids); the symptoms of the virus (pox virus, yellow fever); or, finally, with the names of the persons isolating and identifying the virus (e.g., Epstein-Barr). Using the symptoms caused by a virus can lead to confusion. There are many viruses that cause what we generally call hepatitis. For example, the *picornavirus* hepatitis A causes common jaundice, a self-limiting disease; hepatitis B causes serum hepatitis associated with blood transfusions; and hepatitis C,

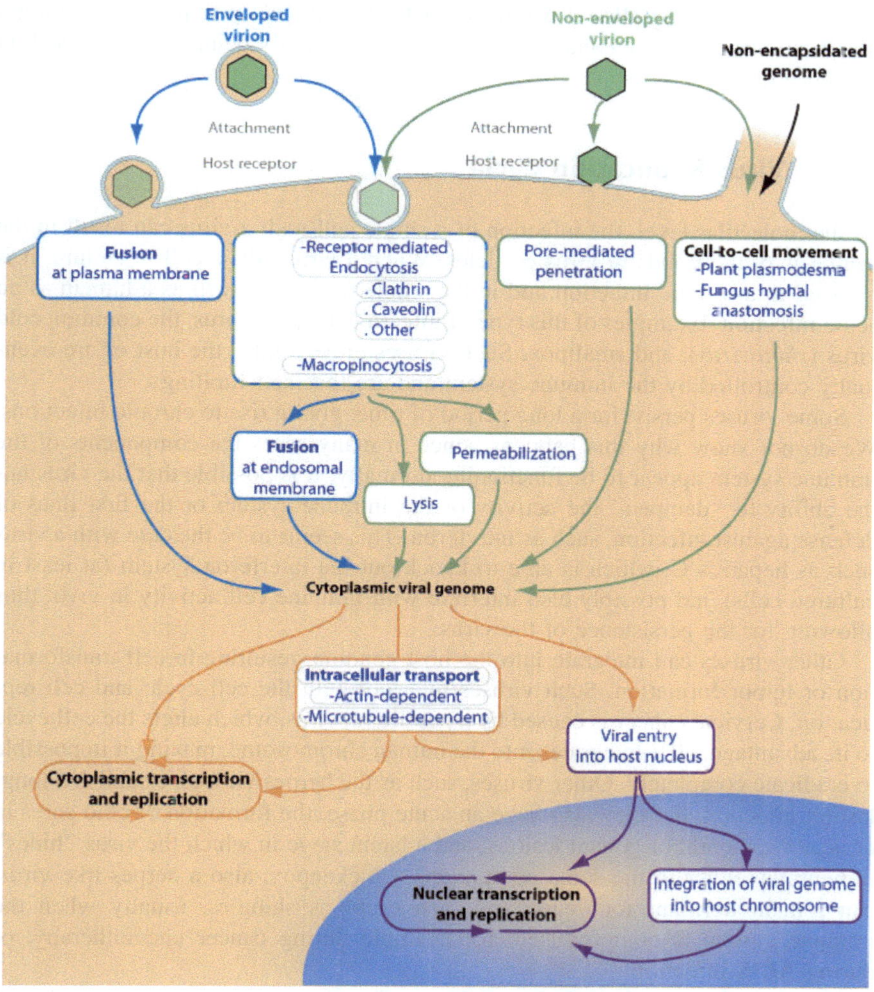

Fig. 2.11 Mode of attachment and uptake of virus into the mammalian cell. Note the difference between enveloped and non-enveloped virus (ViralZone, SIB Swiss Institute of Bioinformatics)

a *flavivirus*, transmitted in blood and blood products, eventually causing liver cirrhosis and liver cancer.

Many attempts have been made to classify viruses, following the classical scheme of Linnaeus, the eminent Swedish botanist. However, since viruses are quite invisible, and we know them from either the diseases they cause or from their shape in electron micrographs, it has been very difficult to relate them to each other. Newer methods of classification depend on the molecular structure of the virus, type of nucleic acid, and mode of replication. More accurate methods in DNA and RNA sequencing allow for virus classification based on sequence homology. The relationship between and among viruses is constantly being revised, based on sequence data.

2.6 Virus Replication Cycle

At the molecular level, the infection of a single cell with a virus can result in the synthesis of thousands of virus particles. When a virus kills a cell in culture, it is referred to as a lytic infection and in the whole organism, such as a human as an acute infection. Examples of this type of infection are poliovirus, the common cold virus (*rhinovirus*), and smallpox. Such viruses either "kill" the host or are eventually controlled by the immune system and are thus self-limiting.

Some viruses persist for a long period of time, giving rise to chronic infections. We do not know why this happens, since in many cases the components of the immune system appear to be functioning normally. It is possible that the virus has the ability to "dampen" the activity of the immune system or the first lines of defense against infection, such as interferon. This seems to be the case with a virus such as hepatitis C, which is able to knock out the interferon system (at least in cultured cells) and possibly also interfere with immune cell activity in vivo, thus allowing for the persistence of the virus.

Other viruses can integrate into the host genome, resulting in cell transformation or tumor formation. Such viruses interfere with the cell cycle and cell replication. Cervical cancer is caused by a papilloma virus, which alters the cell cycle to its advantage. HIV integrates into the human chromosome, making it impossible to eradicate completely. Other viruses, such as the herpes viruses, establish a long-lasting infection, but they also have an acute phase, the formation of cold sores in or near the mouth, or genital lesions, and a latent stage in which the virus "hides" in neurons, either in the soma or in axons. Chickenpox, also a herpes-like virus, can remain dormant for years, and then erupt as shingles, usually when the immune system is compromised, for example during cancer chemotherapy, or during AIDS infection.

The typical virus life cycle is as follows: The virus interacts with its target cell and introduces its nucleic acid into the cell. The interaction occurs through the binding of the virus to a unique receptor domain on the cell membrane (Fig. 2.11). This receptor is a normal component of the cell and may have other functions. Virus proteins are made and modify the host cell metabolism so that host proteins are no longer made, or the host cell replication cycle is affected. All of this is to allow the virus to produce more of itself, thus diverting cell metabolism to virus production. The viral proteins are produced in one part of the cell, the replicated nucleic acid in another, and somehow they find each other, interact, and form virus particles that are expelled from the cell. These viruses then infect nearby cells. Immediately after infection in vivo, the body reacts by producing interferon and other immune enhancing molecules known as cytokines, which are the first line of defense against the virus; this response is called "innate immunity." Adaptive immunity, the production of antibodies for a specific virus, will not occur until much later (10–21 days). Symptoms usually occur a few days after infection, although during this asymptomatic period a person may be infectious. This is particularly true in the case of viruses such as HIV, where obvious symptoms may not occur for quite some time (The role of the immune response is discussed in detail in Chap. 6).

2.7 How Are Viruses Spread?

In discussing the mode of viral transmission, one must consider the tissue or organ tropism of the virus. Viruses bind to receptor sites on the cells of specific tissue. It is the presence of these receptor sites that make an animal or cell susceptible to the virus. The virus contains unique proteins, occasionally as surface projectiles that interact specifically with the receptors. Thus some viruses, such as hepatitis virus, attack the liver and no other tissue. Other viruses, such as the common cold virus (rhinovirus), have adapted to cooler body temperatures and infect the lining of the nose, which is cooler than the rest of the body. Mutations in a receptor—or lack thereof—can lead to resistance to virus infection. Individuals lacking specific chemokine receptor sites, such as CXCR4 and CCR5, are resistant to HIV infection due to the inability of the virus to bind to cells lacking these receptors.

Scientists divide the mode of virus spread into two categories—direct contact and indirect contact. Direct contact includes blood-borne transmission, fecal-oral transmission, airborne spread by droplets or aerosol, contact with bedding, clothes, etc. (fomites), and vector (usually mosquitoes or other insects) [5]. Indirect contact spread includes cases where mucus from a runny nose may get onto the hands, or virus may be left on a surface such as a doorknob, telephone, or countertops, and is picked up by a second individual, who then touches his eyes or nose, resulting in infection. For example, cytomegalovirus (a virus of the herpes class) can be transmitted between children if one child drools on a toy and a second child picks it up and then put his hands in his mouth or eyes. Viruses can be transmitted in droplets; however. the area covered is not very large, usually only a few feet depending on the size of the droplets.

(The exception may be in planes, where the virus droplets may persist and spread in the dry closed confinement.) An example of common vehicle transmission might be where a food contaminated with hepatitis A is consumed by many people who all become sick. Vector transmission is a very common means of transmission; the best studied cases include yellow fever, dengue virus, and West Nile fever—viruses all transmitted by mosquitoes.

The many modes by which viruses are transmitted are summarized in Table 2.1.

Table 2.1 Mode of transfer of viruses

Mode of transfer	Virus	Symptoms
Aerosol and airborne droplets	Rhinovirus, adenovirus, respiratory syncytial virus	Common cold: upper- and lower-tract respiratory infections
Fecal-oral	Poliovirus, hepatitis A, noroviruses	Diarrhea, vomiting, gastric pain
Insect transmission by mosquitoes	Dengue, yellow fever, West Nile fever	High fever, hemorrhagic fever, internal bleeding
Sexual transmission	HIV, HSV-2, herpes 8, hepatitis B, papilloma, EB virus	In case of HIV—immunodeficiency. Kaposi's sarcoma. Others: warts, blisters on genitals and anus
Blood-borne diseases (excluding insects): dirty needles and blood transfusions	Hepatitis C, hepatitis B, HIV	Hepatocellular carcinoma
Routes unknown or mixed: transmitted from other animals (zoonoses)	SARS, MERS, Ebola, Lassa fever, Nipah, hantavirus	Very fatal infections. Usually internal bleeding

2.8 Aerosol and Airborne Spread

In studying the spread of respiratory viruses, scientists have concentrated on respiratory syncytial virus (RSV), influenza, and rhinoviruses, the three viruses most commonly found in homes, old age homes, institutions and schools. RSV is the most common respiratory infection in children worldwide. Attendance at daycare center ensures that a child will be infected with RSV during the first few years of life, and infection results in 0.05–1 % of children being hospitalized with bronchitis.

Although it would be expected that transmission of RSV would be by airborne particles, this does not appear to be the case. Transmission takes place by contact, both direct and indirect. In experiments designed by Hall et al. [6] highly symptomatic infants who were producing abundant secretions were placed in cribs. Three categories of nurse volunteers were brought into the room. "Cuddlers" played with the infant, changed his or her diaper and performed other routine care. "Touchers" did not touch the baby but had extensive contact with the child's environment, which had been heavily contaminated with secretions. "Sitters" sat next to the crib reading a book for 3 h but did not touch anything in the immediate environment. Five of the seven "cuddlers," four of ten "touchers" and not one of 14 "sitters" developed RSV infection. Thus transmission did not appear to be airborne, since none of the nurses sitting beside the children was infected. The other nurses were infected either by direct contact with the child, or from secretions that were on the crib or other objects handled by the child. Children with

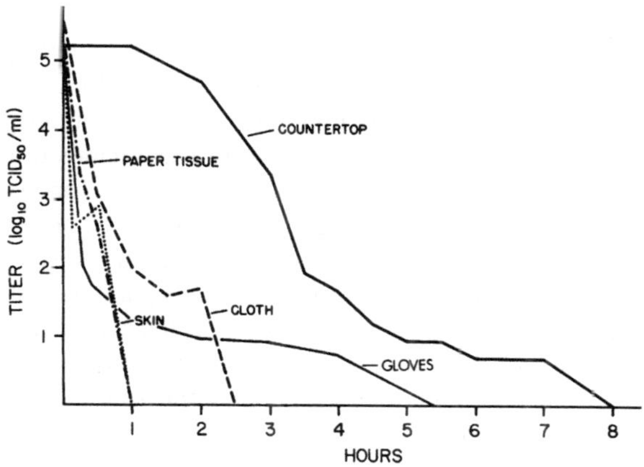

Fig. 2.12 Mean titer and duration of recovery of respiratory syncytial virus (*RSV*) from skin (hands) and four environmental surfaces that were contaminated with RSV in both medium and pooled normal adult nasal secretion [6]

RSV secrete large amounts of virus, which can be viable for up to 7 h on tabletops or other surfaces (Fig. 2.12).

RSV is often spread in hospitals (nosocomial spread) through lack of thorough hand-washing after handling infected children. It is important that all surfaces be washed down with an effective antiviral compound after exposure to children with RSV. Detergents containing alcohol are effective in destroying RSV, and anyone in contact with this infection should wear gloves as a barrier to infection.

Rhinoviruses follow a similar pattern. A 2007 study [7] showed that sick people leave cold viruses on the things they touch, such as door handles, pens and light switches. Adults with naturally acquired rhinovirus colds have been found to contaminate 35 % of environmental surfaces in hotel rooms after overnight occupancy, using reverse transcriptase-polymerase chain reaction (RTPCR) to detect viral RNA. In addition, viral RNA on surfaces (light switch, phone button, and telephone handset) could be transferred to fingertips during activities of daily life. Although RTPCR measures viral RNA contamination, parallel experiments testing for growth of virus in cell culture resulted in similar results. Virus could be found on these objects for as long as 18 h. There are, however, two conflicting sets of data on how rhinoviruses are spread, one maintaining that the virus is predominantly spread by indirect contact through nasal secretions, the other that the major mode of transmission is through aerosol. Studies at the University of Wisconsin have shown that when individuals with a cold are kept in the same room with uninfected individuals, the virus does spread by aerosol, but not as fast as expected. It may take hours of exposure for others to contract the cold. However, in these experiments, virus was not detectable on game chips, suggesting that in the case of contact infection, there is a need for large amounts of fresh secretion [8]. Thus, again, hand-washing should be effective against the spread of this virus.

In the last few weeks of spring 2009, pictures from around the world appeared in the press and on television of people wearing facemasks to prevent the spread of *influenza* (H1N1). Influenza is predominantly spread by droplets that may travel as far as six feet and from surfaces contaminated from nasal secretions. Adults can spread the virus 1 day before becoming sick and up to 7 days after the infection begins. Children can spread the virus for a longer time.

To avoid the spread of the virus, people should stay away from the sick and stay home if sick. It also is important to wash hands often with soap and water; if not available, use an alcohol-based hand rub. Linens, eating utensils, and dishes belonging to those who are sick should not be shared without thoroughly washing them. Eating utensils can be washed either in a dishwasher or by hand with soap and water, and do not need to be cleaned separately. Furthermore, frequently touched surfaces should be cleaned and disinfected at home, at work and at school, especially if someone is ill (http://www.cdc.gov/flu/about/disease/spread.htmCDC).

Secondary infections with pneumonia often occur during an influenza epidemic. During a normal outbreak, influenza targets the elderly. There have been epidemics in which younger members of society are targeted, and the great pandemic of 1918 attacked mostly 20- to 30-year-olds. Experiments indicate that the influenza virus may remain active on various surfaces for a few days, depending on the type of surface. Influenza virus spread can be inhibited to some extent by the use of surgical masks.

2.9 Fecal-Oral Spread

The most prevalent virus spread by this means is the Norwalk (norovirus). These viruses have been in the news recently due to infections that have occurred on cruise ships. The Norwalk virus was named for Norwalk, Ohio, where there was an outbreak in the 1970s. Since there are many species of norovirus, this is now the officially accepted name. This stomach virus ranks second (behind the common cold) in the occurrence of viral illnesses in the United States. The centers for disease control (CDC) reported over 267 million cases of diarrhea in 2000, and estimates about 5–17 % of these may have been caused by a norovirus. Cruise ships are not the only place where one can pick up this nasty bug; of the 348 outbreaks reported to the CDC between 1996 and 2000, only 10 % were in vacation settings. Restaurants, nursing homes, hospitals, and daycare centers are the most likely places to pick up a norovirus infection. Symptoms of norovirus illness usually begin about 24–48 h after exposure to the virus, but they can appear as early as 12 h after ingestion of contaminated food. People infected with norovirus are contagious from the moment they begin feeling ill until at least 3 days after recovery. Some people may be contagious for as long as 2 weeks. Therefore, it is particularly important for people to practice good hand-washing after having recently recovered from a norovirus. The major source of contamination is leafy greens such as lettuce, fresh fruit and shell fish. The most important precautions against norovirus are hand-washing, laundering of contaminated clothing and blankets, and cooking shellfish, to high temperatures.

Another virus spread by this means, from diarrhea and vomiting, is rotavirus. This virus occurs predominantly in children, although there have been outbreaks in adults. It also occurs suddenly with an outbreak of fever, vomiting, and stomach cramps, and usually last a few days. At one time it was the leading cause of death of young children in the U.S. Since the advent of a vaccine in 2005, its seriousness has decreased. Prior to 2005, almost all children in the U.S. were infected with rotavirus before their fifth birthday. In the pre-vaccine period, rotavirus was responsible for more than 400,000 doctor visits; more than 200,000 emergency room visits; 55,000–70,000 hospitalizations; and 20–60 deaths in children younger than 5 years of age (CDC). Many other viruses, such as polio, are spread through the fecal-oral route and are discussed separately.

2.10 Insect Route

Dengue virus is a leading cause of illness and death in the tropics and subtropics. As many as 400 million people are infected annually by dengue virus, which is caused by any one of four related viruses transmitted by mosquitoes. There is no vaccine to prevent infection with dengue virus, and the most effective protective measure is to avoid mosquito bites. When infected, early recognition and prompt supportive treatment can substantially lower the risk of medical complications and death.

An estimated 2.5 billion people live in over 100 endemic countries and areas where dengue viruses can be transmitted. Up to 50 million infections occur annually, with 500,000 cases of dengue hemorrhagic fever and 22,000 deaths—mainly among children. Prior to 1970, only nine countries had experienced cases of dengue hemorrhagic fever (DHF); since then, the number has increased more than fourfold and continues to rise (see map in Fig. 2.13) (World Heath Organization).

During the nineteenth century, dengue was considered a disease that sporadically caused epidemics at long intervals, a reflection of the slow pace of transport and limited travel at that time. Dengue has emerged as a worldwide problem since the 1950s, and in the last 50 years, the incidence has increased 30-fold. Although dengue is rarely found in the continental United States, it is endemic in Puerto Rico and in many popular tourist destinations in Latin America, Southeast Asia and the Pacific islands (Fig. 2.14).

The reason for the increase in epidemic dengue fever has been the lack of eradication of mosquitoes (*Aedes aegypti*). With populations moving from the country to urban areas, there is now a large non-immune population available to the infected mosquitoes. There has been, over time, an increase in the use of plastic containers, which are breeding grounds for mosquito larvae. As has also been shown for yellow fever (see Chap. 10), the availability of old tires filled with rainwater is also a locale for breeding. Air travel has helped spread the mosquito from continent to continent and as a carrier of the disease.

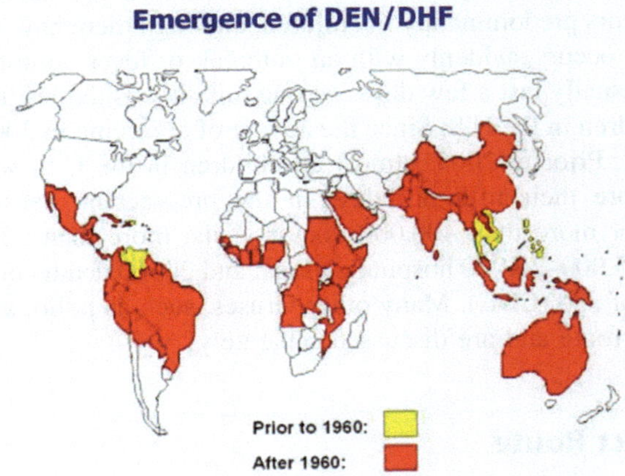

Fig. 2.13 Range of dengue fever infection in the world. Note increase in area after 1960 (World Health Organization)

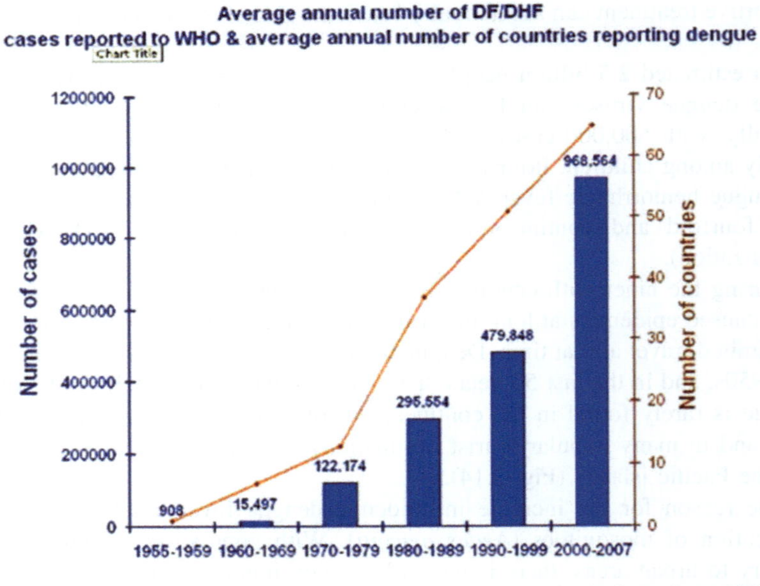

Fig. 2.14 Increase in cases of dengue fever and dengue hemorrhagic fever from 1955–2007 (World Health Organization)

As in the case of other diseases spread by mosquitoes, the virus replicates in the gut of the insect, and then accumulates in the salivary gland waiting to be transmitted at the next meal. The extrinsic period (time in the insect before becoming

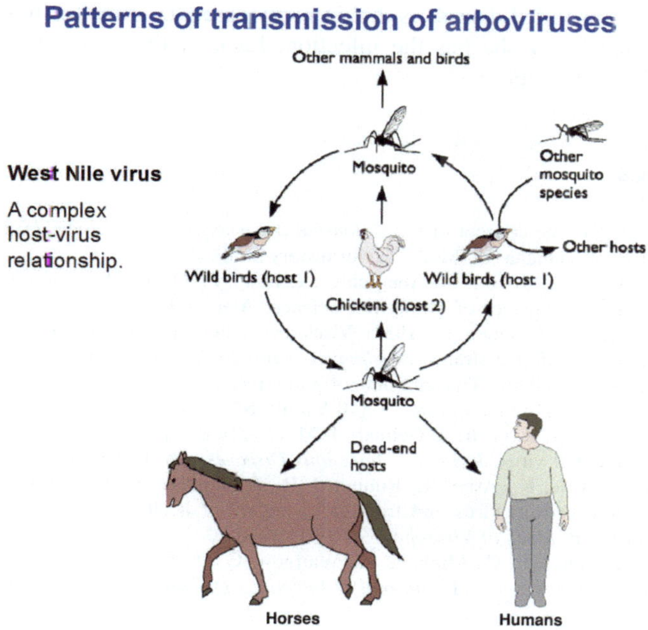

Fig. 2.15 Example of "dead end" infection. Main target of virus/vector is the *bird*; *man* and *horses* are accidental targets

infectious) is approximately 7 days. There is a similar pattern of infection for the yellow fever virus.

West Nile Virus is among the most common mosquito-borne viruses in the U.S., and there is currently no vaccine for it. Only about 20 % of those bitten by virus-loaded mosquitoes develop fever, body aches, headache, and nausea. Another 1 % develop neurological problems that can be serious. Humans are a "dead-end" host for the virus, in that it is not transmitted further. The virus is predominantly found in birds (see http://www.cdc.gov/westnile/resources/pdfs/Bird%20Species%201999-2012.pdf), which lists those species in which the virus is found (Fig. 2.15).

2.11 Sexual Transmission

A large number of viruses can be transmitted sexually. These include Herpes simplex 2, HIV, hepatitis B and papilloma viruses. The mode of transmission is the same in all these cases, and depends on the level of promiscuity, irrespective of whether the sex is heterosexual or homosexual. All can be prevented by the use of condoms or by limiting the number of partners. One of the major problems with the sexually transmitted disease is that in many cases the infected individual does not realize that he or she has the infection. Each of these viruses is discussed separately in later chapters.

References

1. Lwoff, A. (1957). The concept of virus. *Journal of General Microbiology, 17*(2), 239–253.
2. (http://courses.bio.indiana.edu/M430-Taylor/history.html).
3. Raoult, D., Audic, S., Robert, C., Abergel, C., Renesto, P., Ogata, H., et al. (2004). The 1.2-megabase genome sequence of Mimivirus. *Science, 306*(5700), 1344–1350.
4. Fraenkel-Conrat, H., & Singer, B. (1957). Virus reconstitution. II. Combination of protein and nucleic acid from different strains. *Biochimica et Biophysica Acta, 24*(3), 540–548.
5. Goldmann, D. A. (2000). Transmission of viral respiratory infections in the home. *The Pediatric Infectious Disease Journal, 19*(10 Suppl), S97–S102.
6. Hall, C. B., Douglas, R. G, Jr, & Geiman, J. M. (1980). Possible transmission by fomites of respiratory syncytial virus. *Journal of Infectious Diseases, 141*(1), 98–102.
7. Winther, B., McCue, K., Ashe, K., Rubino, J. R., & Hendley, J. O. (2007). Environmental contamination with rhinovirus and transfer to fingers of healthy individuals by daily life activity. *Journal of Medical Virology, 79*(10), 1606–1610.
8. Dick, E. C., Jennings, L. C., Mink, K. A., Wartgow, C. D., & Inhorn, S. L. (1987). Aerosol transmission of rhinovirus colds. *Journal of Infectious Diseases, 156*(3), 442–448.

A History of Cell Culture

3

Abstract

Cells from different tissues of plants and animals can be grown and cultured in artificial media outside the body. Cell culture started off as tissue culture with the development of the "hanging drop technique" by Ross Harrison, whereby a small piece of tissue was placed in a drop of medium (including serum), and cells migrated from the tissue into the surrounding environment. Further advances in tissue culture were made by Alexis Carel and Charles Lindbergh, who grew tissue on glass plates and developed techniques to keep tissue continuously growing. These techniques developed into cell culture, where cells could be collected from a variety of sources, separated by trypsin, and grown into a monolayer if given appropriate medium-containing serum. Such cells of human and murine origin were used for growing viruses. John Franklin Enders and colleagues discovered that poliovirus and other viruses could be grown in cell lines developed from primary cells. These results gave a boost to research in developing a polio vaccine, as well as to basic virology. In the 1940s and 1950s, plaque assays were developed for animal viruses, and quantitative methods of measuring viral growth were developed by Salvador Luria and Renato Dulbecco. In the 1950s and 1970s, methods were developed for clonal growth of cells in culture using newly isolated HeLa cells. This resulted in a new area of research—somatic cell genetics—with the development of stable cell lines. The observation that cells of different species fused in culture led to the development of somatic cell genetics, and eventually to the formation of hybridomas to produce monoclonal antibodies.

M. W. Taylor, *Viruses and Man: A History of Interactions*,
DOI: 10.1007/978-3-319-07758-1_3,

3.1 The Beginnings

Cells from various tissues of plants and animals can be grown and cultured in artificial media outside the body. This technique was developed simultaneously with research on viral diseases, as discussed in the introductory chapter. Cell culture, as it is known today, was not developed until the 1950s and should be differentiated from tissue culture, although both terms are used indiscriminately. Tissue culture is the culturing of pieces of tissue in an artificial environment, in which cells migrate out of the tissue onto an inert surface. Cell culture involves the harvesting of individual cells from a specific tissue and maintaining the cells in an incubator at body temperature (37 °C) in a plastic or glass flask, covered with a rich medium that mimics that of the internal environment of the human body. Both these techniques allow for many more manipulations than possible with living animals, and contribute to progress in understanding pathogens without the disturbing interference of the hormonal and immune systems, and the influences of the whole organism. Vaccines for many diseases owe their existence to the development of cell culture.

3.2 Harrison and Carrel

Cell culture developed from research in embryology, developmental biology and, later, cancer. Early experiments, performed by Ross Harrison (1870–1959) and Leo Loeb (1869–1959) at the beginning of the twentieth century used tissue or pieces of organs kept alive in test tubes using blood clots, agar or salt solutions as the liquid medium. Harrison was the first to grow tissue in the test tube while studying the development of frog nerve fibers. He also pioneered the "hanging drop technique," in which tissue was embedded in plasma on the reverse side of a slide or cover slip. The plasma would form a hanging drop, in which a piece of tissue would be retained. These were tissue or organ slices that produced cells that migrated from the explanted tissue on a solid surface. This was quite an achievement, and depended on progress in elucidating conditions for maintaining and growing isolated fragments of living tissue (Fig. 3.1).

Another pioneer in this field was Alexis Carrel (1873–1944). He advanced and adapted the Harrison methodology and used tissue fragments from chicks and various mammals, as well as tumors. He and his colleague Montrose Burrows (1884–1947) grew these tissues in plasma of the animal of origin in hanging drops,

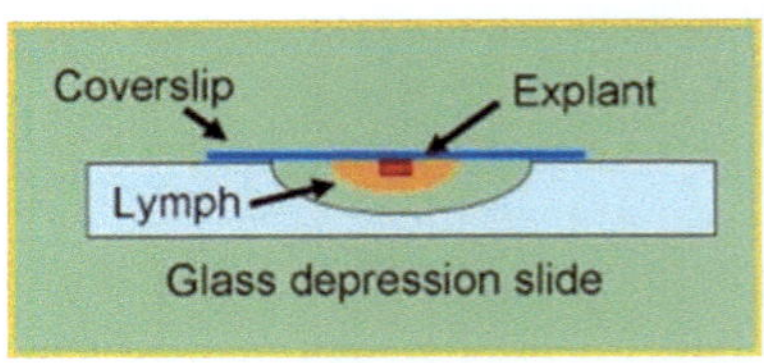

Fig. 3.1 Harrison's hanging drop technique (www.corning.com)

and also on slide covers in glass plates. Cells grew out of the tissue, migrated around the tissue slice and could be transferred serially a number of times to new glass plates [1, 2]. Carrel invented the first tissue culture flasks, called "D-flasks," which were a prototype of the flasks used today for culturing small amounts of cells. He also improved growth media, showed that serum could be used instead of plasma, and that serum from different mammals supported cellular growth. Plasma is derived from blood by centrifuging down the red blood cells, usually in the presence of an anti-coagulant such as heparin, while serum differs from plasma in that some components of plasma, such as the clotting factor, are removed by allowing the blood to clot and then separating the clot from the supernatant which is the serum. Modern-day cell culture is dependent on the use of bovine serum, which is abundantly available. Carrel also emphasized the need for sterile techniques and tested toluene in small amounts as an inhibitor of contaminating bacterial growth. This was in the period before antibiotics and fungicides, which are now added to culture media to prevent bacterial and fungal contamination.

The finding that cells could be grown on a glass surface was a major discovery. This method is used in many laboratories to this day, rather than plastic surfaces, since it is more economical and also environmentally favorable. Glass is easily washed and sterilized. Cells would grow out of the transplants onto most solid surfaces. Carrel also showed that it was possible to grow cells from chick embryos' organs and from skin. He claimed to have kept such cultures alive for long periods, but this is now considered unlikely. It is now thought that these observations were artifactual, that feeding cells "embryo extracts" may have introduced new cells into the population, which would thus give the appearance of continuous growth. Carrel's contribution to the development of tissue culture was enormous and in 1912 he was the first American to receive the Nobel Prize in Medicine.

A major breakthrough, allowing for the growth of large numbers of mammalian cells in culture for virus production, occurred much later with the use of trypsin, an enzyme that splits the peptide bonds linking adjacent cells, thus allowing for the harvesting and transfer of large numbers of cells This is a simple procedure, in which a small amount of trypsin is added to a plate of cells, or tissue clumps, allowed to digest the bonds between the cells for a short time, the plate shaken gently, and the cells are detached from each other and from the glass surface. The cells are then harvested, diluted and re-plated. Modifications of this technique can be used to tease cells from tumors or organs.

One cell line, derived from explanted chick embryo heart fragments in January 1912, was claimed to be successfully sub-cultured hundreds of times. After surviving initial contamination outbreaks that almost wiped them out, they were looked after by Carrel's research assistant. He followed Carrel's methods of subculture and maintained this cell line until 1946, when the cultures were finally terminated. These cultures were widely written about and their anniversary was celebrated each year in the New York World Telegram. The cultures outlived Carrel by almost 2 years [3]! However, it is questionable whether this really was the original culture.

Of interest is the collaboration between Carrel and Charles Lindbergh (1902–1974). Lindbergh's aeronautical feats are well known, but he was also an excellent experimenter interested in advancing medical technology. He developed methods of separating serum from whole blood and also introduced the use of Corning Pyrex tissue culture flasks. One advantage of using Pyrex for the flasks and other equipment is its resistance to high heat, allowing for autoclaving at between 120 and 170 °C for sterilization. Carrel and Lindbergh co-authored many research papers in the 1930s, and their research led to the subsequent explosion of knowledge in cell culture techniques in the 1950s–1960s [4]. Very little changed in the field of tissue culture, as it was then known, for the next 25 years, until the discovery of growth factors and cytokines, proteins of the immune system that enhance or inhibit the growth of cells. These factors can be purified and tested as replacements for the "unknowns" in serum. Many well-known scientists were trained in Carrel's laboratory at the Rockefeller Institute in New York. (See note on Carrel in Appendix and his political views and involvement with Vichy France during World War II).

Among these scientists were Wilton Earle, whose media for growing cells ("Earle's medium") is still very popular, and Raymond C. Parker (1903–1974), whose team at the University of Toronto developed the culture techniques for mass-producing the polio virus, which was used in the first successful polio vaccine in 1954. Testing and perfecting new culture media has made many advances possible. Most importantly, the advances in polio production were due to the introduction of a new medium, M199, resulting in larger yields of polio virus from monkey kidney cells [5]. This medium was originally designed for the growth of chick embryo fibroblasts and did not require serum. If one peruses catalogues of media for cell culture today, there is a plethora of media containing different concentrations of salts, amino acids, buffers, etc., from the 1960s for the growth of different types of cells. It was not until the 1980s that truly defined media were developed, using growth factors instead of serum.

3.3 Growth of Virus in Culture

John Enders (1897–1985) was another pioneer in the field of cell culture and virology. He initially studied the replication of vaccinia virus, the virus used by Jenner in the vaccine against smallpox, and he developed conditions for viral growth in culture. Enders found that by constantly changing the media, he could keep cells alive for an extended period. He utilized the methods developed by Carrel and later by George Gey (1889–1970), in which pieces of minced tissue were cultured on the sides of a rotating tube that was later developed into the "roller bottle." The idea behind rotating the culture was that this would allow for the exchange of nutrients in the media and for respiration during the period that the tissue was not in the media. Virus was added to chick embryonic tissue and incubated for 24 h. The virus was subsequently harvested and titred on rabbit skin, proving that virus replication could occur under these conditions. Moreover, there was no loss of activity of virus even after 2 weeks' storage at −20 °C. The development of these techniques was very important and permitted the growth of virus in vitro rather than in whole animals. Using this approach, various viruses could be cultured, including equine encephalitis, influenza and chickenpox (varicella). This really was the beginning of the cultivation of viruses outside the body. Many of the techniques developed by Enders and his group are still routinely used in the virology lab today.

It was originally believed that poliovirus, a scourge of the early twentieth century, could only be grown in cells of human neural origin. Enders tested a series of human embryonic tissues for their ability to produce virus. By 1948 it was obvious to Enders and his group that a virus found in the feces of man should be able to grow in a variety of tissue. Minced tissue was maintained in culture and poliovirus isolated from mouse brain added to flasks. Samples from the flasks were harvested at various times and tested for their paralytic effect on mice. Enders, Weller and Robbins reported in *Science* in 1949 that poliovirus grew very well on all these tissues over an extended period, and the calculated yield was very high. Enders developed methods for growing human cells from the foreskins of babies circumcised at the local hospital. Quite by accident, an aliquot of poliovirus was added to human foreskin and the virus destroyed the cells, producing a large yield of virus [6, 7]. Foreskin thus became the tissue of choice for basic research with human primary cultures, and a source of cells for growing virus.

In Enders's Nobel Prize lecture in 1954, one feels the excitement from the discovery that poliovirus grew in a number of different cell lines. The finding was pure serendipity, since, as noted, it was firmly believed by virologists previously working in this area that polio could only be produced in human neural tissue. Enders himself describes the approach taken by his group as being similar to a fishing expedition. Finding specific host cells in culture for a specific virus requires the testing of many cell lines, and there is often no obvious logic to the cell line targeted by the virus. For example, HeLa cells (cells derived from Helen Lacks) support the growth of most viruses, both human and non-human (mouse) and, likewise, mouse L-cells can support the growth of viruses from different species. As already pointed out, the idea that polio could only grow in neuronal cells was based on the observation that in vivo the patient's nervous system was affected. Poliovirus is an enteric virus and replicates primarily in the gut, a fact not realized in the 1930s. Enders demonstrated that polio could grow on human embryonic tissue derived from intestines, liver, kidney, adrenal gland, brain, heart, spleen, and lung. His group also studied the cytopathogenic effect of viruses on the culture. It was also possible to demonstrate inactivation of the virus by polio-specific antisera in vitro. Enders and colleagues showed that poliovirus could be grown in most human tissues in large amounts, thus paving the way for the eventual development of the polio vaccines by Salk and Sabin (see Chap. 12 on poliovirus) (Fig. 3.2). John Enders has been called "the father of the modern vaccine."

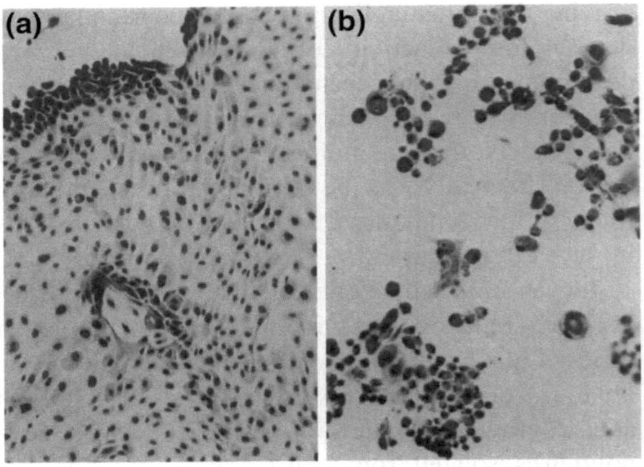

Fig. 3.2 Cytopthogenic effect of poliovirus on monolayer of cells. Preparations from roller tube cultures of human kidney tissue. **a** Control uninoculated culture (×100). **b** Culture 24 h after infection with Type I virus (Pizarro strain) (×185). Note the thickening of the cell nuclei, the numerous single rounding of the cells, and the advanced degeneration of many cells. Hematoxylin and eosin stain. *Source*: Reference [9]

The development of cell culture techniques revolutionized the whole field of virology. Viruses could now be grown in large volume for both molecular studies and the development of vaccines. With the discovery of antibiotics, penicillin and streptomycin were added to synthetic media and prevented bacterial contamination. Later, fungicide would be added when necessary to prevent fungal growth in the rich media. Different media were developed for different cells which all grew with the addition of 5–10 % volume of bovine serum. It became customary to use bovine or calf serum (or fetal calf serum) for this purpose.
One of the major advances in cell culture was the addition of the protein degrading enzyme trypsin to the "tissue" culture. This resulted in the breaking of proteolytic bonds between cells and permitted individual cells to be harvested and re-plated to form a monolayer of cells.

It has proven difficult to replace all the components of serum in cell culture, although various defined media have been developed for different cell lines. These contain different combinations of amino acids, hormones and trace elements.

3.4 Development of the Plaque Assay

Renato Dulbecco (1914–2012), using the techniques developed by his colleague, Salvatore Luria (1912–1991), who had developed quantitative ways of measuring phage by a plaque assay, showed that with certain human viruses it was possible to develop a quantitative plaque assay [8] (Fig. 3.3). When poliovirus was diluted onto a monolayer of cells, the resulting virus infection cleared the area by killing the original infected cell as well as the surrounding cells to form a hole, or a plaque, which could be visualized by staining with a dye such as crystal violet. Such holes in the monolayer were visible by a light microscope or to the naked eye. Viruses are prevented from spreading by adding a layer of semi-solid agar. The advantage of this technique is that all viruses in a single plaque are derived from a single virus particle allowing for the isolation of purified stock. It was

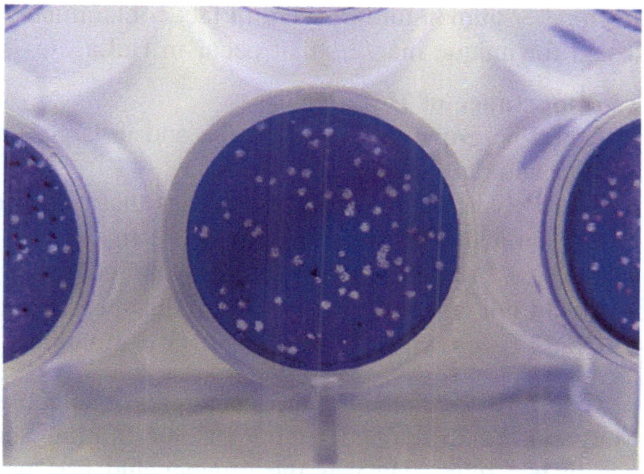

Fig. 3.3 Plaque assay of herpes virus on Vero (monkey kidney) cells (Wikipedia). This could be any virus on a monolayer of cells. Each hole (plaque) represents infection with a single virus and contains thousands of virus particles

possible to quantify virus production, and to select for viral mutants either drug-resistant, or temperature-sensitive. Other types of viruses were shown to infect and transform cells, altering their growth properties and the resulting transformed cell forming colonies in agar. These cells had characteristics associated with cancerous cells, and could in some cases form tumors upon injection into mice. Both Dulbecco and Luria received Nobel Prizes independently for different projects—Dulbecco for his work on tumor viruses, and Luria for his work on phage replication—and both had worked for periods at Indiana University, Bloomington. Jim Watson, the discoverer of DNA structure, was a graduate student of Luria's, who later moved to the University of Illinois, and finally to MIT where he headed the department of microbiology for some time (Fig. 3.3).

3.5 HeLa Cells and Cloning

Over the years, many attempts had been made to culture human tumor cells, but without much success. However, in 1951 the first human tumor cells to grow in culture were isolated from a cervical carcinoma of a young African-American woman by the name of Henrietta Lacks. These cells were isolated by George Gey at John Hopkins University Hospital in Baltimore; they were called HeLa after the donor. HeLa cells grew very aggressively and rapidly in culture. Gey published his results in 1953 and HeLa cells became available to any scientist who requested them [9]. HeLa became the workhorse of the cell culture field, and were used to study cell growth, cell reproduction and, of course, to produce virus. Many viruses that could not be grown before in culture were cultivated in HeLa. Cultures of these cells could be split in a twofold to tenfold dilution and when transferred to new flasks, new monolayers developed very quickly in simple media. Gey distributed these cells to all who requested them, so that almost every tissue culture lab used these cells. The cells grew so aggressively that many cultures derived from other human tumors were in fact contaminated with HeLa, and needed to be tested for unique markers that occur in HeLa.

Importance of HeLa cells

- Could be grown in simple media and in large amounts.
- Had wide host range for viruses.
- Were given freely to all laboratories that requested them.
- Contaminated most other human cell lines.
- Became "work horse" of virology and cell biology.
- Could not be used for genetics studies because of chromosomal aberrations.
- Could be grown from single cell- to form clones.

Gey developed many of the techniques of cell culture used today. With his wife, Margaret, he developed techniques for filming and photographing cell growth within test tubes. He was the founder and first president of the Tissue Culture Association (TCA), an organization founded in 1946 with the purpose of educating scientists in the "art" of cell culture. A new journal, In Vitro, was launched; its purpose was to describe cell culture developments. This education program became so successful that by 1971 the W. Alton Cell Science Center in the Adirondack Mountains of New York State, was opened as a permanent home for TCA's educational programs. This author remembers taking a week-long course in culture techniques at the center, at which Theodore Puck (1916–2005) and Philip Marcus (1927–2013), whose contributions are discussed below, were the instructors. In 1994 the TCA became the Association for In Vitro Biology.

William Earle developed a mouse cell line, similar in growth potential to HeLa, from mouse embryo fibroblasts. These were known as L292 cells, or referred to as L-cells [10]. Ted Puck and Phil Marcus [11, 12] proved in the 1960s that it was possible to dilute HeLa cells so that one could achieve "clonal" growth of the cells—that is, single cells would stick to the glass or plastic surface and daughter cells surround the single progenitor cell, forming a clone of cells all genetically identical [11]. The era of somatic cell genetics was born from the use of this simple technique, which allowed the selection of mutants able to growing in the presence of inhibitors that in turn gave rise to drug-resistant colonies (Fig. 3.4).

Puck also developed a cell line isolated from Chinese hamsters called "CHO," which maintained, unlike other cells in culture, a near diploid chromosome complement. Using selective pressure such as high temperature as well as drugs, Puck was able to isolate mutant cell lines for the analysis of biochemical pathways, in particular mutants of the amino acid pathways [13]. Puck hoped that CHO would become the *E. coli* of mammalian cells, i.e., a model system for molecular genetics, and his hope has been fulfilled. CHO cells are used to manufacture many recombinant drugs and cytokines, including human interferons, and are used in the study of chromosome morphology and cell hybridization. Chromosome behavior could now be studied in a cloned population of cells [14]. In my own laboratory, we isolated mutants of CHO cells in purine metabolism [15, 16]; these mutants

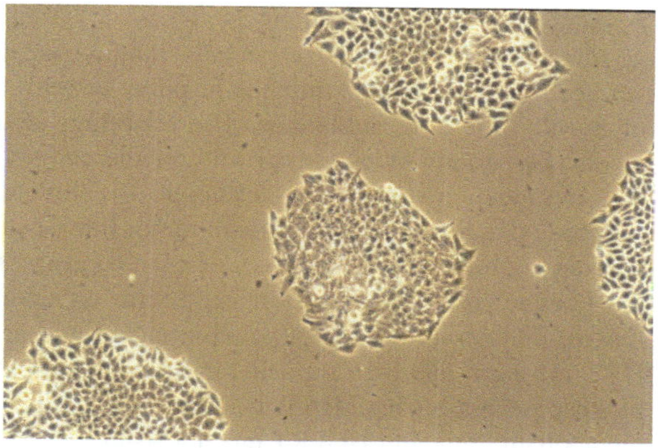

Fig. 3.4 Clonal growth of HeLa cells (www.virology.ws)

mimic human genetic diseases of the same pathway. Thus a new field of research was born, as was a new journal, *Somatic Cell Genetics* (which later became the *Journal of Somatic and Molecular Genetics*).

3.6 Birth of Monoclonal Antibodies

Using such cell culture techniques, it was possible to fuse two different cell lines to form hybrid cells, even between cells of totally different origin (human and mouse). In mixed culture, fusion occurred spontaneously, albeit at a low rate. By adding Sendai virus or, in later experiments, non-biological agents such as polyethylene glycol, the rate of cell fusion was enhanced. Many viruses, such as HIV, herpes and Sendai, cause cell fusion as part of the infectious process. Exploiting this technique, in 1975 Grigori Milstein and colleagues at Cambridge University fused myeloma cells (a tumor cell line) and B-cells from the spleen to produce what they termed hybridomas [17]. These were hybrid cells that had the long life of the tumor cell and produced antibodies derived from the B-lymphocytes in the spleen.

The technique involves a series of steps that will be listed below. But first it is important to emphasize that all the antibodies produced by this method are monoclonal antibodies—that is, each individual hybridoma produces only one type of antibody and not a mixture as in plasma. Monoclonal antibodies are now being used to target specific cancers and are in fact now being tested in clinical trials for the treatment of lymphomas, a colon and breast cancer. In many cases they are combined with chemotherapeutic compounds. The most commonly used monoclonal antibody in these clinical trials is the one for breast cancer therapy; this antibody targets the Her2 protein, and is called Herceptin. This is of special interest because Her2 protein is the human epidermal growth factor receptor-2 [18]. Mutations in this gene result in very aggressive tumors.

And now for the technique:

Mice or other animals (especially rabbits) are injected with a single known antigen that will elicit an immune response characterized by a rapid multiplication of B-cells secreting antibodies. The resulting population of antibodies is polyclonal. Some antibodies are specific for the antigen selected, but most reflect the past immune experience of the animal. B-cells obtained from the spleen of the injected animal and cells of a B-cell tumor line (myeloma) with a selective genetic marker, are mixed in the presence of a fusion agent, such as polyethylene glycol or the Sendai virus. Treatment of cells with either of these agents results in cell fusion. The surviving cell population is selected in selective medium (known as HAT medium) and then screened in 96 well cell culture plates for the production of specific antibodies. HAT medium kills cells that have a defective purine metabolism (deficient HGPRT enzyme)—in this case, the myeloma cells. HGPRT function is supplied by the fused B-cells. Non-fused B-cells die in culture, since they are short lived and therefore only the hybrid myeloma + B-cells responding to the selected antigen survive (Fig. 3.5).

In the 1980s there were major advances in cell culture with the identification of growth factors in serum allowing for the growth of lymphocytes in culture, and the isolation of stem cells. Robert Gallo, one of the co-discoverers of the HIV virus, grew T-lymphocytes in culture by the addition of an interleukin, IL-2 (a molecule that supports the growth of T-cells), thus paving the way for the growth of cells of the immune system and, eventually, embryonic stem cells [19].

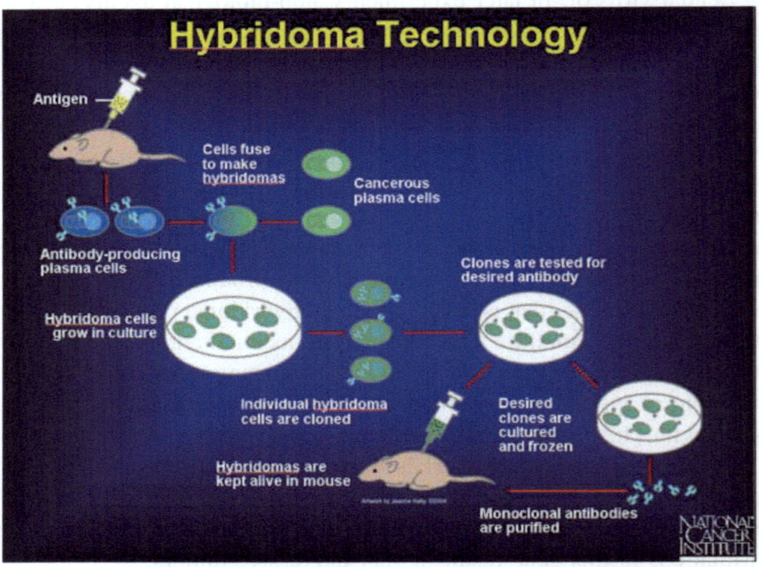

Fig. 3.5 Production of monoclonal antibodies (National Cancer Institute)

Thus from this beginning of "organ" culture, then "tissue" culture, one is able to produce cells industrially, both for the production of vaccines, and also, unfortunately, for virus as weapons of mass destruction. However, cell culture has led to the "stem cell" revolution that is ushering in a new era of research.

3.7 Appendix

In 1935 Carrel published a book titled *L'Homme, cet inconnu* (*Man, The Unknown*), which became a best seller. The book discussed "the nature of society in light of discoveries in biology, physics, and medicine." It contained Carrel's own social prescriptions, advocating, in part, that mankind could improve itself by following the guidance of an elite group of intellectuals, and by implementing a regime of enforced eugenics. Carrel claimed the existence of a "hereditary biological aristocracy" and argued that "deviant" human types should be suppressed, using techniques similar to those later used by the Nazis.

"A euthanasia establishment, equipped with a suitable gas, would allow the humanitarian and economic disposal of those who have killed, committed armed robbery, kidnapped children, robbed the poor, or seriously betrayed public confidence," Carrel wrote in *L'Homme, cet Inconnu*. "Would the same system not be appropriate for lunatics who have committed criminal acts?" he asks.

In the 1936 preface to the German edition of his book, Alexis Carrel added praise to the eugenics policies of the Third Reich, writing that the German government had taken energetic measures against the propagation of the defective, the mentally diseased, and the criminal. The ideal solution would be the suppression of each of these individuals as soon as he has proven himself to be dangerous.

Carrel, although living in the U.S., considered himself French and left his position during World War II to join the Vichy government as head of an institute on eugenics. He died in 1944 (before the end of the war), but he has been accused of being involved in the murder of thousands of mentally defective people and others suffering from disabilities. However, since he died before he could be brought to trial, one has to separate the good of the individual as a scientist from his personal opinions, and the consequences of those opinions. In the case of Carrel, he saved lives by his approach to medicine and supported those who destroyed lives by his belief in eugenics.

References

1. Carrel, A., & Burrows, M. T. (1911). An addition to the technique of the cultivation of tissues in vitro. *Journal of Experimental Medicine, 14*(3), 244–247.
2. Carrel, A., & Burrows, M. T. (1911). Cultivation of tissues in vitro and its technique. *Journal of Experimental Medicine, 13*(3), 387–396.
3. Carrel, A. (1913). Contributions to the study of the mechanism of the growth of connective tissue. *Journal of Experimental Medicine, 18*(3), 287–298.

4. Carrel, A., & Lindbergh, C. A. (1935). The culture of whole organs. *Science, 81*(2112), 621–623.
5. Swim, H. E., & Parker, R. F. (1953). Method for establishing human cells in culture: Susceptibility to poliomyelitis after prolonged cultivation. *Proceedings of the Society for Experimental Biology and Medicine, 83*(3), 577–579.
6. Enders, J. F. (1954). Some recent advances in the study of poliomyelitis. *Medicine (Baltimore), 33*(2), 87–95.
7. Weller, T. H., Robbins, F. C., & Enders, J. F. (1949). Cultivation of poliomyelitis virus in cultures of human foreskin and embryonic tissues. *Proceedings of the Society for Experimental Biology and Medicine, 72*(1), 153–155.
8. Dulbecco, R., & Vogt, M. (1954). Plaque formation and isolation of pure lines with poliomyelitis viruses. *Journal of Experimental Medicine, 99*(2), 167–182.
9. Scherer, W. F., Syverton, J. T., & Gey, G. O. (1953). Studies on the propagation in vitro of poliomyelitis viruses. IV. Viral multiplication in a stable strain of human malignant epithelial cells (strain HeLa) derived from an epidermoid carcinoma of the cervix. *Journal of Experimental Medicine, 97*(5), 695–710.
10. Evans, V. J., Shannon, J. E., Jr., Bryant, J. C., Waltz, H. K., Earle, W. R., & Sanford, K. K. (1953). A quantitative study of the effect of certain chemically defined media on the proliferation in vitro of strain L cells from the mouse. *J Natl Cancer Inst, 13*(4), 773–784.
11. Puck, T. T., & Marcus, P. I. (1955). A rapid method for viable cell titration and clone production with HeLa cells in tissue culture: The use of X-irradiated cells to supply conditioning factors. *Proceedings of the National Academy of Sciences of the United States of America, 41*(7), 432–437.
12. Puck, T. T., Marcus, P. I., & Cieciura, S. J. (1956). Clonal growth of mammalian cells in vitro; growth characteristics of colonies from single HeLa cells with and without a feeder layer. *Journal of Experimental Medicine, 103*(2), 273–283.
13. Kao, F., Chasin, L., & Puck, T. T. (1969). Genetics of somatic mammalian cells, X. Complementation analysis of glycine-requiring mutants. *Proceedings of the National Academy of Sciences of the United States of America, 64*(4), 1284–1291.
14. Puck, T. T. (1972). *The mammalian cell as a microorganism: Genetic and biochemical studies in vitro*. San Francisco: Holden-Day.
15. Taylor, M. W., Souhrada, M., & McCall, J. (1971). New class of purine mutants of Chinese hamster ovary cells. *Science, 172*(979), 162–163.
16. Feldman, R. I., & Taylor, M. W. (1975). Purine mutants of mammalian cell lines. II. Identification of a phosphoribosylpyrophosphate amidotransferase-deficient mutant of Chinese hamster lung cells. *Biochemical Genetics, 13*(3–4), 227–234.
17. Milstein, C. (1982). Monoclonal antibodies. *Cancer, 49*(10), 1953–1957.
18. Ross, J. S., & Gray, G. S. (2003). Targeted therapy for cancer: The HER-2/neu and Herceptin story. *Clinical Leadership & Management Review: The Journal of CLMA, 17*(6), 333–340.
19. Gallo, R. C., Poiesz, B. J., & Ruscetti, F. W. (1981). Regulation of human T-cell proliferation: T-cell growth factor and isolation of a new class of type-C retroviruses from human T-cells. *Haematology and Blood Transfusion, 26*, 502–514.

4 The Discovery of Bacteriophage and the d'Herelle Controversy

Abstract

Bacteriophage (bacterial viruses) were discovered independently by two scientists, Frederick Twort and Felix d'Herelle, in 1915 and 1917. D'Herelle went on to carry out an in-depth study of these viruses, including replication and adaptation, and he proposed their possible use in anti-bacterial treatment. He called this area of research "bacteriophagy." This research should be considered the beginning of molecular biology. D'Herelle was a very controversial character and had difficulty with the scientific establishment accepting his ideas and results, as they doubted the existence of bacteriophage and the concept of phage therapy. The opposite view was that lysis by bacteriophage was the result of autocatalysis. During this period, an institute of phage therapy was established in the Soviet Republic of Georgia by Giorgi Eliava, a colleague of d'Herelle's at the Pasteur Institute. D'Herelle was invited to work at the institute but Eliava was executed during a purge shortly after its opening in 1937.

4.1 Discovery of Bacteriophage

That viruses are ubiquitous and can infect ALL living things was reinforced with the discovery of bacteriophage. That bacteria can be hosts for viruses was corroborated by the discovery of bacterial viruses, later termed "bacteriophage." These were first observed by Frederick Twort in England [1] and Felix d'Herelle, [2] a French-Canadian working at the Pasteur Institute in Paris. Twort was attempting to grow vaccinia virus in an artificial medium, without a host cell, when he noted a micrococcal contamination in the growth tubes. He describes an unusual effect on these bacterial colonies "*…inoculated agar tubes often showed watery-looking areas, and in cultures that grew micrococci it was found that some*

M. W. Taylor, *Viruses and Man: A History of Interactions*,
DOI: 10.1007/978-3-319-07758-1_4,

of these colonies could not be subcultured, but if kept they became glassy and transparent". He attempted to isolate the substance that produced this effect, but was unable to detect anything under a microscope. He found that by touching this transparent area with a loop and transferring it to other cultures, it had a similar effect, and he thought of this as an infection of the culture. Even years afterwards, it was not clear to Twort whether this was a viral infection of the culture or an enzyme that caused autocatalysis of the bacterium. The fact that his interpretation was unclear had serious consequences later in a controversy that developed on whether bacteriophage were viruses or an autocatalytic enzyme. In his paper he concludes, "*…from these results it is difficult to draw definite conclusions… [This] may be living protoplasm that forms no definite individuals or an enzyme with the power of growth…In any case, whatever explanation is accepted, the possibility of its being an ultra-microscopic virus has not been definitely disapproved because we do not know for certain the nature of such a virus… On the whole it seems probable, though by no means certain, that the active transparent material is produced by the micrococcus, and since it leads to its own destruction and can be transmitted to fresh healthy cultures it might almost be considered as an acute infectious disease of micrococci*" [1, 3].

Twort did not have the funds available to pursue this matter further and his findings lay in obscurity for a number of years. His paper was not rediscovered until much later, during the controversy surrounding the work of d'Herelle described below. The evidence indicates that d'Herelle was not familiar with Twort's work when he later published his discovery of bacteriophage in 1917.

Felix d'Herelle was born in Montreal in 1873, but educated in Paris, his family having moved back to France when he was 7 years old. On completing high school in Paris, he moved to Bonn where he attended classes for a few months, but never completed a university degree; he was thus a self-taught microbiologist. He studied microbiology from books, and built a home laboratory. Through a family friend he received a commission from the Canadian Government to study the fermentation and distillation of maple syrup. Together with his brother he invested money in a chocolate factory that soon went bankrupt. During this period, d'Herelle published his first paper in *Le Naturaliste Canadien* on the formation of carbon by plants. Although the claims in the paper were erroneous, it did show some scientific aptitude. Always an adventurer, he then took a position as a bacteriologist at the General Hospital in Guatemala City. Although not a physician, some of his work was on the major diseases of that time—yellow fever and malaria. In 1907 he received an offer from the Mexican government to continue his studies on fermentation, a topic that interested him from his early bacteriology experiments. He and his family moved to a sisal plantation near Merida, in the Yucatan, and he developed a method of producing schnapps from sisal. As far as I am aware, this was never a commercial success.

During his stay at the sisal plantation in Merida, he attempted to stop a plague of locusts. He noted that the locusts appeared to be infected by an organism that caused diarrhea (dysentery). He cultured the causative organism, a bacillus from the insects. During growth of the bacterium, he noted areas of clearing in the culture but did not realize the importance of the discovery until a few years later. However, he did attempt to stop the locust plague using the bacillus he had isolated from their guts. He used this bacillus again against locusts a few years later in Argentina. This was probably the first case of attempted biological control of an insect pest.

He was asked to investigate an outbreak of severe dysentery that occurred among a group of soldiers in Paris during World War I. He isolated and characterized a unique strain of dysentery bacillus and during culture noted holes in the bacterial lawn. D'Herelle realized then that the areas of clearing were plaques caused by a filterable agent (he passed the solution of "phage" through filters), and called this a bacteriophage, a bacteria-eating entity. He claimed that this was a virus that infected bacteria, and he also noted that the culture of bacteria growing in liquid, and which was usually turbid, cleared when this filtrate was added to the culture and bacteria did not survive. Emile Roux, his supervisor at the Pasteur Institute, read d'Herelle's paper on the discovery of bacteriophage at a meeting of the Académie des Sciences on September 3, 1917 (for details on d'Herelle's life, see Summers [4], and Hausler [5]).

4.2 Bacteriophagology

D'Herelle himself published his research on bacteriophage in a monograph, "The bacteriophage and its behavior," as well as many subsequent papers and books [6–9]. This study was the basis of what he called "bacteriophagology." He introduced new terminology to describe the events happening after infection and describes in some detail the purification and titration of bacteriophage cultures. He also noted that different strains of bacteria (even of the same genus) produce different bacteriophage and grow under different conditions and were thus very host-specific. His initial approach was basic: showing that actively growing, viable bacteria were optimal for producing phage and that one could titer (calculate the number of) phage by serial dilution. He also made the important point previously discovered by Martinus Beijerink for plant viruses, such as tobacco mosaic virus, that bacteriophage were obligate parasites and could not be grown without a host bacteria. His monograph reads like a lab manual, including techniques for plaque assays and isolation of what he called the corpuscles. He showed that each plaque was derived from a single phage infection and that plaques contained thousands of new phage.

D'Herelle also describes increases in virulence with passage of the phage in its or other host cells as well as decreases in phage "activity" by high temperatures. Unaware of its importance, he describes adaptation of strains (he calls them races) of bacteriophage to different strains of bacteria as a result of numerous passages. It is possible that he was isolating mutants of the phage and resistant strains of bacteria. Later in life, as a faculty member at Yale, he did isolate specific mutants of phage and their hosts. He also suggested that bacteriophage were particulates and developed antisera in rabbits to specific phage. He presented his work with great enthusiasm at international meetings, and was ready to meet criticism, which he ran into—head on. It is fair to suggest that he was the founder or "father of modern molecular biology," since much of the pioneer work done by him was later repeated by the Cold Spring Harbor phage group in the 1950s.

d'Herelle and his many collaborators published numerous papers on the characteristics of bacteriophage, and later on their possible clinical applications. He proposed that phage produced a sort of immunity to bacterial diseases and that many bacterial diseases, such as cholera, bubonic plague, and dysentery could be controlled by using phage in a therapeutic fashion. Unfortunately, he was not interested in the basic science of virology, but only in clinical adaptation of his work, although during his time at Yale he did basic research on bacterial mutants or, as he called it, adaptation. He concluded that bacteriophage are viruses and are alive, a controversy that later dogged virology for many years after the crystallization of the tobacco mosaic virus by Wendell Stanley. In his chapter called "*The criteria of life,*" he argues that a major component of living is the ability to modify the environment to make more of oneself. He proposes two fundamental properties of life, the assimilation of heterologous media to make more of oneself different from the original medium, and the power of adaptation. We have no doubt that a bacterium is living, since it uses a defined (or undefined) medium to make more of itself—in the same way the phage uses the bacteria as its medium, to reproduce more of itself. He argues that what is important in this definition is the heterologous medium. "Assimilation in a heterologous medium implies the power of modifying the heterologous substances contained in this medium rendering them comparable to the distinctive substances of the being which brings about the transformation." Thus, phage use bacteria as their heterologous substance and convert it to more phage, in the same way that sheep convert grass to more of themselves.

"The power of adaptation may be defined as the faculty possessed by a being to respond to a specific opposing reaction to variations in the conditions of the medium." D'Herelle was a disciple of Jean-Baptiste Lamarck, and believed that organisms adapted to their environments rather than the environment selecting for mutants. Although discredited for over 200 years, we now believe that some aspects of Lamarckism, such as changes in DNA in vivo due to environmental pressures, may explain genetic adaptations. At the chemical level, this would be due to methylation or other alterations in DNA (coined epigenetics).

Of course d'Herelle had no idea that phage actually contained its own genetic material and thus instructed the cell to make more phage using pools of nucleotides and amino acids derived from the host. The concept of a gene at that time was a very hazy one.

D'Herelle also made the observation that certain strains of bacteria were "lysogenic," i.e., they carry bacteriophage in a latent, non-infectious fashion. This observation was also made by some of his rivals and led to further controversy as to whether such an organism as a bacteriophage really exists. Without discussing the mechanisms of lysogeny, d'Herelle implies that one can isolate strains of bacteria that seem to carry "contaminating" phage (We will discuss the phenomenon of lysogeny in more detail later; see following chapter). However, d'Herelle's opponents used this as proof of their theory that bacteriophage were "enzymes" produced under certain circumstances by the host cell.

4.3 Phagotherapy

D'Herelle's primary interest was in the clinical relevance of bacteriophage. He attempted to correlate the pathogenicity and severity of a disease with the presence of bacteriophage specific for diseases causing bacterium. He isolated phage from patients and assayed them on both laboratory indicator bacterial strains and bacterial isolates of the patients. Such studies were performed with isolates from patients with dysentery, typhoid fever, and later, on a much larger scale, with isolates derived from patients suffering from bubonic plague and cholera. He and his colleagues claimed that in the majority of cases, it was the presence of the bacteriophage and the subsequent destruction of the pathogenic strain of bacteria that resulted in the alleviation of symptoms. He also described cases of resistant bacteria and described situations in which the bacteriophage is less active and the condition of the disease more severe. Similar studies were done with the agents of bubonic plague (*Yersinia pestis*) and staphylococcus infections. D'Herelle even proposed that this occurs in plants, as in the case of legumes infected with *Bacillus radicicola* and its phage. He wrote, "The bacteriophage is as much an infectious agent as the bacterial it is using as a host, and there is a constant struggle between the bacterial host and the phage." If the phage "wins" this struggle, the disease is attenuated or resolved. Using this rationale it should be possible to use bacteriophage to treat or, as d'Herelle calls it, "immunize" against a bacterial infection. Although the work of many investigators did not show any improvement in experimental infections in rabbits using phage, d'Herelle states that he is uninterested in these experimental systems, since the bacteriophage "immunity" only works in the natural infection. He proposed that during a natural infection, where phage are present, the infection would be self-limiting due to the destruction of bacterial cells; if phage were not present, then the infection would be more severe, and in most cases, fatal. He tested this hypothesis by adding phage to three very different infections: avian typhosis, which is an enteric disease of chickens; septicemia of cattle (barbone of buffalo); and a toxic disease, shigella, in rabbits. These treatments were all carried out in the field and were successful.

Claims were made that a single injection of the appropriate phage cured bacillary dysentery and staphylococcus infections, although d'Herelle suggested that a cocktail of phage types may be necessary to overcome resistance (and I assume to find the right phage), an approach that was later used in clinical practice in the Soviet Union. The results obtained in the treatment of bubonic plague were encouraging, as they were for other bacterial diseases. D'Herelle termed this "phagotherapy." This novel approach to curing disease was portrayed in in 1925 in Sinclair Lewis's book *Arrowsmith*, in which a physician experiments with bacteriophage to cure bubonic plague in the West Indies.

4.4 D'Herelle and the Phage Controversy

Without a doubt, there was a great deal of controversy surrounding d'Herelle's discovery and his interpretation of data. This in part originated from the character of d'Herelle himself, as well as from bitter scientific rivalries. D'Herelle was an adventurer who travelled to many countries on many continents. He had no formal education in microbiology, although his self-learned techniques were exemplary. This lack of a formal education and thus an academic degree meant that he did not have a stable position at any institute. He was given space in different laboratories at the Pasteur Institute, but did not really have a lab or financing of his own until later in life. His discovery of bacteriophage led to the controversy of whether this was a virus (as he proposed) or some catalytic enzyme produced by the host cells that led to their own destruction. His major adversary was the Nobel Prize-winning immunologist Jules Bordet, who refused to believe that bacteriophage were independent viruses, but tried to fit d'Herelle's data into his own interpretation, based on the interaction of anti-sera and complement leading to lysis of the bacterial cell. He proposed (as did Twort earlier, and even some of d'Herelle's own colleagues) that the destruction of the bacterial cells resulted from autocatalysis [10]. It appears that there were stormy sessions at various scientific meetings between d'Herelle and his supporters on the one side, and Bordet and other immunologists on the other. Even as late as 1934, a paper by Monroe Eaton and Stanhope Bayne-Jones [11] in the Journal of the American Medical Association still questioned the concept of bacteriophage as viruses and listed a long list of failures of bacteriophage therapy in experimental animals, although there was some success in the treatment of staphylococci infections and enteric infections. This paper is particularly biased against the use of bacteriophage as curative agents and even doubts the existence of bacteriophage. Why there was so much bitterness towards the idea of bacteriophage as a clinical cure is not clear; one of the reasons may have been the "commercialization" of bacteriophage treatment by pharmaceutical companies, without strict quality control—or simply antagonism to d'Herelle the man.

By the 1940s, d'Herelle's idea that bacteriophage were particulate viruses began to attract more support. New methods and new investigators sided with d'Herelle, although there was still some opposition. The development of the electron microscope, and the work of Helmut Ruska in 1939 and Salvador Luria in 1942 [12] showed phage particles attached to bacterial cells. By 1942, d'Herelle's concept had won. It was obviously a long, difficult fight to persuade some immunologists and other microbiologists about the existence of phage (Fig. 4.1).

D'Herelle was obviously a very colorful and controversial character. Quoting an excellent paper by Donna Duckworth [13] everywhere that d'Herelle went there were fireworks. He was well travelled, gave many lectures on his bacteriophage phenomenon, and speculated on the origins of life and the role of viruses in this origin. He also spent considerable time in the Soviet Union, not something to endear him to American scientists. He was nominated for the Nobel Prize countless times, Hausler [5] but unfortunately until very recently [4] was absent

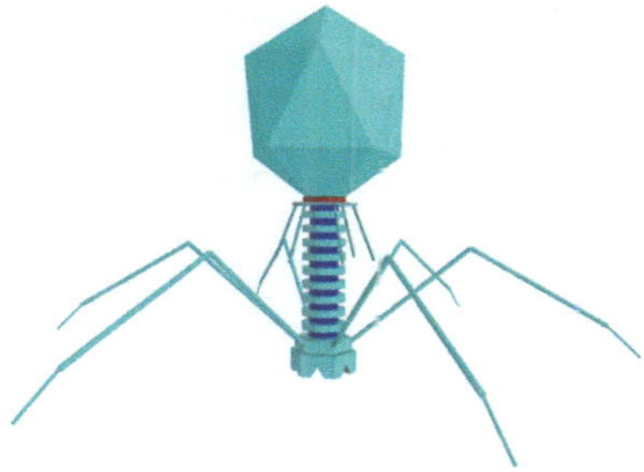

Fig. 4.1 Artist's rendition of bacteriophage T4. (With permission, www.cronodon.com)

from the history of microbiology. He was awarded three honorary degrees in medicine, which allowed him to use the title "doctor," although after his appointment to the faculty in Yale he became professor. He also had problems at Yale because of his arrival many months after the beginnng of his appointment, and also because of illness, suggested by some to be due to hypochondria, although in truth he had caught a number of infectious diseases in the tropics (he suffered from malaria and amoebiasis). In addition, a self-inoculation of tetanus vaccine led to some paralysis. At Yale he was a subject of controversy, travelling and giving talks most of the time rather than working in his laboratory and teaching [4].

4.5 Phage Institute in Tiflis (Tbilisi)

D'Herelle was at Yale from 1928–1933. He had many arguments with the chairman of the microbiology department, and when his colleague from Paris, George Eliava, offered him a position in Tiflis (now Tbilisi) in the Soviet Republic of Georgia, he jumped at it. During this period there were some suspicions at Yale that he was running a commercial business from Paris, where he spent his summers (at the Laboratoire du Bacteriophage, which supplied phage stocks commercially.) In 1923 Eliava returned to Tiflis from Paris to work on bacteriophage therapy, and he established the Tiflis Bacteriological Institute, to study the use of bacteriophage in clinical and epidemiological outbreaks. He received encouraging support from the Soviet government and established the All-Union Bacteriophage Institute, today known as the Eliava Institute. (This was during a period when the Soviet Union encouraged novel research in medicine.)

When d'Herelle decided to leave Yale, Eliava offered him everything he wanted: a new institute with colleagues interested in phage therapy, a government interested in his work, and the added attraction, the ideology of "scientific socialism." D'Herelle arrived in Tbilisi in 1933, leaving an economically depressed USA for what appeared to be a prosperous USSR. His laboratory was well funded, and he was given good living conditions and even a chauffeur. Unfortunately, this situation did not last long, since Eliava fell victim to the Stalinist purges. He was arrested personally by Stalin's chief of police, Lavrentiy Beria, and executed immediately, although it is not clear why. He was a bit of a playboy and Beria may have been

Fig. 4.2 D'Herelle (seated) and Eliava (standing to the right) in Tblisi, just before Eliava was arrested and executed (from *Science* 298. 2002)

"in love" with his wife or mutually attracted to another woman, possibly the librarian of the institute [14] (Fig. 4.2). D'Herelle then returned disappointed to Paris, where he continued to carry out research in the private company he had established, but continued to collaborate with colleagues at the Tiflis institute. This company supplied phage for therapeutic purposes. The laboratory in Paris produced phage against carbuncles, dysentery, and wound infections.

This was a very productive period both for d'Herelle and, in particular, for Professor Alexandr Petrovich Tsulukidze, who produced 14 papers on the subject between 1935 and 1957. Tsulukidze remained in Russia and headed a team that tested bacteriophage for war wounds during the Finnish campaign. By 1946, phage therapy had become firmly established in the Soviet Union. During World War II, d'Herelle had moved to Vichy, but his son-in-law remained in Paris throughout the war, in charge of the research laboratory and business. D'Herelle and his group examined phage preparations prepared in other laboratories from other countries and concluded that most were not "capable of effecting recovery from infectious disease." In part. he blamed this on capitalism and the profit motive, which he claimed never entered into his work.

There was active research and clinical use of bacteriophage in Russia and in Poland. Since d'Herelle's phage was used with mixed results to treat various infections by the German and Soviet armies during World War II, and also extensively explored in the U.S. before the advent of antibiotics, with the production of antibiotics in the 1940s, particularly penicillin and streptomycin, the use of phage as a therapeutic agent disappeared from medicine in the U.S. and western Europe. With the observations today, in the twenty-first century, of antibiotic-resistant strains of bacteria, particularly among pathogenic organisms such as staphylococcus, bacteriophage therapy is again being explored. It has been reported to be very successful, and is now being studied in the West. Although the clinical trials in eastern Europe may not have met western (U.S. and U.K.) standards, they did show a high rate of response, and if the reports are to be believed, many outbreaks of bacterial infections were controlled by the judicious use of phage. Most of this work was done at the Eliava Institute of Bacteriophage and Microbiology in Tblisi in the ex-Soviet Republic of Georgia, where to this day there is a large collection of phage that could be used clinically—in particular against cholera.

References

1. Twort, F. W. (1915). An investigation on the nature of ultra-microscopic viruses. *Lancet, 2*, 1241–1243.
2. d'Herelle, F. (1917). An invisible microbe that is antagonistic to the dysentery bacillus. *Comptes Rendus Academie Sciences Paris, 165*, 373–375.
3. Twort, F. W. (1922). The bacteriophage; the breaking down of bacteria by associated filter passing lysins. *British Medical Journal, 2*, 293–296.
4. Summers, W. C. (1999). *Felix d'Herelle and the origins of molecular biology*. New Haven: Yale University Press.
5. Hausler, T. (2006). *Viruses versus superbugs*. New York: Macmillan.
6. d'Herelle, F. (1921). Sur l'historique du bacteriophage. *Comptes Rendus Seances Societe Biologie Paris, 84*, 863–864.
7. d'Herelle, F. (1921). The nature of bacteriophage. *British Medical Journal, 2*, 289–293.
8. d'Herelle, F. (1922). *The bacteriophage: its role in immunity*. Baltimore: Williams and Wilkins.
9. d'Herelle, F. (1926). *The bacteriophage and its behavior*. Baltimore: Williams and Wilkins Co.
10. Bordet, J., & Ciuca, M. (1921). Remarques sur l'historique des recherches concernant la lyse microbienne transmissible. *Comptes Rendus Seances Societe Biologie Paris, 84*, 745–747.
11. Eaton, M. D., & Bayne-Jones, S. (1934). Bacteriophage therapy. *Journal of the American Medical Association, 103*, 1769–1776.
12. Luria, S. E., & Anderson, T. F. (1942). The identification and characterization of bacteriophages with the electron microscope. *Proceedings of the National Academy of Sciences of the United States of America, 28*(4), 127–130.
13. Duckworth, D. H. (1976). Who discovered bacteriophage? *Bacteriol Reviews, 40*(4), 793–802.
14. Chanishvili, N. (2012). Phage therapy–history from Twort and d'Herelle through Soviet experience to current approaches. *Advances in Virus Research, 83*, 3–40.

The Beginnings of Molecular Biology 5

Abstract

This chapter describes the beginnings of molecular biology with the discovery of bacterial transformation and identification of DNA as the genetic material. The role that bacteriophage and the Cold Spring Harbor Phage group played in this development are emphasized. The discovery of transducing phage and the regulation of gene expression in the temperate phage lambda led to models of transcriptional and translational control. The discovery of lysogeny, and the integration of the bacteriophage into the host chromosome, ultimately led to the idea of cancer viruses being integrated into the chromosome and the oncogene model of cancer.

5.1 Early Studies with Bacteriophage

Virology began its venture toward the understanding of molecules with the crystallization of the tobacco mosaic virus by Wendell Stanley in 1935 [1]. Working with highly purified TMV, he identified conditions where virus particles were arranged in a lattice so that X-ray beams could elucidate their molecular features by diffraction, and Stanley received the Nobel Prize in Chemistry in 1946 for his groundbreaking work. Even at that late stage in the 1930s, proteins were considered to be the genetic material, although the virus was shown to contain small amounts of nucleic acids, not thought to be of great importance.

Although the importance of bacteriophage as therapeutic agents waned in the 1930s, they became of interest as models for the study of viruses and their structure. The discovery that it was possible to crystallize a virus [1] had an enormous impact on the field. In the 1930s, Max Schlesinger reported that bacteriophage could be studied by chemical and physical-chemical techniques. He was the first to obtain a pure culture of the bacteriophage, WLL (later identified as one of the T-even phages)

M. W. Taylor, *Viruses and Man: A History of Interactions*,
DOI: 10.1007/978-3-319-07758-1_5,

of *Escherichia coli* [2] by passing a suspension of the bacteriophage through a colloidal filter with a graduated range of pores and then precipitating it in an ultracentrifuge. He also reported that the phage preparation contained high levels of a Feulgen staining material, now known to be nucleic acid [3].

A group at the California Institute of Technology, including physicist Max Delbruck, were interested in studying the origins of life and the mechanisms of replication of all organisms. They initially studied embryology and explored problems of importance in biology to the physicist. For this purpose, they decided to study viral replication, and the simplest available system was the bacteriophage. In order to do this, quantitative techniques were essential. In 1939, Ellis and Delbruck [4] developed the one-step growth curve for bacteriophage. (In fact, d'Herelle had performed similar experiments but these appear to have been forgotten.) The question posed was whether one could infect a bacterial cell with a single phage particle and measure both the changes occurring inside the cell, and the production of bacteriophage from that cell under experimental conditions. Luckily these experiments were performed with a virulent phage, isolated from sewage ultra-filtrates that grew rapidly on the common enteric bacterium, *E. coli*. This particular phage formed large visible plaques on a lawn of bacterial cells. Delbruck, who rediscovered the "Poisson distribution" originally published in 1837 that dealt with random sampling, developed a mathematical approach to this problem. From such experiments, the burst size from a single phage infection, and the time of the burst (extent of the replication cycle), were calculated.

5.2 The Phage Group

In 1940 the phage group that was formed for the study of the replication of bacteriophage consisted of Alfred D. Hershey, Salvador Luria and Max Delbruck. This group was later expanded to include other scientists interested in the physical structure of bacteriophage, and was centered in Cold Spring Harbor, on Long Island. In 1945 Delbruck began his annual phage course, which "spread the word" about the usefulness of bacteriophage to study what was eventually to become molecular biology. It helped that these meetings were held at Cold Spring Harbor, a beautiful spot on Long Island. The meetings became known not only for bacteriophage biology but for many topics in molecular biology, and, later on, topics in genetics and medicine. The last evening of the meeting, with a fancy lobster dinner, was always an added attraction, and became a tradition of the Cold Spring Harbor meetings.

It was not until further research with bacteriophage, and simultaneously with bacteria, [5] that it could be shown that the nucleic acid carried the genetic information. These experiments included DNA transformation in bacillus, and later experiments with bacteriophage, in which it could be shown using radiolabeled material that the DNA of the bacteriophage entered the bacterial cells, replicated and produced progeny phage, and that the protein did not carry the

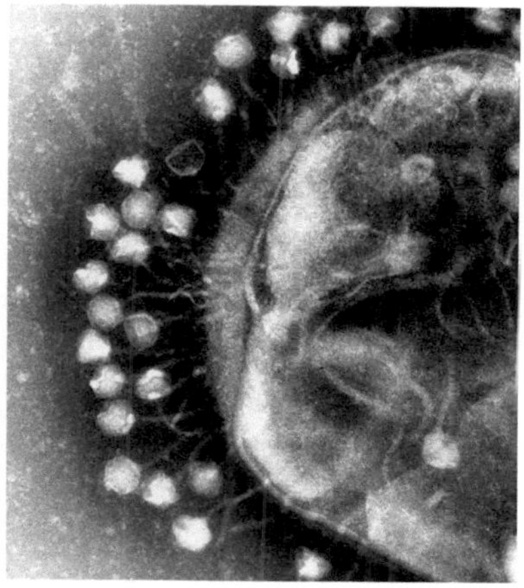

Fig. 5.1 Phage attached to the outside of a bacterial cell (Wikipedia; photograph Graham Coln 2008)

genetic information as previously thought [6]. The protein coat of the phage remained attached to the outside of the cell. This was a seminal discovery in molecular biology (see Fig. 5.1). DNA was the genetic material, the basis of inheritance, and not protein as was thought until the mid-1940s and early 1950s.

These experiments defined an eclipse period, when no phage is detected even inside the infected cell. The importance of this discovery should not be underestimated since unlike other organisms, the absence of phage during the first period after infection indicated that events took place in the cell without the intact phage being present. In fact, phage structural components could not be detected in the infected cell for quite some time, suggesting that replication occurred in a fashion quite different from that seen in other organisms. Similar results were later reported with animal viruses.

One of the most important advances in virology was the development of the electron microscope (EM), already referred to in Chap. 2. Ernst Ruska and Max Knoll invented the EM in Germany in 1931–1933, and the Siemens Company patented the microscope and then developed its application to biological samples, initially to visualize poliovirus. Ruska was a German physicist who received the Nobel Prize in Physics in 1986 for his work on developing the EM. Ruska, Edgar Pfankuch and Gustav Kausche (1940) were the first to study bacteriophage preparations with the EM, and in 1941 Ruska described sperm-shaped particles. He was uncertain whether these were virus or components of the host bacteria [7]. The first paper published in English on this subject (others were in German), was that of Luria and Anderson [8] on the structure of bacteriophage as seen by electron microscopy. They described various isolates of phage, showing the presence of a head and tail-like structure, the attachment of the phage to the bacterial host, and also variations in morphology with different phage, some having long tails with brush-like appendages, others just straight tails or stubs (Figs. 5.2, 5.3). This report, however, was met with skepticism in some circles; one interesting question was whether the tails were flagella-like and gave motility

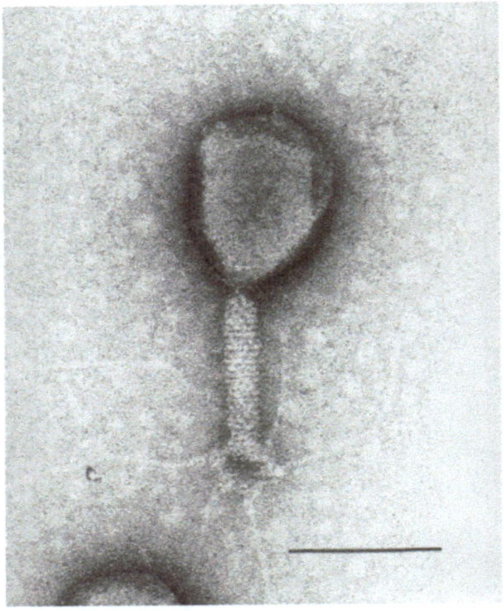

Fig. 5.2 Bacteriophage T4 (electron micrograph from ICTV)

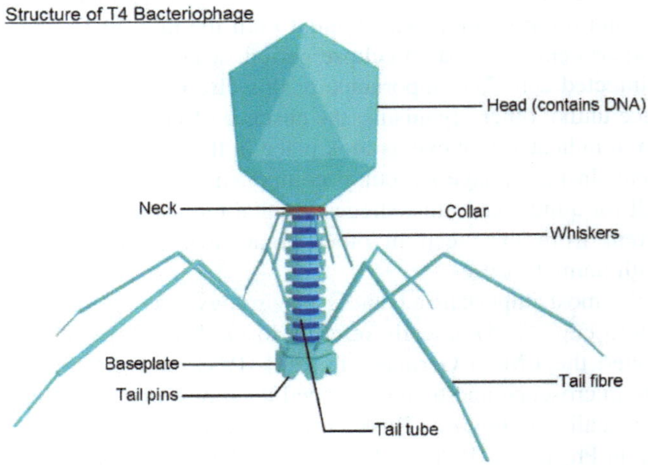

Fig. 5.3 Schematic structure of phage T4 with parts labeled (www.cronodon.com with permission). The tail fibers attach to the cell wall of the bacterium. The tail pins also attach to the cell wall and bring the tail tube in contact with the surface. This causes the tail tube to contract like a hypodermic syringe, and the DNA in the head is pushed into the bacterial cell where it undergoes replication, transcription and translation to produce thousands of new phage particles

to the phage particle. It was later shown that the tails were not flagella, but instruments of delivery of the nucleic acid. These were not simple structures, but quite complicated appendices and appeared to act like syringes after attaching to the outside of the bacterial host.

An important observation made in these EM studies was that the phage particles themselves appeared to "stick" to the bacterial cell wall and not enter the cell. This led to the discussion of whether this was analogous to monospermy, in which only one sperm enters the egg to fertilize it, a characteristic of the higher bisexual world. Was it possible that only one phage particle entered the cell and this event blocked other phage particles from entering the cell? We now know that many phage particles can enter the cell; however, if a cell is infected with too many particles, it may "lyse" from without. Later work using radioactive tracers would show that the protein components of the phage (the head and tail) never entered the cell, only the nucleic acid component not visible by electron microscopy [6]. This is not true of animal viruses, which are completely engulfed by the cell.

Delbruck insisted that his group work specifically with phage that affected *E. coli* strain B and not others, and to concentrate only on these; this proved to be a significant and important decision. These were the T-phages that were "well-behaved" in that they were easy to quantitate in a plaque assay. They became known as phage T1-T7, in order of isolation, and later were divided into the T-even phages and T-odd phages. Some of them, such as T1, formed huge plaques that for anyone working with other phaged were a menace since they easily contaminated cultures. These T-phages became the "workhorses" of virology and later of molecular biology. The experiments of Hershey and Chase described above [6] are among the most important in the birth of molecular biology. The identification of nucleic acid as the basis of genetic information opened a new area of biology: molecular genetics. Studies on mutations in bacteria and bacteriophage would eventually lead to the deciphering of the genetic code [9–12].

Studies by Doermann and others indicated that infection with two different phages often led to recombinant phage, suggesting that the DNA of the phage was mixed and compartmentalized during replication in the cell, underwent recombination by a mechanism of strand switching, and was packaged into the phage head [13, 14]. The isolation of temperature-sensitive (ts) mutants and advances in electron microscopy allowed one to study the intracellular events after phage infection. It was thus possible to visualize components of the phage as well as accumulation of nucleic acid (black spots in the EM) and the stages of assembly. Thus, phages were assembled inside the cell, either under instruction from a specific gene, or from interactions between proteins (self-assembly). The DNA was packaged into the phage head (see Fig. 5.5). Utilizing different mutants, phage could be self-assembled from two mutants in vitro by mixing the components. Using temperature, it was possible to look at this assembly as if in slow motion photography. Using ts mutants, one could visualize the components of the phage, including tail structures and heads [14, 15] (Fig. 5.4).

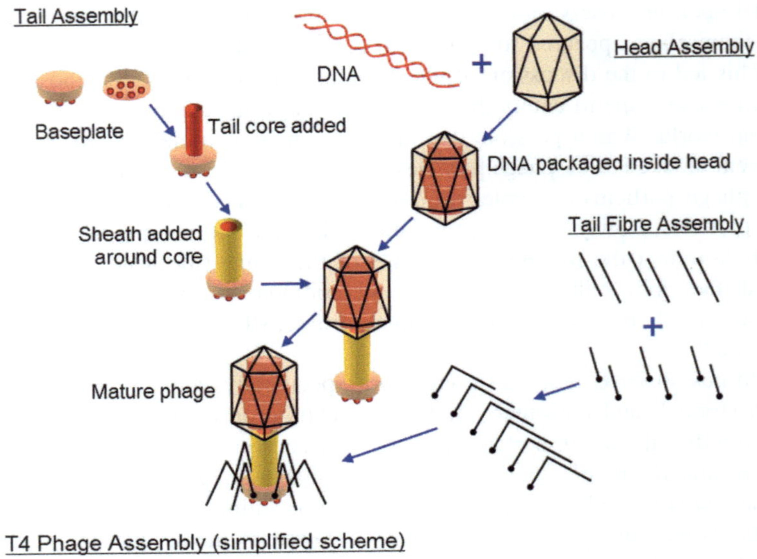

Fig. 5.4 T4 phage assembly (www.cronodon.com with permission)

5.3 Lysogeny and Phage λ

D'Herelle had noted that certain strains of bacteria appeared to be lysogenic for bacteriophage (see Chap. 4). Bacterial cells were occasionally lysed by "carrier" phage, which were not always present in an infectious form. This was a puzzling phenomenon that caused quite a bit of consternation and was behind the controversy surrounding the discovery of phage and whether they were viruses or autocatalytic molecules. The French scientist André Lwoff at the Pasteur Institute decided to study this phenomenon [16]. He studied single bacterium (*Bacillus megatarium*) in micro-drops of media under the microscope, watching for lysis and the production of phage (by titration). Only occasionally did a bacterium lyse and produce phage; one can imagine how tedious and bored he must have been. Thus he hypothesized that certain bacteria carried within them an infectious entity that was normally not replicated; this entity was termed a "prophage." When lysogenic bacteria were placed in culture, there was occasional lysis and production of phage. Lwoff and others assumed this to be triggered by changes in the media by bacterial products. What these factors were was never discovered.

After months of experimentation, it was decided to irradiate a bacterial culture with UV radiation. This really did not make sense, since the bacteria themselves would be sensitive to the UV irradiation. However, to Lwoff's surprise (these experiments were done with Siminovitch and Kjedgaard, both later eminent phage

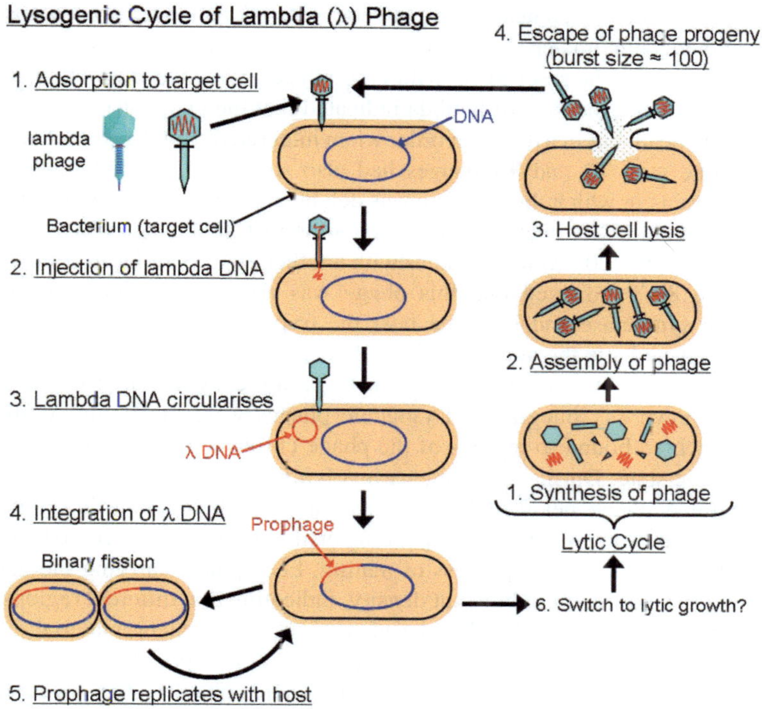

Fig. 5.5 Lysogenic cycle of phage lambda (www.cronodon.com, with permission)

geneticists), the culture lysed with the production of thousands of phages; thus, lysogenic phage induction was discovered. A model was proposed in which the phage DNA was repressed by a protein produced by the prophage; the UV irradiation destroyed this repressor (or converted it to an inducer), freeing the phage DNA to replicate. The prophage was considered innocuous, but when induced, the "vegetative" phase of phage replication begins, and this leads to the destruction of the host cell and production of thousands of viral particles. Lysogeny was used as a model for developing theories of cancer formation by viruses (see Fig. 5.5).

While studying phage λ in the 1950s, it was discovered that the phage picked up a host gene (a gene for galactose metabolism, referred to as the "gal gene") and transmitted it to gal-deficient *E. coli* [17]. Further work indicated that the prophage integrated into the bacterial chromosome near or next to the gal gene. When induced, the phage existed independently and expressed this gene. Such a phage was known as a "transducing phage." Possible mechanisms were either the phage genome synapses (pairs) to the bacterial chromosome and undergoes recombination, or, as proposed by Alan Campbell, the phage genome circularizes, and recombines with the bacterial chromosome near the gal gene at a specific attachment site. Upon being induced by UV light or spontaneously induced, the phage occasionally incorporated nearby host genetic material, and in the process lost some of its own genetic material [18, 19]. This transducing phage was replication defective but could be rescued by wild type phage. Evidence from other research using three genetic markers of lambda indicated that the phage genome was linear in the phage head, but in a different order when integrated, and thus circular. Similar studies done with T2 and T4 phages had also suggested a circular genome. In bacterial strains in which the gal gene was deleted, the phage could transduce other nearby genetic markers such as biotin, indicating that the lambda phage integrated into a specific site [20]. Similar studies with another phage similar in structure to λ, called "phi 80," indicated that this phage was able to transduce the gene for tryptophan synthetase and integrate near the tryp gene [21]. Thus, not only did lysogenic phage have the ability to circularize and integrate into the chromosome, they had specific integration sites (att) and specific enzymes involved in recombination, known as "integrase." Lysogenic phages were under the control of a repressor, which bound to regions of the phage chromosome and maintained it in the lysogenic state. During experiments involving mating of two bacterial strains, the integrated phage behaved like a bacterial gene, being transferred with the bacterial chromosome—although depending on the strains used, it could be induced to the lytic state during the process of mating. Each phage carrying the gal locus was found to be different in buoyant density, indicating that different regions of the phage genome were deleted and replaced with bacterial genetic material.

Using bacteriophages such as T2 and lambda, it was possible to dissect each stage of virus replication, the role of nucleic acid, quantify the appearance of mutation and mechanism of mutation, and develop a model from phage lambda for gene repression and regulation. The number of scientists involved is too large to list, but the main players all constituted the phage group headed by Max Delbruck, that met at the Cold Spring Harbor Laboratories. They were later joined by James Watson and Francis Crick of DNA double helix fame.

Studies with bacteriophage led to a greater understanding of how cells regulate protein synthesis, and to the recognition of what has been termed the "trinity of molecular biology"—namely, DNA–RNA–protein. In a landmark paper in 1961, Sydney Brenner, François Jacob, and Matthew Messelson, using bacteriophage T4, showed that infection with the phage led to cessation of host protein synthesis, that phage protein synthesis occurred on organelles called ribosomes, and that there was an RNA attached to such ribosomes, termed "messenger RNA [22] ." This RNA was short lived and appeared to be involved in protein synthesis. Prior to this, it was believed (or proposed) that the information for the protein was carried by the ribosome itself: "one gene–one ribosome–one protein." However, this theory was already doubtful because of the homogeneity of the ribosomal population. Jacob and Jacques Monod proposed that the ribosome was not the intermediate, but played a role in the transfer of an unstable message (mRNA) copied from the DNA that carried the instructions for the genetic information, and that this "message" interacted with the ribosome. Transfer RNA carrying the respective amino acid would interact with the mRNA to form a protein on the ribosome. Host and phage RNA could be distinguished using radioactive density, labeled

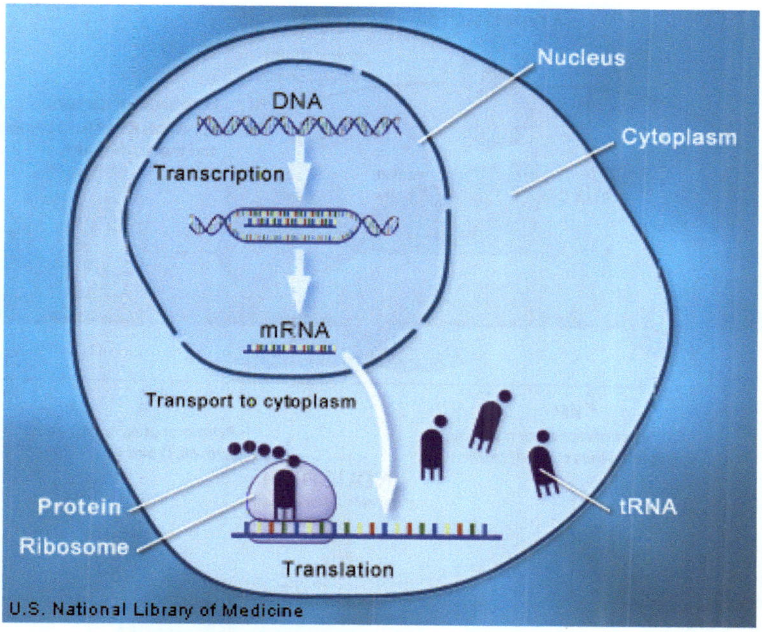

Fig. 5.6 Mechanism of transcription and translation in simplified fashion (U.S. National Library of Medicine)

DNA and RNA. Host protein synthesis stopped a few minutes after infection, and the new proteins being made were 60 % or more phage structural proteins. The newly synthesized RNA was short lived and similar in sequence to phage DNA. This RNA interacted with ribosomes to form polysomes, the machinery for protein synthesis. From phage research the idea of mRNA was born, as well as the process of transcription and translation of genetic material. Information is transferred from the DNA by means of mRNA to protein [23] (Fig. 5.6).

Continuing studies with bacteriophage, particularly with the lysogenic phage λ, Mark Ptashne and others described the intricacies of gene regulation. This developed from a study of λ "immunity," i.e., the maintenance of the phage in the lysogenic state, the destruction of the repressor by UV treatment leading to induction and replication of the phage [24, 25]. Looking back, it is surprising that studies of a single bacteriophage such as this *E. coli* phage would result in discoveries that were applicable to the whole field of gene regulation and even development in higher organisms. The idea of repressors acting on specific DNA sequences known as operators developed from the work of Jacob and Monod with gene regulation in bacteria in the 1960s. It was suggested that there were proteins that interacted with sequences of a gene that either shut the gene down (repressed) or were in turn modified and turned the gene on (activated). Mark Ptashne, as a student at Harvard, used this model in studying lambda phage. In 1967 he

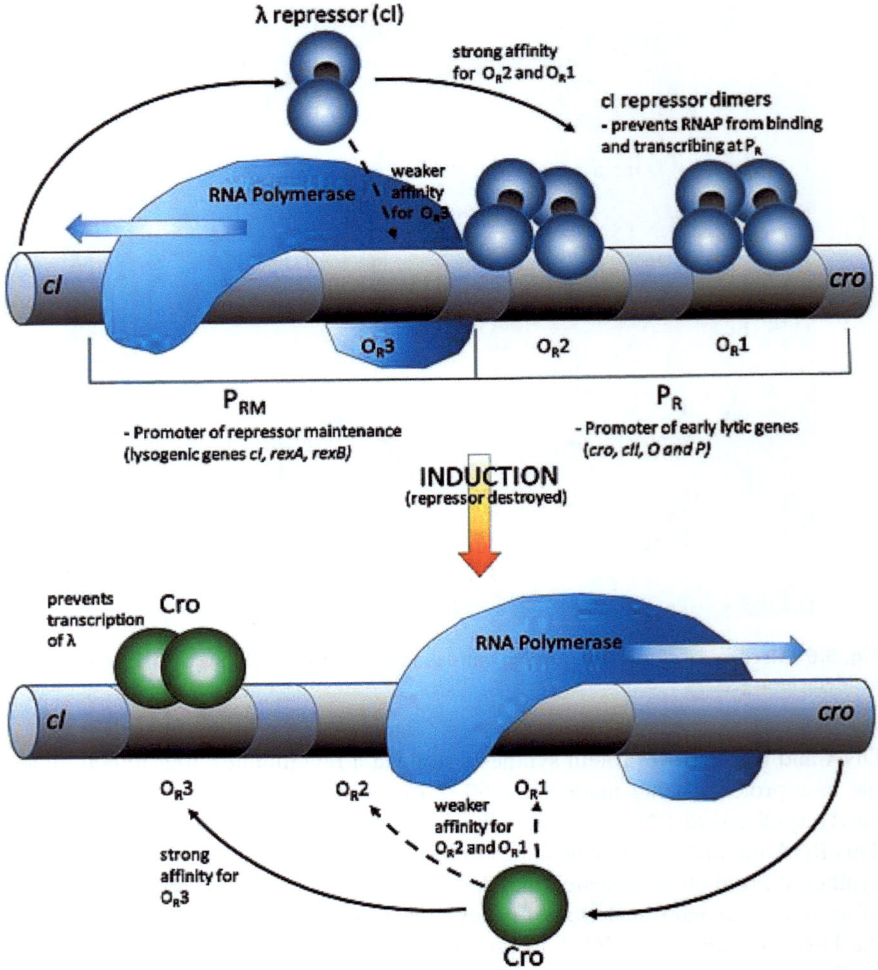

Fig. 5.7 Transcriptional state of the P_{RM} and P_R promoter regions during a lysogenic state versus induced, early lytic state. (http://en.wikipedia.org/wiki/Lambda_phage)

proposed that the phage was kept in its lysogenic state by a repressor that interacted with a specific sequence of phage DNA (promoter region). This repressor was made from a phage gene (C1), analogous to similar repressors hypothesized to regulate certain bacterial genes (lac repressor). The repressor was identified by using ts mutants of the phage; this repressor could also interact and prevent the replication of other superinfecting phages of the same type by binding to the same sequence of DNA. Lysis occurred when the repressor was destroyed and an activator protein bound to a nearby promoter and started the lytic cycle (see Fig. 5.7). This model, with variations, could be applied to any gene, and it gave rise to the idea that gene regulation in general, including the various steps in embryonic development, was controlled by repressor proteins and activator proteins that bound to different regions of DNA. RNA transcription would be controlled in the same fashion. This is of course a simplified representation, since we now know that transcriptional regulation involves many proteins interacting with each other at each promoter region, one or more of which binds to specific DNA sequences.

5.4 The Oncogene Theory of Cancer

The discovery of lysogeny and transduction led to the development of models explaining how animal viruses could cause tumors. This model was first proposed by Robert Huebner and George Todaro [26], two scientists working with Rous sarcoma virus and other retroviruses and expanded upon by Gross, who worked with leukemia viruses. At this time (the 1960s–1970s), it was believed that viruses were the major cause of cancer. Earlier work (referred to in Chap. 1) had shown that viruses could cause leukemia and solid tumors in chickens. In 1933 Shope discovered a papilloma virus [27] that caused tumors in cottontail rabbits, and in 1936 John Joseph Bittner identified a mouse mammary tumor virus [28]. In the 1960s, the Epstein-Barr virus was associated with the human Burkitt's lymphoma and other viruses, such as polyoma virus, SV-40, adenovirus; adenovirus and papillomavirus transformed cells in culture, indicating that they might be cancerous for humans. The discovery of a series of retroviruses and the reverse transcriptase enzyme, an enzyme that could copy RNA into DNA, reinforced the idea that viruses were the major cause of human cancers. Viral transformed cells, when injected into mice or hamsters, nearly always developed into tumors. The oncogene theory proposed that all cells carry genes that could be altered by environmental factors and turned into cancer-causing genes; these genes were known as "proto-oncogenes" and the cancerous form as "c-oncogenes." When cells are infected with a tumor virus (retrovirus), the oncogenes integrate into the chromosome by recombination, replicates with the cell, and occasionally—if activated—a tumor develops. The virus itself would be silent and not expressed, although the genetic material would be present in all cells (see Chap. 8 on viruses and cancer). An adaptation of this theory, proposed by Gross, was that the viruses involved were normal components of the cell and such viruses would be transmitted vertically (from one generation to the next at the cellular level and thus in individuals), and occasionally would cause a tumor (or leukemia) as a result of some action of the viral promoter on an nearby oncogene. Large numbers of cancer-causing retroviruses were discovered in rodents and fowl (chickens) and most did carry an oncogene. As in the case of transducing phages, the viruses were defective and had lost part of their genetic material with the insertion of the proto-oncogene. With further research, the number of identified oncogenes increased, as these are genes involved in transcriptional regulation, cell growth and cell division (see http://www.cancerquest.org/oncogene-table.html).

Other tumor viruses, such as adenovirus, or SV-40, or polyoma, contained viral genes that functioned to regulate cell growth, and thus acted in a fashion similar to oncogenes. Such gene products interacted with host proteins, resulting in unlimited growth and tumor formation. The type of tumor virus was categorized as either fast, when the virus carried an oncogene, or slow, as in the case of viruses causing leukemia, where the virus integrated into a site close to a proto-oncogene and acted on this gene, usurping its normal regulation.

Although this model is correct and has held up for experimental tumors in rodents, most human tumors are not caused by oncogenic viruses (see Chap. 8).

References

1. Stanley, W. M. (1935). Isolation of a crystalline protein possessing the properties of tobacco-mosaic virus. *Science, 81*(2113), 644–645.
2. Schlesinger, M. (1933). *BiochemZ, 264*(6).
3. Schlesinger, M. (1936). The Feulgen reaction of the bacteriophage substance. *Nature, 138*, 508–509.
4. Ellis, E. L., & Delbruck, M. (1939). The growth of bacteriophage. *Journal of General Physiology, 22*(3), 365–384.
5. Avery, O. T., Macleod, C. M., & McCarty, M. (1944). Studies on the chemical nature of the substance inducing transformation of pneumococcal types: Induction of transformation by a deoxyribonucleic acid fraction isolated from pneumococcus type III. *Journal of Experimental Medicine, 79*(2), 137–158.
6. Hershey, A. D., & Chase, M. (1952). Independent functions of viral protein and nucleic acid in growth of bacteriophage. *Journal of General Physiology, 36*(1), 39–56.
7. Ruska, E. (1987). Nobel lecture. The development of the electron microscope and of electron microscopy. *Bioscience Reports, 7*(8), 607–629.
8. Luria, S. E., Anderson, T. F. (1942). The identification and characterization of bacteriophages with the electron microscope. *Proceedings of the National Academy of Sciences of the United States of America, 28*(4):127–130, 121.
9. Yanofsky, C., Carlton, B. C., Guest, J. R., Helinski, D. R., & Henning, U. (1964). On the colinearity of gene structure and protein structure. *Proceedings of the National Academy of Sciences of the United States of America, 51*, 266–272.
10. Nirenberg, M. W. (1963). The genetic code II. *Scientific American, 208*, 80–94.
11. Nirenberg, M. W., Matthaei, J. H., Jones, O. W., Martin, R. G., & Barondes, S. H. (1963). Approximation of genetic code via cell-free protein synthesis directed by template RNA. *Federation Proceedings, 22*, 55–61.
12. Brenner, S., Stretton, A. O., & Kaplan, S. (1965). Genetic code: the 'nonsense' triplets for chain termination and their suppression. *Nature, 206*(988), 994–998.
13. Broker, T. R., & Doermann, A. H. (1975). Molecular and genetic recombination of bacteriophage T4. *Annual Review of Genetics, 9*, 213–244.
14. Edgar, R. S., & Lielausis, I. (1964). Temperature-sensitive mutants of bacteriophage T4d: Their isolation and genetic characterization. *Genetics, 49*, 649–662.
15. Wood, W. B. (1980). Bacteriophage T4 morphogenesis as a model for assembly of subcellular structure. *The Quarterly Review of Biology, 55*(4), 353–367.
16. Lwoff, A. (1953). Lysogeny. *Bacteriological Reviews, 17*(4), 269–337.
17. Morse, M. L., Lederberg, E. M., & Lederberg, J. (1956). Transduction in *Escherichia Coli* K-12. *Genetics, 41*(1), 142–156.

18. Campbell, A. (1965). The steric effect in lysogenization by bacteriophage lambda. I. Lysogenization of a partially diploid strain of *Escherichia coli* K-12 *Virology, 27*(3), 329–339.
19. Campbell, A. (1965). The steric effect in lysogenization by bacteriophage lambda. II. Chromosomal attachment of the b2 mutant. *Virology, 27*(3), 340–345.
20. Campbell, A. (1958). The different kinds of transducing particles in the lambda-gal system. *Cold Spring Harbor Symposia on Quantitative Biology, 23*, 83–84.
21. Taylor, M. W., & Yanofsky, C. (1966). Chromosomal relocation of prophage-associated bacterial genes. *Journal of Bacteriology, 91*(4), 1469–1476.
22. Brenner, S., Jacob, F., & Meselson, M. (1961). An unstable intermediate carrying information from genes to ribosomes for protein synthesis. *Nature, 190*, 576–581.
23. Jacob, F., & Monod, J. (1961). Genetic regulatory mechanisms in the synthesis of proteins. *Journal of Molecular Biology, 3*, 318–356.
24. Ptashne, M. (1967). Specific binding of the lambda phage repressor to lambda DNA. *Nature, 214*(5085), 232–234.
25. Ptashne, M. (1967). Isolation of the lambda phage repressor. *Proceedings of the National Academy of Sciences of the United States of America, 57*(2), 306–313.
26. Huebner, R. J., & Todaro, G. J. (1969). Oncogenes of RNA tumor viruses as determinants of cancer. *Proceedings of the National Academy of Sciences of the United States of America, 64*(3), 1087–1094.
27. Shope, R. E. (1932). A filtrable virus causing a tumor-like condition in rabbits and its relationship to virus myxomatosum. *Journal of Experimental Medicine, 56*(6), 803–822.
28. Bittner, J. J. (1936). Some possible effects of nursing on the mammary gland tumor incidence in mice. *Science, 84*(2172), 162.

The History of Immunology

6

Abstract

Immunology begins with Edward Jenner's discovery that vaccination with cowpox protects against smallpox. That there was an immune response was confirmed by the observations of many scientists that the same disease did not return a second time to a recovered individual. With the recognition by Friedrich Henle that germs caused disease, and the isolation of infectious bacteria by his pupil Robert Koch, the stage was set to examine how an immune response was achieved. Modern immunology begins with the research of Metchnikoff, who discovered the phenomenon of phagocytosis in starfish and extrapolated it to macrophages in humans as cells that engulf infectious agents; this was the beginning of cellular immunology. Paul Ehrlich investigated the formation of antibodies recognized as later as proteins that destroyed infectious agents. However, an explanation of how antibodies were formed and selected was puzzling. Did the body have enough genes to code for every type of antibody, and did specific cells produce antibodies, or did each cell have the ability to produce antibodies to any challenging molecule? Following the work of Karl Landsteiner, Felix Haurowitz, Niels Jerne and others, the "clonal selection theory" was proposed by MacFarlane Burnett. This theory states that each B-cell produces one type of antibody, and once activated, it expands and produces memory cells. Meanwhile, work on cellular immunity and innate immunity recognized the role of various types of T-cells, dendritic cells and cytokines in the immune response. New classes of T-cells and cytokines are constantly being found, and there is an intricate connection between these three branches of the immune system.

M. W. Taylor, *Viruses and Man: A History of Interactions*,
DOI: 10.1007/978-3-319-07758-1_6,

6.1 What Is Immunology?

Since the aim of a pathogenic virus or bacteria is to requisition the body completely for the replication of itself, why does one not always die from infection? In the course of millions of years of evolution, defense mechanisms arose to overwhelm such pathogenic attacks. One can view this as a constant struggle between pathogen and host, so that eventually a "truce" is achieved in which both co-exist (lysogeny in bacteria or chronic disease in mammals) or one (the host) vanquishes the invading pathogen or the pathogen is triumphant and kills the host. All organisms, from the most primitive bacterium, to plants to man, developed defenses against invasion and infection. In bacteria, this system is referred to as "restriction", since enzymes (restriction enzymes) are produced that destroy the DNA of an invading bacteriophage, or the genetic material of an invading foreign bacterium. Likewise, plants have developed primitive immune systems to protect themselves against both viral and bacterial pathogens, the result of convergent evolution rather than descent from a common ancestor. Plants, however, lack specialized mobile immune cells as occurs in animals [1]; instead, every plant cell is thought to be capable of launching an effective immune response. Somewhere along the time scale of evolution, vertebrates have developed a two-tier system, one a general immediate action against a foreign invader—the innate immune system, the other a delayed response, but a very specific reaction against the invader—the adaptive immune response. This consists of two components, plasma cells and antibodies, and a complex cellular system of "immune" cells that interact with antibody-producing cells.

In order to understand the complexities of immunology, and it is indeed very complex, and to gain a better insight into the theories and progress made over the last 150 years or so, it is worth considering the progress made in this science in chronological order.

6.2 Henle's Contribution to Immunology

As discussed in Chap. 1, Friedrich Gustave Jakob Henle (1809–1885) proposed the germ theory of disease around 1840. Henle was a major figure in both medicine and anatomy. In 1852–1873 he published a *Handbook of Systematic Human Anatomy*, which at that time was the most complete and comprehensive text of its kind. It was remarkable not only for its anatomical descriptions, but also for the number and quality of its illustrations. Henle wrote, "I have shown, then, that the infective material must be independently alive, and thought to consist of organisms—either plant or animal—or of parts of animals which have achieved a limited independence". (*Pathologische Untersuchungen*). Henle was a very colorful character, typical of the Enlightenment, born into a Jewish family in Fürth,

Fig. 6.1 Friedrich Gustav Jakob Henle (from Wikipedia)

Germany, interested in music and the arts as well as medicine. Among his circle of friends was the composer Felix Mendelssohn. Henle was a superb lecturer, a poet and a fine musician, playing the violin, viola and cello Fig. 6.1. He was very close to his most famous student, Robert Koch, who later on was to provide definitive proof of Henle's belief in infectious agents as independent organisms.

6.3 Metchnikoff and Ehrlich

Modern immunology begins with the research of Emil Metchnikoff (1845–1910), a Russian microbiologist and embryologist who recognized the importance of phagocytosis as a general biological phenomenon. He was a professor at Odessa University, but later established a private laboratory in Messina, Italy, to study comparative embryology, in particular of marine organisms. Tauber and Chernyak [2] he demonstrated, analogous to his work with invertebrates, that certain cells of the blood also digested (phagocytized) bacterial particles. Eminent scientists such as Louis Pasteur (1822–1895) and Emil Von Behring (1854–1917) were very skeptical at the time.

In 1888 Pasteur invited him to the Pasteur Institute in Paris, where he spent the remainder of his life. Of importance are his drawings of cells that he called "macrophages", taking up foreign particles, and showed that these cells were important in the clearance of disease. He proposed that such cells were the first line of defense against infection by a bacterium rather than soluble factors (antibodies), as proposed by Paul Ehrlich (1854–1915) and others. He became a leading proponent of the "cellularists", who believed that phagocytes, rather than antibodies, played the leading role in immunity. It is now thought that many of the cells described by Metchnikoff were actually dendritic cells rather than macrophages (see Figs. 6.2, 6.3). Metchnikoff received the Nobel Prize in Medicine in 1908, sharing it with Ehrlich. Of course both Metchnikoff and Ehrlich were correct, and each system depends on the other.

The humoral (antibody-mediated) theory of immunity really begins with the experiments of Emil von Behring and Shibasaburo Kitasato (1853–1931) with diphtheria and tetanus toxin. Diphtheria was rampant during the end of the nineteenth century, with a high rate of fatality, and tetanus occurred predominantly in the military. Von Behring and Kitasato showed that it was possible to passage the diphtheria toxin through horses (or other animals) and obtain serum-containing anti-toxin. This serum, when injected into other animals, protected them against the diphtheria toxin. In his Nobel Prize address in 1901, Von Behring equated his serum therapy with the presence of antibodies formed against diphtheria toxin. Certainly unknown at the time was the fact that diphtheria toxin was actually a viral (bacteriophage) product and not a direct product of the bacillus.

Building on the results of the research on diphtheria toxins, Erhlich proposed that specific cells produced antibodies. According to Ehrlich's side-chain theory, immune cells were dotted with a vast array of receptors, each specific to a particular substance (Fig. 6.4). When an antigen interacted with the relevant receptor, the cell would be activated and would react by producing more receptors, which would then be released into the bloodstream as antibodies to neutralize the toxin. This hypothesis, and other research including developing treatment against syphilis and techniques for the staining of various types of blood cells, earned Ehrlich a Nobel Prize in Medicine in 1908. His side-chain antibody theory was later discredited, although it did form the basis for other theories of antibody formation. Ehrlich also proposed the idea of "self versus non-self" as a characteristic of the immune system.

In summary, Metchnikoff proposed a cellular basis for the immune response, whereas Ehrlich proposed a humoral response. Until the 1940s, the antibody theory of the immune response was dominant. By 1915, Ehrlich's side-chain theory had been largely rejected, in part because it stipulated that every cell expressed all of the receptors it might need to neutralize toxins or bacteria or virus.

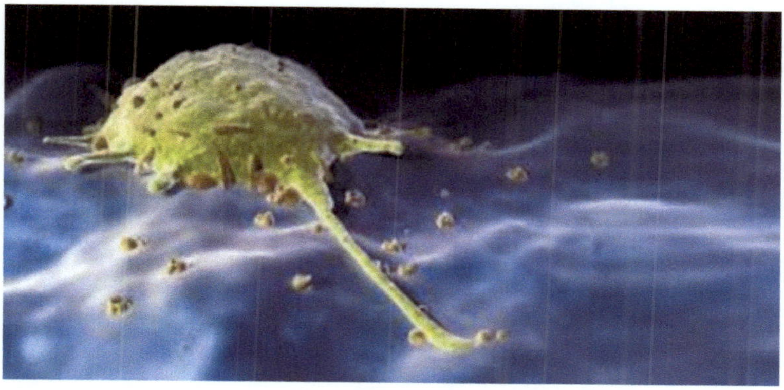

Fig. 6.2 Macrophage having a meal (novimmune.com)

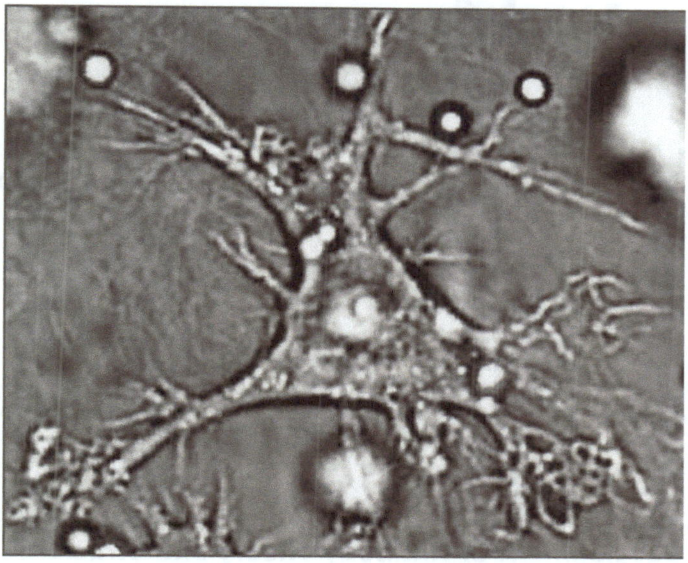

Fig. 6.3 Dendritic cell. *Source* http://www.plospathogens.org/article/info:doi/10.1371/journal.ppat.0030013

Other researchers “began finding antibodies against everything; against normal, non-dangerous proteins, like egg albumin” http://www.the-scientist.com/?articles.view/articleNo/36175/title/Side-Chain-Theory--circa-1900/. With that kind of diversity, it was inconceivable that a cell could carry all possible receptor configurations.

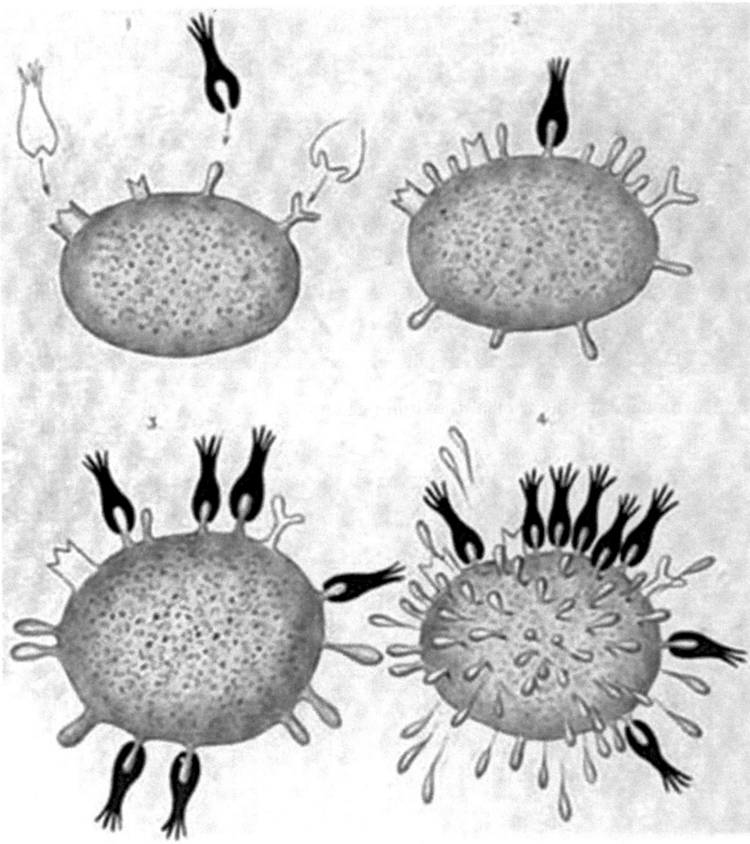

Fig. 6.4 Ehrlich's drawing of antibodies (magic bullets) binding to cellular receptors

6.4 Development of Antibody Theory and Structure

Michael Heidelberger (1888–1991) is considered the "father of American immunology" (or immune-chemistry). His background, although he was born in the USA, is similar to that of Ehrlich and Henle. He came from a relatively poor family, but one in which there was an emphasis on education. He was also interested in music, and considered it as a possible career, and he even performed with an active trio. Following his education and post-doctoral studies in Germany, he was offered a position at the Rockefeller Institute in Simon Flexner's group, which was working on polio virus. At the same time, Karl Landsteiner

(1868–1943) had been working in Germany and discovered the ABO blood groupings and isolated the polio virus from rhesus monkeys. In 1921 he was also invited by Flexner to join the Rockefeller Institute in New York. When Karl Landsteiner left Vienna in 1922 at the age of 54 for the Rockefeller Institute, he had already done much of his most important work. In the early years of the century he had discovered a simple technique of agglutination, whereby human blood was divided into four groups, a discovery that made safe blood transfusions possible. This work in turn led him to the study of hereditary differences in the antigens of blood groups, and thus to the birth of serological genetics. A succession of discoveries followed in the 1920s and 1930s, after he had moved to the Rockefeller Institute: the discovery of blood factors (M, N, P, and Rh factors), and investigations of the chemical specificity of serological reactions, which became a basis of immunochemistry. Landsteiner was awarded the Nobel Prize in Physiology or Medicine in 1930 (biographical data on Karl Landsteiner from the American Philosophical Society http://www.amphilsoc.org/mole/view?docId=ead/Mss.B.L23m-ead.xml).

Heidelberger collaborated with Landsteiner on studies of the antigenic properties of different types of hemoglobin. Throughout his life, Heidelberger was proud to say that he first learned immunology from Landsteiner. He researched the structure of pneumococcal antigens and the antibodies they elicited. In the 1920s, Michael Heidelberger and Oswald Avery (1877–1955) observed that antigens could be precipitated by antibodies, and they went on to show that antibodies were made of protein [3, 4].

The chemistry of antigen-antibody reactions was uncovered largely by the development of the quantitative precipitin reactions between antibody and antigen in collaboration between Heidelberger and Elvin Kabat (1914–2000), his graduate student. Kabat [5] later was instrumental in the discovery of the structural and genetic basis of the remarkable specificity of antibodies. He first demonstrated that antibodies are gamma globulins, and he characterized the size and stunning heterogeneity of antibody-combining sites [6]. The heterogeneity extended over the whole length of the antibody molecule. These were in both the heavy and light chain and extended to genetic markers. There were multiple types of antibodies, derived from five different classes of human immunoglobulin. The question was how to explain this variability at the genetic level. The structure of a typical antibody is presented in Fig. 6.5. Each antibody is made up of two identical heavy chains and two identical light chains, shaped to form a "Y". The sections that make up the tips of the Ys' arms vary greatly from one antibody to another; this is called the "variable region". It is the amino-acid sequence in the antigen-binding site that allows the antibody to recognize an antigen. The constant region is the same for all antibodies of the same type of immunoglobulin.

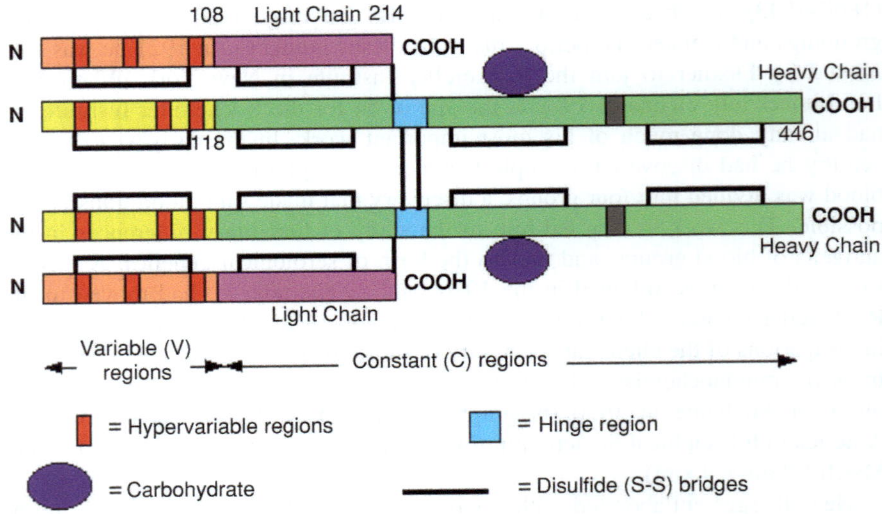

Fig. 6.5 Schematic of typical antibody molecule. Variable region is the antigen binding region (http://biology-pages.info, John Kimball)

6.5 The Clonal Theory of Antibody Production

The large variety of antibodies formed, and the fact that T-cells, which will be discussed later, can also recognize any antigen, has led to multiple hypotheses on how such diversity arose. From the knowledge of genetics at that time, it seemed inconceivable that there should be genes for the production of each type of antibody; this would require many more genes than we possess. Since antibodies are formed in response to infection, and take some time for processing (a minimum of 5 days), there has to be a mechanism for selection. Are they preformed and floating around in our blood, or are they formed anew each time we are infected? Immunologists were preoccupied with these conflicting ideas for a long time.

Landsteiner [7] showed that antibodies were not produced against the whole bacterium or infecting agent, but against small regions of this agent, the "determinant group", a short polypeptide that we now call "epitope". Initially he suggested that antigens might act as mutagens, altering the cell in some fashion to produce more antibodies that react with the specific antigen. Other hypotheses were based on the idea of a template in which the antigen entered the antibody-producing cell and directed the formation of the antibody. This theory assumes that antibodies can be produced only by cells in which the antigen is present. This theory is known as the "antigen-template theory." The specific affinity of an antibody molecule toward the antigen is due to complementarity in sequence of amino acids. The antigen thus serves as a template for the formation of the antibody. These ideas were presented and discussed by renowned immunologists such as Felix Haurowitz (1896–1987), Niels Jerne (1911–1994), and Linus Pauling (1901–1994). A genetic mechanism for this could not be found.

The 1950s were a period of intellectual ferment in immunology. How can we explain the large number of antibodies that are produced to "everything" that infects us? MacFarlane Burnett (1899–1954) and Frank Fenner (1914–2010) [8, 9] proposed an alternate hypothesis known as the "modified enzyme" hypothesis. They proposed that the introduction of an antigen into cells containing enzymes that normally degraded foreign material induced the formation of "enzymic units" programmed to destroy the antigen. Renewed contact with the antigen stimulates the replication of these "enzymic units", thus explaining memory in antibody production. Circulating antibodies are partial replicas of the modified enzymic units, carrying specificity but lacking enzymic action.

Jerne [10] proposed that the body already has the antibodies, made in specialized cells, that it needs to fight any infection. The interaction of the antibody with the antigen results in the stimulation of more antibodies. This was called the "natural selection hypothesis". Jerne rejected the idea that the antigen is a template or an enzyme modifier. He proposed that globulin molecules are being synthesized in a large number of different shapes and structures and that these interact with the antigens. This interaction in turn stimulated the production of more antibodies. This would explain the fact that antibodies can be found in the circulation after an infection is over. Such alternate antibodies would also be able to shut down the production of antibodies and that antibodies against self would be eliminated in the organ where antibodies mature. The selection theory was later advanced to include not only antibodies, but also T-cell receptors that recognized foreign antigens.

An alternative hypothesis, based on the ideas of Jerne, was the "clonal selection hypothesis". This proposes that antigens react with sequences on the surface of antibody producing cells (or later expanded to T-cells with their unique receptors), and this interaction leads to expansion of these specific antibody producing B-cells (plasma cells) and T-cells. This interaction between the antigen and the receptor results in the clonal expansion of such cells. Since there are obviously thousands of different types of antibodies, these result from random recombination within segments of the immunoglobulin genes, which gives rise to thousands of different antigen-receptor structures. The mechanism at the gene level was discovered much later. Thus the process of antibody formation is basically a random process, akin to Darwinian selection. This is now the accepted theory for explaining the diversity of antibodies and T-cell receptors (see Fig. 6.6).

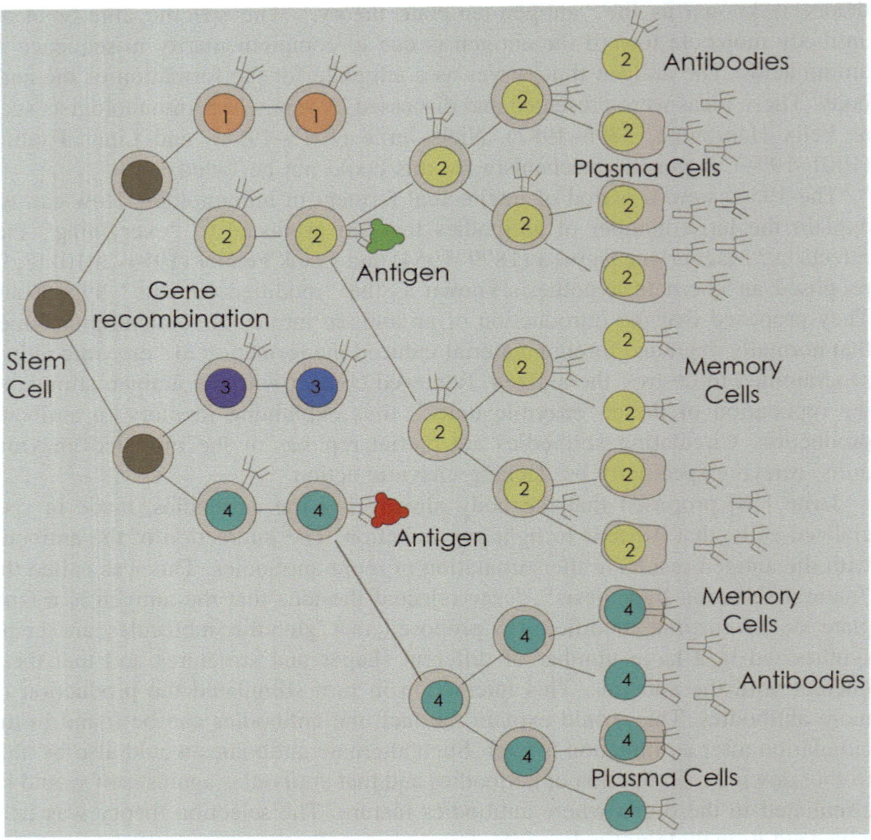

Fig. 6.6 Clonal selection with two different antigens. The clonal selection theory of lymphocytes. Schematic of clonal selection hypothesis of antibody production by lymphocytes, illustrating the idea that each naïve lymphocyte has a different receptor specificity, each of which can bind a different antigenic determinant. When a pathogen is recognized by the cells, in this case by two different antigenic determinants, then the cells that bind to these determinants are selected to proliferate or undergo clonal expansion, and then differentiate into effector cells that either secrete antibodies or mediate various effector mechanisms of cell-mediated immunity [11]

To summarize the clonal selection theory:

1. Animals contain numerous cells called lymphocytes.
2. Each lymphocyte is responsive to a particular antigen by virtue of a specific surface receptor molecule.
3. Upon contacting its appropriate antigen, the lymphocyte is stimulated to proliferate (clonal expansion) and differentiate.

4. The expanded clone is responsible for the secondary response (more cells to respond), while the differentiated ("effector") cells secrete antibody or antibodies.
5. A small number of cells remain to form memory cells.
   See excellent review of this history [12].

6.6 Genetic Basis of Antibody Formation

There are hundreds of genes for immunoglobulins, the proteins that give rise to antibodies. If each antibody were coded for by a separate gene, one would need millions of such genes. Instead, the DNA for such genes is reorganized by splicing mechanisms and by recombination by enzymes known as recombinases. All cells of the same clonal origin produce the same antibody and have identical rearrangements. Each gene contains exons, the parts that code for proteins, and introns, non-coding components; such introns are removed by a splicing mechanism. This splicing mechanism is not unique to immunoglobulin genes, but to almost all genes. In the case of immunoglobulin, each exon would code for a different region of the antibody; some regions code for the variable region, while others code for the constant region. Each immunoglobulin (antibody) has a constant region and variable region coded for by different chromosomes.

There are four key antigen receptors: the T-cell, MHC-1, MHC-II and B-cell receptor. The antigen interacts with the variable domain of the protein, which will differ from cell to cell. A typical receptor gene is shown in Fig. 6.7.

6.7 Cell-Based Immunity

While major studies in immunology during the first half of the twentieth century were predominantly on elucidating the structure of antibodies and how and where they are made, research on humoral immunity, abandoned for some time, picked up speed in the second half of the century. The early work on cellular immunity really derives from transplantation experiments of grafting tissue or tumors from one mouse to another. Grafts between inbred strains of mice were accepted, whereas those between non-inbred strains or different genetic strains were not accepted. This resulted in studies of self versus non-self in tissue transplants. The work of George Snell (1903–1996) and colleagues [13] illustrated two principles: First, tissue rejection depends on genetic differences between donor and host. Second, the processes involved in tissue rejection are immunological in nature, and specific genes are involved in the rejection process. In humans, when an organ is transferred from one patient to another, this will be rejected if the immune system detects it as "non-self". Recognition depended on specific genes later called "major histocompatibility genes" (MHC). An MHC protein serves as a

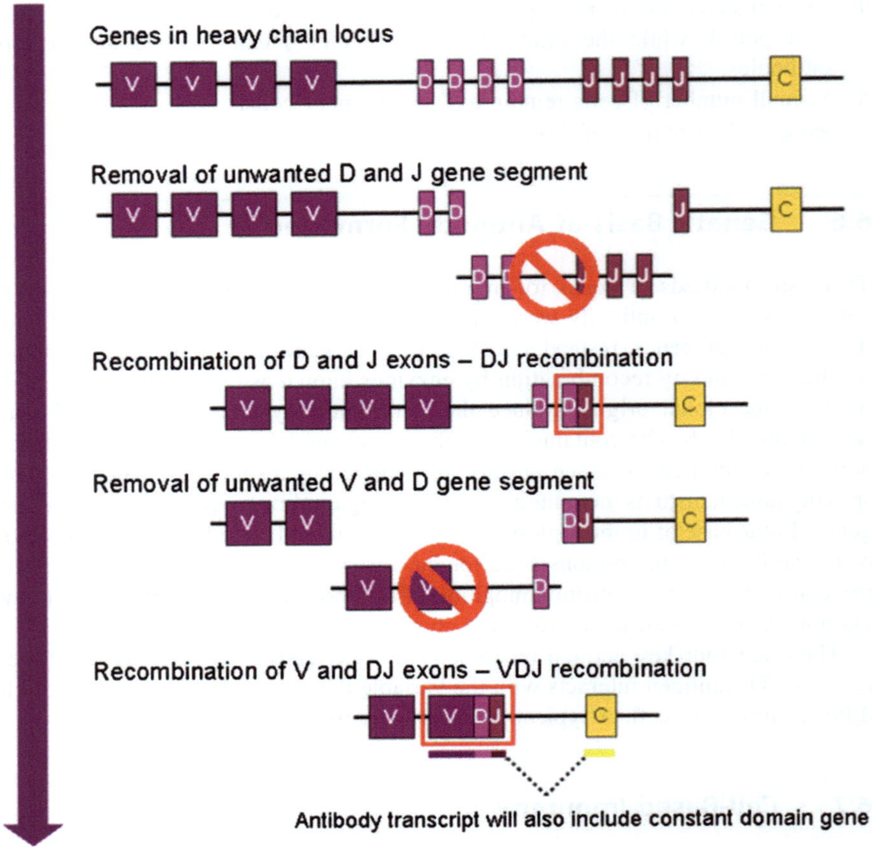

Fig. 6.7 Simplistic overview of V(D)J recombination of heavy chain (from Wikipedia)

recognizable scaffold that presents peptides of a foreign protein (antigenic) to immune cells. An empty "foreign" MHC scaffold itself can act as an antigen when donor organs or cells are introduced into a patient's body. These MHC self-marker scaffolds are also known as a patient's "tissue type", or as human leukocyte antigens (HLA) when a patient's white blood cells are being characterized. (See http://www.cancer.gov/cancertopics/understandingcancer/immunesystem/page4.)

Every cell in the human body is covered with these MHC self-marker proteins, and—except for identical twins—individuals carry different sets. MHC marker proteins are as distinct as blood types and come in two categories–MHC Class I: humans bear 6 markers out of 200 possible variations; and MHC Class II: humans display 8 out of about 230 possibilities (NCI Web site). All lymphocytes carry either MHC1 or MHCII proteins on their surfaces. The MHC I complex occurs in all nucleated cells, whereas type II occurs in dendritic cells, macrophages, B-lymphocytes and cells of the thymus.

The role that the MHC loci played in immunology was further explored by studying the role of T lymphocytes (T-cells) in the immune response to viral meningitis. Virus-infected mouse cells were mixed with T-lymphocytes from infected mice of various strains. T-lymphocytes did destroy the virus-infected cells, but only if the infected cells and the lymphocytes came from a genetically identical strain of mice. Lymphocytes would ignore virus-infected cells that had been taken from another strain of mice; [14] thus MHC proteins were strain-specific and involved in T-cell recognition.

The definitive proof that cells played a role in the immune response came from the classic experiments of Landsteiner and Merrill Chase (1905–2004). Cells from guinea pigs, which had been immunized with *Mycobacterium tuberculosis* or hapten, were transferred into naive guinea pigs. (A hapten is a small molecule that can elicit an immune response only when attached to a large carrier such as a protein.) When an antigen or hapten was injected into these naïve guinea pigs, they elicited an immune response. This did not happen when the serum fraction alone was transferred [15–18]. Thus the immune response required both a cellular fraction as well as a serum fraction. The need for both cell-mediated and antibody-mediated immunity had been established by 1940; however, the type of cells involved was still unknown.

The findings that there are two classes (or many classes) of lymphocytes came from an unexpected source. Bursectomy—the removal of the bursa from the "tail" of an adult chicken—had no noticeable morphologic or functional effects on the immune system. Bursectomy before or immediately after hatching produced major immunological impairments in the chick: specific antibody responses were abolished, serum antibody levels were very low, and plasma cells and germinal centers for the production of antibodies were absent. The rejection of foreign grafts, however, was normal. Thymectomy (removal of the thymus) had the opposite results; in both chicks and mice it impaired the host versus graft reaction as well as decreasing antibody production [19]. These experiments pointed to two types of lymphocytes, those derived from the bursa, coined "B-lymphocytes", and those from the thymus, coined "T-lymphocytes". Since man does not have a bursa, the B-cells are produced elsewhere [20]. A series of experiments [21–25] demonstrated that antibody-producing cells in man were made in the bone marrow and stored in the spleen, and that for full antibody production one required cells from the thymus, now termed T-cells and also macrophages. Thus T-cells were involved in two activities: one in helping antibody production and the other in the graft-host reaction.

By the 1970s it became clear that there were at least two classes of T-lymphocytes. These could be distinguished by antibodies made to surface antigens known as "CD antigens" (clusters of differentiation) in humans, or Ly antigens (lymphocytes) in mice. T-cells could be divided into helper T-cells, i.e., those that were involved in "helping" antibody production and "cytotoxic" or "killer T-cells" that were involved in the tissue-rejection phenomenon.

It was originally thought that T-cell receptors would also be immunoglobulin molecules, but this proved to be wrong. In 1983 Philippa Marrack and colleagues [26] showed that the receptor consisted of two polypeptide chains of ~45 KDa, with both constant and variable regions, and was similar, but not identical, to immunoglobulin. A year later, cDNAs for the T-cell receptor were cloned and the gene was organized in a similar manner to the antibody gene. The T-cell receptor, unlike the B-cell receptor, only binds protein antigens and does not bind antigens directly, but only after the antigen has been internalized by other cell types and re-expressed on the cell surface in association with MHC molecules; such cells are known as "antigen-presenting cells" [27].

In summary, the inter-relationship between adaptive and innate immunity as defined in the 1970s and 1980s is presented in Fig. 6.8. When an antigen enters the mammalian body, there is an instant reaction with the mobilization of dendritic cells, macrophages and natural killer (NK) cells. Macrophages present virus antigens on the cell surface and activate T-c cells. This is done in conjunction with cytokines produced by the macrophage. NK cells are activated by interferons and also destroy virus-infected cells. Interferon plays a major role in the activation of the first line of defense; this is described in Chap. 7.

B-cells recognize antigens and expand with the help of Th cells, resulting in plasma cell formation and soluble antibody production. Th cells and Tc cells expand, and Tc cells destroy cells that contain viruses.

This is a very simplified presentation, and we now recognize that many different cytokines and many sub-sets of T-cells are involved in the immune response. Fig. 6.9

6.8 T-helper Cells

T-cells are now divided into a number of different classes, based on CD identification and different functions. New classes are being discovered continuously.

T-h or T-helper cells, also known as CD-4 cells, help with the maturation of B-cells into plasma cells, which produce antibodies and also mature into B-memory cells. T-helper cells can also activate T-cytotoxic cells and macrophages. Helper T-cells become activated when they are presented with peptide antigens in the presence of MHC class II molecules, expressed on the surface of antigen-presenting cells. These cells can differentiate into different types of T-cells, including T_H1, T_H2, T_H3, T_H17, T_h9, or T_{FH}. The classification of T-cells depends on the type of cytokines they produce. Proliferating helper T-cells that develop into effector T-cells differentiate into two major subtypes of cells known as T_h1 and T_h2 cells. T_h1 helper cells are triggered by the cytokines IL12 and IL2 and produce as activator IFN-gamma. The main effector cells of T_h1 immunity are macrophages, CD8 cells, and IFN-gamma producing CD4 cells and B-cells. T_h2 helper cells are the major effectors against multicellular

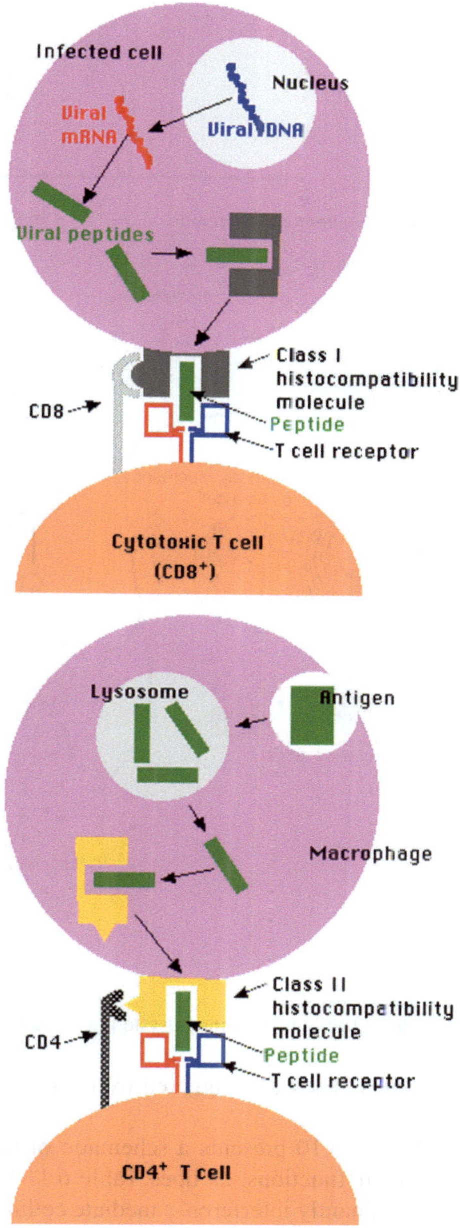

Fig. 6.8 Interaction of MHC1 and MHC2 as a link between antigen-presenting cells (macrophage and CD8 and CD4 T-cells). MHC are transmembrane proteins that hold the epitope that is presented to the T-cell, or other cell types (http://biology-pages.info Kimballs biology pages)

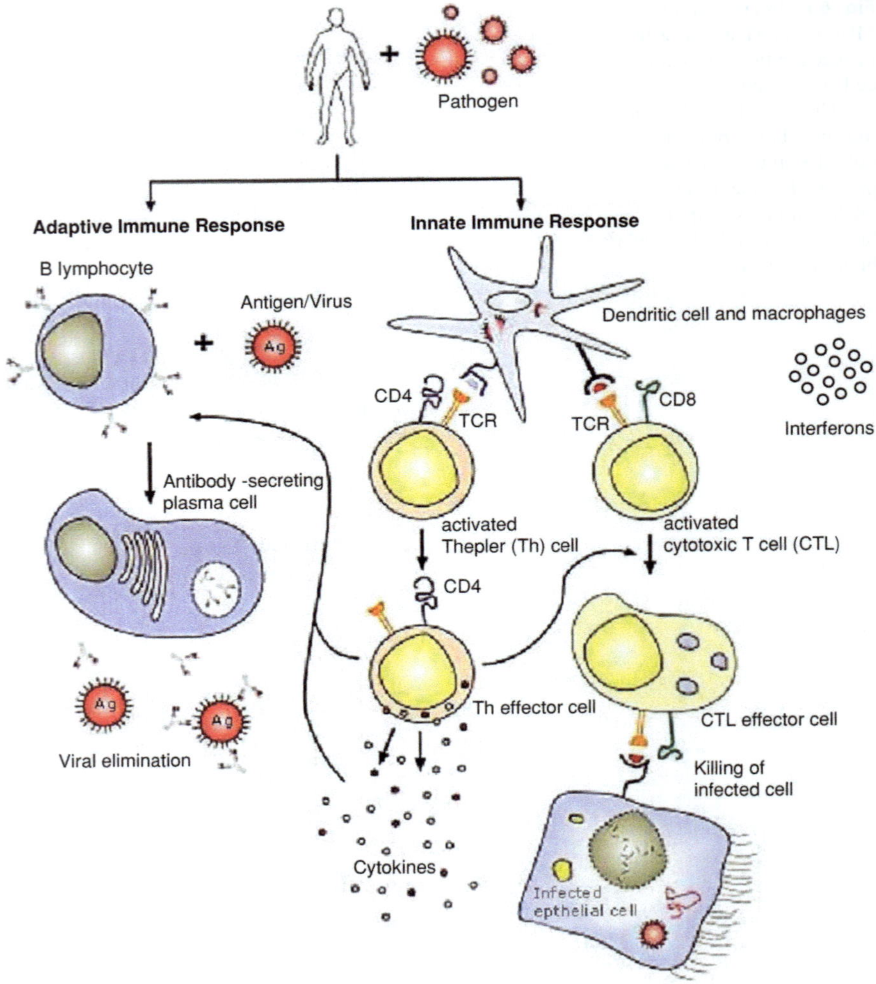

Fig. 6.9 The human immune system as visualized in 1990

parasites. They are triggered by IL-4, and their effector cytokines are IL-4, IL-5, and IL-13.

Figure 6.10 presents a schematic of the differentiation into T_h1 and T_h2 cells and their functions, as does Table 6.1. T helper cells type 1 (T_h1s) that produce predominantly interferon-γ mediate cellular immune responses and are involved in the development of chronic inflammatory conditions, whereas T_h2s, which produce large amounts of IL-4 and IL-5, up-regulate IgE production and are prominent in the pathogenesis of allergic diseases. The precise factors determining whether T_h1- or T_h2-mediated immune responses preferentially occur at a peripheral site of antigen exposure are largely unknown. Chemokines, a

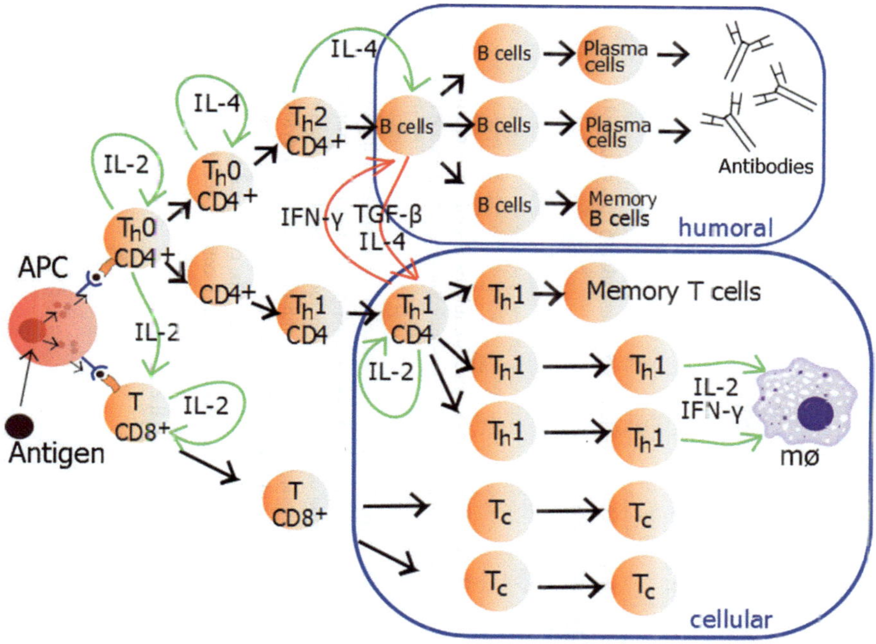

Fig. 6.10 T_h1/T_h2 Model for helper T-cells. An antigen is ingested and processed by an antigen presenting cell. It presents fragments from it to T-cells. The upper, T_h0, is a T helper cell. The fragment is presented to it by MHC2. IFN-γ, interferon γ; TGF-β, transforming growth factor β; mø, macrophage; IL-2, interleukin 2; IL-4, interleukin 4 (from Wikipedia)

superfamily of polypeptide mediators, are a key component in the leukocyte recruitment process [28].

T_H3 cells are involved in mucosal immunity and protecting mucosal surfaces in the gut from nonpathogenic, non-self antigens. They mediate this non-inflammatory environment by secreting TGF-beta and IL-10 (Fig. 6.11).

Since all these cells contain CD4, CXCR4 and CXCR5 and are thus susceptible to HIV infection, the balance of immune cells is altered following HIV infection. A recent paper suggests that HIV infection may drive the T-helper cell population towards T-reg cells, thus suppressing the function of other T-cells [29].

6.9 Complement Cascade

This is a system in which antibody antigen complexes are destroyed by interaction with specific proteins of the "complement cascade". It is mostly functional against bacteria, and is not a predominant anti-viral system. The complement system is

Table 6.1 The difference in functions of T_h1 and T_h2 is presented in the table below (adapted from Wikipedia)

	Type 1/T_h1	Type 2/T_h2
Main partner cell type	Macrophage	B-cell
Cytokines produced	Interferon-γ and TGF-beta. (Interleukin-2 was classically associated with T_h1 cells, but this association may be misleading; IL-2 is produced by all helper T cells early in their activation.) Interleukin-10 production has been shown to be induced in activated T_h1 cells	Interleukin-4, interleukin-5, interleukin-6, interleukin-10, interleukin-13
Immune stimulation promoted	Cellular immune system. Maximizes the killing efficacy of the macrophages and the proliferation of cytotoxic $CD8^+$ T-cells. Also promotes the production of opsonizing antibodies	Humoral immune system. Stimulates B-cells into proliferation, to induce B-cell antibody class switching, and to increase neutralizing antibody production
Other functions	The Type 1 cytokine IFN-γ increases the production of interleukin-12 by dendritic cells and macrophages, and via positive feedback, IL-12 stimulates the production of IFN-γ in helper T-cells, thereby promoting the T_h1 profile. IFN-gamma also inhibits the production of cytokines such as interleukin-4, an important cytokine associated with the Type 2 response, and thus it also acts to preserve its own response	The Type 2 response promotes its own profile using two different cytokines. Interleukin-4 acts on helper T-cells to promote the production of T_h2 cytokines (including itself; it is auto-regulatory), while interleukin-10 (IL-10) inhibits a variety of cytokines including interleukin-2 and IFN-γ in helper T-cells and IL-12 in dendritic cells and macrophages. The combined action of these two cytokines suggests that once the T-cell has decided to produce these cytokines, that decision is preserved (and also encourages other T-cells to do the same)

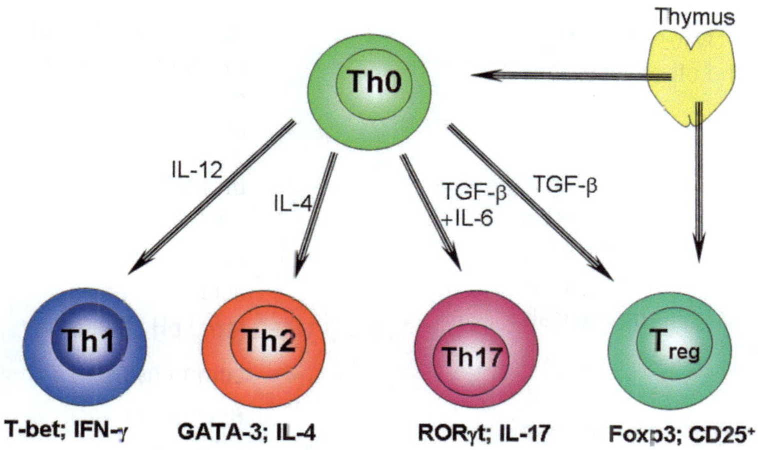

Fig. 6.11 T_h17 cells are so called because they produce IL17. They appear to be involved in the autoimmune response. T regulatory cells (Treg) are a component of the immune system that suppresses immune responses of other cells. This is an important "self-check" built into the immune system to prevent excessive reactions (Wikipedia [33])

made up of about 25 proteins that aid in the destruction of bacteria tagged by antibodies. The complement proteins circulate in the blood in an inactive form, with each component taking its turn in a precise chain of steps known as the complement cascade. There are at least three different cascades that inactivate the invading bacterium by puncturing holes in the cell wall. The complement cascade may play a role in specific virus infections such as in measles and retroviruses. However, compared to interferon and cytokines, its role in anti-viral innate response is minimal.

6.10 Skin as a Component of the Immune System

A major line of defense against all infections is the skin of an animal or human. We have become so preoccupied with defining the functions of the various cells in the immune response that we are inclined to forget the role that this organ plays in defense. The first lines of defense against infection are the external "organs" of the body, which include the skin, tears produced by the eye, and cilia of the ears, nose and bronchi. Mucus in other parts of the body may have antimicrobial activity. A lower temperature in the nose, and low physiological pH in the gut (acidity) all act as deterrents to infection. That the skin is in fact an active component of the immune system has recently been demonstrated by the use of patches to elicit an immune response against anthrax and other infectious agents, by introducing genes from such organisms into *Escherichia coli* and applying the recombinant *E. coli* directly to the skin [30, 31] (see Fig. 6.12).

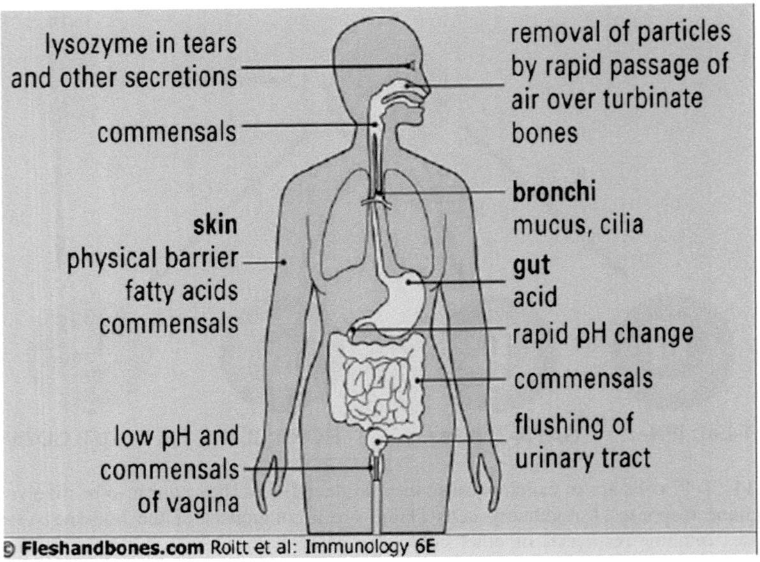

Fig. 6.12 Major first-tier defense mechanism against invading organisms (Fleshand bones.com with permission)

When the skin is broken by a cut or wound, it then becomes a site of entry for viruses or bacteria. Cases of HIV and HCV have been reported by physicians after an accidental cut or mistaken injection with contaminated blood. HIV in particular enters through wounds in the skin or any of the membranes of the genital tract or anus; a herpes infection causing genital wounds is also a site of entry for other viruses.

6.11 Toll-Like Receptors and Pattern Recognition

Once the virus penetrates the skin or mucous membrane, it attaches to cells with specific receptors. Within the cell there has to be a signaling system for recognition of a pathogen or foreign agent. Recently, components of the cell, known as "toll-like receptors", which are either membrane-bound or cytoplasmic and recognize pathogens or components of pathogens such as the double-stranded RNA of a virus, have been discovered and induce the production of cytokines and interferons to counteract the invader. There are many different toll-like receptors; some are specific for viral RNA, others for bacterial surface proteins, lipids, toxins, etc.; these are also known as pathogen-recognition receptors. The component of the virus that is recognized by such molecules is called the pathogen-associated molecular pattern. There is a whole series of pattern recognition receptors (Table 6.2).

Table 6.2 List of known toll-like receptors as of 2013. The TLR family discriminates between specific patterns of microbial components [32]

Name of pattern recognition receptor	Molecule-recognized
TLR1	Triacylated lipoprotien
TLR2	Specific glycoproteins
TLR3	Viral sequences, dsRNA
TLR4	Specific glycoproteins, lipopolysaccharide
TLR5	Flagellin
TLR6	Diacylated lipoprotein
TLR7	Viral sequences
TLR8	Vial sequences
TLR9	Viral sequences. CpG DNA
RIG-1	Short dsRNA sequences
MDA-5	Long dsRNA sequences
C-type lectin receptor	
DC-SIGN	Viral envelope glycoproteins

The results of activation of the toll-like receptors is complicated and involves the activation of a series of kinases, enzymes that phosphorylate other enzymes, and activate them. The end result is the production of a cascade of intermediate molecules that in turn interact with signals on DNA and produce interferon and other cytokines. (This is discussed in part in the chapter on interferon, Chap. 7.) There are as many as 80 known cytokines, and these control the innate immune response.

References

1. Spoel, S. H., & Dong, X. (2012). How do plants achieve immunity? Defence without specialized immune cells. *Nature Reviews Immunology, 12*(2), 89–100.
2. Tauber, A. I., & Chernyak, Leon. (1991). *Metchnikoff and the origins of immunology: From metaphor to theory*. New York: Oxford University Press.
3. Avery, O. T., & Heidelberger, M. (1923). Immunological relationships of cell constituents of Pneumococcus. *Journal of Experimental Medicine, 38*(1), 81–85.
4. Heidelberger, M., Avery, O. T., & Goebel, W. F. (1929). A soluble specific substance derived from gum arabic. *Journal of Experimental Medicine, 49*(5), 847–857.
5. Kabat, E. A., & Heidelberger, M. (1937). A quantitative theory of the precipitin reaction: V. The reaction between crystalline horse serum albumin and antibody formed in the rabbit. *Journal of Experimental Medicine, 66*(2), 229–250.
6. Kabat, E. A. (1970). Heterogeneity and structure of antibody-combining sites. *Annals of the New York Academy of Sciences, 169*(1), 43–54.
7. Landsteiner, K., & van der Scheer, J. (1940). On Cross Reactions of Egg Albumin Sera. *Journal of Experimental Medicine, 71*(4), 445–454.

8. Burnett, F. M. (1949). *Production of antibodies*. London: Macmillan & Co.
9. Burnet, F. M., & Fenner, F. (1948). Genetics and immunology. *Heredity, 2*(3), 289–324.
10. Jerne, N. K. (1955). The natural-selection theory of antibody formation. *Proceedings of National Academy Science USA, 41*(11), 849–857.
11. Abbas, A. K., & Janeway, C. A, Jr. (2000). Immunology: improving on nature in the twenty-first century. *Cell, 100*(1), 129–138.
12. Hodgkin, P. D., Heath, W. R., & Baxter, A. G. (2007). The clonal selection theory: 50 years since the revolution. *Nature Immunology, 8*(10), 1019–1026.
13. Snell, G. D. (1979). Recent advances in histocompatibility immunogenetics. *Which Publication?, 20*, 291–355.
14. Zinkernagel, R. M., & Doherty, P. C. (1974). Restriction of in vitro T cell-mediated cytotoxicity in lymphocytic choriomeningitis within a syngeneic or semiallogeneic system. *Nature, 248*(450), 701–702.
15. Landsteiner, K., & Chase, M. W. (1941). Studies on the sensitization of animals with simple chemical compounds: IX. Skin sensitization induced by injection of conjugates. *Journal of Experimental Medicine, 73*(3), 431–438.
16. Landsteiner, K., & Chase, M. W. (1940). Studies on the sensitization of animals with simple chemical compounds: VII. Skin sensitization by intraperitoneal injections. *Journal of Experimental Medicine, 71*(2), 237–245.
17. Landsteiner, K., & Chase, M. W. (1939). Studies on the sensitization of animals with simple chemical compounds: VI. Experiments on the sensitization of guinea pigs to poison ivy. *Journal of Experimental Medicine, 69*(6), 767–784.
18. Landsteiner, K., & Chase, M. W. (1937). Studies on the sensitization of animals with simple chemical compounds: IV. Anaphylaxis induced by picryl chloride and 2:4 dinitrochlorobenzene. *Journal of Experimental Medicine, 66*(3), 337–351.
19. Glick, B., Chang, T. S., & Jaap, R. G. (1956). The bursa of fabricius and antibody production. *Poultry Science, 35*, 224.
20. Miller, J. F., Brunner, K. T., Sprent, J., Russell, P. J., & Mitchell, G. F. (1971). Thymus-derived cells as killer cells in cell-mediated immunity. *Transplantation Proceedings, 3*(1), 915–917.
21. Claman, H. N. (1966). Human thymus cell cultures-evidence for two functional populations. *Proceedings of the Society for Experimental Biology and Medicine, 121*(1), 236–240.
22. Claman, H. N., Chaperon, E. A., & Triplett, R. F. (1966). Thymus-marrow cell combinations. Synergism in antibody production. *Proceedings of the Society for Experimental Biology and Medicine, 122*(4), 1167–1171.
23. Mosier, D. E. (1967). A requirement for two cell types for antibody formation in vitro. *Science, 158*(3808), 1573–1575.
24. Mitchell, G. F., & Miller, J. F. (1968). Cell to cell interaction in the immune response. II. The source of hemolysin-forming cells in irradiated mice given bone marrow and thymus or thoracic duct lymphocytes. *Journal of Experimental Medicine, 128*(4), 821–837.
25. Mitchell, G. F., & Miller, J. F. (1968). Immunological activity of thymus and thoracic-duct lymphocytes. *Proceedings National Academy Sci USA, 59*(1), 296–303.
26. Marrack, P., Hannum, C., Harris, M., Haskins, K., Kubo, R., Pigeon, M., et al. (1983). Antigen-specific, major histocompatibility complex-restricted T cell receptors. *Immunological Reviews, 76*, 131–145.
27. Davis, M. M., Chien, Y. H., Gascoigne, N. R., & Hedrick, S. M. (1984). A murine T cell receptor gene complex: isolation, structure and rearrangement. *Immunological Reviews, 81*, 235–258.
28. Bonecchi, R., Bianchi, G., Bordignon, P. P., D'Ambrosio, D., Lang, R., Borsatti, A., et al. (1998). Differential expression of chemokine receptors and chemotactic responsiveness of type 1 T helper cells (T_h1s) and T_h2s. *Journal of Experimental Medicine, 187*(1), 129–134.

29. Favre, D., Mold, J., Hunt, P. W., Kanwar, B., Loke, P., Seu, L., et al. (2010). Tryptophan catabolism by indoleamine 2,3-dioxygenase 1 alters the balance of TH17 to regulatory T cells in HIV disease. *Science Translational Medicine, 2*(32), 32–36.
30. Shi, Z., Curiel, D. T., & Tang, D. C. (1999). DNA-based non-invasive vaccination onto the skin. *Vaccine, 17*(17), 2136–2141.
31. Tang, D. C., Shi, Z., & Curiel, D. T. (1997). Vaccination onto bare skin. *Nature, 388*(6644), 729–730.
32. Takeda, K., & Akira, S. (2004). TLR signaling pathways. *Seminars in Immunology, 15*(1), 3–9.
33. Rang, H. P. (2003). *Pharmacology* (p. 223). Edinburgh: Churchill Livingstone. ISBN 0-443-07145-4.

Interferons

7

Abstract

Interferon was discovered by Alick Isaacs and Jean Lindenmann in 1957. It was originally thought that interferon could be used as a general anti-viral agent and in anti-cancer therapy. There are many different types of interferons, now known as interferons "alpha," "beta," "gamma" and "lambda," with different cellular receptors and modes of action, and there are possibly 24 different types of alpha interferon. Independently and simultaneously, a group of Japanese scientists found an "interfering factor," which upon subsequent analysis turned out to be interferon, probably of the alpha type. The interferon alpha gene was the first mammalian gene to be cloned in a bacterial system and became the prototype for gene cloning technology. Until the cloning of the interferons in *Escherichia coli*, and expression of the interferon genes in mammalian cells in culture, it was impossible to obtain enough material for clinical use. Interferon today is predominantly used in the treatment of hepatitis B and C, leukemia and Kaposi's sarcoma. As an anti-viral agent, interferon has not lived up to its initial promise, since in vitro most viruses block the activity of interferon and clinical trials have given inconclusive results with severe side effects. Interferon induces hundreds of genes in vivo and in vitro, each interferon producing overlapping and distinct gene profiles. The mechanism of both interferon induction and anti-viral response is complicated and involves the interaction of many regulatory molecules. Interferon is now known to be a component of the large family of cytokines or interleukins.

M. W. Taylor, *Viruses and Man: A History of Interactions*,
DOI: 10.1007/978-3-319-07758-1_7,

7.1 Discovery of Interferon

No history of virology would be complete without a discussion of interferons and how they led to the discovery and identification of cytokines (small proteins that influence the activity of the immune system and nearby cells), their function in innate immunity, and their pharmaceutical properties as anti-viral and anti-cancer agents. The cloning of the interferon gene and its production in *E. coli* initiated the biotechnology revolution. As was the case of many other major discoveries in science, interferon was a fortuitous discovery.

In 1957, Alick Isaacs (1921–1965) and a post-doctoral Swiss student, Jean Lindenmann, were studying the phenomenon of "viral interference"—the ability of one virus to inhibit the replication of another virus. When 10-day-old chick chorioallontoic membranes from chick embryos were infected with heat or UV inactivated influenza virus, a material was produced that interfered with subsequent viral replication. The experimental procedure is illustrated in Fig. 7.1. Influenza virus production (or inhibition) was measured by hemagglutination, the ability of the virus to interact and agglutinate red blood cells. They termed the interfering substance "interferon".The end point of the titration was the identification of that well (on a plate of small wells) with partial agglutination; the reciprocal of the influenza dilution thus observed was taken as the interferon titer (concentration).

The first reports of interferon were published in 1957 in the prestigious *Proceedings of the Royal Society*, [1, 2] and a more detailed description in the *British Journal of Pathology* [3]. Using standard biochemical techniques, it was shown

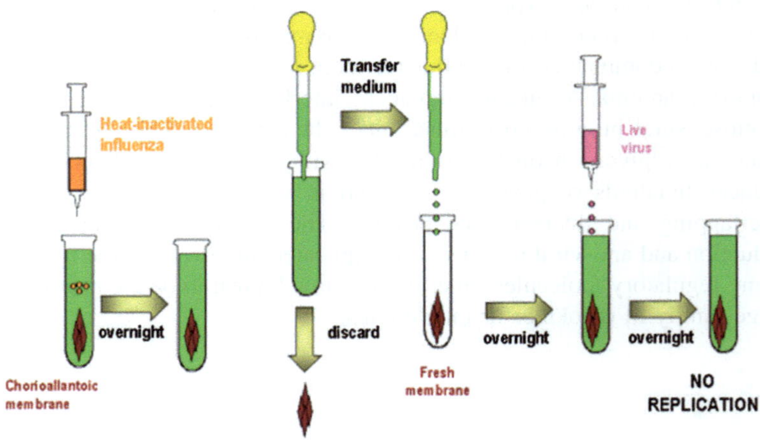

Fig. 7.1 The Discovery of intereferon. From Isaacs and Lindenmann, Proc. Roy Soc B, 1957 (adapted from http://www.microbiologybook.org/book/virol-sta.htm)

that interferon was a protein, was pH 2 resistant, could be precipitated by ammonium sulfate, destroyed by ether, and digested by trypsin—all characteristics of proteins. The stability at pH 2 was an important characteristic, later used to distinguish the original interferon (called IFN-alpha) from other later discovered interferons. Attempts to initially purify the interferon were unsuccessful. A very low concentration was induced and the investigators at that time were unaware that multiple types of interferon existed, so that even with the techniques of column chromatography available at that time, it proved impossible to purify a single type.

Jean Lindenmann returned to Zurich and did not continue to work on interferon but continued research on influenza virus, discovering a strain of mice resistant to the virus. This resistance did not appear to be related to interferon, but to the Mx protein coded by an autosomal gene, *MX*. However, in later years Lindenmann and colleagues found that the Mx protein was interferon-inducible, and that this protein was a component of the cascade of genes induced by interferon.

Rather surprisingly, interferon was not viral-specific; it not only inhibited influenza, the inducing virus, but also unrelated vaccinia and other viruses [4]. That interferon was an agent with a very wide inhibitory range suggested that interferon could be used as a general anti-viral agent, much as the recently discovered antibiotics. As a result, three pharmaceutical companies, Glaxo Laboratories, ICI Pharmaceuticals, and Burroughs Wellcome (later to become the Wellcome Foundation) supported the research. The primary purpose was to produce enough interferon for use in clinical trials, and a second was to keep the patent for interferon production in the U.K., since there was the feeling that penicillin, a British discovery, had been appropriated by the U.S. (See Burke [5] for a review of the early days of interferon research.)

7.2 Inhibitory Factor

In parallel with the research described above, but quite independently, in 1954 a group of researchers in Japan characterized what they called a "virus inhibitory factor" [6]. They injected ultra-violet irradiated vaccinia (inactivated virus) into the backs of rabbits and subsequently inoculated them with live vaccinia at various times after the initial inoculation. The vaccinia replication was inhibited by the pre-treatment with the inactivated virus. They isolated this substance and called it "inhibitory factor" (IF). Further experimentation was difficult because of the system used. These observations were presented at the 1956 Annual Meeting of the Japan Society for Viral Research and later at the meeting of the Japan-France Biology Society held in 1957. This was really the first report on interferon but was not recognized as such until much later. The IF was later tested for interferon activity using rabbit standard interferon, designated as such by the International Committee on Interferons, and it was found to have an interferon titer of 300,000 IU, an astonishingly high level of activity. It is now clear that Yasuichi Nagano and Yasuhiko Kojima observed the production of interferon in response to an inactivated DNA virus in the whole animal. Because this work was carried out in Japan, and originally published in a French journal, it was not given the recognition it deserved. (A fuller description of this study can be found in the review by Ozato and colleagues [7].)

7.3 First Clinical Trials

An ambitious first "clinical trial" was performed at the Salisbury Common Cold Center in England in 1962. Its purpose was to examine whether interferon inhibited the production of vaccinia virus following injection of the virus into the arms of individuals who had not been previously vaccinated with vaccinia. The result was limited inhibition of vaccinia replication; it was also obvious that much more interferon was needed if it was to be of clinical value, but it was not until the 1980s that enough interferon could be produced in a pure state for this purpose. Prior to gene cloning, it was extremely difficult to produce enough interferon for clinical use, even when attempting to produce it in cells in culture. In the 1960s there was a large demand for interferon in order to test it as an anti-cancer drug. Kari Cantell in Finland produced leukocyte interferon, in large vats, in quantities sufficient for some limited clinical trials. He harvested leukocytes from many sources and infected them with the Sendai virus, which greatly stimulated the production of interferon. However, even with these facilities, it was impossible to produce quantities sufficient for all the desired clinical trials (Fig. 7.2).

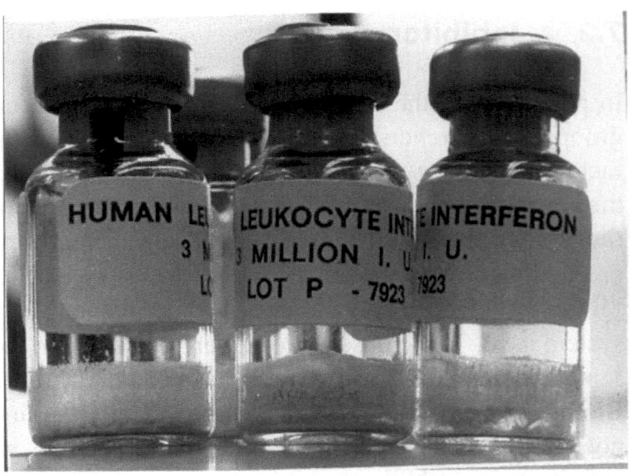

Fig. 7.2 Vials of leukocyte interferon

7.4 How Does Interferon Protect Cells Against Virus Infection?

During the 1960s, many different viruses were shown to induce interferon in cell culture, and interferon in turn—when added to cells in culture—could inhibit most viruses, particularly at high concentrations, but not all, since it was realized much later that some viruses had developed methods of neutralizing interferon activity. Interferon was measured by its anti-viral activity, one unit of interferon being sufficient to inhibit viral growth by 50 %, usually measured by a plaque assay. Since there was some variability depending on the cell line and virus used, standard lots were titered or measured the NIH and supplied as "standards" to laboratories on request. When interferon was added directly to virus, there was no effect; it was the addition of interferon to the cells that inhibited virus production. A virus-infected cell produces interferon, which protects nearby cells; thus. the nearby cells are in an anti-viral state. (This is illustrated in Fig. 7.3).

Interferons are not only induced by viruses, but by viral intermediates such as double-stranded RNA, by synthetic double-stranded RNA such as PolyI:PolyC, some species of bacteria, endotoxins, and other cytokines. By the 1970s, a number of different types of interferon were characterized. Differences were found on the basis of cell types protected, and the stability of the interferons at different pH (Table 7.1). The binding of interferon to specific receptors on the cell triggers a large number of biochemical reactions, leading to the inhibition of virus replication or maturation and the induction of further interferon and other cytokines. The system is much more complex than the early researchers of this field could ever have imagined.

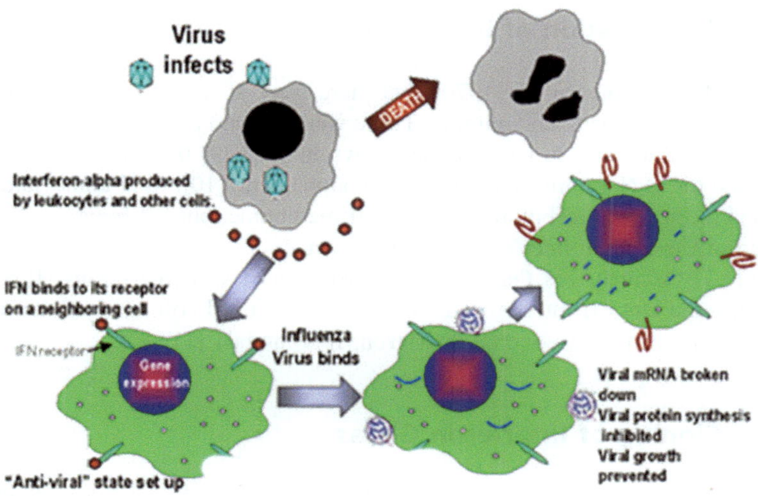

Fig. 7.3 How interferon affects neighboring cells. Adapted from Hunt (http://www.microbiologybook.org/book/virol-sta.htm)

Table 7.1 Classification and characteristics of human interferons

Characteristics	Interferon-alpha	Interferon-beta	Interferon-gamma	Interferon-lambda
Other designation	Intron-A, pegasys. consensus	IFN-b2. At one time mistakenly called IL-6	Macrophage activating factor: Immune-interferon	IL28A, IL28 B IL29, IFNA14
Number of genes	24 (+)?	1	1	3 (+)
Chromosomal location	9p22	9p21	12q14	19q13.13
Introns in gene	None	None	Yes	Possibly yes
Cell of origin	Leukocytes	Fibroblasts	Lymphocytes, macrophages, NK cells, dendritic cells	Epithelial cells
Inducers	Virus, dsRNA	Virus, dsRNA	Antigens, mitogens, other interferons, cytokines, IL2, NK receptors	Virus

Using mice as a model, experiments were initiated to examine whether interferon produced in mouse cell culture culture could be used to inhibit virus infection in mice, If given in large daily doses, interferon did inhibit the pathological effects of the EMC virus, a virus causing paralysis and death in mice [8]. Antibodies were made to mouse and human interferon in rabbits that neutralized the anti-viral effect in the mouse and the anti-viral effect of human interferon-alpha.

7.5 Classification of Interferons

It was rather obvious by the 1970s that there were at least three major types of interferons: alpha, beta, and gamma. These were defined as interferons because of their anti-viral activity in vitro. It was shown in 1984 [9] that there were separate receptors for IFN-gamma and interferon alpha/beta [10]. IFN-alpha and -beta attached to the same receptor(s) on the cell membrane, but the biological activity of each of these three interferons was different.

Table 7.1 summarizes the differences between class I interferons (IFN-alpha and beta) and class II interferon, IFN-gamma. The third column presents what is known of a recently discovered interferon, IFN-lambda [11, 12].

7.6 Cloning of Interferon Genes

In 1980 the interferon alpha gene was cloned into *E. coli* and the methodology used became standard for cloning mammalian genes in this bacterium. Double-stranded complementary DNA (cDNA) was prepared from random mRNA from interferon-producing leukocytes (stimulated with Sendai virus) and cloned into an *E. coli* plasmid, known as pBR 322. Five thousand bacterial clones were screened by extracting plasmid DNA; they were expressed in frog oocytes, and those synthesizing biologically active interferon isolated. mRNA coding for a polypeptide with IFN activity was isolated, and the IFN was shown to have normal antiviral activity [13]. This research was done in the laboratory of Charles Weissmann at the University of Zurich, and supported by a new biotech company, Biogen. Fifteen percent of the shares of Biogen were held by Schering-Plough, which eventually sold the interferon as Intron A in the U.S. market. This major breakthrough in the cloning of IFN-α2a led to the cloning of other interferons. Meanwhile, other biotech and pharmaceutical companies entered into this very competitive market. The Wellcome Research Laboratories, one of the initial companies involved in interferon research in the U.K., extracted interferon from virus-transformed lymphoblastoid cells, and Searle Laboratories, also in England, produced interferon from fibroblasts. Biogen went on to sell interferon-beta 1a (Avonex) for the treatment of multiple sclerosis, and eventually joined up with a company specializing in monoclonal antibodies, Idec, to form one of the largest biotech companies in the world. Interpharm in Israel began the manufacture and production of IFN-beta in Chinese hamster ovary cells in culture, and finally interferon gamma was cloned in 1981 at the Genentech Company in San Francisco. Thus interferon could now be produced in *E. coli,* yeast, and mammalian cells [14, 15]. Interferon cloning led to the development of the biotech industry, and for this alone it is historically significant.

7.7 Interferon Alpha and Beta

Class I interferons have been studied more than class II, reflecting the fact that they were discovered first, cloned early, and used in the clinic. There appear to be many types of IFN-alpha based on gene homology, [16] but only one type of IFN-beta. Since the genes for all class 1 interferons reside on the same chromosome, it is assumed that they arose as duplications of a single interferon gene. Using high performance (or pressure) liquid chromatography (HPLC), it became possible to purify the various interferons and assay them with a biological assay. Using fluorescent tags, the amino acid content and sequence of each type of interferon was determined. There are as many as 24 different interferons that are classified as IFN-α. Each one has a slightly different amino acid sequence, different specific activity and antiviral spectrum [16]. Different cell types make different amounts of each type; different genes codes for each one. Very little is known of the biological bases of the multiple species and their activities, since most research has only been done with one of these, termed IFNα2. This may not be the most active of the alpha interferons; all the alpha interferons bind to the same receptor, and when one looks at gene (protein) induction as measured by a few key induced proteins, the same level of induction occurs in all cases, suggesting that a minimal amount of interferon triggers activity [16]. Interferon-alpha is induced in various types of leukocytes.

There is only one known species of IFN-beta, which shares 25 % amino acid homology with IFN-α. IFN-β is produced by fibroblasts and epithelial cells in response to virus and double-stranded RNA.

In the early 1990s a group of scientists at the Amgen biotech company attempted to go one better than nature and create what they termed a "consensus interferon," which was made by comparing the most common amino acids occurring in the then-available 20 different species of IFN-alpha, constructing a " gene" from these amino-acids introducing this new "gene" into *E. coli*, and producing an active molecule. A number of such molecules were produced, one of which had very high specific activity. This molecule had higher specific activity in anti-viral activity, and anti-tumor activity, as measured by the death of hairy cell leukemia cells, and the activation of natural killer cells [17, 18]. Eventually, consensus interferon was approved for the treatment of hepatitis C. A well-planned clinical trial was never performed to examine whether consensus interferon was superior for the treatment of hepatitis C than IFN-α2a (intron A) or IFN-α2b (Pegasys). However, in clinical trials of HCV patients who were not responsive to IFN-α2a or IFN-α2b, there was a 30–50 % response rate.

7.8 Interferon-Gamma

Cells of the immune system produce interferon-gamma: dendritic cells, natural killer cells (NK), both CD4 and CD8 T-lymphocytes, and macrophages. Interferon-gamma is also induced by cytokines such as IL-1, IL-2, and by many growth factors. It is also induced during viral infections in vivo as a part of the immune system. IFN-gamma stimulates the expression of class I and class II MHC molecules and promotes the differentiation of naïve helper T-cells into Th1 cells, activates dendritic and cytotoxic T-cells, and increases the cytotoxicity of NK cells. Thus it is a key player in the immune response. (For an excellent short summary of IFN-gamma activities, see http://www.bio.davidson.edu/courses/immunology/students/spring2006/v_alvarez/ifn-gamma.html). IFN-gamma is an important component of cellular immunity and plays a significant role in cell differentiation, cell growth, and cell survival. In fact, if IFN-gamma were to be re-discovered today, it would not be called an "interferon" but rather an "interleukin."

IFN-gamma binds to specific receptors on the cell membrane known as IFN-γR. The bound receptor interacts with enzymes, JAK kinase (Just Another Kinase), which phosphorylates STAT proteins, which in turn interact with specific DNA sequences upstream of inducible genes that have sequences in their upstream DNA, known as "GAS sequences" (gamma activated sequences—see Fig. 7.4). The spectrum of genes induced by IFN-gamma overlap, but with some differences from the genes induced by interferon alpha and beta [19]. One of the most studied genes induced leads to the production of a protein, IDO (indoleamine 2,3-dioxygenase). This enzyme catalyzes the rate-limiting step in tryptophan (an essential

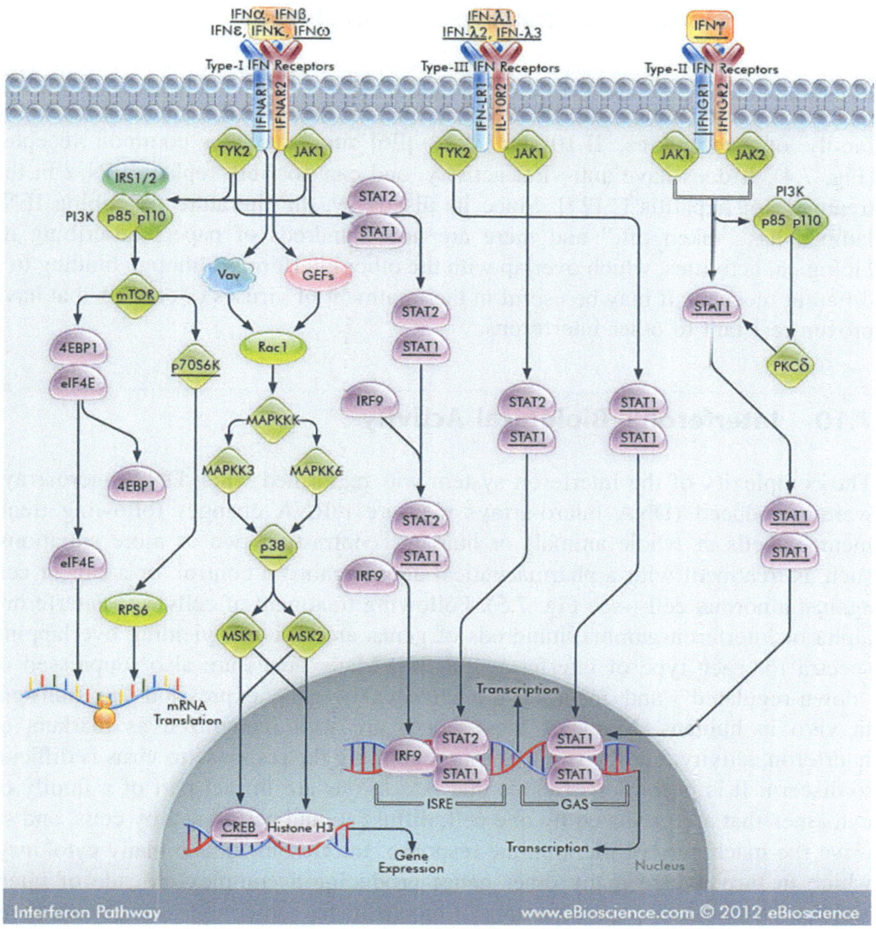

Fig. 7.4 Simplified schematic of how IFN-alpha//beta, IFN-gamma and IFN-λ stimulate gene induction (Courtesy of eBioscience, an Affymetrix company)

amino acid in humans) catabolism. In culture, cells die in the presence of interferon gamma, unless supplemented with tryptophan [20, 21]. It has been suggested that this enzyme is responsible for T-cell tolerance [22], is induced during pregnancy, and is important in the maintenance of pregnancy in humans [23]. However, this latter hypothesis is controversial [24, 25]. IDO may also have detrimental effects, being higher during septic shock and in certain types of cancers. Thus, the function of this enzyme, and many other proteins induced by the interferons, are still unknown.

7.9 Interferon Lambda: Type III Interferons

More recently, another group of interferons, called "IFN-lambda," has been discovered. Recent studies with IFN-lambda indicate that it is close in structure to the family of interleukins, IL10 and IL22 [26] and shares a common receptor (Fig. 7.4). It does have anti-viral activity, and can possibly replace IFN-α in the treatment of hepatitis C [27]. Since its discovery, the literature describing IFN-lambda has "taken off," and there are now hundreds of papers describing its biological activities, which overlap with the other interferons although binding to a different receptor. It may be useful in the treatment of viruses or cancers that have proven resistant to other interferons.

7.10 Interferon's Biological Activity

The complexity of the interferon system was reaffirmed when DNA microarrays were introduced (DNA micro-arrays measure mRNA changes following treatment of cells or whole animals or humans, contrasting two or more situations, such as treatment with a pharmaceutical agent against a control, or a cancer cell against a normal cell—see Fig. 7.5). Following treatment of cells with interferon-alpha or interferon-gamma, hundreds of genes are induced, yielding overlapping spectra for each type of interferon [28–30]. Many genes are also suppressed or "down-regulated" and most of the changes in gene expression are transient in vivo in humans. Many of these genes are well recognized as markers of interferon activity, but their function in regulating the response to virus is difficult to discern. It is becoming obvious that interferons are in fact part of a family of cytokines that are produced by one cell, diffuse to and act on nearby cells, and so drive the machinery of the immune response. Interferons induce many cytokines, which in turn induce many other genes producing a complex cascade of interactions. Cytokine biology is an area of biochemistry too complex to discuss here, but it must be emphasized that it has become an important component of immunology.

There are many questions that need to be answered in studying interferon at the cellular level. What happens after interferon binds to the cell receptor and triggers an anti-viral response? How does this response inhibit virus production? Has the virus developed methods of overcoming the interferon response? Are there specificity and differences among the various types of interferon? In answering these questions, one delves into the basis of the innate immune response and the relationship between the innate immune response and the induced cellular immunity. Answers to some of these questions are found in Chap. 6 (on immunology), and in Figs. 7.4 and 7.6.

As stated, interferons bind to specific receptors on the cell membrane. IFN alphas and betas both bind to the same receptors, IFNAR1, IFNAR2, and IFN-gamma bind to separate receptors, IFGNR1 and IFNGR2, and IFN-lambda to a

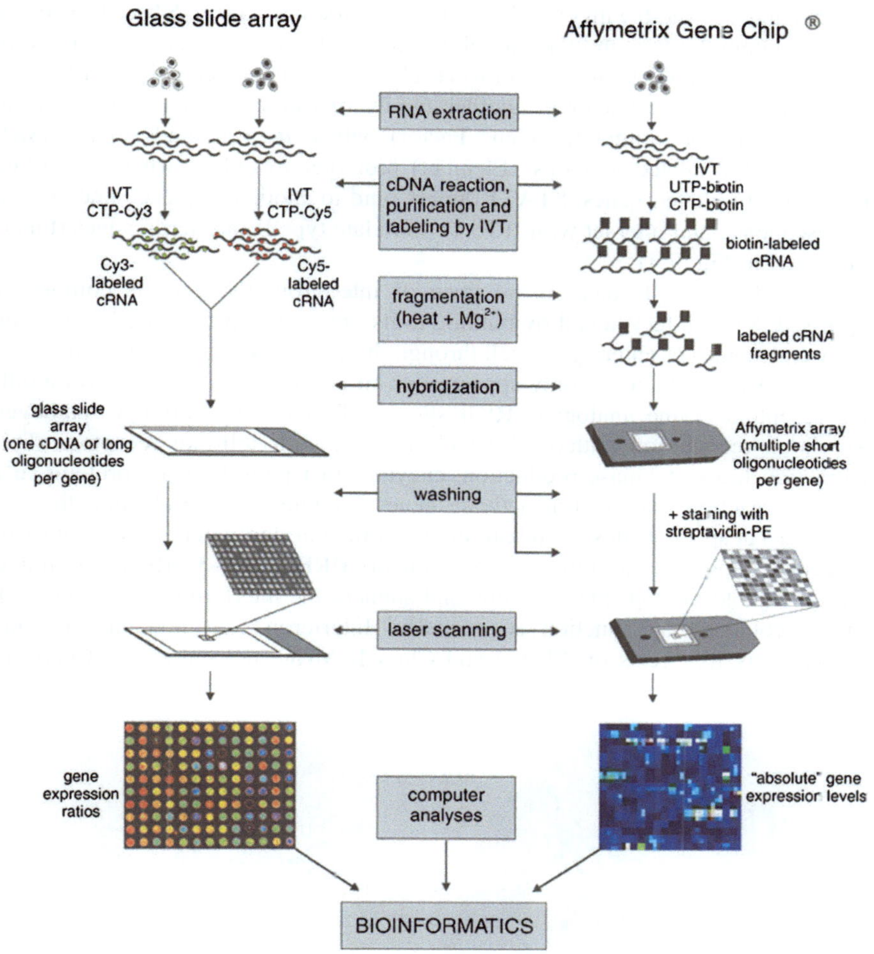

Fig. 7.5 Hybridization to detect differential gene expression by DNA microarrays (Affymetrix)

third set of receptors which are shared with interleukins IL10, IL28A, IL28B, and IL29 [31] (see Fig. 7.4). Interestingly, virus infection also induces these interleukins. Binding of the interferons activates a number of kinases, including a series of proteins termed JAK kinases (Janus Kinase) which phosphorylate tyrosine on a group of proteins known as STATS (Signal Transducer and Activator of Transcription). The phosphorylation activates and leads to dimerization of these STAT proteins. There are seven STAT proteins that have been recognized, each related to a different function in either the immune response or metabolic response. The function of these STAT proteins has been elucidated from "knock-out" mice, i.e., mice with specific genes (in this case STATs), deleted. STAT1 is involved with the IFN-α/β and IFN-γ signaling pathway, STAT 2 also with IFN-α/β pathway, STATs

3, 4, 5A and 5B, and 6 with other interleukins (cytokines). These STAT proteins in turn form complexes with other proteins that bind to specific DNA sequences. In type I interferon signaling, STAT1–STAT2 heterodimers combine with IRF9 (Interferon Response Factor 9, another family of transcription factors) to form ISGF3 (Interferon Stimulated Gene Factor), which in turn binds to the ISRE (Interferon Stimulated Response Element) promoter to induce what are termed interferon-stimulated genes. STAT1 dimers bind to another sequence known as a GAS sequence and interact with IRF1 to stimulate type II interferon genes (this is illustrated in Fig. 7.4).

In uninfected cells there is no trace of interferon; the gene is completely repressed. Virus infects a cell by binding to its unique receptor. After being taken up by endosomes or entering the cell through the cell membrane, viral components or viral dsRNA, which can only appear after initial viral replication, activate toll-like receptors or the analogous RIGI system. This is done through "pathogen associated recognition patterns," which interact with toll-like receptors. This interaction leads to kinase production, enzymes that phosphorylate proteins in a complex reaction that in turn activate genes for interferon production through complexes at specific sites on the promoters of the interferon genes, by a family of molecules known as interferon response factors (IRF-1, IRF-3, IRF-7). A similar family of proteins stops the induction and silences the interferon genes (Fig. 7.6). These proteins may function sequentially. Interferon can modulate immune responses by its effects on Class I and Class II MHC molecules (see Chap. 6).

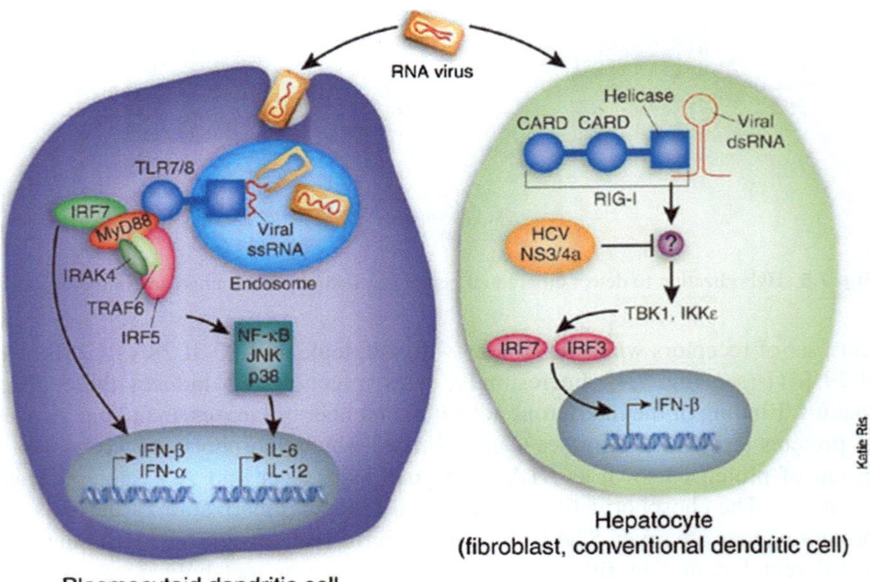

Fig. 7.6 Toll-like receptor—and RIG-I-dependent—induction of type I interferon during RNA virus infection. Adapted from *Nature Medicine* **11**, 929–930 (2005) [32]

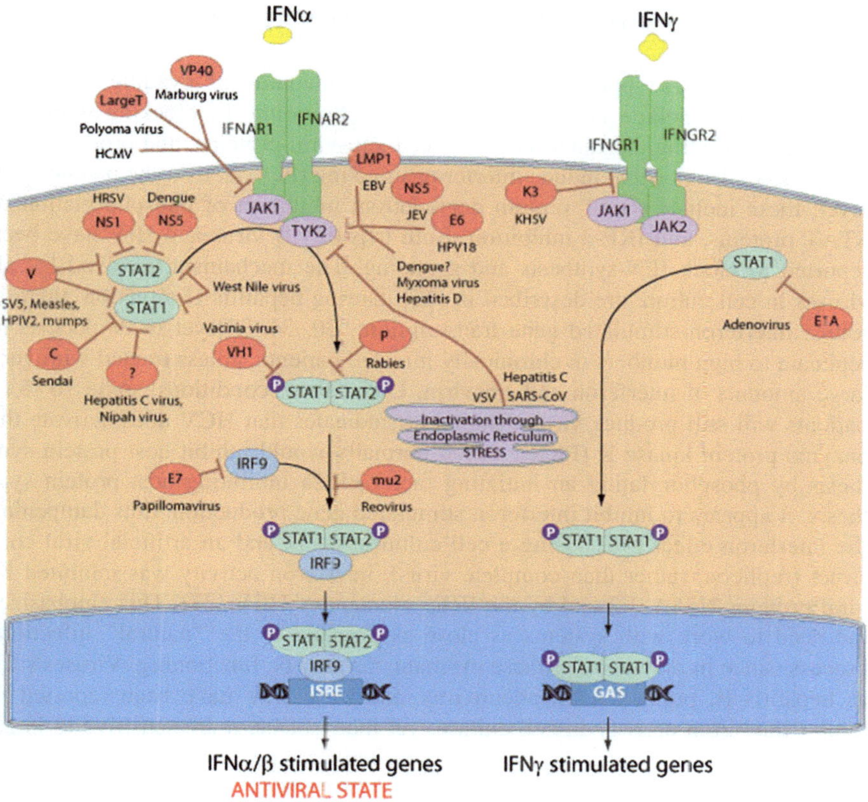

Fig. 7.7 Antiviral pathway and steps at which viruses block IFN activity (Source: ViralZone www.expasy.org/viralzone, Swiss Institute of Bioinformatics)

IFN-alpha, IFN-beta and IFN-gamma increase expression of Class I molecules on all cells, thereby promoting recognition by cytotoxic T-cells that can destroy virus-infected cells. IFN-gamma can also increase expression of Class II MHC molecules on antigen-presenting cells, resulting in better presentation of viral antigens to CD4^{+} T helper cells. Furthermore, IFN-gamma can activate NK cells and dendritic cells, which can kill virus-infected cells. As stated above, interferon induces many genes that are involved in gene regulation, cell differentiation, other cytokine induction and anti-viral activity. As discussed below, many viruses block interferon activity, as shown in Fig. 7.7.

Figure 7.6 illustrates the pathway of an RNA virus. Viral double-stranded RNA, an intermediate of virus replication, binds to proteins known as toll-like receptors, which are proteins that alert the cell to the presence of a pathogen. A complex forms on the toll-like receptor or RIG I, another toll-like receptor molecule, and interacts with a number of proteins that migrate to the nucleus and activate the genes for IFN, which in turn then activate nearby cells.

7.11 How Do Viruses Escape Interferon Activity?

Almost all viruses appear to have developed methods of evading interferon activity. These include inhibition of interferon biosynthesis, blocking interferon signaling, inhibiting the function of induced anti-viral proteins, and production of decoys to molecules that induce interferon signaling (Fig. 7.7). At the biochemical level, these include STAT protein degradation, inhibition of phosphorylation of STAT proteins, and IRF-3 inhibition. Both hepatitis B virus and HIV have been reported to block IFN-synthesis and signaling. The mechanism involved for all viruses in cell culture are described in [33]. During hepatitis C infection, there is robust interferon-stimulated gene transcription, [30, 34, 35] yet HCV persists to replicate to high numbers in chronically infected patients, unless treated with very large amounts of interferon and ribavirin. Under these conditions, some 70 % of patients will still produce virus. One report indicates that HCV does activate the enzyme protein kinase R (PKR), which normally would inhibit host protein synthesis by phosphorylating an initiating factor eIF2a inhibiting host protein synthesis, it appears to inhibit interferon stimulated gene production, thus dampening the interferon effect [36]. Using a cell culture system and an artificial viral construct (replicon, rather than complete virus), interferon activity was inhibited by viral proteins NS5A, E2 and by the IRES element of HCV [37]. This emphasizes the need to work with systems as close as possible to the "natural" infectious process, since in man none of these mechanisms may be functioning. Viruses such as hepatitis B, polio, SARS, adenovirus, and even HIV have been reported to inhibit interferon activity in cell culture, yet many of them are sensitive to interferon in man.

7.12 Clinical Studies of Interferons with Virus

Since 2002 interferon has been PEGylated to increase stability, with the addition of a polyethylene glycol molecule. The major use of interferon in the clinic has been in the treatment of hepatitis B and hepatitis C. In hepatitis B patients, the response rate is between 15–40 % of treated patients responding with long-term remission. IFN-alpha is used with a reverse transcriptase inhibitor, Lamivudin.

Until recently, the standard treatment for hepatitis C in patients was a 48-week regimen of interferon-alpha and the anti-viral drug ribavirin. The percentage of patients with no detectable virus after completion of treatment was approximately 30 %, depending on the type of interferon used, the antigenic type of the virus, and the race of the patients [38]. This protocol has recently been modified and usually includes a viral protease inhibitor.

Interferon has also been used in many clinical trials with inconclusive results. A small group of asymptomatic HIV-infected individuals were treated with IFN-α2b: 41 % had decreased viral titer, and no patients in the IFN-α group developed AIDS-defining opportunistic infection, compared with 5 patients in the placebo group ($P = 0.02$) [39]. However, 35 % of the patients in the treatment group withdrew from the study because of the severity of the side effects. Other clinical trials have not been so successful. Attempts to treat children with Japanese encephalitis with IFN-alpha were unsuccessful, even though this virus and other flaviviruses are sensitive to interferon in cell culture [40]. During the SARS epidemic of 2002–2003, interferon was used among many other treatments, but the data are inconclusive [41]. Overall, the effect of interferon on virus infections in the clinical setting has been disappointing. Although interferon is active in inhibiting many viruses in cell culture, it does not carry over to the clinic, and there are many reasons for this. Most viruses induce interferon at an early stage of infection, so that the effect of interferon on the virus is established during the course of the infection, and additional amounts may not make much of a difference.

Even in the case of chronic infections, such as hepatitis C, interferon is induced and present during the infection [34]. It requires extremely high doses to have an effect on the virus, and at these levels IFN is quite toxic. In most clinical trials, a percentage of the patients will drop out of the trial. The experience with Japanese encephalitis and HIV indicate that once a viral infection has been established, it is difficult to eradicate the virus with interferon. Even in cells in culture, the timing of the addition of the interferon is crucial in inhibiting the virus. Long-term exposure to interferon—as in the clinical treatment of hepatitis C—may cause a decrease in white blood cells (leukopenia), leaving the patient susceptible to infections. Apart from the problems with the treatment, interferon therapy is expensive, i.e., $10,000–$20,000 for a 48-week regimen of interferon plus ribavirin.

7.13 Interferon as an Anti-cancer Agent

Although originally characterized as an anti-viral agent, to everyone's surprise interferon was also an anti-cancer agent. The first experiments performed were on mouse leukemia known to be of viral origin, including Friend's leukemia (identified by Charlotte Friend in 1957) and Rauscher's leukemia; both tumors later identified of retrovirus origin. However, interferon was not only active against tumors caused by viruses but against a large number of transplantable mouse tumors of different origins. In these experiments, tumors were injected intraperitoneally or intramuscularly, so that the interferon could be administered directly into the tumor.

Interferon later became a standard treatment for a number of types of human cancers, including hairy cell leukemia, Kaposi's sarcoma in AIDS patients, chronic myelogenous leukemia (CML), and papilloma infections (warts) [42].

By 1982 a phase 1 trial was conducted in a large group of patients with various cancers. There was a variety of side effects now known to occur with interferon therapy, including nausea, fatigue, headache, muscle pain, and occasionally elevated liver enzymes. However, there was objective evidence of antitumor activity in non-Hodgkin's lymphoma, chronic lymphocytic leukemia (CML), Hodgkin's disease, breast cancer, and melanoma [43]. One of the first clinical uses of interferon (IFN) was in the treatment of a rare leukemia known as hairy cell leukemia, which results from over-production of mature (and abnormal) B-cells. The cells under the microscope have a hairy morphology, hence the name. These cells in culture are very sensitive to interferon [44]. There are approximately 500 new cases of hairy cell leukemia each year in the U.S. Interferon (alpha, or beta, which was discovered later) is the therapeutic of choice, and about 80–90 % of patients will respond to treatment with remission for a period. Interferon therapy did extend the life span in early-stage melanoma and early-stage CML; however, in clinical trials it is inefficient against solid tumors. In Kaposi's sarcoma, as occurs in AIDS patients, it was effective if injected directly into the lesion, but not systemically. A major problem is the large dose of interferon required in any treatment as well as the severe side effects.

IFN-beta has been approved for the treatment of multiple sclerosis. The response to the drug varies and in some cases is ineffective, depending on the source of the interferon and the type of multiple sclerosis. Several studies have found IFN-β beneficial in reducing rates of relapse, whereas others have reported no benefit in this regard [45].

7.14 The Interferon Society

The first meeting of those interested in interferon was held in 1964 in Bratislava (at that time in Czechoslovakia but today in Slovakia). There were sporadic meetings during the 1960s and early 1970s, but without an official organization. A report in 1974 claiming that interferon produced by Cantell could be used to treat cancer motivated a meeting in New York, organized by Mathilde Krim to stimulate interest in interferon and its anti-cancer activities. Dr. Krim was an activist in medical research, and was a faculty member of the Sloan Kettering Institute for Cancer Research. She was a well-known "socialite"in New York and her party for President Kennedy's 45th birthday was quite famous. She later became an AIDS activist, recognizing early on the problems and ethical dilemmas of the disease. (http://en.wikipedia.org/wiki/Mathilde_Krim). Bill Stewart, one of the early pioneers of interferon research, founded the Interferon Society in 1982. The first formal meeting and the publication of the *J. Interferon Research* occurred in 1983. The society met once a year with the presentation of papers on the three major classes of interferons (α, β, and γ). The *Journal of Interferon Research*, only for papers on interferon, was established (although papers on interferon were published in other journals related to virology).

By 1989 the importance of the cytokines and their relationship to interferon was beginning to be realized, and the journal changed its name to the *Journal of Interferon and Cytokine Research* to encompass papers dealing with "other interferons," which were not called interferons but a variety of names based on their activities, such as the "tumor necrosis factor" (which was initially called lymphotoxin), IL-1, IL-2, etc. A separate Society of International Cytokine Research (ICS) was established in 1989. There is also a *Journal of Cytokine*

Research, which publishes papers on interferon research. We do not know how many cytokines (initially lymphokines, now interleukins) exist. The International Society for Interferon Research joined with the Cytokine Society in 2012 to form one organization for scientists working in both areas, which were obviously now interconnected. This would be known as the International Cytokine and Interferon Society. Many of the other cytokines (interleukins) are being tested for their effects on diseases and cancer.

7.15 Conclusions

The days of interferon as a clinical entity may be over. It has been replaced in the treatment of viral infections by small molecules that inhibit specifically viral enzymes, and such molecules may have fewer side effects. Interferon is still an important molecule to study since it elucidates the workings of the immune system. It is an important "backup" in the event of a sudden outbreak of an unknown virus epidemic; this was demonstrated during the SARS epidemic, when there was no alternative but interferon. There is not yet sufficient knowledge of the new interferons recently discovered, or whether they will have clinical applications.

References

1. Isaacs, A., Lindenmann, J. (1957). Virus interference. I. The interferon. *Proceedings of the Royal Society of London Series B, Containing papers of a Biological character Royal Society, 147*(927), 258–267.
2. Isaacs, A., Lindenmann, J., Valentine, R. C. (1957). Virus interference. II. Some properties of interferon. *Proceedings of the Royal Society of London Series B, Containing papers of a Biological character Royal Society, 147*(927), 268–273.
3. Lindenmann, J., Burke, D. C., & Isaacs, A. (1957). Studies on the production, mode of action and properties of interferon. *British Journal of Experimental Pathology, 38*(5), 551–562.
4. Burke, D. C., & Isaacs, A. (1958). Further studies on interferon. *British Journal of Experimental Pathology, 39*(1), 78–84.
5. Burke, D. C. (1987). Early days with interferon. *Journal of Interferon Research, 7*(5), 441–442.
6. Nagano, Y., Kojima, Y., & Sawai, Y. (1954). Immunity and interference in vaccinia; inhibition of skin infection by inactivated virus. *Comptes Rendus des Seances de la Societe de Biologie et de Ses Filiales, 148*(7–8), 750–752.
7. Ozato, K., Uno, K., & Iwakura, Y. (2007). Another road to interferon: Yasuichi Nagano's journey. *Journal of interferon & cytokine research: the official journal of the International Society for Interferon and Cytokine Research, 27*(5), 349–352.
8. Gresser, I. (2007). Interferon: an unfolding tale. *Journal of Interferon & Cytokine Research: The Official Journal of the International Society for Interferon and Cytokine Research, 27*(6), 447–452.
9. Orchansky, P., Novick, D., Fischer, D. G., & Rubinstein, M. (1984). Type I and Type II interferon receptors. *Journal of Interferon Research, 4*(2), 275–282.
10. Novick, D., Cohen, B., & Rubinstein, M. (1994). The human interferon alpha/beta receptor: characterization and molecular cloning. *Cell, 77*(3), 391–400.

11. Sheppard, P., Kindsvogel, W., Xu, W., Henderson, K., Schlutsmeyer, S., Whitmore, T. E., et al. (2003). IL-28, IL-29 and their class II cytokine receptor IL-28R. *Nature Immunology, 4*(1), 63–68.
12. Kotenko, S. V., Gallagher, G., Baurin, V. V., Lewis-Antes, A., Shen, M., Shah, N. K., et al. (2003). IFN-lambdas mediate antiviral protection through a distinct class II cytokine receptor complex. *Nature Immunology, 4*(1), 69–77.
13. Nagata, S., Taira, H., Hall, A., Johnsrud, L., Streuli, M., Ecsodi, J., et al. (1980). Synthesis in E. coli of a polypeptide with human leukocyte interferon activity. *Nature, 284*(5754), 316–320.
14. Derynck, R., Leung, D. W., Gray, P. W., & Goeddel, D. V. (1982). Human interferon gamma is encoded by a single class of mRNA. *Nucleic Acids Research, 10*(12), 3605–3615.
15. Derynck, R., Singh, A., & Goeddel, D. V. (1983). Expression of the human interferon-gamma cDNA in yeast. *Nucleic Acids Research, 11*(6), 1819–1837.
16. Rubinstein, M. (1987). Multiple interferon subtypes: the phenomenon and its relevance. *Journal of Interferon Research, 7*(5), 545–551.
17. Blatt, L. M., Davis, J. M., Klein, S. B., & Taylor, M. W. (1996). The biologic activity and molecular characterization of a novel synthetic interferon-alpha species, consensus interferon. *Journal of Interferon & Cytokine Research: The Official Journal of the International Society for Interferon and Cytokine Research, 16*(7), 489–499.
18. Ozes, O. N., Reiter, Z., Klein, S., Blatt, L. M., & Taylor, M. W. (1992). A comparison of interferon-Con1 with natural recombinant interferons-alpha: antiviral, antiproliferative, and natural killer-inducing activities. *Journal of Interferon Research, 12*(1), 55–59.
19. Sanda, C., Weitzel, P., Tsukahara, T., Schaley, J., Edenberg, H. J., Stephens, M. A., et al. (2006). Differential gene induction by type I and type II interferons and their combination. *Journal of Interferon & Cytokine Research : The Official Journal of the International Society for Interferon and Cytokine Research, 26*(7), 462–472.
20. Feng, G. S., & Taylor, M. W. (1989). Interferon gamma-resistant mutants are defective in the induction of indoleamine 2,3-dioxygenase. *Proceedings of the National Academy of Sciences, 86*(18), 7144–7148.
21. Taylor, M. W., & Feng, G. S. (1991). Relationship between interferon-gamma, indoleamine 2,3-dioxygenase, and tryptophan catabolism. *The FASEB Journal, 5*(11), 2516–2522.
22. von Bubnoff, D., Hanau, D., Wenzel, J., Takikawa, O., Hall, B., Koch, S., et al. (2003). Indoleamine 2,3-dioxygenase-expressing antigen-presenting cells and peripheral T-cell tolerance: another piece to the atopic puzzle? *The Journal of allergy and clinical immunology, 112*(5), 854–860.
23. Mellor, A. L., Chandler, P., Lee, G. K., Johnson, T., Keskin, D. B., Lee, J., et al. (2002). Indoleamine 2,3-dioxygenase, immunosuppression and pregnancy. *Journal of Reproductive Immunology, 57*(1–2), 143–150.
24. Entrican, G., Wattegedera, S., Rocchi, M., & Wheelhouse, N. (2009). Pregnancy, indoleamine 2,3-dioxygenase (IDO) and chlamydial abortion: an unresolved paradox. *Veterinary Microbiology, 135*(1–2), 98–102.
25. Clark, D. A., Blois, S., Kandil, J., Handjiski, B., Manuel, J., & Arck, P. C. (2005). Reduced uterine indoleamine 2,3-dioxygenase versus increased Th1/Th2 cytokine ratios as a basis for occult and clinical pregnancy failure in mice and humans. *American Journal of Reproductive Immunology, 54*(4), 203–216.
26. Gad, H. H., Dellgren, C., Hamming, O. J., Vends, S., Paludan, S. R., & Hartmann, R. (2009). Interferon-lambda is functionally an interferon but structurally related to the interleukin-10 family. *Journal of Biological Chemistry, 284*(31), 20869–20875.
27. Lasfar, A., Abushahba, W., Balan, M., & Cohen-Solal, K. A. (2011). Interferon lambda: a new sword in cancer immunotherapy. *Clinical & developmental immunology, 2011*, 349575.
28. de Veer, M. J., Holko, M., Frevel, M., Walker, E., Der, S., Paranjape, J. M., et al. (2001). Functional classification of interferon-stimulated genes identified using microarrays. *Journal of Leukocyte Biology, 69*(6), 912–920.

29. Taylor, M. W., Grosse, W. M., Schaley, J. E., Sanda, C., Wu, X., Chien, S. C., et al. (2004). Global effect of PEG-IFN-alpha and ribavirin on gene expression in PBMC in vitro. *Journal of Interferon & Cytokine Research: The Official Journal of the International Society for Interferon and Cytokine Research, 24*(2), 107–118.
30. Taylor, M. W., Tsukahara, T., Brodsky, L., Schaley, J., Sanda, C., Stephens, M. J., et al. (2007). Changes in gene expression during pegylated interferon and ribavirin therapy of chronic hepatitis C virus distinguish responders from nonresponders to antiviral therapy. *Journal of Virology, 81*(7), 3391–3401.
31. Yang, L., Luo, Y., Wei, J., & He, S. (2010). Integrative genomic analyses on IL28RA, the common receptor of interferon-lambda1, -lambda2 and -lambda3. *International Journal of Molecular Medicine, 25*(5), 807–812.
32. Benedict, C. A., & Ware, C. F. (2005). RIGing a virus trap. *Nature Medicine, 11*(9), 929–930.
33. Bose, S. (2006). *Viral Defense Mechanisms against interferon*. Weinheim, Germany: Wiley_VCH Verlag GmbH and Co.
34. Bolen, C. R., Robek, M. D., Brodsky, L., Schulz, V., Lim, J. K., Taylor, M. W., et al. (2013). The blood transcriptional signature of chronic hepatitis C virus is consistent with an ongoing interferon-mediated antiviral response. *Journal of Interferon & Cytokine Research: The Official Journal of the International Society for Interferon and Cytokine Research, 33*(1), 15–23.
35. Taylor, M. W., Tsukahara, T., McClintick, J. N., Edenberg, H. J., & Kwo, P. (2008). Cyclic changes in gene expression induced by Peg-interferon alfa-2b plus ribavirin in peripheral blood monocytes (PBMC) of hepatitis C patients during the first 10 weeks of treatment. *Journal of Translational Medicine, 6*, 66.
36. Garaigorta, U., & Chisari, F. V. (2009). Hepatitis C virus blocks interferon effector function by inducing protein kinase R phosphorylation. *Cell Host & Microbe, 6*(6), 513–522.
37. Wohnsland, A., Hofmann, W. P., & Sarrazin, C. (2007). Viral determinants of resistance to treatment in patients with hepatitis C. *Clinical Microbiology Reviews, 20*(1), 23–38.
38. Conjeevaram, H. S., Fried, M. W., Jeffers, L. J., Terrault, N. A., Wiley-Lucas, T. E., Afdhal, N., et al. (2006). Peginterferon and ribavirin treatment in African American and Caucasian American patients with hepatitis C genotype 1. *Gastroenterology, 131*(2), 470–477.
39. Lane, HC., Davey, V., Kovacs, J.A., Feinberg, J., Metcalf, J.A., Herpin, B., Walker, R., Deyton, L., Davey, R.T., Jr., Falloon, J. et al. (1990). Interferon-alpha in patients with asymptomatic human immunodeficiency virus (HIV) infection. A randomized, placebo-controlled trial. Annals of internal medicine, *112*(11), 805–811.
40. Solomon, T., Dung, N. M., Wills, B., Kneen, R., Gainsborough, M., Diet, T. V., et al. (2003). Interferon alfa-2a in Japanese encephalitis: a randomised double-blind placebo-controlled trial. *Lancet, 361*(9360), 821–826.
41. Stockman, L. J., Bellamy, R., & Garner, P. (2006). SARS: systematic review of treatment effects. *PLoS Medicine, 3*(9), e343.
42. Strander, H., & Cantell, K. (1974). Studies on antiviral and antitumor effects of human leukocyte interferon in vitro and in vivo. *In vitro Monograph, 3*, 49–56.
43. Sherwin, S. A., Knost, J. A., Fein, S., Abrams, P. G., Foon, K. A., Ochs, J. J., et al. (1982). A multiple-dose phase I trial of recombinant leukocyte a interferon in cancer patients. *JAMA, 248*(19), 2461–2466.
44. Reiter, Z., Ozes, O. N., Blatt, L. M., & Taylor, M. W. (1992). Cytokine and natural killing regulation of growth of a hairy cell leukemia-like cell line: the role of interferon-alpha and interleukin-2. *Journal of Immunotherapy: Official Journal of the Society for Biological Therapy, 11*(1), 40–49.
45. Nikfar, S., Rahimi, R., & Abdollahi, M. (2010). A meta-analysis of the efficacy and tolerability of interferon-beta in multiple sclerosis, overall and by drug and disease type. *Clinical Therapeutics, 32*(11), 1871–1888.

Cancer and Viruses

8

Abstract

Cancer morbidity decreased in the last few years of the twentieth century, in large part due to early detection as well as the campaign against tobacco. The relationship between viruses and cancer, and the elaboration of the oncogene theory—from the early work on phage lambda and lysogeny—to oncogenes is presented. The majority of cancers result from somatic mutations and chromosomal rearrangements, which may cause proto-oncogene activation. Mutations in tumor suppressor genes also end in tumor formation. The function of these different genes is discussed, as are hereditary cancers. Viruses—such as papilloma, hepatitis C, hepatitis B, human herpes 8, Epstein-Barr virus and HTLV-1—induce cancers by varied mechanisms.

8.1 The War on Cancer

Cancer is the second leading cause of death in the U.S. (after heart disease). In 2010, over 500,000 people died of cancer in the U.S. Millions of dollars have been spent on finding a cure, and President Nixon proposed a "War on Cancer," saying on December 23, 1971, "**I will also ask for an appropriation of an extra $100 million to launch an intensive campaign to find a cure for cancer. The time has come in America when the same kind of concentrated effort that split the atom and took man to the moon should be turned toward conquering this dread disease. Let us make a total national commitment to achieve this goal.**"

According to Vincent DeVita, M.D., director of The National Cancer Institute from 1980–1988, the "War on Cancer" "…**did everything it was supposed to do. It supported basic and applied research**…" The frequency of cancer and mortality from cancer started dropping in the U.S. in 1990 and has continued to drop every year. "**And the morbidity from cancer, comparing 1971–2005, is like night and day, so every benchmark has been hit.**"

M. W. Taylor, *Viruses and Man: A History of Interactions*,
DOI: 10.1007/978-3-319-07758-1_8,

Both frequency and death rates from all cancers combined declined significantly in the last twenty years for both men and women and for most racial and ethnic populations. Between 1990/1991 and 2009, the most recent year for which data is available, overall death rates decreased by 24 % in men, 16 % in women, and 20 % overall. These decreases were mostly in both occurrence and death rates for the three most common cancers in men (lung, prostate and colorectal cancer), and for two of the three leading cancers in women (breast and colorectal). Overall, the decline was significant for most cancers, although on examining the data it was quite modest. These data are more the result of early screening and the campaign against smoking than treatment. Some cancers and mortalities have remained static. (A breakdown of the prevalence and results of treatment of each cancer can be found at http://seer.cancer.gov/). There were 232,340 new cases of breast cancer in the U.S. in 2013, with an estimated 39,620 deaths. The percentage living five years after diagnosis was 89.2 % for 2003–2009. It is estimated that 12.3 % of women living in the U.S. will be diagnosed with breast cancer sometime in their lives; there are currently 2,829,094 women living with breast cancer in the U.S.

New cancer drugs—that certainly in the case of breast cancer have extended the life of the patients—have come on the market. Lists of new drugs (as well as old) and ongoing clinical trials for different types of cancer can be found at http://www.cancer.gov/. The development of the monoclonal antibody Herceptin, against Her2+(ERb2 oncogene) receptor in breast cancer combined with other therapies, such as taxol, has been quite effective in metastatic breast cancer. This is the first drug derived from monoclonal antibody research. The cost of treatment with this and other drugs of this class is extremely high, reflecting the cost of research in this area. Another drug, Gleevac (Imatinibmesylate), inhibits a key enzyme (tyrosine kinase) in cell replication and has been successfully used in the treatment of chronic myelogenous leukemia (CML) and gastrointestinal stromal tumors. This is the first of a new class of anti-cancer molecules derived from small molecules that target specific metabolic enzymes involved or regulating cell replication.

Despite the glowing report of Dr. DeVita, there is some disagreement in the cancer research community as to whether the touted success is justified. Many oncologists and researchers feel that the NCI has stifled research (as has the American Cancer Society) by funding research on chemotherapeutic agents and radiation rather than more novel approaches to treatment, such as the use of immune-therapeutic, naturally occurring agents, viral oncolysis and gene therapy. Another complaint is that not enough time and money were spent on attempting to find the root causes of cancer. It is now generally accepted that cancer is a genetic disease in which many gene products are affected, either by a primary mutation in one key gene, or by numerous mutations in an array of genes. Thus instead of treating cancers by organ pathology, it is essential to identify which gene(s) are mutated or activity-enhanced and to tailor treatment accordingly. This will require a new approach to cancer diagnosis as well as treatment. Platforms such as DNA—micro-arrays in which the expression of many genes can be examined at the same time—should eventually lead to individualized treatment. A microarray may show numerous changes or differences between a normal tissue and a cancerous one, but it is nearly impossible to identify with certainty the relevant gene; even so, great progress has been made in this area and it is discussed in more detail later in this chapter.

Chemotherapy is poisonous and in many cases does not extend life. In large part, the treatment of cancer has been driven by the pharmaceutical industry, which has developed hundreds of compounds that do eliminate cancer cells, but also at the expense of destroying healthy cells. In the same fashion as the development of new anti-virals, such as protease or replicase inhibitors, it should be possible to target specific enzymes that are unique to the cancer cell, or to surface markers on the cancer cells. Large numbers of monoclonal antibodies have been produced both against surface markers and even internal proteins of cancer cells. A list of such monoclonal antibodies is available at http://en.wikipedia.org/wiki/List_of_therapeutic_monoclonal_antibodies. Many of these are currently undergoing clinical trials. Within the next few years this should be a standard mode of treatment.

8.2 Introduction to Cancer at the Molecular Level

Although Peyton Rous discovered that avian tumors could be transmitted from chicken to chicken by injection of tumor material at the turn of the twentieth century, its applicability to human cancer was not realized. It was thought, at the time, that this was an oddity of the avian world; that viruses might be involved in human cancer was not investigated until the 1950s with the discovery that viruses caused tumors in mice and other rodents. Over the next 50 years, the emphasis in cancer research was on the interrelationship between viruses, particularly retroviruses and cancer. However, a viral origin for all cancer is no longer acceptable; it is only one component of a very complicated biological story. It is estimated that only 20 % of cancers are related to virus infection; these include Epstein-Barr virus, human herpes 8 (Kaposi's sarcoma), hepatitis B, hepatitis C, papilloma and retroviruses.

The more cancer was studied, the more it was realized that the majority of cancers are genetic defects of the cell resulting from mutation—either spontaneous or as a result of environmental insult that damages DNA. Cancer cells also arise as a result of chromosomal breakage and rearrangement. Many other cancers are inherited as a result of a mutation carried in families of specific ethnic groups, such as the BRCA-1 and BRCA-2 mutations that cause breast and other cancers in Ashkenazi Jewish women and occasionally result in cancer in men.

Cancers of the same tissue may differ and result from multiple mutations in different genes. Hundreds of genes have been identified as being involved in cancer, either as the primary mutation or as a secondary defect due to the activation (disruption) of multiple biological pathways. Some of these mutations result in defects in apoptosis, the name given to the process of natural cell death; other mutations result in disruption in the regulation of cell division. Cells usually stop dividing when they reach a specific stage of differentiation, i.e., kidney cells will stop dividing when the kidney is formed. Likewise, blood cells of various types have a definite half-life, stop dividing, and only divide for replacement. If new cells are needed, these are provided by stem cells that differentiate to the required tissue. Cancerous cells, on the other hand, continue to divide—ignoring signals to stop dividing, and cancer cells grow much faster than normal cells since signals that naturally would slow or stop growth are now continuously "firing." Cancer cells have acquired the property of immortality, and since they are not foreign cells, they may not be recognized by the immune system, although it is hypothesized that there is a general surveillance system that will eliminate cancerous cells. This may be one of the roles of the "natural killer cells" that circulate in the blood.

Diagnostically, cancers fall into two classes: benign or malignant. These are clinical terms to distinguish the aggressiveness of the tumor. Box 1 lists definitions used in describing cancer. A benign tumor is one in which the tumor, a mass of cells, has been contained at the site of origin and has not spread through the blood and lymphatic system to other sites, whereas a malignant tumor is one where the cancerous cells have spread systemically and colonized other areas of the body. This results in the establishment of the cancer at another site and is known as "metastasis" and is a hallmark of malignancy. This is how cells of prostate origin can be found in bone marrow, or breast cancer cells at sites other than the mammary glands. Another term often used to refer to tumor initiation is "oncogenesis." As will be discussed later, genes that appear to be primarily implicated in cancer are referred to as "oncogenes." The examples given above, of two drugs that affect cancerous growth—Herceptin and Gleevec—are examples of drugs that affect oncogenes and their biological pathways.

Box 1. Terminology

Some basic terminology:

Oncogenesis = process of initiation of tumours (cancer) in an organism (onkos = mass; genesis = birth)

Tumor = tissue composed of cells that deviate from normal program of cell division and differentiation.

Benign tumor = tumor cells remain together in a single mass and do not invade and disrupt surrounding tissues

Malignant tumor = tumor cells invade and disrupt surrounding tissues.

Metastasis = spread of malignant tumor cells throughout the body typically through the blood and lymphatic system

8.3 Oncogenes

Normal cell growth is controlled by growth factors, proteins that bind to specific receptors on the cell surface, or on the nuclear membrane. These signals appear to be of two classes—those that stimulate cell division, and those that inhibit cell division. These transcription factors migrate to the nucleus of the cell and activate genes to produce proteins (oncoproteins) that stimulate cell division, and suppressor proteins that suppress cell division (Figs. 8.1, 8.2, 8.3).

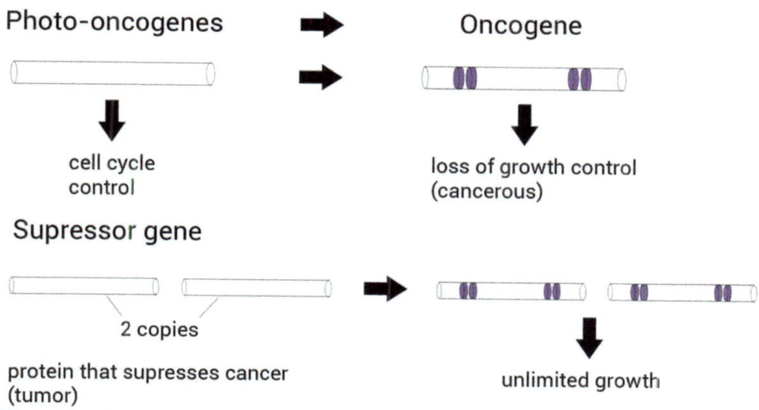

Fig. 8.1 Proto-oncogene and suppressor genes. Mutants in proto-oncogenes have a gain of function; those in suppressor have a loss of function. Both enhance growth of the tumor

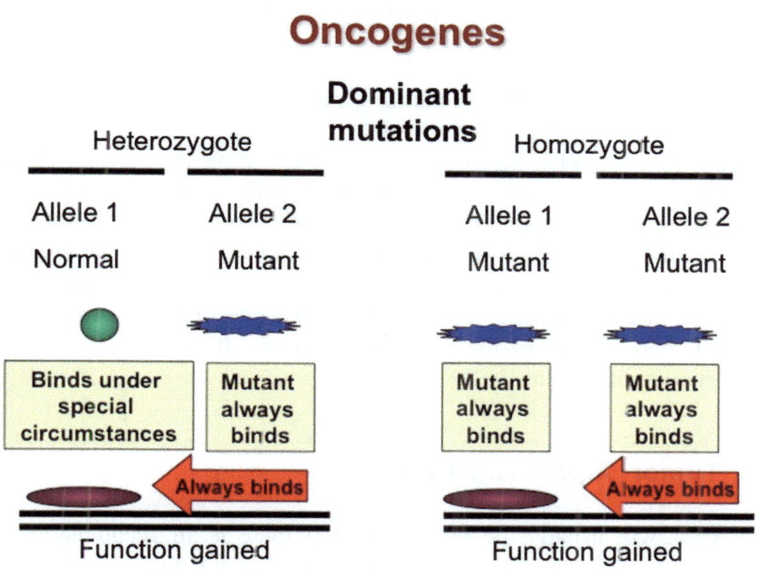

Fig. 8.2 Gain of function by mutation in proto-oncogene. Constant binding leads to cell growth

The fact that a single mutation in an oncogene may have multiple consequences downstream from the primary event makes it difficult to identify the primary event and to develop small molecules that may target the oncogene. Some DNA and RNA viruses can trigger cancerous growth, by mechanisms similar to those discussed above, by affecting an oncogene or, in some cases, inhibiting tumor

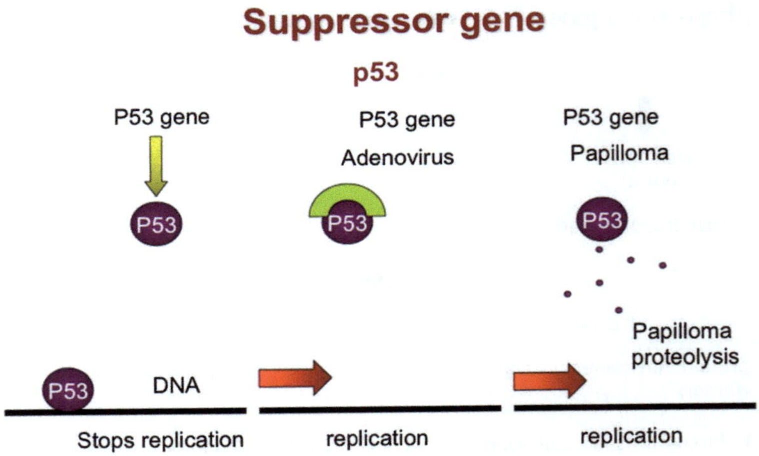

Fig. 8.3 Function of tumor suppressor gene p53. Adenovirus binds to p53 and prevents its functioning; papilloma virus protease degrades p53. Both cases lead to tumor formation. Adenovirus does not cause tumors in man but does lead to tumors in hamsters and transforms cells in culture

suppressors—or in others, by activating genes generally silent (this will be discussed later in this chapter).

Figure 8.2 shows what happens if a regulatory gene is mutated. The normal protein binds only under special circumstances, such as when a growth signal is received at the cell plasma membrane. In the heterozygote, the mutant oncogene protein can bind to the DNA (in the absence of a signal) and produce growth. Thus the mutation is dominant over the wild type, as there will be mutant protein in the heterozygote that can always signal growth. The rate of growth may be further enhanced if both copies of the gene are mutated, although in most cases one mutation may be sufficient.

8.4 Tumor Suppressor Genes

Healthy cells will usually die by apoptosis if there is an accumulation of any injury, particularly DNA damage. In cancer cells, proteins that usually function in the process of repair (for example, a protein known as "p53") do not recognize DNA damage and the cell continues to replicate. A protein such as p53 that removes DNA damage and restores normal growth is known as a "suppressor" gene product, since it suppresses tumor growth. A mutant form of the TP53 gene occurs in about 50 % of human tumors, and in about 80 % of colon cancers. p53 normally inhibits DNA replication if the DNA is damaged. Transgenic mice with deletion in the TP53 gene are viable, and appear normal, but close to 100 % develop cancer within 10 months.

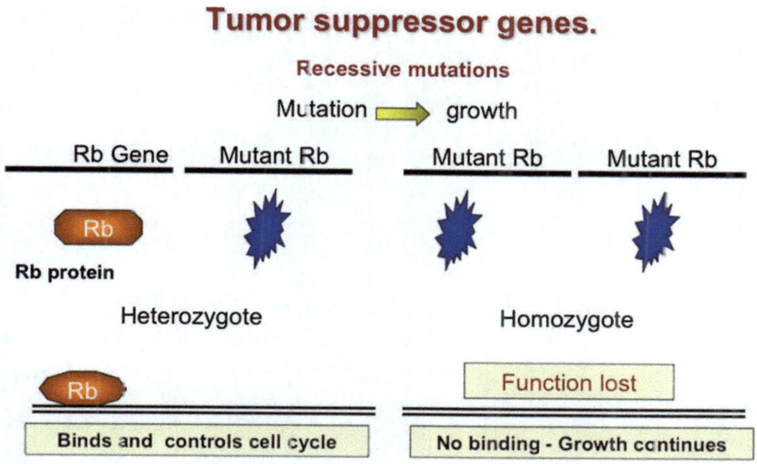

Fig. 8.4 Mechanism of action of tumor suppressor gene as exemplified by the retinoblastoma gene (Rb). Both alleles must be mutated to give rise to cancerous cells

Cells in culture can easily be transformed into tumor cells by infecting them with a series of DNA viruses such as polyoma, SV-40, and adenovirus. Viral proteins interact with host proteins that usually suppress the cell cycle and inhibit these genes, resulting in uncontrolled growth of the transformed cell (Fig. 8.3). Another major human suppressor gene is the Rb or retinoblastoma gene that is expressed in all cells and is involved in cell cycle control. In humans, a mutation in both copies of this gene results in tumors in the eye, hence the original name "retinoblastoma" [1] (Fig. 8.4). Suppressor genes require two mutations, one in each allele, to be totally non-functional.

8.5 Chromosomal Aberrations

An alternative theory of cancer genesis not involving viruses has become more acceptable in the last few years. It has been proposed that cancers occur as a result of multiple mutations possibly due to a master "mutator" gene, such as a gene involved in DNA repair, resulting in a thousand-fold increase in the mutation rate, most of which will occur in oncogenes. A related hypothesis is that cancer arises from chromosomal breakage, i.e., translocation of one segment of a chromosome to another chromosome, or due to aneuploidy (an aberrant chromosome number) resulting from breakage of chromosomes during cell division. Such aneuploidy may increase with time, resulting in a tumor with a very abnormal chromosome composition. Many types of tumors, particularly leukemia, show this phenomenon, as do cultured cells. This mechanism of cancer formation has been proposed by Peter Duesberg for many years [2] (see Fig. 8.5). The aneuploidy theory of cancer is a very old one that was first proposed by the German embryologist Theodor

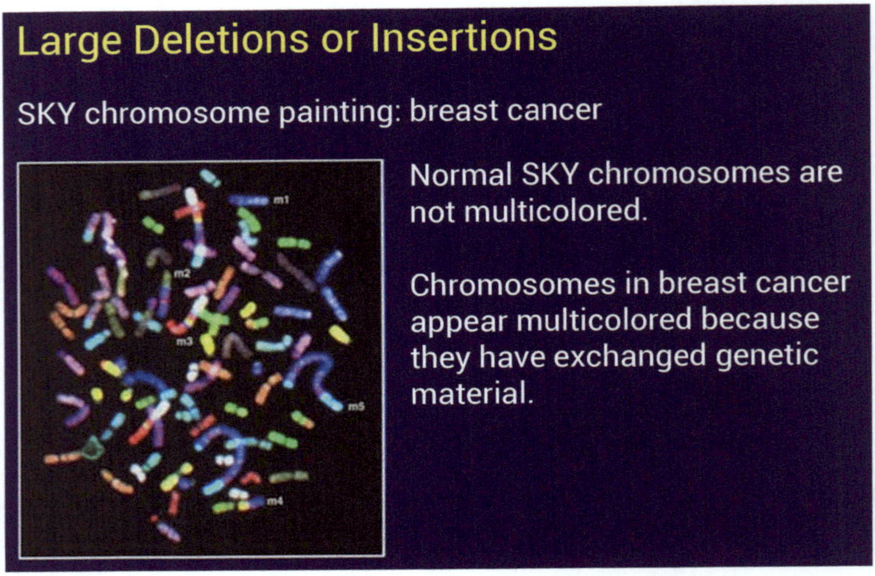

Fig. 8.5 Chromosome aberration in breast cancer (artwork originally created for the National Cancer Institute. Reprinted with permission of the artist, Jeanne Kelly. Copyright 2014)

Boveri in 1902 in a classic paper that was translated and annotated by Dr. Henry Harris in 2008 [3]. Interestingly, this paper touches upon many of the topics still discussed today in relation to cancer, problems of malignancy, and the relationship of disjunction of chromosomes and tumor development. Boveri worked with sea urchin eggs and often noted that cells (embryos) with abnormal chromosomes did not develop normally; many died, but a few did survive. He wrote, "But the essential element in our hypothesis is not the abnormal mitosis, but always a specific abnormal chromosome constitution. However this might arise, a particular tumor will be the result" (Fig. 8.6).

8.6 Viral-Oncogene Theory of Cancer

It was an accepted theory during the last half of the twentieth century that viruses caused most cancers. The observation that retroviruses carried "oncogenes" resulted in a burst of research activity resulting in Nobel Prizes for the viral origins of cancer. Some viruses are still implicated in cancer, in particular the DNA viruses such as papilloma (cervical cancer in particular), human herpes virus-8, Epstein-Barr virus, the causative agent of a number of types of lymphomas, hepatitis B and C, HTLV-1 and, just recently implicated, a polyoma virus [4]. The viral-oncogene theory of cancer derived from ideas explaining lysogeny in bacterial systems. There are two slightly different oncogene theories, one proposed by George Todaro and Huebner [5], and the other by Ludwig Gross [6].

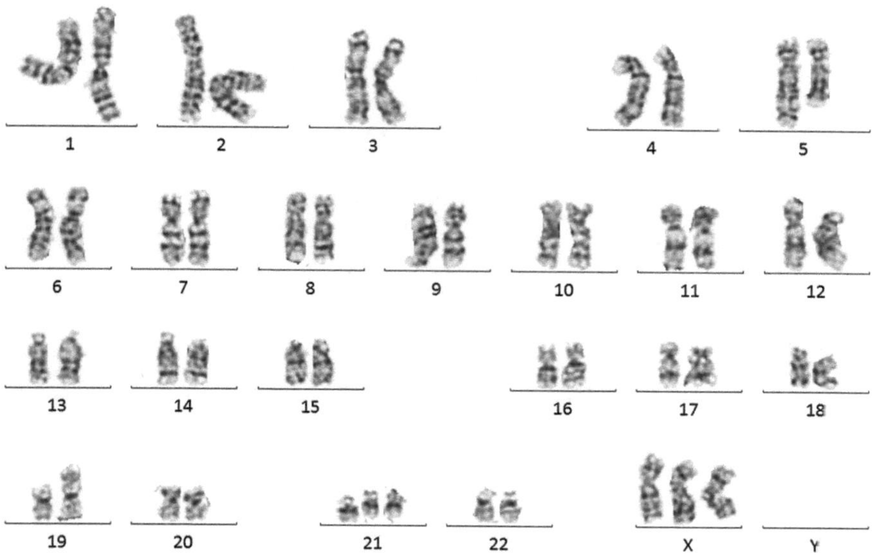

Fig. 8.6 Chromosome from female myeloid tumor. Note extra X chromosome and extra chromosome 21. Part of chromosomes 19 and 5 are missing (courtesy of Hana Aviv, Ph.D., Director of Cytogenetics, Rutgers-Robert Wood Johnson Medical School., N.J. USA)

The replication cycle of viruses has been discussed elsewhere in this volume. Most viruses replicate and produce more of themselves through the lytic cycle. In the 1950s, it was noted that under specific circumstances, some bacteriophage do not undergo the normal lytic cycle, but rather integrate into the bacterial host chromosomal DNA and reside in the bacterial chromosome. Such bacteriophage could be activated from a dormant state to a lytic state by treating the cells with UV light, or during the transfer of the bacterial chromosome from one host to the other (a type of mating that can occur between related species of bacterium) resulting in viral replication. Such bacteriophage were called "temperate phage" and are discussed in detail in Chap. 5.

A similar model was proposed to explain the origins of tumor viruses (called C-type particles) and cancer [7]. Based on work with cells in culture, particularly cells of rodent origin, Todaro and Huebner suggested that all cells contained genes that were the result of viral infections far back in evolutionary time, i.e., possibly millions of years ago. Thus the genes of retroviruses are present in all mammalian cells, but expression is repressed. These genes could be activated by treatment with UV radiation, or potential mutagens or compounds that interfered with DNA synthesis. Thus new viruses could be formed from the component parts, which did not have to be on the same chromosome, and occasionally the virus could incorporate an oncogene and express this in the cell-triggering tumor formation; these were known as "C-type particles" (Fig. 8.7).

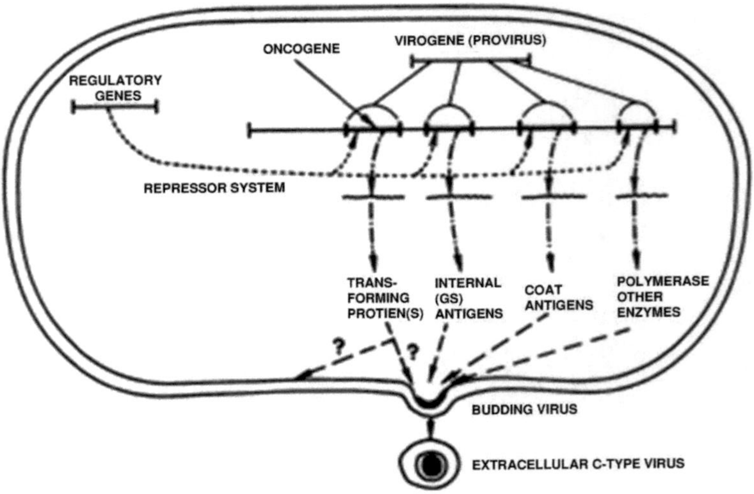

Fig. 8.7 Model of the proposed oncogene theory (from Huebner and Todaro)

Although parts of this model have proven to be correct, there is no evidence that C-type particles (retroviruses) are formed in this way. Retroviruses do carry oncogenes, which can be inserted into the chromosome, and since they are under the control of viral promoters, they are expressed. If a retrovirus is integrated next to an oncogene, that oncogene can be activated and expressed abnormally, resulting in leukemia. The fact that all cells would carry viral genes and that such genes would be transmitted to further generations was termed "vertical transmission" rather than "horizontal transmission" when the virus would spread from cell to cell. Todaro and Huebner [7] point out that this differs from the phage lambda model, in that the phage can be lytic and is not necessarily carried for lengthy periods as part of the host genetic material.

An alternative model was proposed by Ludwig Gross [6] to explain the vertical transmission and familial incidence of leukemia in mice and other animals. In this model, C-type particles exist in all cells, including embryonic stem cells, and are transmitted vertically to daughter cells. This virus entered the cell (animal) millions of years ago and may exist, latent and harmless, for long evolutionary periods. Activation of the virus, leading to tumor formation, may occur because of environmental insults or hormonal changes and leukemia or some other cancer occurs. Gross predicted that such viruses would eventually be found in humans. Apart from HTLV-1 and HTLV-2, the first associated with T-cell leukemia, [8] such viruses have not been found, but of course this does not mean that such viruses do not exist in the human population. It is hypothesized that such viruses, if they do exist, cause leukemia by integrating near an oncogene, which is now under the control of the viral long terminal repeat (LTR), equivalent of a strong promoter region rather than its own promoter. That such retrovirus sequences exist in our DNA and that of other mammals has been confirmed.

This model is applicable to retroviruses, but not to other viruses that cause cancers. Most retroviruses integrate their nucleic acid into the host chromosome, resulting in cell transformation in culture, and tumor formation in vivo. Such viruses produce low levels of new viruses, often defective and needing a "wild type" of virus to complete their replication.

This model explains the integration of HIV into the host chromosome, since HIV is in many respects a typical retrovirus, but more complicated, having genes other than the usual gag-pol-env genes found in most retroviruses. Neither HIV nor HTLV-1 are defective viruses and cause cancer by other mechanisms.

8.7 Retroviruses and Cancer

Retroviruses are RNA viruses that undergo a DNA stage during viral replication and integrate into the host chromosome. David Baltimore and Howard Temin independently found the enzyme "reverse transcriptase," [9, 10] and both received the Nobel Prize in Physiology or Medicine for this work in 1975. The reverse-transcriptase enzyme copies RNA into DNA by a complex mechanism. Integration of the viral chromosome into the host chromosome occurs by an intricate mechanism involving two enzymes, a recombinase and integrase enzyme, which together open up the double-stranded DNA, and integrate the viral genes into the host chromosome.

A schematic presentation of a typical retrovirus and the gene order is presented in Figs. 8.8 and 8.9. Most retroviruses contain three blocks of genes: "gag," "pol," and "env." (gag stands for group antigen, the genes that make up the capsid; pol. for polymerase, in this case the reverse transcriptase and integrase; and env. for the lipid membrane that surrounds the capsid or core Fig. 8.10). Examples of retroviruses include Rous sarcoma virus, the first tumor virus discovered; mouse mammary tumor virus, a virus vertically transmitted from mother to daughter; HTLV-1, known to cause a rare leukemia in man; HTLV-2, which is associated with hairy cell leukemia; and HIV-1 and 2, the causative agent of AIDS and "slim disease" respectively. Since the viral RNA is copied into DNA and is integrated into the chromosome (possibly at random or at a few specific sites), it replicates as part of the cell chromosome and its genes are expressed in these cells. If the virus contains an oncogene, this is also expressed during replication and can result in unlimited growth of the cell. The oncogene would replace part of one of the viral genes.

The genome of a typical retrovirus includes a long sequence of RNA at both ends. In the integrated form, it is known as the "LTR" (Long Terminal Repeat) (Fig. 8.10) and includes sequences for regulation of transcription and a site that primes the reverse transcriptase (PBS). This is the promoter region of the viral chromosome. Next are the sequences gag-pol-env, which code for the structural components of the virus as well as the reverse transcriptase and outer envelope. This is followed by a repeat of the LTR. This infers that after insertion of the virus

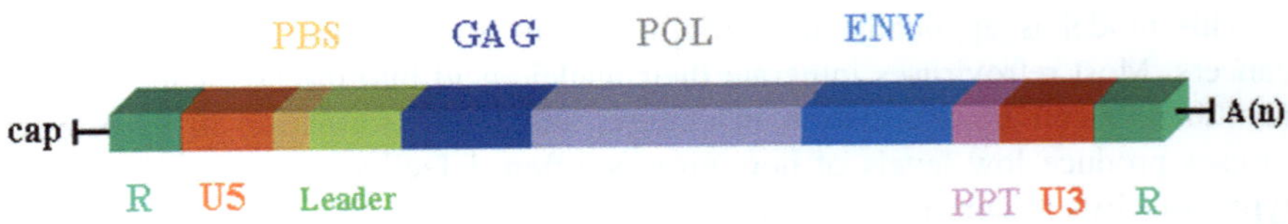

Fig. 8.8 Gene order of a typical unintegrated retrovirus (Hunt, University S. Carolina with permission). PBS is a primer binding site for the initiation of reverse transcriptase

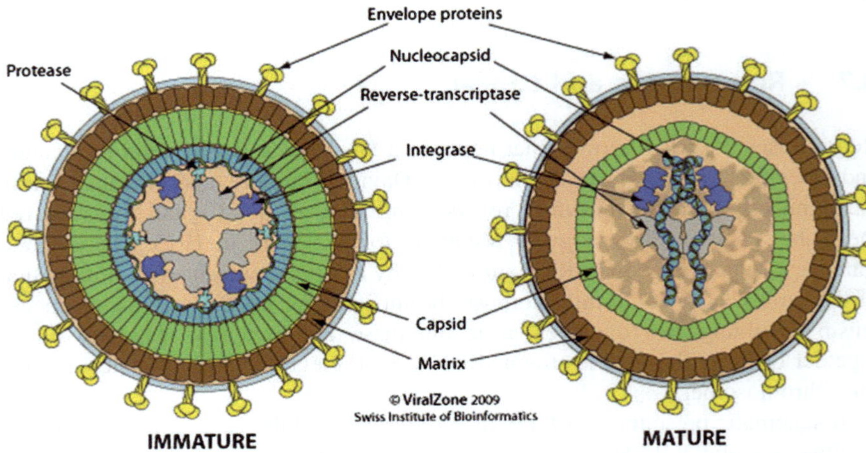

Fig. 8.9 Schematic of typical retrovirus (HTLV-1). (Viralzone, Swiss Institute of Bioinformatics)

Fig. 8.10 Integrated retrovirus genes (from Hunt, University of S. Carolina, with permission)

into the chromosome, the end LTR could be next to an oncogene, a gene regulating cell division. The presence of this LTR sequence or promoter is sufficient to "drive" this gene rather than the native promoter. This is the basis of certain viral leukemias, where an oncogene is not transposed by the virus, but the LTR is next to a proto-oncogene; this effect on a nearby gene is known as "insertional oncogenesis" (Fig. 8.11), which illustrates the incorporation of an oncogene into a series of retroviruses.

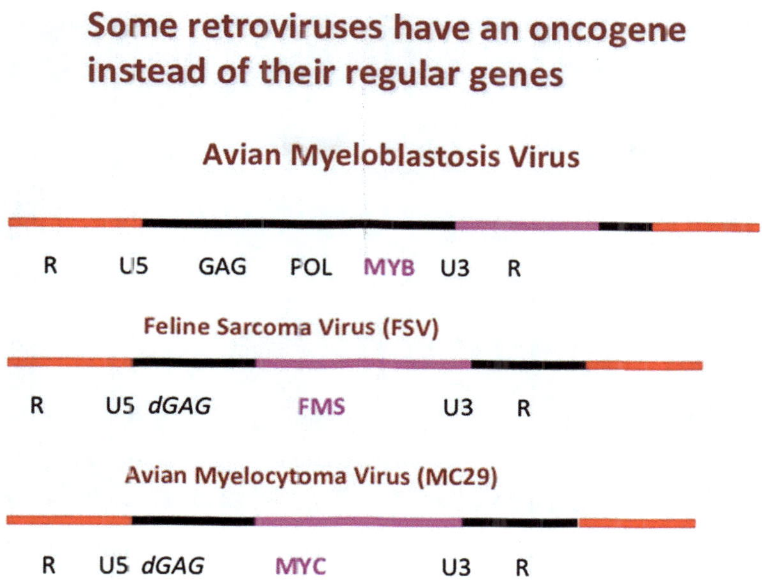

Fig. 8.11 Oncogenes integrated into different retroviruses (*mys* in AMV, *fms* in FSV and *myc* in AMV)

Using a different approach to identify genes involved in cancer, Robert Weinberg and colleagues at MIT [11] identified cancer-causing genes by transforming a mouse cell line known as 3T3, with DNA from various human cancers. They selected the transformed cells using a selective gene and re-isolated the human DNA from the mouse background. Using this approach, they identified mutant genes involved in cancer formation. These were mostly genes involved in gene regulation and gene transcription, and thus a list of potential oncogenes was created. These were in many cases identical to genes shown to be carried by retroviruses. This list is available at http://www.cellsignal.com/reference/oncogene_table.html.

In fact, the more cancers are examined at the molecular level, the more mutations in different genes are found, and each mutation may lead to differences in the growth rate of the tumor and affect the response to anti-tumor drugs. Papilloma virus initiates tumor formation by one of the viral proteins interacting with p53 and preventing its activity.

8.8 The BRCA1 and BRCA2 Genes

Other suppressor genes are involved in hereditary breast cancer. BRCA1 and BRCA2 produce tumor suppressor proteins that help repair damaged DNA. About 5 % of breast cancers are hereditary. If DNA is not repaired, there is an increased

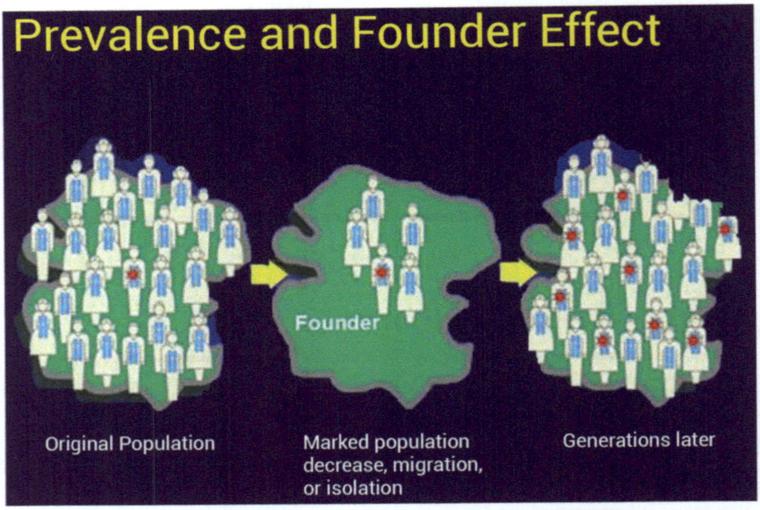

Fig. 8.12 Founder effect of mutation in isolated population (artwork originally created for the National Cancer Institute. Reprinted with permission of the artist, Jeanne Kelly. Copyright 2014)

likelihood of mutation. Together, BRCA1 and BRCA2 mutations account for about 20–25 % of hereditary breast cancers [12] and about 5–10 % of all breast cancers [13]. In addition, mutations in BRCA1 and BRCA2 account for around 15 % of ovarian cancers overall [14]. Either sex can transmit defects in the BRCA1 and BRCA2 genes.

The link between BRCA1 and BRCA2 mutation and breast cancer was discovered in 1994. Follow-up studies done with DNA samples from individuals with Tay-Sach's disease, a disease of Ashkenazi Jews, showed a high prevalence of mutations in these genes in the Ashkenazi Jewish population, although mutations in these genes are not confined to this group. Because of reproductive isolation, later generations of an isolated population will have a higher frequency of mutation than in the original population (Fig. 8.12). Today, one person in 40 among Ashkenazi Jews carries a 185delAG mutation in BRCA1, which places them at higher-than- average risk for breast and ovarian cancer.

Three specific mutations in the BRCA1 and BRCA2 genes account for 90 % of the mutations identified in Ashkenazim. A different spectrum of mutation has been found in other groups; thus, the treatment for breast cancer has to be individually catered, since even though the same gene may be involved, the mutation may be at a different site within the gene and result in a differential response to chemotherapeutic agents [15, 16] (Figs. 8.13, 8.14). These mutations are carried in every cell of the body, and predispose towards cancer (Fig. 8.15).

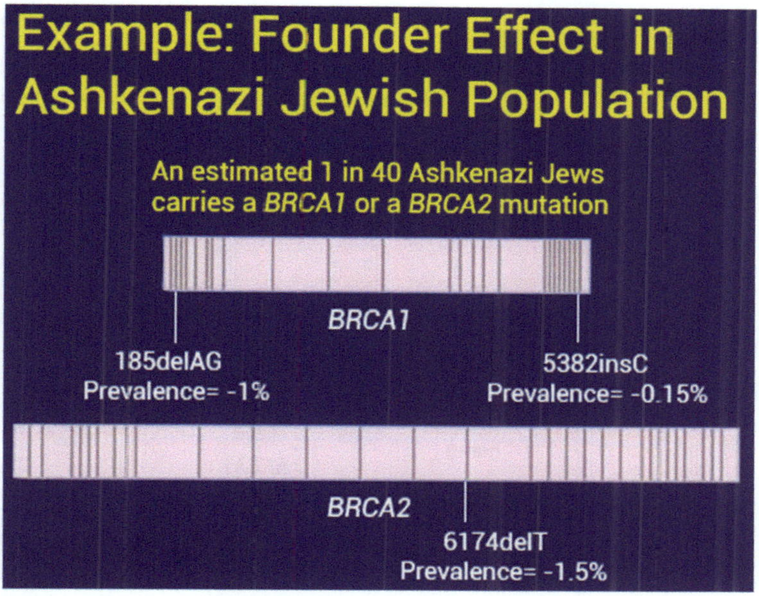

Fig. 8.13 Location of mutations in BRCA1 and BRCA2 genes (artwork originally created for the National Cancer Institute. Reprinted with permission of the artist, Jeanne Kelly. Copyright 2014)

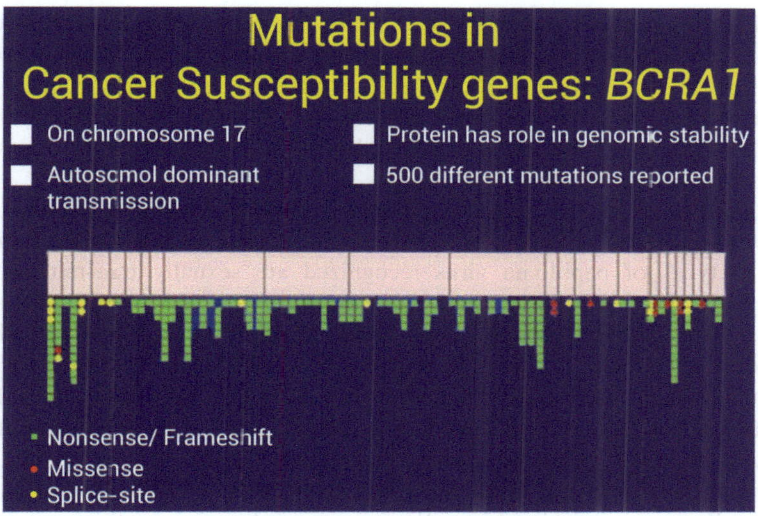

Fig. 8.14 Mutations in the BRCA1 gene. Note large number of mutants (artwork originally created for the National Cancer Institute. Reprinted with permission of the artist, Jeanne Kelly. Copyright 2014)

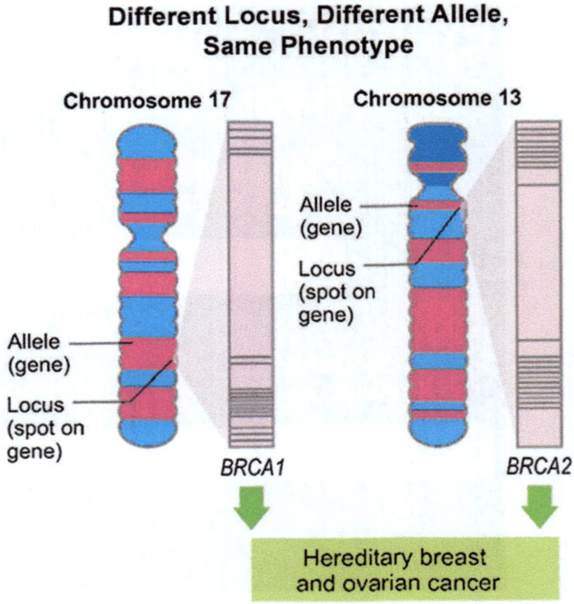

Fig. 8.15 Chromosomal location of BRCA1 and BRCA2 mutations. From National Cancer Institute

8.9 DNA Tumor Viruses

In culture, most DNA tumor viruses readily transform cells. Such transformed cells have all the characteristics of tumor cells: unlimited growth, lack of contact inhibition, and the ability to grow freely in agar and give rise to tumors when injected into nude mice (i.e., mice lacking an immune system). These viruses include polyoma, SV-40, and adenovirus. Hepatitis B and papilloma viruses grow very poorly in culture.

Papilloma virus: The human papilloma virus (HPV) is one of the most common virus groups to affect the skin and mucosal areas of the body. More than 40 of the 150 types of papilloma virus recognized are sexually transmitted. Some antigenic strains of this virus cause cervical cancer in women, while other strains cause a variety of warts, mostly benign. Approximately 42 % of all women have a papilloma infection at any point in time; 10 % of all human cancers and 16 % of all female cancers may be HPV-linked. Two of the papilloma viruses, types 16 and 18, cause almost all cancers. Squamous cell carcinomas of the larynx, esophagus, and lung are histologically similar to cervical carcinomas.

Tumors are mostly of epithelial origin (skin cells). The epithelial surfaces include all areas covered by skin and/or mucosa, such as the mouth, throat, tongue, tonsils, vagina, penis, and anus. Different strains of the human papilloma virus are known to infect different parts of the body. Papilloma virus also causes a variety of warts on the hands and feet and penile warts in men (Fig. 8.16).

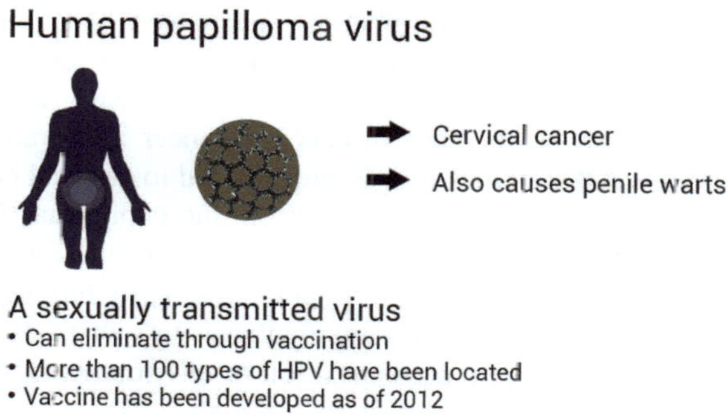

Fig. 8.16 Papilloma virus effect

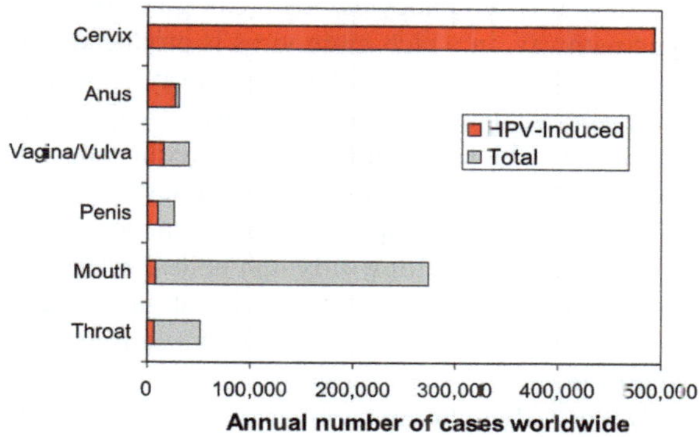

Fig. 8.17 Number of cases of cancers caused by HPV in 2006, before the introduction of the vaccine (WHO)

HPV infection is among the most common sexually transmitted diseases. About 50 % of sexually active people in the USA (both sexes) have HPV at some time in their lives, although the virus is often transmitted without the person knowing that he or she is infected. It can be transmitted during vaginal or anal intercourse, or by oral sex. The Pap smear, which should be taken every year in sexually active women, tests for the presence of early stages of cervical cancer. Newer tests detect the presence of papilloma virus DNA. A vaccine against the tumor-causing strains has recently been developed and approved. The vaccine (Gardasil) consists of four inactivated papilloma virus types (6, 11, 16 and 18), all known to cause cervical cancer and penile warts; types 6 and 11 are the most common that infect humans. The vaccine has not yet been licensed for use in men although this is under study, since it may prevent penile warts and certainly would inhibit the spread of the disease. The application of the vaccine, however, has run into controversy. The CDC has recommended vaccination for girls aged 11–26, but the question has arisen whether the use of this vaccine, particularly among teens, will encourage promiscuity. Since the vaccine is new, it is not known how long immunity will last. Some of these questions are the same as will arise if and when there is a vaccine against HIV. Figure 8.17 illustrates the number of cases of papilloma cancers world-wide. Note that it is predominantly cervical and anal cancer.

Herpes virus. A number of cancers have been associated with herpes virus (see Chapter on herpes), including Burkitt's lymphoma, nasopharyngeal carcinoma, and Kaposi's sarcoma. Herpes virus is usually a passenger in many tumors, as appears to be the case in cervical cancer and brain tumors. At one time it was believed that herpes type 2 infection led to cervical cancer, but as discussed above, this cancer is known to result from the papilloma virus.

Epstein-Barr virus (EBV) virus is an interesting virus in that it causes Burkitt's lymphoma in some populations, nasopharyngeal cancer in others, and the self-limiting disease of mononucleosis in a third population. Thus, there seems to be either a genetic predisposition or environmental factor related to the infection. EBV is very common, with more than 90 % of individuals in the US testing positive for EBV antigens. The virus is ubiquitous and latent in most individuals, residing in B-lymphocytes. EBV can go through two alternate pathways, either a lytic pathway, in which more virus is produced (although the cells do not lyse but rather bud out virus) or a latent pathway in which a subset of genes is active in repressing the lytic response. The virus is capable of immortalizing B-lymphocytes in culture, which is similar to a leukemic effect in vivo. In mononucleosis, the virus multiplies in B-cells, causing a severe immune reaction with expansion of CD8 T-cells. The symptoms of mononucleosis (also called glandular fever) are sore throat with swelling of lymph nodes, spleen, and liver. Epstein-Barr virus has also been associated with "chronic fatigue syndrome" and rheumatoid arthritis. However, there is no firm evidence of a causative link between this virus and these conditions.

Burkitt's lymphoma is widespread in areas of East and Equatorial Africa and occasionally outside them. In East Africa, it accounts for 74 % of childhood cancers. Children in this area often have malaria, which appears to predispose them to Burkitt's lymphoma. Most children by age three are EBV-positive and carry a latent virus. Recent research has shown that *Plasmodium falciparum* (the agent responsible for malaria) membrane protein-1 can activate a latent EB virus in B-cells. The tumor invariably has a translocation of chromosome 8 with transposition of the myc gene, an important oncogene to another chromosome. The disease characteristically involves the jaw or other facial bones, and sometimes ovaries, kidneys or breasts. That there is a relationship to malaria in African cases is noted by the low rate of Burkitt's lymphoma in Africans with sickle-cell anemia, which protects against malarial infection and low frequencies in areas of Africa where there is little malaria. However, the intricacies of this relationship with malaria are still not clear but appear to be very complex [17].

The sporadic type of Burkitt's lymphoma (also known as "non-African") is another form of non-Hodgkin's lymphoma. The tumor cells have a similar appearance to the cancer cells of classic African or endemic Burkitt's lymphoma. Non-Hodgkin's, which includes Burkitt's, accounts for 30–50 % of childhood lymphomas. One finds the same chromosomal translocation, the translocation of the myc gene. Another significant difference is the age of onset of the lymphoma; in East Africa the onset occurs in young children around age three, while in other locations it is in older persons.

Kaposi's sarcoma. What was until recently a rare cancer initiated by a herpes virus has been discovered in patients in the early stages of AIDS. Kaposi's sarcoma was one of the initial opportunistic infections first characterized in early AIDS patients, [18] later shown to be of viral origin [19]. The virus triggering Kaposi's sarcoma is **human herpesvirus-8**. During the early stages of the AIDS epidemic, more than 50 % of the patients had Kaposi's sarcoma, a very high rate for a malignant disease. This was one of the distinguishing "marks" of AIDS, since the tumor causes "raised purple" marks on the skin. At that time, it was proposed that this was due to a sexually transmitted virus that appeared only in immune-compromised individuals. KS also occurs at a high rate in Sub-Saharan Africa independent of AIDS, and in these cases it is not sexually transmitted. The virus can exist in a latent stage or lytic stage. How the virus is transmitted is still unclear, although there are indications that it can be spread in saliva. In the latent stage, viral genes interact with tumor suppressor genes p53 and Rb. HHV8 is not integrated into the chromosome, but exists as an episome, an independent self-replicating circular fragment of DNA.

8.10 Hepatocellular Carcinoma

There are as many as 2.7 million cases of hepatitis C in the U.S., and about 15 million worldwide. The rate varies from country to country, with Egypt, and possibly China, India and Japan having the highest incidence. HCV is spread by intravenous drug users sharing needles, thus the percentage of infection is high among drug addicts and prisoners. It is also spread by blood transfusion, which is performed in many third world countries under less-than-sterile conditions. There is little evidence of sexual transmission. Unlike the other viruses discussed above, HCV is a single-stranded RNA virus.

In a few individuals, the disease is self-limiting (acute infection), whereas in the majority of cases there is chronic infection. About 1 % of chronic infected individuals will go on to develop liver cancer. Although work is ongoing, a vaccine is not yet available. New drugs—mostly viral protease inhibitors—were approved within the last few years by the FDA. Small molecules that might inhibit the viral replicase are also being explored. The current treatment for chronic HCV is a combination of protease inhibitors and ribavirin, an older antiviral. Until very recently, the treatment consisted of a course of pegylated interferon-alpha and ribavirin. The success rate for this protocol, as measured by a decrease in virus in the blood to non-detectable limits, is about 30–50 % in Caucasian and Oriental subjects, but considerably lower in African or African-American subjects [20, 21]. This is in part due to a genetic polymorphism in the IL28 or interferon λ gene [22]. With the new drug regimen, the rate of "cure" is considerably higher, and with fewer side effects. As in the case of HIV/AIDS, a change in behavior could affect the HCV rates of infection and reduce the burden on society. The disease is progressive and eventually affects the liver, in most cases leading to cirrhosis of the liver and the need for a liver transplant. There is no evidence of integration of the viral genome, and since this is an RNA virus, it is located in the cytoplasm or internal cell membranes. It is hypothesized that carcinomas develop as a result of stress on the liver, with constant destruction and regeneration of liver tissue. Complete liver transplant is becoming more common as a treatment; however, with the advent of new drugs, this disease should be under control in a few years.

Hepatitis B is a DNA virus that also affects the liver. There are possibly 10 times more cases of hepatitis B than hepatitis C in developing or undeveloped countries, but not in the U.S. The mode of spread is through sexual transmission, needle-sharing and blood transfusions. This virus is unique in that it goes through a DNA-RNA-DNA stage, coding for and carrying its own reverse transcriptase. The hepatitis B DNA can be integrated into the chromosome. It has been suggested that an HBV protein (X protein) acts on the p53 suppressor, and leads to liver cancer. However, it is just as likely that cancer is a result of constant damage to the liver from viral replication—as indicated above for hepatitis C. Unlike the case of hepatitis C, there is now a very effective vaccine against hepatitis B. This will certainly cut down on the number of impending cases, although at the moment both hepatitis C and B present significant medical problems for the future. One can expect to see large increases in liver cancer.

References

1. Knudson, A. G, Jr. (1974). Heredity and human cancer. *American Journal of Pathology, 77*(1), 77–84.
2. Duesberg, P. (2007). Chromosomal chaos and cancer. *Scientific American, 296*(5), 52–59.
3. Harris, H. (2008). Concerning the origin of malignant tumours by Theodor Boveri. Translated and annotated by Henry Harris. Preface. *Journal of Cell Science, 121 Suppl 1*, v–vi.
4. Nakajima, H., Takaishi, M., Yamamoto, M., Kamijima, R., Kodama, H., Tarutani, M., et al. (2009). Screening of the specific polyoma virus as diagnostic and prognostic tools for Merkel cell carcinoma. *Journal of Dermatological Science, 56*(3), 211–213.
5. Todaro, G. J., & Huebner, R. J. (1972). N.A.S. symposium: New evidence as the basis for increased efforts in cancer research. *Proceedings of the National Academy of Sciences of the United States of America, 69*(4), 1009–1015.
6. Gross, L. (1974). Facts and theories on viruses causing cancer and leukemia. *Proceedings of the National Academy of Sciences of the United States of America, 71*(5), 2013–2017.
7. Huebner, R. J., & Todaro, G. J. (1969). Oncogenes of RNA tumor viruses as determinants of cancer. *Proceedings of the National Academy of Sciences of the United States of America, 64*(3), 1087–1094.
8. Rho, H. M., Poiesz, B., Ruscetti, F. W., & Gallo, R. C. (1981). Characterization of the reverse transcriptase from a new retrovirus (HTLV) produced by a human cutaneous T-cell lymphoma cell line. *Virology, 112*(1), 355–360.
9. Temin, H. M., & Mizutani, S. (1970). RNA-dependent DNA polymerase in virions of Rous sarcoma virus. *Nature, 226*(5252), 1211–1213.
10. Panet, A., Baltimore, D., & Hanafusa, T. (1975). Quantitation of avian RNA tumor virus reverse transcriptase by radioimmunoassay. *Journal of Virology, 16*(1), 146–152.
11. Murray, M. J., Shilo, B. Z., Shih, C., Cowing, D., Hsu, H. W., & Weinberg, R. A. (1981). Three different human tumor cell lines contain different oncogenes. *Cell, 25*(2), 355–361.

12. Easton, D. F. (1999). How many more breast cancer predisposition genes are there? *Breast cancer research: BCR, 1*(1), 14–17.
13. Campeau, P. M., Foulkes, W. D., & Tischkowitz, M. D. (2008). Hereditary breast cancer: new genetic developments, new therapeutic avenues. *Human Genetics, 124*(1), 31–42.
14. Pal, T., Permuth-Wey, J., Betts, J. A., Krischer, J. P., Fiorica, J., Arango, H., et al. (2005). BRCA1 and BRCA2 mutations account for a large proportion of ovarian carcinoma cases. *Cancer, 104*(12), 2807–2816.
15. Lancaster, J. M., Carney, M. E., & Futreal, P. A. (1997). BRCA 1 and 2–A Genetic Link to Familial Breast and Ovarian Cancer. *Medscape Womens Health, 2*(2), 7.
16. Robson, M., Dabney, M. K., Rosenthal, G., Ludwig, S., Seltzer, M. H., Gilewski, T., et al. (1997). Prevalence of recurring BRCA mutations among Ashkenazi Jewish women with breast cancer. *Genet Test, 1*(1), 47–51.
17. Burkitt, D. P. (1983). The discovery of Burkitt's lymphoma. *Cancer, 51*(10), 1777–1786.
18. Gessain, A., Sudaka, A., Briere, J., Fouchard, N., Nicola, M. A., Rio, B., et al. (1996). Kaposi sarcoma-associated herpes-like virus (human herpes virus type 8) DNA sequences in multicentric Castleman's disease: Is there any relevant association in non-human immunodeficiency virus-infected patients? *Blood, 87*(1), 414–416.
19. O'Neill, E., Henson, T. H., Ghorbani, A. J., Land, M. A., Webber, B. L., & Garcia, J. V. (1996). Herpes virus-like sequences are specifically found in Kaposi sarcoma lesions. *Journal of Clinical Pathology, 49*(4), 306–308.
20. Conjeevaram, H. S., Fried, M. W., Jeffers, L. J., Terrault, N. A., Wiley-Lucas, T. E., Afdhal, N., et al. (2006). Peginterferon and ribavirin treatment in African American and Caucasian American patients with hepatitis C genotype 1. *Gastroenterology, 131*(2), 470–477.
21. Howell, C. D., Dowling, T. C., Paul, M., Wahed, A. S., Terrault, N. A., Taylor, M., et al. (2008). Peginterferon pharmacokinetics in African American and Caucasian American patients with hepatitis C virus genotype 1 infection. *Clinical gastroenterology and hepatology: the official clinical practice journal of the American Gastroenterological Association, 6*(5), 575–583.
22. Abe, H., Hayes, C. N., Ochi, H., Maekawa, T., Tsuge, M., Miki, D., et al. (2011). IL28 variation affects expression of interferon stimulated genes and peg-interferon and ribavirin therapy. *Journal of Hepatology, 54*(6), 1094–1101.

Smallpox

9

Abstract

It is estimated that smallpox arose in the human population either 16,000 or 68,000 years ago from a rodent source, possibly as two different diseases. The virus mutated to greater pathogenicity in the sixteenth and seventeenth centuries, with epidemics in most urban centers of the world. It was a disease that was endemic in European and Asian populations, but extremely virulent among the non-immune native populations of the New World. Ninety percent of the American Indians died as a result of smallpox, leading to the easy colonization of the American continent by the Spaniards, and later by other European settlers. The decimation of the Amerindians led to the introduction of slavery to work the silver and gold mines. Smallpox found an immune "virgin population" in the Americas, and in other isolated populations. Reasons for this difference in response to the infection are discussed. Smallpox is the first human infectious viral disease to have been eradicated from the earth; its eradication is a fascinating history—from variolation to development of a vaccine in the eighteenth century to complete elimination in 1977.

9.1 Origins of Smallpox

Using DNA-sequencing technology and mutational analysis, researchers at the Lawrence Livermore Laboratory in California and the Centers for Disease Control in Atlanta have calculated that the original smallpox virus evolved as a human disease either 160,000 or 68,000 years ago, depending on whether the African or Asian virus is used as the basis of the calculation [1]. This virus would have evolved from an African rodent virus, possibly taterapox. These conclusions are based on the rate of change that occurred in the viral DNA with time. The wide discrepancy in dates is due to the use of different clades (sub-groups of virus): one clade from variola major, the smallpox that occurred in Asia, and the other from a

M. W. Taylor, *Viruses and Man: A History of Interactions*,
DOI: 10.1007/978-3-319-07758-1_9,

Fig. 9.1 The earliest evidence of smallpox is believed to be the vesicular skin lesions seen on the mummy of Ramses V, who died in Egypt in 1157 B.C. (Courtesy of the CDC)

West African virus, which included alastrim, a less virulent strain of the smallpox virus. This is much further back in time than expected. Researchers hypothesized that the virus may have arisen from an intermediate animal host that has disappeared completely, possibly an African rodent. The pox viruses are among the largest known animal viruses, and there has been speculation as to whether they evolved as other smaller viruses, possibly from the host cell, or are a degeneration of a more complex organism akin to a bacterium or a protozoan. Since these are DNA viruses, they undergo mutation at a very slow rate.

Although the time and place of man's first encounter with the smallpox virus will remain speculative, it is believed that the disease originated in agricultural settlements in Northeast Africa, China, or the Indus River Valley. It is generally assumed that viral diseases developed with the formation of agricultural communities after the hunter-gatherer stage of civilization. Writings dating from 3700 B.C. (Egypt) and 1100 B.C. (China), and mummies dating from the 18th Egyptian Dynasty (1580–1350 B.C.) show evidence of smallpox (Fig. 9.1). DNA sequence analysis of material from the pockmarks confirmed their origin. Interestingly, smallpox is not mentioned in the Bible or in the writings of Hippocrates or in other Greek or Roman texts, (unless we assume that the plagues mentioned in Chap. 1 were due to smallpox), suggesting that the disease was unknown or not serious in the Middle East a few thousand years ago, except in Egypt. Smallpox likely spread throughout Southwest Asia by means of trade, conquest, and exploration. Smallpox became endemic in India in the first millennium B.C.; from India it spread into China in the first century A.D. and reached Japan by the sixth century. It reached Europe during the Middle Ages, most likely when the armies of Islam passed through North Africa into Spain, or later with the crusaders returning from the Holy Land.

In China, Ho-Kong (d. 340 A.D.) gave the first recognizable description of smallpox. He describes pockmarks that appear on the body and face, and writes that these last for a long time even if the person recovers. If the disease is not treated early. the patients die. However the description of the treatment is rather puzzling. To quote from a translation from the Chinese, "If the people boiled edible mallows, mixed them with garlic, and ate the concoction, the epidemic would stop. If when contracting the disease one ate this concoction with a small amount of rice to help it down, this too would effect a cure." The disease was called "the Barbarian (or Hun) Pox," since it came from the West after an attack on Nan-Yang, in central China. "The Marsh-mallow" (not to be confused with the confectionery marshmallow), an edible weed, has been used for thousands of years in medicine, and can soothe irritated mucous membranes. Mallows have been used in the treatment of many illnesses in folk medicine. Adding garlic might have a medicinal effect, and of course may also prevent contact between people. In 569 A.D., Marium, Bishop of Avenches, named the agent causing smallpox "variola," from the Latin *varius*, meaning "pimple" or "spotted."

It was difficult at that time to distinguish between measles and smallpox, both fatal diseases. Rhazes (Mohammad Ibn Zakariya al-Razi), a ninth-century Persian physician, was the first to differentiate between the two infections. Rhazes was a physician, philosopher, and mathematician who wrote many books on medicine and philosophy and is considered the forerunner of modern empiric medicine. He describes the rash stage of smallpox thus: "Smallpox appears when blood 'boils' and is infected, resulting in vapours being expelled. Thus juvenile blood (which looks like wet extracts appearing on the skin) is being transformed into richer blood, having the color of mature wine. At this stage, smallpox shows up essentially as 'bubbles found in wine' (as blisters) … this disease can also occur at other times (i.e., not only during childhood). The best thing to do during this first stage is to keep away from it; otherwise this disease might turn into an epidemic." From the description of Rhazes and other physicians of the period, it seems that smallpox was not a dangerous disease. By 1000 A.D. smallpox had spread across most of Europe and Asia. From the twelfth to fourteenth centuries there were large movements of peoples, particularly during the Crusades, when armies from Europe (England, France, and Spain) journeyed to the Middle East and back, thus spreading disease. We don't know exactly when smallpox reached Europe—but we know that "pox houses" were built in the era of the Crusades, along the routes from Europe to the Holy Land. It's likely that the movement of armies across Europe in these years hastened the spread of the disease [2]. Some historians believe smallpox in the medieval period was just a minor variant of the disease—a childhood illness like measles, but less severe. It certainly wasn't feared like the bubonic plague, but in the 200 years that followed the bubonic plague (Black Death), smallpox grew more and more deadly. By the sixteenth century it was a major killer [3].

Table 9.1 Royal Houses of Europe affected by smallpox

Queen Mary II of England (1662–1694)	Wife of William of orange	Died at age 32
King Louis I of Spain (1707–1724)	Son of Philip V	Reigned only 8 months. Died at age of 17
Tsar Peter II of Russia (1715–1730)	Grandson of Peter the Great	Ruled only 3 years
Emperor Joseph I of Austria, Holy Roman Empire (1678–1711)	King of Hungary, Austria, Germany	Hapsburg king
Queen Ulrika Eleonora 1688–1741	Queen of Sweden	
Louis XV of France (1710–1774)	Known as Louis the beloved	Became king at age 5

Even Girolamo Fracastoro (1478–1553) describes it as an endemic disease, not one with severe symptoms (it was predominantly a childhood disease). The virus must have mutated to one with increased virulence sometime in the sixteenth or seventeenth century. Smallpox was "endemic" throughout Europe—meaning that it was a constant presence. When the virus reached a dense population with low immunity, it became "epidemic," and a major outbreak occurred. Epidemics were cyclical; with each passing generation, as immunity levels dropped, epidemics recurred.

9.2 The Effect of Smallpox on European History

Smallpox in Europe (as elsewhere) attacked the rich as well as the poor. It drastically altered European history by destroying the line of succession of many royal houses (listed in Table 9.1) and it drastically affected everyday life. In 1800, English historian Thomas Macaulay wrote that the disease "was always present, filling the churchyard with corpses, tormenting with constant fears all whom it had not yet stricken … making the eyes and cheeks of a betrothed maiden objects of horror to the lover." Between 1719 and 1746, London suffered five major smallpox outbreaks. As the eighteenth century drew to a close, smallpox was killing 400,000 Europeans each year and had spread from Eurasia to Australia, South Africa, and North and South America. It replaced the "black death" as the major scourge of Europe and the rest of the world.

9.3 Effect of Smallpox on the New World

Smallpox had a devastating effect in the Americas where it occurred as a continuous deadly epidemic. The first recorded smallpox outbreak in Latin America occurred in 1507, when Spanish explorers brought the disease to Hispaniola (today Haiti and the Dominican Republic). The native population lacked immunity to smallpox and entire nations were decimated. The population of Hispaniola, estimated anywhere between 300,000 and 1,000,000 in 1492, had decreased to less than 1000 by 1541. The first outbreak of smallpox occurred in 1518 among the Taino people, who inhabited the Caribbean Islands, including Cuba and Puerto Rico. By 1550, the Tainos were extinct. In their diaries, both Bartolomé de las Casas and Columbus describe the Taino as a very gentle, peace-loving, honest people, not given to the evils of the Spaniards. Bartolomé de las Casas, a Spanish priest living in Hispaniola at the time (1542) wrote [4]:

> There were 60,000 people living on this island [when I arrived in 1508], including the Indians; so that from 1494 to 1508, over three million people had perished from war, slavery and the mines. Who in future generations will believe this? And of all the infinite universe of humanity, these people are the most guileless, the most devoid of wickedness and duplicity, the most obedient and faithful to their native masters and to the Spanish Christians whom they serve. They are by nature the most humble, patient, and peaceable, holding no grudges, free from embroilments, neither excitable nor quarrelsome. These people are the most devoid of rancors, hatreds, or desire for vengeance of any people in the world. And because they are so weak and complaisant, they are less able to endure heavy labor and will soon die of no matter what malady … Yet into this sheepfold, into this land of meek outcasts there came some Spaniards who immediately behaved like ravening wild beasts, wolves, tigers, or lions that had been starved for many days. And Spaniards have behaved in no other way during the past 40 years, down to the present time, for they are still acting like ravening beasts, killing, terrorizing, afflicting, torturing, and destroying the native peoples, doing all this with the strangest and most varied new methods of cruelty, never seen or heard of before, and to such a degree that this Island of Hispaniola once so populous (having a population that I estimated to be more than three million), has now a population of barely two hundred persons … The island of Cuba is nearly as long as the distance between Valladolid and Rome; it is now almost completely depopulated. San Juan [Puerto Rico] and Jamaica are two of the largest, most productive and attractive islands; both are now deserted and devastated.

As is obvious from the above description, not only did the native peoples die from smallpox, but also from the hardships and cruelty imposed upon them by the Spaniards. The description of Bartolomé de las Casas is unique in its honesty of the sufferings imposed on the Indians and of the greed of the Spaniards.

9.4 Smallpox in Mexico

Bernard Diaz del Castillo describes in great detail in his book *The True History of the Conquest of New Spain* [5] the war between the Aztecs of Montezuma and the Spaniards, led by Cortes. As in the case of the writings of Bartolomé de las Casas, the picture of Montezuma and the Indians is a very sympathetic one. Following the death of Montezuma, the Spaniards were forced to flee the city and take refuge in Tlaxcala, and signed a treaty with that tribe to help conquer Tenochtitlan, offering the Tlaxcalans freedom from any kind of tribute and the control of Tenochtitlan. In the ensuing battle between the Aztecs and the Spaniards, it is possible that smallpox was transmitted to the Aztecs, who then died by the thousands. This included Montezuma's brother, the new king. The smallpox was possibly brought to Mexico by a local slave attached to the army of Panfilo de Narvaez, who headed the expedition against Cortes. The disease spread from town to town, resulting in the abandonment of farms and the starvation of the population. On August 21 the Spaniards stormed the city, only to find that some greater force had already pillaged it. "I solemnly swear that all the houses and stockades in the lake were full of heads and corpses," Cortes's chronicler, Bernal Diaz, wrote of the scene. "It was the same in the streets and courts … We could not walk without treading on the bodies and heads of dead Indians. I have read about the destruction of Jerusalem, but I do not think the mortality was greater there than here in Mexico … Indeed, the stench was so bad that no one could endure it … and even Cortes was ill from the odors which assailed his nostrils."

Historians cite from 2 to 15 million Aztec deaths due to smallpox. It was claimed that the population of Mexico had shrunk from a population of 25.2 million in 1518 to 1.1 million by 1605 [6]. However, other historians have disputed these numbers as exaggerated. From Mexico, the disease spread both north and south to the Incas. It appeared in Guatemala in 1520. The disease reached Peru by 1525 and killed the emperor of the Incas, Huayna Capac, and most of his family. Following the death of the emperor and his sons, civil war broke out in the Incan Empire, which was quickly conquered by Pizarro. The smallpox epidemic was followed by measles, spreading through Mexico and Peru from 1530 to 1531. Another epidemic (perhaps typhus) followed in 1546, which was followed by an influenza epidemic on both sides of the Atlantic in 1556–1560. Reports indicate that as many as 20 % of the population of Europe and the Americas died during this epidemic. It may have been similar to the great influenza epidemic of 1918. By 1630, the Andean population was only about 7 % of what it had been in 1524. Not only was the population decimated by disease, but also by the conditions in the silver mines, where harsh conditions led to death. Smallpox moved northward from Mexico and westward from the eastern United States and Canada. Probably by the time most settlers arrived in what is now the U.S. and Canada, a large proportion of the native populations had died off because of smallpox. We do know that after the French set up a post in Nova Scotia in 1616, a great pestilence (smallpox?) swept through the Massachusetts Bay area. Thus both the Indians and the Europeans believed that the "God of the White Man" prepared the way for the white settlers, By the mid-1800s, smallpox had decimated many Native American tribes in the United States, Canada, and Alaska (see Table 9.2) [7].

9.5 Smallpox and Slavery

The historical importance of these events should not be underestimated. This disease "depopulated" most of the Americas and certainly helped in the settlement of the country by "white men." Estimates of the total population of North America before the arrival of Columbus vary from 50 to 100 million. There was also systematic killing of the native population, either by terrible working conditions in

Table 9.2 History of smallpox by date as related to the New World

1492	Discovery of America
1495	First major epidemic: 75–80 % of inhabitants of Santa Domingo die
1515	2/3 of Indians of Puerto Rico die
1518	Introduction of slave labor, including slaves from Africa
1518	Outbreak among Taino
1519–1521	Battle between Cortes and Aztecs
1520–1527	Millions of inhabitants of Mexico die: death of Incan leader and 200,000 people; destruction of Incan empire
1577	Smallpox kills 1/3 of population of Venezuela
1587	90 % of Indians of Bogota die from smallpox
1590–1610	2 million Indians of South America die from smallpox
1616–1619	Algonquin nation in Massachusetts exterminated. About 90 % of Indian population dies
1620	With arrival of Pilgrim Fathers in New England, spread of smallpox along East Coast into Carolinas
1720	New Epidemic kills millions in Peru
1633–1764	Epidemics in Boston and surrounding areas
1662–1669	Epidemic among Iroquois and Huron: high death rate
1699	Disease hits Mohawk Indians
1703–1718	Epidemics in Quebec, among Hurons and Iroquois
1730	Smallpox spreads to West Coast Indians
1731	Epidemic among Seneca Indians
1738–1757	Epidemic among the Miami and Cherokee, and west of the Dakotas
1778	Smallpox in Louisiana and Texas
1837	Smallpox epidemic among the Great Plains Indians, Pacific Coast Indians
1830s	Reduced Dena'in a Athabaskan population in Cook Inlet region of south-central Alaska by half. Smallpox also devastated Yupik Eskimo populations in Western Alaska

Adapted from *Texas Department of State Health Services, History of Smallpox.* http://www.dshs.state.tx.us/preparedness/bt_public_history_smallpox.shtm and http://www.genealogyinc.com/encyclopedia/epidemics/

the mines, or by outright massacres. The smallpox epidemics may have brought about the importation of slaves from other continents, in particular from Africa since there was no native population to work the land, and the gold and silver mines of South America. It has been suggested that Africans were more resistant to smallpox and it is possible that African children were exposed to smallpox as a childhood infection, survived the infection and thus were immune for life.

Fig. 9.2 Indians and Spaniards, unknown seventeen century

Borges, the famous Argentinian writer, wrote:

> In 1517, the Spanish missionary, Bartolomé de las Casas, taking great pity on the Indians who were languishing in the hellish work pits of the Antillean gold mines, suggested to King Charles V of Spain, a scheme for importing blacks, so that they might languish in the hellish work pits of the Antillean gold mines. To this odd philanthropic twist we owe, all up and down the Americas, endless things.

Although Bartolomé de las Casas did initially advocate the importation of black slaves to replace the local peoples, he later regretted this decision and was opposed to slavery. He sympathized with the local Indian population and tried to persuade the Spanish king to abolish the Enconomendia, which resulted in maltreatment and slavery of the Indians. He was later appointed Bishop of Chiapas.

One other important result of the smallpox epidemics and plague was the conversion of the remaining Indians to Christianity. The Aztecs and Incas saw that the "white man" seemed to be resistant to the smallpox, whereas their people died by the thousands. Thus the Spaniards' god seemed much more powerful than the Aztecs' or Incas' gods. Missionaries soon followed the conquistadores and converted a weak and disillusioned population to Christianity. The native Amerindians could not understand what was happening; they were downtrodden, defeated and humiliated. They readily accepted Christianity, or submitted without resistance, and the religious among them felt that their gods had abandoned them. The Spaniards and other colonizers did not deliberately, in most cases, introduce smallpox and other diseases. These diseases, mostly viral, found "virgin soil", i.e., a population without immunity and the diseases spread with extreme devastation. This continued for many generations, although there must have been some selection for eventual resistance, particularly following intermarriage between Spanish settlers and local Amerindians (Fig. 9.2).

There was at least one case in which smallpox was used as a biological weapon against an Indian tribe (Fig. 9.3). General William Amherst of the British army ordered blankets known to be contaminated with pus from smallpox to be given to the Indian tribe. Lord Amherst was the commanding general of the British forces

Fig. 9.3 Image of the "great dying" from smallpox. Unknown, probably 17th century

during the French and Indian War (1754–1763); this was the war that acquired Canada for Britain. Apparently after the defeat of the French, one Indian tribe did not accept British rule (they expected supplies from the British, as given to them by the French). Below is a link to a copy of the letter written by Amherst supporting the use of small pox blankets to get rid of this tribe (see Chap. 22 on biological warfare). http://www.nativeweb.org/pages/legal/amherst/34_41_114_fn.jpeg.

9.6 Why Were Native Peoples More Susceptible to the Virus?

Why were the native populations so susceptible to smallpox and measles? One hypothesis is that there had been no selection for immunity in the native populations. Europeans had lived with plagues and viruses for centuries and had developed an immunity that allowed for the survival of the fittest. Native peoples did not live in proximity to their animals (they had very few domestic animals) and thus had not experienced viral diseases, which were transmitted from domestic animals to humans. By all accounts, the native peoples were very healthy, even those living in urban areas. In the case of measles, Europeans had developed immunity since most (99 %) were exposed to this virus in childhood; it was much more lethal to adults than to children, and when brought to the New World, or introduced into native populations, it wrought havoc on the adult population.

However, in cases of both measles and smallpox, the mortality rate among the native peoples was around 90 % while in the European peoples around 10 % [7]. This might imply non-pathogenic exposure of Europeans to the virus early in life, with resulting immunity—as is the case with many other virus infections.

An explanation for the lack of immunity among the Amerindians may be simply that their ancestors crossed the Bering Straits from Asia before smallpox was established as an infectious disease in the human population. It is estimated that the crossing occurred 21,000–40,000 years ago. This would predate the development of the smallpox virus, or at least be about the time that the virus became a human infection. As stated earlier, it is thought that the virus mutated and became more virulent in the sixteenth century, so that its introduction to an immune naïve population was devastating. The lethality may have been much higher than in previous outbreaks. Smallpox epidemics in North America were not confined to the native population; from 1666 until 1903, numerous outbreaks of smallpox occurred in the Eastern seaboard cities of the United States. Massachusetts had five epidemics between 1617 and 1722, and smallpox afflicted George Washington's troops in 1775 during the Revolutionary War. Historians claim that the Revolutionary Armies were unable to conquer Canada from the French due to epidemics of smallpox among the troops [8]. In fact, Washington recalled his troops from the lines of battle so that they might be variolated before joining a fighting regiment. The British Army also suffered from smallpox during this period but they practiced variolation on a larger scale than the Americans.

Abraham Lincoln delivered the Gettysburg Address while suffering from the early symptoms of smallpox and the disease was rampant among both Union and Confederate solders. Smallpox continued to be a considerable problem up until the mid-twentieth century in large parts of Africa and in India. The World Health Organization estimated in 1967, when the campaign to eradicate small pox was launched, that there were about 15 million individuals affected annually worldwide. However, at no time did the death rate in these countries reach the level that it did among native peoples of the Americas [2].

New research indicates that the Amerindians might be lacking a component of the adaptive immune response. From studies done on newly discovered tribes of the Amazon, it appears that the initial immune response to virus might be normal, but that there is not enough heterogeneity in T-cell or B-cell responses to counteract the invasion of the virus. Thus, there was not enough selection against the virus, which could then overcome the antibody response by selection for virus variants. The constant exposure of European and Asian populations to the virus (and other infectious diseases) may have led to selection for a more heterogeneous immune response, and that genes involved in resistance were lost in isolated populations, where these diseases were not rampant. This genetic basis is partly supported by the observation that genetically mixed children of Spaniards and native peoples are more resistant to smallpox [9]. On the other hand, the smallpox virus is relatively stable and does not mutate at a high frequency, so there should be little selection.

9.7 Outbreaks of Smallpox Among Other "Virgin Populations:" Iceland and the Shetland Islands

One of the best-documented studies of a smallpox epidemic occurred among the small population of Iceland in the eighteenth century. This epidemic reached Iceland in 1707, killing about a quarter of the population over the next 2 years. A previous epidemic had occurred 35 years earlier; thus, all born since then had no immunity to the disease. The epidemic occurred when an Icelander visiting Denmark had died of the infection while at sea, and his clothes were sent back to his home on the island. The smallpox spread very rapidly over the southern and western parts of the island, and in the course of the epidemic 7,300 inhabitants died. By 1708 the infection spread to the rest of Iceland and the Isle of Grimsey, where 33 of 87 inhabitants died. By the spring of 1709 the epidemic was over. It is estimated that approximately 12,000 inhabitants died of smallpox between 1707 and 1710. The population of the island might have been 50,000 before the epidemic began, thus about 25 % of the population died from the viral infection.

The epidemic of 1707 is an example of what transpires in a non-immune population. Previous epidemics had left a proportion of the population immune, and thus most deaths occurred in the under-35 age bracket. Most of the victims were adults of various ages and very few children were affected, but the reason for this is unknown. This was different from most European epidemics where young children were the only ones susceptible to the virus, since the disease was endemic in the population, or in the case of the Amerindians, which can be described as "virgin soil," where there was no history of previous infection. Iceland was relatively isolated with little contact with the outside world. It also had a small population, not large enough to sustain a full-blown endemic outbreak. There had been sporadic outbreaks in the previous two centuries [10].

A curious factor of the Icelandic epidemic was that young children had a milder form of the disease than young adults. With time, the virus appeared to be less lethal. The outbreak of 1707 had economic consequences: fewer farmers, and a resulting loss of livelihood for landlords. Variolation was introduced in Iceland in 1802 by government decree [11].

That smallpox might have become more virulent in the sixteenth and eighteenth centuries is also suggested by the high mortality that occurred in 1720 on the small island of Foula in the Shetland Islands of Scotland, with the mortality rate approaching 90 %. This could have been the result of an isolated population lacking immunity, as in the case of the New World Indians, or because of increased virulence of the virus [12].

9.8 Religion, Tradition and Smallpox

As did Europeans until the nineteen-century, Amerindians believed that evil spirits and/or witches instigated smallpox, and other diseases, for sins committed in the community against their gods, a superstition that did not differ much from those

Fig. 9.4 Chinzei Hachiro Tametomo (1139–1170), a skillful archer, was exiled to the island of Oshima. He is reputed to have prevented a smallpox demon from landing there. His image was hung on the walls of Japanese homes to help to protect them against smallpox. (Wikipedia PD-1923) [2]

found among other peoples. One of the approaches to warding of the disease was the construction of the "sweat lodge." which is similar to Finnish or Russian saunas. It is not clear whether this was a truly Amerindian custom or learned from the immigrants. The sweat lodge was used both to cleanse the soul and the body, and attendance at the lodge was accompanied by incantations to ward off the smallpox.

As in the case of the Amerindians, other cultures tried to explain the occurrence of smallpox through religious or superstitious belief. According to legend in Japan, a smallpox demon or devil, who returned to earth to seek revenge, was blamed as the cause of the disease. Smallpox devils were said to be afraid of red, and thus dolls with red clothing were placed near the smallpox victim. In Japan, a red picture of Tametomo, a twelfth century hero who was reputed to have thwarted a smallpox demon, was often hung in the rooms of smallpox victims to aid their recovery. The color red had an ancient and persistent association with smallpox and was supposed to promote recovery. Many Japanese textbooks on dermatology stated that exposure to red light alleviated the symptoms of smallpox (Fig. 9.4).

On the island of Okinawa, special music was composed for the *shansin*, a precursor of the Japanese *samishan* (a three-stringed instrument similar to a banjo). Special songs or poetry were written to appease the smallpox devil, written in the local language of the Ryukyu Islands and special dances were also created. As in many other cultures, the color red was associated with "curing" smallpox http://en.wikipedia.org/wiki/Smallpox_demon.

In Europe there are records that show that Queen Elizabeth 1 of England, and King Charles V of France wore red during the period they had smallpox. Queen Elizabeth was wrapped in a red blanket. This was common in China, India, Turkey and Georgia; even in Western Africa, the Yoruba god of smallpox, Sopona, was associated with the color red.

In parts of India, China, Africa and Latin America, sacrifices were made to appease the gods of smallpox. In medieval Europe, prayer and pious living were recommended as one way to guard against sickness. St. Nicaise of Rheims, a martyr from the time of the Huns (fifth century) was said to have recovered from smallpox and became the patron saint of the smallpox victim.

9.9 Variolation

In China, and later in Turkey and the rest of the Ottoman Empire, people discovered that exposure to smallpox prevented subsequent infection. Material from the smallpox wound (pox) was dried and inhaled into the nose. Later, rather than inhaling through the nose, pus from the open sores of the smallpox victim was scarified onto the arm or another part of the body. In most cases this resulted in a mild infection of smallpox, making the person "immune" to subsequent disease, although in 2–3 % of the cases, it could be harmful. In 1713, Emanuel Timoni, a physician, published an account of the practice in the *Philosophical Transactions of the Royal Society* in London, and in 1716, Jacob Pylarini, a Venetian physician and consul in Smyrna, Turkey, published a further account in the same journal. He variolated the children of the British Ambassador and was one of the first to practice variolation outside Turkey. Lady Mary Wortley Montagu, wife of the British ambassador to Turkey, who herself had been a victim of smallpox at the age of 26, brought the practice of variolation to Britain and the British Royal Court (Fig. 9.5). In 1717, while in Turkey, she learned about variolation and initially had her 6-year-old son and 4-year-old daughter "variolated" despite opposition from her husband and the British consulate, who thought it a "heathen" act. Mary Montagu's children did not come down with smallpox.

Variolation was much discussed in London, and this discussion must have encouraged Lady Montagu to take a bold step. She convinced the English court, including Caroline of Ansbach, Princess of Wales (later Queen of England during the reign of George II) to variolate her daughter. By 1721, variolation (then called then "engrafting") began to be widely practiced in Europe as protection against smallpox. On returning to Britain, she urged the British court to try variolation by introducing pus from a pock onto the skin of a group of six prisoners and

Fig. 9.5 Lady Mary Montagu as painted by Charles Jarves (Wikipedia)

abandoned children as a trial run. Dr. Charles Maitland, the court surgeon, performed variolation under the watchful eye of the Royal Society. Subsequently exposing these subjects to the smallpox virus showed that variolation worked, as they did not become ill with the virus [13]. Maitland subsequently published a book describing these and other successful variolations. Such experiments could not be carried out today in most countries, and in fact it would be difficult to develop a vaccine against smallpox under present-day restrictions on the use of human subjects for medical purposes since there is no animal model for smallpox (Fig. 9.5).

Variolation in the U.S. followed a different pathway. During an outbreak of smallpox in Boston in 1706, Cotton Mather learned from one of his slaves (Onesimus) about variolation. He then read Dr. Timoni's paper published in the *Philadelphia Transaction* [13]. He published and introduced the procedure into the colonies and then tried to convince physicians of the need to perform variolation in Boston. One of them, Zabdiel Boylston of Brookline, variolated his 6-year-old-son, his 30-year-old slave, and the slave's 2-year-old son. In 1721 Boston was hit with an epidemic that afflicted about 50 % of the city's 11,000 inhabitants and killed 844. By 1722 Boylston had variolated 242 of his patients, of whom only six died during the epidemic. This was a mortality rate of 2.5 % compared to 15–20 % among the non-variolated population. Boylston reported these results in the *Boston Gazette*. However, variolation was opposed by the medical establishment and the clergy, and even the general public, and Boylston had to go into hiding [14]. (Boylston Street, a major thoroughfare in Boston, is *not* named after him, but after one of his descendants, a wealthy businessman and philanthropist.)

Both the British Army and the Revolutionary Forces during the American War of Independence widely used variolation, since both sides suffered terribly from outbreaks of smallpox in their respective armies. Documents indicate that in 1777 George Washington ordered the return to Philadelphia of solders in his army to receive smallpox variolation since smallpox was rampant in the revolutionary army. The British had vaccinated their troops, and the story is that they sent variolated civilians to mix with the colonists in the hope of spreading the disease. In 1776, the British did it again as the Americans besieged Quebec City. It is told that this time they variolated prostitutes and sent them among the troops; half of the 10,000 Americans fell ill, and after burying their dead in mass graves, they retreated in disorder from the colony, which remained in British hands [14].

One of the puzzling aspects of variolation is why taking material from pus and infecting an individual with this was not as severe as the natural infection. However, variolation could result in devastating effects, and about 2–3 % of those treated in this fashion did die, yet this was ten times less than the effect of smallpox. It could be that due to scarification of the skin—rather than inhalation—the disease was less severe.

9.10 Jenner's Vaccination

Local English farmers had observed for some time that milkmaids often had sores (pockmarks) on their hands, which prevented smallpox from developing. This appeared to be due to "cowpox" transmitted from cows during milking. In fact, many farmers inoculated their families with cowpox, and the story went around that this could protect against smallpox. In 1792, as a country physician in Gloucestershire, Edward Jenner noted this phenomenon and decided to "scarify" James Phipps, a local 8-year-old boy, with the cowpox, and subsequently expose him to pus from a smallpox lesion; James proved resistant to the subsequent

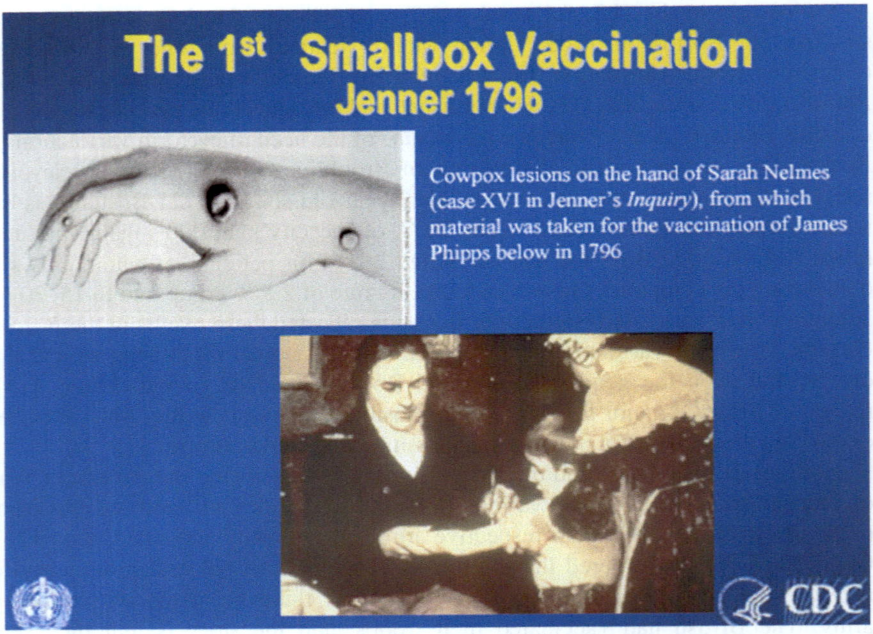

Fig. 9.6 Cowpox lesion on hand of milkmaid and Jenner's first smallpox vaccination on James Phipps (Courtesy of the CDC)

smallpox infection. This method of protection was called "vaccination" from the Latin "*vacca*" meaning "cow" (Fig. 9.6). Jenner was also a naturalist and studied cuckoo behavior; he published a paper on the natural history of the cuckoo, and was a member of the British Royal Society. Despite this, the society refused to publish his observations on cowpox as a preventative of smallpox. Luckily for mankind, Jenner independently published a pamphlet "An inquiry into the causes and effects of the variola vaccine, a disease discovered in some of the western counties of England, particularly Gloucestershire, and known by the name of cowpox." In this, he described the results from 20 patients who developed immunity to smallpox after cowpox inoculation.

Although some accepted this work, the medical establishment ridiculed it—in particular the clergy who proclaimed it an interference with God's way. Others claimed that it was not mentioned in the Bible (neither was smallpox), and that it was a filthy habit. Despite this campaign, Jenner's pamphlet was widely read and the practice of vaccination spread all over the world. He was honored with letters and awards from President Thomas Jefferson, from the Chiefs of the Five Nations of the North American Indians, and from Napoleon, who released prisoners at Jenner's request [15]. (More details can be found in Chap. 19 on vaccines.)

The cowpox virus used was later called "vaccinia." This method, and possibly the same original virus with some modification (vaccinia), was used until very recently to prevent smallpox epidemics. Such experimentation would never pass an institutional medical review board today, nor would the subsequent experimentation on prisoners confirming that smallpox vaccination worked. Although vaccination was effective, it took about another 200 years for it to be universally accepted. Smallpox was only totally eradicated in 1977.

The acceptance of vaccination was not easy. Many people fought against it, claiming that it was interfering with the work of God; others made fun of the idea that material from a cow could protect humans. There are many cartoons from the eighteenth century that ridicule the concept of variolation and vaccination. Anti-vaccination societies became especially vocal during the late nineteenth century, and many anti-vaccinators believed that vaccination was, as George Bernard Shaw put it, a "filthy piece of witchcraft" that did more harm than good.

9.11 Eradication of Smallpox

The eradication of smallpox is one of the all-time greatest achievements in medicine. A Russian epidemiologist, Victor Zhdanov, the vice-minister of the Soviet Ministry of Health proposed in 1958 at the UN and World Health Assembly the total eradication of the smallpox virus. Initially there was much skepticism expressed as to whether this was achievable. The WHO, a branch of the United Nations, selected Dr. Donald A. Henderson of the CDC—Atlanta to co-ordinate the effort and head the WHO's Smallpox Eradication Unit from its inception in 1966 until the final case in 1977. Henderson was the driving force behind the whole program. Initially very little funding was allocated to this project, since large amounts of funding went to an anti-malaria program that seemed more promising. In fact, the anti-smallpox program may have been the result of rivalry between the Soviet Union and the U.S. at the United Nations. Despite this, in 1966 a medical team under the supervision of the WHO set out for Central and Western Africa, armed with smallpox vaccine and special jet injectors. It was to become history's most extensive microbe hunt, encompassing India, Nepal, Indonesia, Bangladesh, Pakistan, Afghanistan, Yemen and South America. Teams were dispatched to known sources of epidemics to carry out mass vaccination, but instead of mass vaccination, a program of surveillance and eradication was introduced. Infected individuals were tracked down and anyone having contact with them was vaccinated. This was more efficient than mass vaccination, although the approach was initially controversial. This idea was the brainchild of Dr. William Foege, who had been working as a medical missionary for the Lutheran Church in Eastern Nigeria when the eradication program began; he now works for the Carter Center in Atlanta. Dr. Foege was the architect of Surveillance containment isolating areas of outbreak and then vaccinating there. This freed health workers from the burden of mass vaccination and preserved the often limited supplies of vaccine.

By 1975, it appeared that the campaign was successful and that smallpox had been eradicated. The last naturally occurring case of the disease was traced to Somalia on October 26, 1977, to Ali Maow Maalin from the town of Merka; he survived the infection. The total eradication of smallpox was not easy in the field. Many believed that we could not eliminate an organism without affecting the total ecology of human life. There were many practical problems; one was the surveillance and reporting of cases, particularly in Third World countries where there was little means of communication. Some countries posed particular problems since there were on-going civil wars, such as in the Congo. The size of the area covered meant that local teams had to communicate and travel long distances. Even after eradication, one had to follow up rumors of possible outbreaks and test for the presence of virus and cases.

Unfortunately, the case of Ali Maow Maalin was not the last. There was a localized outbreak of smallpox in Birmingham, England, in 1978 due to a lab accident a lab technician—not associated with the smallpox research laboratory—died. Janet Parker, a 40-year-old medical photographer who was working on the floor above a smallpox laboratory contracted the disease. It has never been conclusively determined how the virus reached her, but an official inquiry cited an air duct between the two floors as the most likely conduit. Unbelievably, another medical photographer working in the same room had come down with smallpox 12 years earlier. As a result of Janet Parker's illness, close to 300 people who'd come in contact with her had to be quarantined. Among those, only her mother contracted smallpox and she survived.

Despite this, in 1980 the World Health Assembly recommended that all countries cease vaccination, and declared that smallpox had been eradicated. A WHO expert committee recommended that all laboratories destroy their stocks of variola virus or transfer them to one of two WHO reference laboratories—the Institute of Virus Preparations in Moscow, or the Centers for Disease Control and Prevention (CDC) in Atlanta. The decision was to be made in the future whether to destroy all stockpiles or retain samples at these sites for future research.

Why was it possible to eradicate smallpox so successfully but not other viruses? The principal reason was that smallpox infects only one host, and that is *man*. There is no smallpox reservoir in other animals. The vaccine is easy to prepare and store and does not require refrigeration. Many other vaccines (such as the polio vaccine) require refrigeration, which is not possible in many counties and certainly not in the "bush." The smallpox infection is easy to diagnose and is only infectious during the eruption phase, unlike measles and other viruses that can be infectious before the rash appears; also, the vaccine confers immunity for approximately 10 years. Today the population of the world has no immunity to this disease. However, with the virus having been eliminated, this should not have posed a problem, but it is now known that there is danger lurking. The stocks of the virus sit in two freezers and can be restored to life. We do not know whether they are not stockpiled elsewhere and could be used by a fanatical terrorist. As discussed below (see Chap. 22), smallpox virus was prepared in liter amounts for biological warfare purposes, despite a treaty that was supposed to prevent this development.

9.12 Natural Resistance to Smallpox

In any model of epidemiology, one would expect a resistant population to arise. Smallpox has been endemic for close to 16,000 years. In most of the world, children have been the main victims, hence there should have been some selection among the survivors for genetic markers that endow resistance. Yet, the evidence for this is scant. Although AIDS has only been with us for a relatively short time, we already know that individuals lacking certain chemokine receptors are resistant to the virus. Some research has been done in this area [16]; in 1965–1966, scientists studied the incidence and survivors among 415 unvaccinated survivors of smallpox in a Bengali population. They reported that people who were blood type A were seven times more likely to be infected by smallpox, three times more likely to have a severe infection, and two times more likely to die from the disease than those of other blood types. However, these results have not been confirmed by other researchers.

It has been suggested [17] that the high frequency of the ccr-5-Δ32 deletion that confers resistance to HIV infection (CCR-5 is a co-receptor for HIV) in the European population may have been selected in the human population as a result of conferring resistance to smallpox. Obviously there has not been enough time for HIV selection in the human population, since HIV is a relatively new disease. Others have argued that this allele may have resulted in resistance to the bubonic plague (Black Death). However, given that smallpox is an older infection, Alison Galvani and Montgomery Slatkin [17] make the case for smallpox as the selective agent. In fact, the evidence contradicts the hypothesis that there is a relationship between bubonic plague and the incidence of ccr-5 deletion. Since one cannot do experiments on a non-human system, it is impossible to state categorically which of these hypotheses is correct.

9.13 Smallpox as Weapon of Mass Destruction

Unfortunately, smallpox has been considered again as a possible biological weapon. Populations today are no longer vaccinated and are thus susceptible to infection. Smallpox vaccination was stopped in the U.S. and most of Europe in 1980. Ken Alibek, a defector from the Soviet Union, reported that beginning in 1980, the Soviet government embarked on a successful program to produce the smallpox virus in large quantities and adapted it for use in bombs and intercontinental ballistic missiles; the program had an industrial capacity capable of producing many tons of smallpox annually [18]. If used as a biological weapon, smallpox would represent a serious threat to civilian populations because of its case-fatality rate of 30 % or more among unvaccinated persons. There is no proven therapy against the virus. It may be necessary to revaccinate the entire population, since there are no drugs to counteract a smallpox epidemic. This virus is also a potential weapon in the hands of a terrorist, even one contemplating a painful suicide. The few side effects and few cases of smallpox resulting from vaccination do not justify the complete stoppage, since very shortly (if not already), the total population will have lost immunity and may share the fate of the American–Indian if an attack with the virus occurs. The population of the world is on "virgin soil"—as far as the infection is considered.

After the terrorist attacks of September 11, 2001, the U.S. government took further actions to improve its level of preparedness against a biological terror attack. One of many such measures—designed specifically to prepare for an intentional release of the smallpox virus—included updating and releasing a smallpox response plan. In addition, the U.S. government has prepared enough vaccine to vaccinate every person in the United States in the event of a smallpox emergency. In the event of a single terror attack, one would need a hundred million doses of vaccine. Until recently, injecting the "vaccinia" virus into the bellies of calves and harvesting material from the developing lesions produced the vaccine. This vaccine is derived from the New York City Board of Health strain of vaccinia, and then essentially freeze-dried for storage. It was licensed by the FDA in 1931 but is no longer manufactured. ACAM2000, a "second-generation" smallpox vaccine, is derived from a clone of the original vaccinia virus, purified, and produced, using modern cell culture technology. Large amounts of the vaccine can be produced as required. The question of whether to keep or destroy the remaining vials of smallpox still haunts us, and the argument in favor is the possible need to study the pathology of the virus. However, vaccinia—the virus used in inoculation—which is probably not the original cowpox isolated by Jenner, could function as an alternative model.

9.14 The Pox Viruses

The variola virus causes smallpox. It only infects humans and has no known animal reservoir. Transmission is by airborne droplets and contact with the pus from lesions; under some circumstances, coughing can spread the virus. The maximum infectious period is 7–10 days after the appearance of the rash. It is highly contagious, and is carried in clothing and blankets. Figure 9.7 illustrates the progression of the virus, with the development of blisters all over the skin. Immunity is relatively long-lasting although there is some controversy as to how long it lasts. The vaccine is easy to produce, and relatively inexpensive, and can also be stored in its dry form without the need for refrigeration.

Variola is one of the largest known viruses. It has a "strawberry-like" shape with an internal core often described as dumbbell-shaped (Fig. 9.8). It contains enough DNA to code for hundreds of genes. Many of the viral genes code for proteins that mimic interferon, cytokines, cytokine receptors, complement proteins, and components of the normal immune system; these "decoys" interrupt normal cellular processes. This virus, unlike most other viruses, contains genes that code for enzymes involved in its own DNA and RNA synthesis. After attachment, the virus undergoes phagocytosis, and the outer membrane is removed in the cytoplasm of the infected cell. Unlike most DNA viruses, smallpox replicates in the cytoplasm, since it has genetic information for its own replicating enzymes.

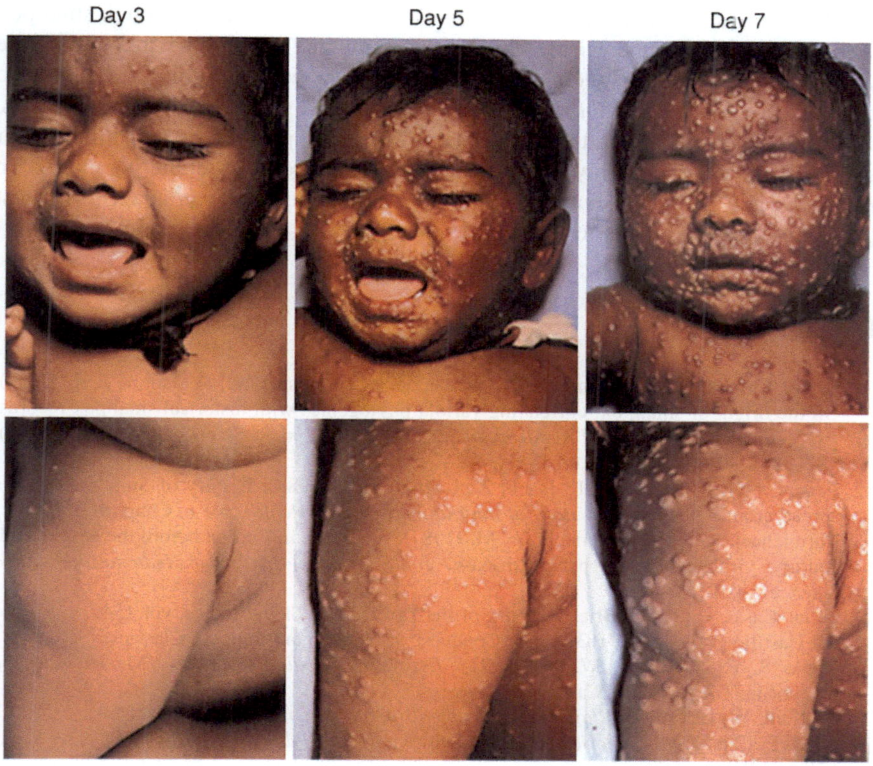

Fig. 9.7 Progression of smallpox in a child. Note the increase and development of poxmarks up to day 7 (Courtesy of the CDC)

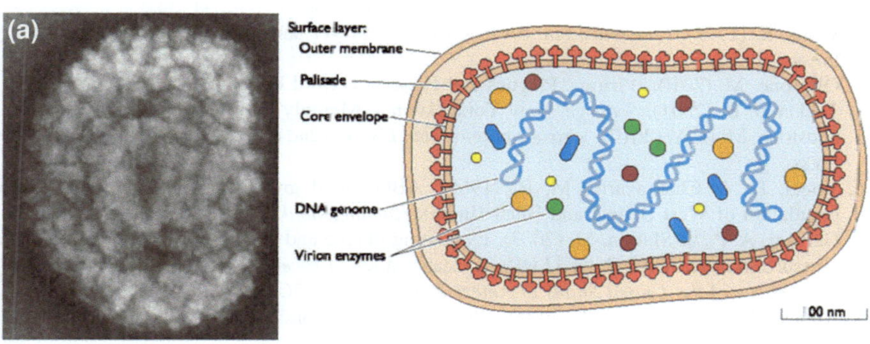

Fig. 9.8 The smallpox virion (Courtesy of the CDC). Note shape and size of smallpox virus. Large inner core containing DNA, outer shell with attachments

The poxviruses form a large class of viruses that affect man, although smallpox is the major agent. Other viruses include vaccinia and cowpox (perhaps the same virus originally used in vaccination), monkeypox (which has symptoms similar to smallpox in humans), *molluscum contagiosum*, which gives rise to benign tumors, and orf, a disease transmitted from sheep. There are poxviruses that infect other animals, one of which, myxoma, was used to control the rabbit population in Australia, and is a good example of how a pathogenic virus can adapt to its host (and how the host adapts to the virus).

References

1. Li, Y., Carroll, D. S., Gardner, S. N., Walsh, M. C., Vitalis, E. A., & Damon, I. K. (2007). On the origin of smallpox: correlating variola phylogenics with historical smallpox records. *Proceedings of the National Academy of Sciences of the United States of America, 104*(40), 15787–15792.
2. Fenner, F. H. D., Arita A. I., Jezek, Z., Ladnyi, I. D. (1988). History of Smallpox and its spread around the world. *History of International Public Health*, vol 6, Chap. 5.
3. Carmichael, A. G., & Silverstein, A. M. (1987). Smallpox in Europe before the seventeenth century: virulent killer or benign disease? *Journal of the History of Medicine and Allied Sciences, 42*(2), 147–168.
4. de las Casas, B.(2007). Brief Account of the Devastation of the Indies, vol 1542.
5. del Castillo B. D. (1963/1632). The Conquest of New Spain: Penguin Classics.
6. Cook, S. F., & Borah, W. W. (1971). *Essays in population history*. Mexico and the Caribbean: University of California Press.
7. McNeill, W. H. (1977). *Plagues and people* (3rd ed.). New York: Random House.
8. Fenn, E. (2001). *Pox Americana: The Great Smallpox Epidemic of 1775–1782*. New York: Hill and Wang. (2201).
9. Black, F. L. (1994). An explanation of high death rates among New World peoples when in contact with Old World diseases. *Perspectives in Biology and Medicine, 37*(2), 292–307.
10. Hays, J. N. (2005). *Epidemics and pandemics: their impact on human history*. CA: Santa Barbara.
11. Steffensen, J. (1977). Smallpox in Iceland. *Nord Medicinhist Arsb*, 41–56.
12. Watts, S. (1997). *Epidemics and history: disease*. Power and Imperialism: Yale University Press.
13. Oldstone, M. (2010). *Viruses*. Plagues and History: Oxford Univeristy Press.
14. Tucker, J. B. (2001). *Scourge*. New York: Atlantic Monthly Press.
15. Oldstone, M. B. A. (2010). *Viruses, plagues and history* (2nd ed.). Oxford: Oxford University Press.
16. Vogel, F., & Chakravartti, M. R. (1966). ABO blood groups and smallpox in a rural population of West Bengal and Bihar (India). *Humangenetik, 3*(2), 166–180.
17. Galvani, A. P., & Slatkin, M. (2003). Evaluating plague and smallpox as historical selective pressures for the CCR5-Delta 32 HIV-resistance allele. *Proceedings of the National Academy of Sciences of the United States of America, 100*(25), 15276–15279.
18. Alibek, K., & Handelman, S. (1999). *Biohazard*. New York: Random House.

10 Yellow Fever

Abstract

Other than smallpox, perhaps no disease has influenced major events shaping the development of North and Central America than yellow fever. This fatal disease—with 50 % mortality—destroyed the economies of many Southern American cities. Yellow fever inhibited the attempts of the French to establish an empire in the Western Hemisphere, using Haiti as a base. This resulted in the sale of the Louisiana Territories to the U.S., resulting in doubling the size of the country. Attempts to build a canal across the Isthmus of Panama failed because of disease, and yellow fever killed more troops in Cuba than were casualties of hostilities in the Spanish–American War. These events led to the formation of a U.S. commission, headed by Walter Reed, to investigate the causes of yellow fever. Carlos Finlay, a Cuban scientist, had suggested some 20 years before the establishment of the commission, that mosquitoes transmitted yellow fever. This was tragically confirmed when two members of the commission allowed themselves to be bitten by yellow fever-loaded mosquitoes. One died and the other suffered chronic illness. Reed established a camp with volunteers to test the hypothesis and proved without a doubt that the insects spread the disease. Later commissions confirmed these results and a virus was eventually isolated, although not without other casualties among the researchers. Max Theiler, at the Rockefeller Institute, developed a vaccine which is used to this day. Outbreaks of yellow fever still occur in Africa and areas of South America.

10.1 Introduction

Other than smallpox, perhaps no disease has influenced major events shaping the development of North and Central America than yellow fever. High fever, chills, headache, back pain, nausea and general fatigue accompany acute yellow fever. In more severe cases there is bleeding, jaundice, and eventual shock and death, with

M. W. Taylor, *Viruses and Man: A History of Interactions*,
DOI: 10.1007/978-3-319-07758-1_10,

the mortality rate about 20–50 %. This disease has struck within North America numerous times, and its origins were blamed on everything from rotting coffee beans to bad sanitation and filth.

The disease was also known as "Yellow Jack" because of the flying of the yellow flag of quarantined ships. The yellow was associated with a custom of placing patients in jackets with yellow patches to separate them from other patients in a hospital in Greenwich, London. A yellow flag was also hoisted where there were cases of yellow fever [1, 2]. Was this the disease described in the "Rime of the Ancient Mariner?" Yellow fever disrupted the economic development of the American South, almost ruining the growth of cities like New Orleans and Memphis. The focus here is on America, since it was here that the origins of the disease were first discovered, but it also affected many European seaports and had a particularly devastating effect on African countries.

In the Americas, the disease was the precursor to historical events of major consequence: The slave revolt in Haiti led to yellow fever infections among the French troops sent to suppress the revolt to such an extent that Napoleon did not believe it possible to build an empire in North America. The disastrous failure of the initial attempts to build the Panama Canal due to the lethal effects of this illness on engineers and workers alike, led to its abandonment and later in the twentieth century to the U.S. building of the canal. The deaths of thousands of American troops in Cuba led to the establishment of a commission of investigation and to the discovery of the mosquito as the vector of the disease [3]. The story of the research to discover the means of transmission of this disease reads like a detective novel in which many of the major players are victims of the disease itself.

10.2 Origins of Yellow Fever

Whether yellow fever originated in Africa, or was transmitted to Africa from tropical America, remains controversial. Evidence from pre-Columbian texts indicates that a disease related to yellow fever occurred in tropical America before 1491. However, it was more likely another hemorrhagic fever—of which there are many—and not yellow fever. A plague ascribed to yellow fever destroyed the Mayans of the Yucatan Peninsula in 1648. However, again there is disagreement as to whether this was yellow fever or some other disease with similar symptoms. The first cases recorded in Africa also date to the seventeenth century [4]. After that date, yellow fever and smallpox replaced the Black Death as the major diseases afflicting the Americas and parts of Europe and Africa. The *Aedes aegypti* mosquito is of African origin, since no related species are found in the Western hemisphere; this strain of mosquito probably arrived in the Americas in slave ships from West Africa. The virus can persist in mosquito eggs for a long time.

Recent phylogenetic studies using RNA sequence analysis of the pre-membrane gene and part of the envelope gene of a collection of yellow fever viruses, isolated over time and from various locations, indicate that the virus evolved in Africa about 1,500 years ago [5]. This study indicated that the virus began in East Africa, moved to West Africa, and eventually crossed the ocean. Four unique "clades" of the virus were discovered, two in Africa (East and West) and two in the Americas. The rate of nucleotide change is as expected for a typical RNA virus. Estimates indicate that the virus was introduced into the Americas about 300–400 years ago, confirming its African origin. It remained a major scourge of the West Indies, both protecting them from invasion by outsiders as well as killing off large segments of the community; between 1853 and 1900 it killed 36,000 people in Havana alone. To this day, yellow fever is endemic to Central and South America and West Africa, and it exists among tree-living monkeys in the forests as "Sylvan (jungle) Yellow Fever," and is spread by different strains of mosquitoes (*Aedes* in Africa, and *Haemagogus* in South America). In urban settings it is spread mostly by *Aedes aegypti.* This species of mosquito is very widespread, and prefers an urban environment, living in pools of water near homes, in abandoned tires, and anywhere that water accumulates. The World Health Organization estimates that there are about 200,000 cases of yellow fever annually, with 30,000 deaths, 90 % of which occur in Africa. West Africa is particularly badly hit due to a breakdown in vaccination and mosquito control. It is now considered a re-emerging virus, with East Africa as well as West Africa being in danger.

10.3 Yellow Fever and Slavery

One of the suggested reasons for the slave trade from Africa to the Americas was the resistance of the African to yellow fever. However, there is no evidence of immunity in the African population. Yellow fever outbreaks have occurred a number of times in West Africa in the last century and well into this century. In 1986, an epidemic broke out among a widely scattered population in Eastern Nigeria; some 10,000 people, out of an estimated regional population of 200,000, suffered "clinical" illness, with more than 5,000 deaths. In 1987 another major epidemic broke out in Western Nigeria: there were 120,000 cases and 24,000 deaths. In 1995, yellow fever re-emerged in epidemic form in Senegal with a case fatality rate of 18.9 %.

Since 2001 there have been large outbreaks in five major African cities. The origin was sylvatic—that is, spread from monkeys or other primates to humans through mosquitoes. Thus in modern times, Africans are as susceptible to yellow fever as other races, although the death rate is less than in the Caucasian population. If Africans were immune a few hundred years ago, at the time of the slave trade, this immunity would have been transmitted to future generations. Perhaps a more reasonable explanation is that yellow fever is endemic to West Africa, and children are exposed to the virus at a young age, with resulting acquired immunity. As in the case of polio, childhood yellow fever may be asymptomatic in the majority of cases, and thus the adult population would be immune. Evidence for this hypothesis is that a high percentage of children in some areas of Africa have antibodies to yellow fever, indicating early exposure to the virus. There is no evidence of genetic selection for resistance, although that may have occurred in the case of other diseases, such as resistance to malaria (sickle-cell hemoglobin). In fact, one can look at the importation of Black Africans as a decision imposed by the economy of the colonial plantations. Sheldon Watts argues that the idea that Africans were immune to yellow fever derives from racial stereotypes, developed in the plantations of North and South America, since there was little yellow fever among plantation workers [6]. If there was exposure to the virus while the child was young, there may have been acquired immunity. Yellow fever was predominantly an urban disease, and not a disease usually found in the countryside [7].

10.4 Yellow Fever in North America

Table 10.1 presents the dates and duration of yellow fever outbreaks since the earliest dates recorded. Individual outbreaks are discussed.

10.4.1 Philadelphia

In the U.S. and Canada, yellow fever struck major cities every few years. This included northern cities as far north as Halifax, and Savannah and New Orleans in the south. Philadelphia suffered outbreaks in 1699, 1741, 1747, 1762 and the 1793 epidemic, which killed about 10 % of the population. The city at that time had a population of about 50,000, and was the functioning seat of government (Washington as the capital was in the planning stages). The outbreak correlated with the influx of large numbers of refugees from the Caribbean fleeing the local epidemic on the islands. It was a hot summer, and there were lots of mosquitoes. The outbreak came to an end with the first frost, which, in hindsight, should have alerted the medical establishment to the fact that the disease correlated with an insect or other living form unable to over winter.

Dr. Benjamin Rush (1746–1813), one of the signatories of the Declaration of Independence and an important physician, urged people to leave the city. For most of his life, he practiced and taught in Philadelphia and helped establish the city as a center for medical training. Rush set up the Philadelphia Dispensary for medical relief for the poor, which was the first institution of its kind in the United States. He also helped found the College of Physicians of Philadelphia.

For an interesting account of the horrors of the yellow fever epidemic of that year and the role of African Americans (and the accusations made against them), see *A Narrative of the Proceedings of the Black People, during the Late Awful Calamity in Philadelphia, in the Year 1793: And a Refutation of some Censures, Thrown upon Them in Some Late Publications* (Philadelphia: Printed for the authors, by William W. Woodward, 1794) by Absalom Jones (1746–1818) and

Table 10.1 Major outbreaks of yellow fever in the New World

Date	Estimated duration in months	Place	Estimated mortality
1672–1673	24	Cuba	2,000
1793	5	Philadelphia	5,000
1798	4	New York	2,100
1802–1803	14	Saint Domingue (Haiti)	29,000–55,000
1853	4	New Orleans	7,800
1853	3	Norfolk, Virginia	3,000
1873	2	Philadelphia	5,000
1873	6	Memphis	2,000
1878	6	New Orleans	4,046
1878	6	Mississippi Valley (Memphis)	13,000

From [8, 9]

Allen Richard (1760–1831) http://www.ncbi.nlm.nih.gov/bookshelf/br.fcgi?book=history&part=Jones.

This booklet describes the attention taken by a group of black men and women in caring for the sick in Philadelphia during the epidemic. Many of them also caught the disease, although it was believed by the white population that they were immune. The booklet was a response to accusations made in a publication that they had "over-charged" for their services. From this publication, we get a picture of the panic and horrors of this epidemic, where people died faster than their bodies could be picked up, the neglect of children, and abandonment of the dying. African-Americans died from the disease at the same rate as Americans of European descent. The following is a description from Samuel Beck, a merchant of the city.

Yellow Fever Attacks Philadelphia, 1793.
Eye Witness to History, www.eyewitnesstohistory.com

"I had scarcely become settled in Philadelphia when in July, 1793, the yellow fever broke out, and, spreading rapidly in August, obliged all the citizens who could remove to seek safety in the country. My father took his family to Bristol on the Delaware, and in the last of August I followed him… I was compelled to return to the city on the 8th of September, and spend the 9th there. Everything looked gloomy, and forty-five deaths were reported for the 9th. And yet it was nothing then to what it became three or four weeks later, when from the first to the twelfth of October one thousand persons died. On the twelfth a smart frost came and checked its ravages.

The horrors of this memorable affliction were extensive and heart rending. Nor were they softened by professional skill. The disorder was in a great measure a stranger to our climate, and was awkwardly treated. Its rapid march, being from ten victims a day in August to one hundred a day in October, terrified the physicians, and led them into contradictory modes of treatment. They, as well as the guardians of the city, were taken by surprise. No hospitals or hospital stores were in readiness to alleviate the sufferings of the poor. For a long time nothing could be done other than to furnish coffins for the dead and men to bury them. At length a large house in the neighborhood was appropriately fitted up for the reception of patients, and a few pre-eminent philanthropists volunteered to superintend it. At the head of them was Stephen Girard, who has since become the richest man in America.

In private families the parents, the children, the domestics lingered and died, frequently without assistance. The wealthy soon fled; the fearless or indifferent remained from choice, the poor from necessity. The inhabitants were reduced thus to one-half their number, yet the malignant action of the disease increased, so that those who were in health one day were buried the next. The burning fever occasioned paroxysms of rage which drove the patient naked from his bed to the street, and in some instances to the river, where he was drowned. Insanity was often the last stage of its horrors.

10.4.2 Memphis

After the abolition of slavery, and the cessation of the slave trade in the North, yellow fever retreated from the northern cities, where it had previously been endemic. However, it continued in the southern states, perhaps with more fury. One of the best reported epidemics occurred in Memphis, Tennessee, in 1878 [7]. That year there had been a very wet winter in West Africa, with an outbreak of yellow fever among the monkeys of the area. Ships sailed to New Orleans, not necessarily with slaves, but with other precious cargoes and mosquitoes. That year, a year of the El Niño cycle, also brought wet weather and hot temperatures to the south—ideal weather for breeding mosquitoes. Memphis was at the peak of its prosperity, being a major port for cotton and a major port for the importation of slaves. However, it was one of the filthiest cities in the Americas, situated in a swamp, lacking sanitation, and a non-functioning city government (http://historic-memphis.com/memphis/yellow-fever/yellow-fever.html). Ships from West Africa would stop at New Orleans and then continue on to Memphis. Although the city of Memphis quarantined any ship suspected of carrying passengers with yellow fever, some did get through by bypassing the quarantine. There had been many outbreaks of yellow fever in the city, but the one of 1,878 took the highest toll on lives (Table 10.1). Whoever could afford it, abandoned the city. Twenty-five thousand people fled, and of the remaining 17,500, over 5,000 died. The city government collapsed, the police force was decimated, and there was a shortage of food.

The only people who went out were the collectors of the dead. With horse and wagon, they shouted 'Bring out your dead!' And then they loaded and took all to the cemetery for a hasty burial. They believed that corpses spread the disease so they tried to get them in the ground as quickly as possible. They also thought the disease was spread by bad air. So with temperatures close to 100, they boarded up their windows and kept fires burning to ward off the outside air. When people died, their clothing and beds were dragged into the streets and burned. Names of the dead were written in ink in leather-bound ledgers. There was an average of 200 deaths per day and corpses were everywhere. Half of those who died were the Irish. Sixteen Catholic priests and 30 sisters died in their heroic battle to tend the sick. (http://historic-memphis.com/memphis/yellow-fever/yellow-fever.html)

Table 10.2 Number of deaths and cases of yellow fever in Memphis

1828	First yellow fever epidemic	650 cases	150 deaths
1855	Second yellow fever epidemic	1,250 cases	220 deaths
1867	Third yellow fever epidemic	2,500 cases	550 deaths
1873	Fourth yellow fever epidemic	5,000 cases	2,000 deaths
1878	Fifth yellow fever epidemic	17,000 cases	5,000 + deaths
1879	Sixth yellow fever epidemic	2,000 cases	600 deaths

The 1878 epidemic—the last great one on American soil—changed the face of Memphis forever. A new and different population replaced those who had left the city, and the city lost its place as the major southern port after the epidemic [7] (Table 10.2).

10.5 The Louisiana Purchase

Perhaps the greatest effect of yellow fever, and an indirect one, was the events leading to the purchase of the Louisiana Territories by the United States. These territories included much more than the present state of Louisiana—they included a large swath of land stretching from New Orleans to Montana, along the Mississippi River (see Fig. 10.1). France had surrendered its North American possessions to Spain at the end of the French and Indian War. New Orleans and Louisiana west of the Mississippi were transferred to Spain in 1762, and French territories east of the Mississippi, including Canada, were ceded to Britain the next year. In 1798 the French retook the island of Saint Domingo (Haiti) from the British. The British had occupied the Saint Domingo in 1793 in part to put down a slave rebellion and to control buccaneers preying on European ships. Between 1793 and 1798, about 70 % of the British troops succumbed to disease, predominantly yellow fever. This was the primary reason for abandoning the island, as well as the victory of Toussaint L'Ouverture and the slave uprising, which abolished slavery in Haiti (More details on the Louisiana Purchase can be found at http://www.monticello.org/site/jefferson/louisiana-purchase).

Napoleon, who came to power in 1799, aimed to restore France's presence on the continent. Napoleon aspired to develop a French empire on North and Central American territory, based on the island of Haiti. He accepted the Louisiana Territory from Spain in exchange for Spain's gaining a province in Italy. However, the port of New Orleans remained in Spanish hands, and during the Spanish period the U.S. used the Mississippi River and New Orleans quite freely. A second slave revolt broke out in Haiti in 1802, led by Toussaint L'Ouverture. France sent a large army of as many as 80,000 men, including a 5,200-man Polish Legion to suppress the revolt. The French also sent a large fleet, one of the largest ever assembled. However, although they were initially victorious over the slave revolt, the French

Fig. 10.1 Area of the Louisiana Purchase

troops began to die in droves from yellow fever. Estimates are that as many as 50,000 French troops died from the epidemic in one year, in 1802. L'Ouverture tied the French troops down in the swampy areas, aware that there was a high incidence of disease in these areas particularly affecting newly arrived French troops. The new leader of the slave revolt, Jean-Jacques Dessalines, declared the island independent and changed its name to Haiti, Taino for "Land of High Mountains," on the defeat of the French forces in 1803 [8].

Napoleon realized that his dream of building a western empire was impractical, since he required large sums of money for his fight against Britain and his wars in Europe. He felt that these colonies could never be settled—because of disease. Thus in 1803 he offered the territories to the United States, and the U.S. government was taken by surprise. Jefferson had authorized a negotiating team that included James Monroe, to negotiate the purchase of New Orleans and adjoining land for $10,000,000. However, the French were willing to part with all of the Louisiana territory for $15,000,000. This was an area about half the size of the U.S.: 827,000 square miles. Although there was some opposition by the Federalists, congress approved the purchase in 1803 (Fig. 10.1).

10.6 The Spanish-American War

In 1880, the Department of the Navy appointed a commission to look into the cause of an outbreak of yellow fever on board the USS Plymouth. This was a case where multiple outbreaks occurred over the span of a year onboard ships. The commission concluded that the disease was caused by a germ, possibly picked up in the ports of call (St. Thomas, Santa Cruz) and brought on board by supplies purchased in these ports. The commission also concluded that although yellow fever was a germ, it was not infectious in the sense that smallpox was contagious, but appeared to be related to specific places and climates. They proposed that it might be related to certain plant growths. Although the relationship between malaria and mosquitoes was studied at this time, there was no attempt to relate the disease to an insect vector. One has to ask why for some 20 years or even earlier, no attempt was made to confirm the theory of Carlos Finlay, that the mosquito was involved in the transmission of the disease. Was this because Congress was unwilling to spend money on research into the cause of the disease until there was a military threat to the U.S.?

During the Spanish-American War of 1898, more soldiers died of disease, including typhoid and yellow fever, than from military action. The U.S. declared war against Spain following an accidental explosion on the USS Maine that killed 268 American sailors. The U.S. government blamed the Spanish government for the explosion and launched the Spanish-American War. Cuba, a Spanish colony, was at that time a major source of disease. Epidemics of yellow fever in Cuba date back to 1649 when approximately 30 % of the population died from the disease. There were subsequent outbreaks every summer. U.S. troops should never have been sent into the tropics, certainly not in the summer. They landed in June of 1898. Although then-Lieutenant Colonel Theodore Roosevelt was aware of the danger, and warned his commanders, "If we are kept here it will in all human possibility mean an appalling disaster, for the surgeons here estimate that over half the army, if kept here during the sickly season, will die." Despite the warning, troops stayed in Cuba until the defeat of the Spaniards in July of that year. Fewer than 400 American soldiers died in combat during the war, but more than 2,000 contracted yellow fever during the campaign. On July 6, in the town of Siboney, the first case of yellow fever among U.S. troops occurred. Yellow fever began to spread, with officers and doctors blaming it on infected buildings in Siboney, the town in which the American troops were garrisoned. In response, General Nelson Miles ordered Siboney evacuated and burned on July 11. An infantry unit of black soldiers, thought to be immune to yellow fever (see above), was brought in to tend to the afflicted. More than a third of their regiment died from yellow fever or malaria. Over 50,000 U.S. troops remained in Cuba after Spain surrendered and although massive efforts were made to clean up the island and to introduce "modern" sanitation, epidemics continued. Finally Dr. George Sternberg, the Surgeon-General of the Army from 1893 until 1902, established a commission in 1898, headed by Dr. Walter Reed, an eminent physician, to investigate the causes of Yellow Fever (http://www.pbs.org/wgbh/amex/fever/peopleevents/e_cuba.html).

10.7 Carlos Finlay

Although Dr. Carlos Finlay of Cuba (1833–1915) had proposed that mosquitoes were the source of the disease 20 years before the commission appointment, this idea was looked upon with great skepticism by the medical profession at the time.

Fig. 10.2 Carlos Juan Finlay (courtesy of the USPHS)

The American research on the causes of malaria, the American physician and founder of the republic Benjamin Rush suspected that the mosquito might be involved in yellow fever as early as 1793.

Carlos Finlay (Fig. 10.2), of Scottish–French parentage, was born in Cuba and was sent to France for schooling. While there he became ill with chorea (a disease characterized by jerking movements) and typhoid fever and returned to Cuba. Unable to get into the medical school of Havana, Finlay attended Jefferson Medical College in Philadelphia. There he learned about the germ theory of disease, and became an advocate for this hypothesis. He was keenly interested in bacteriology and epidemiology, and proposed that water was the source of cholera outbreaks on the island of Cuba. He also suggested that mosquitoes of specific strains were the vectors of yellow fever, an idea he insisted on, based on his experimentation for 20 years before the Reed Commission was established. He published his first paper on yellow fever in 1872 and in 1881 he proposed that the *Culex* mosquito be "hypothetically considered as the agent of transmission of yellow fever." Between 1881 and 1902 he carried out experiments on more than 100 mosquito-bitten individuals, most of whom did not contract yellow fever (although a few did). He was unaware of the extrinsic period required for the virus to replicate in the mosquito (see below) before it became infectious [10], which cast doubt on his theory.

Although most histories credit Finlay with the hypothesis that the *Culex* mosquito was the agent of yellow fever, in a monograph published in 1991, François Delaporte [11] takes issue with this and suggests that Finlay's hypothesis was based on the work of Patrick Manson, who proposed in his paper of 1898 that the filaria responsible for elephantiasis required time outside the mosquito body to develop into the form that invades humans. Delaporte states that Finlay's writings deliberately ignored Manson's work. Finlay must have been aware of this work, which was extensively reported, and in fact Manson proposed that perhaps yellow fever was spread by the mosquito, although exactly how was unknown. Whatever the truth, Finlay helped the Reed Commission with its work by supplying mosquitoes infected with the virus.

Other physicians previously hypothesized that mosquitoes could spread disease. In 1807 John Crawford of Baltimore published a paper stating that mosquitoes were responsible for malaria and yellow fever, among other diseases [1]. Crawford (1746–1813) was a physician who had spent considerable time on board ships, and traveled in the East. He was at odds with most of his medical colleagues and believed that "animacules" and other living creatures caused disease. "Crawford preached investigation and observation, directly applying those principles to the pathogenesis of infestation by parasitic living microorganisms. Nature and his Creator hold the keys, and responsible medical reasoning must observe and interpret accordingly. Well ahead of his time with his thinking, his theories unfortunately would not receive proper acceptance until well after his death" (http://archive.hshsl.umaryland.edu/bitstream/10713/11/1/crawford_bio.pdf).

10.8 The United States Army Yellow Fever Commission

All documents relating to the commission, including personal letters sent home from the members of the commission, can be found in the University of Virginia archival collection http://yellowfever.lib.virginia.edu/reed/collection.html, (Figs. 10.3 and 10.4).

"Here I have been sitting reading that most wonderful book—*La Roche on Yellow Fever*—written in 1853. Forty-seven years later it has been permitted to me and my assistants to lift the impenetrable veil that has surrounded the causation of this most dreadful pest of humanity and to put it on a rational and scientific basis." Walter Reed wrote this to his wife on December 31, 1900. The commission consisted of James Carroll, Aristides Agramonte, and Jesse Lazear and, of course, Walter Reed, and assistance was received from Dr. Carlos Finlay. Walter Reed was born in Virginia in 1851, received his MD at age 17 from UVA, and then moved to Bellevue Hospital in New York, today affiliated with New York University. He was appointed an official of the New York Board of Health. From 1874 to 1890 he served in the U.S. Army as a physician, with a sabbatical at Johns Hopkins Medical School, where he trained in bacteriology. In 1893 he was appointed professor of bacteriology at the Army Medical School. Apart from his work with yellow fever, he established the importance of human–human transmission of typhoid fever.

Aristides Agramonte was born in 1868 in Camaguey, Cuba was who received his MD from Columbia University and worked for the New York Health Department. He was thought to be immune to yellow fever since childhood and thus performed most of the autopsies in Cuba.

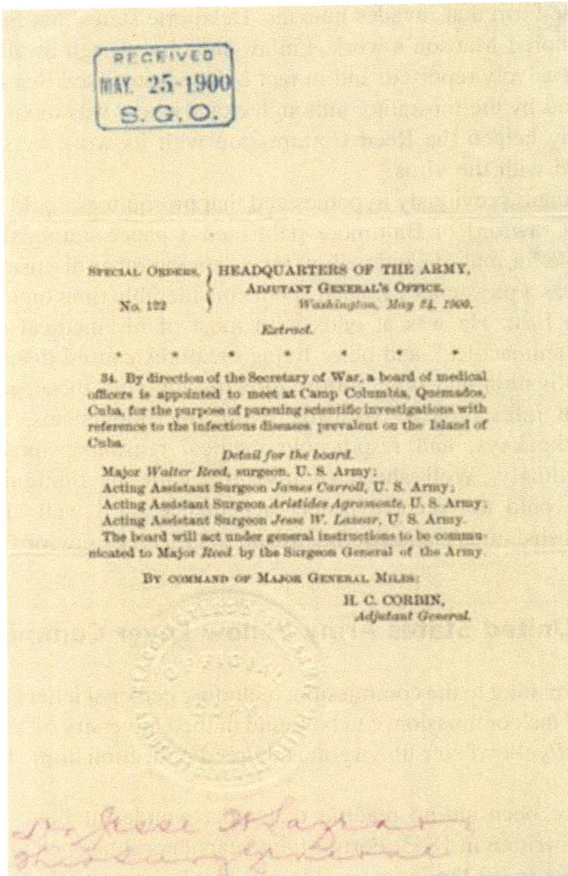

RECEIVED
MAY. 25 1900
S.G.O.

SPECIAL ORDERS, No. 122 } HEADQUARTERS OF THE ARMY, ADJUTANT GENERAL'S OFFICE, *Washington, May 24, 1900.*

Extract.

* * * * * * *

34. By direction of the Secretary of War, a board of medical officers is appointed to meet at Camp Columbia, Quemados, Cuba, for the purpose of pursuing scientific investigations with reference to the infectious diseases prevalent on the Island of Cuba.

Detail for the board.

Major *Walter Reed*, surgeon, U. S. Army;
Acting Assistant Surgeon *James Carroll*, U. S. Army;
Acting Assistant Surgeon *Aristides Agramonte*, U. S. Army;
Acting Assistant Surgeon *Jesse W. Lazear*, U. S. Army.

The board will act under general instructions to be communicated to Major *Reed* by the Surgeon General of the Army.

BY COMMAND OF MAJOR GENERAL MILES:

H. C. CORBIN,
Adjutant General.

Fig. 10.3 A copy of the military orders establishing the Reed Commission (courtesy of the Historical Collections and Services, Claude Moore Health Sciences Library, University of Virginia)

Fig. 10.4 Members of the Reed Commission (courtesy of the Historical Collections and Services, Claude Moore Health Sciences Library, University of Virginia)

James Carroll was born in England on June 5, 1854. He was educated at Albion House Academy, Woolwich, England, graduating in 1869. He graduated from the Medical Department, University of Maryland in 1891, and from a post-graduate course in pathology at Johns Hopkins Hospital in 1891.

Jesse Lazear (1866–1900) was born in Baltimore joined the Yellow Fever Commission in 1900. Previously, he had worked at Johns Hopkins Hospital in Baltimore, and had become familiar and interested in the work of Sir Ronald Ross on the role of the *anopheles* mosquito in the spread of malaria. Prior to the mosquito hypothesis, a great deal of effort was placed on trying to prove or disprove a suggestion from an Italian scientist, Giuseppi Sanarelli, that *bacillus icteroides* was the causative agent of the disease. However, the team proved conclusively that this bacterium was secondary to the infection and was not related to yellow fever. The bacterium could be isolated from many patients and thus was a candidate for the disease. It produced a toxin that gave rise to symptoms of yellow fever in many animal models. Agramonte showed that it was not present in all yellow fever victims and could be isolated from uninfected individuals and was thus a "contaminating" agent (Fig. 10.5).

A second bacterium—a spirochete—was also claimed to be the cause of yellow fever, It could be isolated from mosquitoes, and caused symptoms similar to yellow fever in guinea pigs and other animals. However, when pursued further, it was found in patients with a disease similar to yellow fever, but not the true condition. To quote from Reed,

> In a word, then, the careful bacteriological study which the commission had made in cases of yellow fever had given no indications as to the presence of the specific agent of this disease. The same may be said concerning the result of numerous microscopic examinations of fresh and stained specimens of blood which we had in the meanwhile studied with the view of finding possibly some intracellular or extracellular body.

Reed then proposed human experimentation on a large scale in a letter he wrote to George Miller Sternberg (Fig. 10.6):

> "There is plenty of material in Havana, with every probability of its rapid increase—our last case here died on Monday—we will therefore expect to transfer our field of work to Military Hospital No. 1—Lazaer, Carroll and Agramonte are all deeply interested in the problem, Personally, I feel that only can experimentation on human beings serve to clear the field for further effective work—with one or two points cleared up, we could then work to so much better advantage.

> The final determination to investigate the mosquito theory was arrived at during an informal meeting of the Board (Dr. Agramonte being absent) at Columbia Barracks on the evening before Dr. Reed's departure for the United States early in August 1900... The proposal to submit ourselves to inoculation was made by myself, twice, before it was brought up by Dr. Reed, for the first time, at the meeting above mentioned, where it was finally decided upon by actual vote.
>
> (letter from James Carroll to Robert M. O'Reilly, August 29, 1906)

Reed, Caroll & Agromonte & Lazear disprove
B. icteroides as cause of Yellow Fever

TABLE II.

No. of Case.	Day of Disease.	Time of Autopsy.	Source of Culture.	B. Icteroides.
1	7th.	2 hours after death.	Blood, liver, spleen, kidney.	Neg'tive
2	6th.	13 hours after death.	Blood, liver, spleen, kidney.	"
3	4th.	8 hours after death.	Blood, liver, spleen, kidney.	"
4	8th.	4 hours after death.	Abdominal cavity, blood, liver, spleen, kidney, bile. duodenum.	"
5	4th.	4 hours after death.	Blood, liver, spleen, kidney, bile, duodenum.	"
6	6th.	6½ hours after death.	Abdominal cavity, blood, pericardial fluid, lung, spleen, kidney, liver, bile, duodenum.	"
7	6th.	50 minutes after death.	Blood, lung, liver, spleen, kidney, bile, jejunum.	"
8	6th.	½-hour after death.	Blood, lung, liver, spleen, kidney, urine, small intestine.	"
9	4th.	2 hours after death.	Liver, spleen, small intestine.	"
10	5th.	7 hours after death.	Liver, kidney, spleen, small intestine.	"
11	3rd.	½-hour after death.	Liver, kidney, spleen.	"

Our failure to isolate B. icteroides in these eleven autopsies of yellow fever patients was a result which we had not anticipated. One

Proceedings 28th Annual Meeting of the American Public HealthAssociation, Indianapolis IN, October 22-26, 1900

Fig. 10.5 Results of the tests for *B. icteroides* as the cause of yellow fever

There follows a long correspondence between Carrol, Lazear, their respective wives, and Walter Reed. Some of this is reproduced below (Fig. 10.7).

- Lazear experiments on himself and eight volunteers—none become ill as mosquitoes have not undergone extrinsic incubation period.
- 8/27/1900 Lazear places a mosquito on Carroll's arm that had fed 12 days earlier on a yellow fever patient. 8/29 Carroll develops symptoms.
- However Carroll travels off post to Havana before developing yellow fever, so that when he develops yellow fever it is not conclusive that it was from the mosquito. He recovers by 9/7/1900.

Both Carroll and Lazear allowed themselves to be bitten by mosquitoes to prove their point: that the insect carried the virus. To justify the experimentation on others, Carroll voluntarily submitted to the bite of contaminated mosquitoes that were previously linked to three well-documented cases of yellow fever. Within four days he became ill and suffered a severe attack of the disease. He recovered, but with significant damage to his heart. He died in Washington, D.C., in 1907—probably as a result of the experiment. Jesse Lazear became familiar with the work of Finlay and received mosquitoes from him and allowed them to

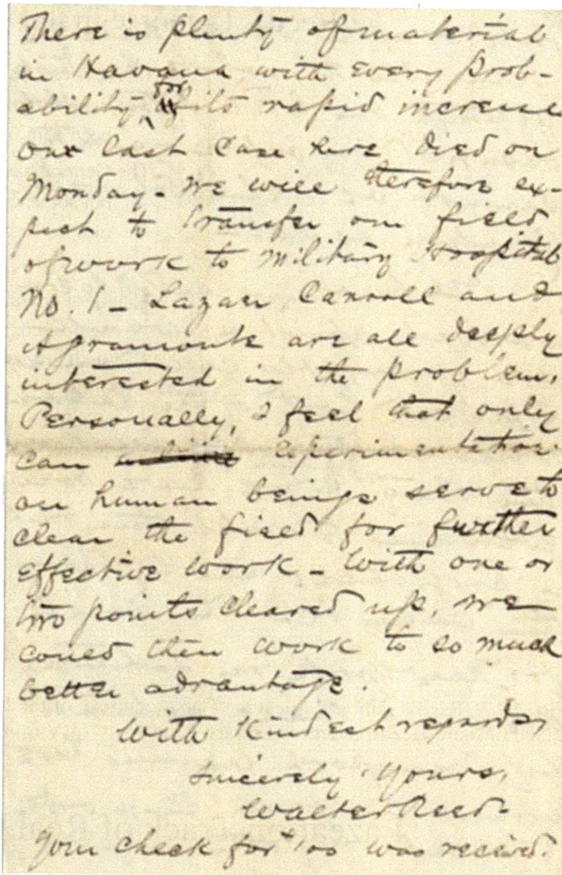

There is plenty of material
in Havana with every prob-
ability of its rapid increase
Our last case here died on
Monday. We will therefore ex-
pect to transfer our field
of work to Military Hospital
No. 1 - Lazear, Carroll and
Agramonte are all deeply
interested in the problem,
Personally, I feel that only
can experimentation
on human beings serve to
clear the field for further
effective work - With one or
two points cleared up, we
could then work to so much
better advantage.
With kindest regards,
Sincerely Yours,
Walter Reed.
Your check for $100 was received.

Fig. 10.6 July 1900 The letter and its contents are presented below (courtesy of the Historical Collections and Services, Claude Moore Health Sciences Library, University of Virginia)

gorge themselves on yellow fever victims. It is suggested in his notebook that he allowed himself to be bitten by such a mosquito, and became ill with a Severe and sudden case of yellow fever from which he never recovered. His death was one more proof that mosquitoes carried this deadly disease (Fig. 10.8).

Progress of Lazear's infection with Yellow Fever

- September 13, 1900. Lazear allows himself to be bitten.
- September 18, 1900. Develops fever.
- September 22, 1900. Black vomit.
- September 25, 1900. Lazear dies.
- Carrol wrote "I will never forget the expression of alarm in his eyes when I last saw him alive on the 3rd or 4th day of his illness."

Letter from Reed to Mrs. Emilie Reed: "Dr. Lazear contracted the disease at the Yellow Fever [Hospital in Havana] by letting an infected mosquito bite him—He saw the insect on his hand & deliberately let it get its fill of blood in order to test our theory—Five days later

Reed's Remorse in the US

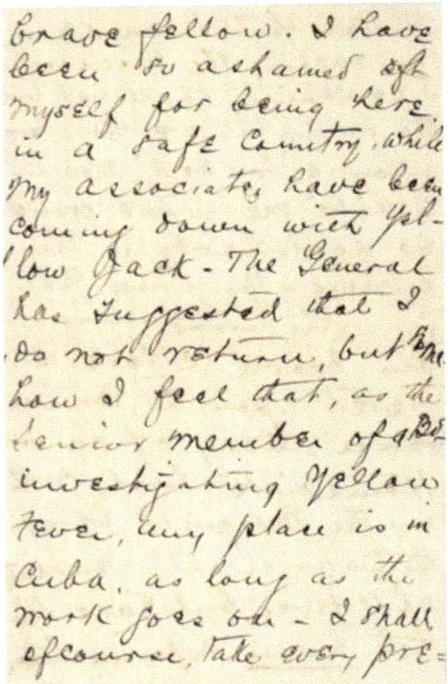
brave fellow. I have been so ashamed of myself for being here in a safe Country, while my associates have been coming down with yellow Jack - The General has suggested that I do not return, but some how I feel that, as the Senior member of a Bd- investigating Yellow Fever, my place is in Cuba, as long as the work goes on - I shall of course, take every pre=

I have been so ashamed of myself for being here in a safe country, while my associates have been coming down with yellow Jack

The General has suggested that I do not return, but somehow I feel that, as the Senior member of a Bd- investigating yellow Fever, my place is in Cuba, as long as the work goes on-

I shall, of course, take every precaution that I can against contracting the disease, and I certainly shall not, with the facts that we now have allow a "loaded" mosquito to bite me! That would be fool-hardy in the extreme- (Reed to Keane, 9/25/1900)

Fig. 10.7 Reed's remorse at leaving Cuba (courtesy of the Historical Collections and Services, Claude Moore Health Sciences Library, University of Virginia)

Lazear on Track of Real Germ

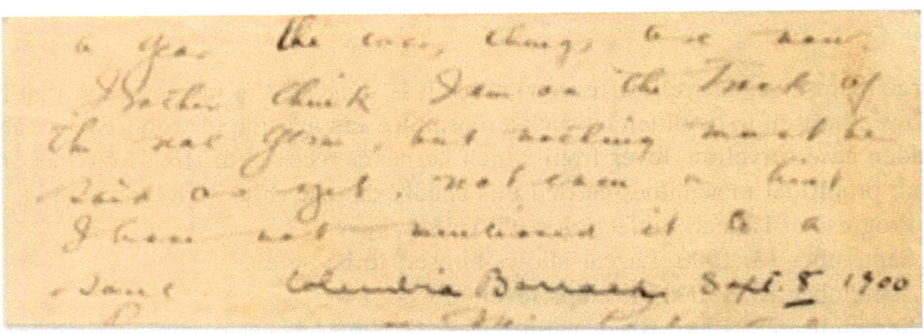
I rather think I am on the Track of the real germ, but nothing must be said as yet, not even a hint. I have not mentioned it to a soul. Columbia Barracks Sept. 8 1900

I rather think I am on the track of the real germ, but nothing must be said as yet, not same a hint I have not mentioned it to a soul.

Columbia Barracks, Sept. 8 1900 Letter from Dr. Jesse Lazear to his wife.

Fig. 10.8 Lazear's note to his wife (courtesy of the Historical Collections and Services, Claude Moore Health Sciences Library, University of Virginia)

he had his chill, followed by high fever—His case was a very severe one from the beginning, his death occurring on the 6th day there after—He was a splendid, brave fellow & I lament his loss more than words can tell; but his death was not in vain—His name will live in the history of those who have benefited humanity."

The emphasis in the commission then changed from what was the cause of the disease to what was the vector, since experiments showed that the disease could not be spread by fomites (clothing) or by personal contacts. Both the mosquitoes used to bite Lazear and a volunteer soldier, William Dove, and Dr. Carroll, had obtained a meal of blood 12 days before the bite. The fact that there was an extrinsic incubation period in the mosquito was an accidental discovery, but an important one, since it explained why most of Finlay's experiments did not work. Following the death of Lazear and the positive results with Carroll and Dove, a camp with volunteers was established. That the disease was not of bacterial origin was also proven by the observation that it was carried in "infected blood" that had passed through a Berkefeld filter, known to retain bacterium.

Reed, who in general received most of the acclaim for proving that yellow fever was transmitted by mosquitoes, acknowledges in his letters that most of the work was done by Finlay in isolation before the days of the commission. When the commission decided to test the mosquito theory, Finlay provided the mosquitoes, and with the commission's first scientifically valid success, Reed wrote triumphantly, "The full run of experiments at Camp Lazear vindicated Finlay's two-decade-long struggle. In the light of this success," Reed acknowledged, "it was Finlay's theory, and he deserves much for having suggested it."

Camp Lazear was established, and volunteers gave informed consent to be bitten by mosquitoes. This is probably the first case of "informed consent" of human subjects in medical experimentation. These were mostly army personnel and new Spanish immigrants to Cuba. Some were immune and others non-immune to yellow fever. Reed's idea was to establish conditions suspected up to now of being involved in the disease transmission (unsanitary conditions, etc., but not mosquitoes), and separate facilities that were clean but allowed access to mosquitoes. A description of the camp and informed consent by the volunteers can be found in the collection of papers at the Philip S. Hench Walter Reed Yellow Fever Collection at the University of Virginia (http://yellowfever.lib.virginia.edu/reed/story.html).

The results of these experiments left no doubt that mosquitoes were the vectors of yellow fever. As for the causative agent, the question was left unsolved until much later, when it was shown that it was a virus. It is worth reading Reed's reports on the experiments carried out to prove the mosquito link. These papers describe in detail the conditions taken to isolate some men from mosquitoes, yet expose them to yellow fever-contaminated blankets, towels, and even clothing. In one of the set-ups, the disease could only be transmitted by mosquitoes, or by direct injection of blood from a patient with the disease. This proved without a doubt that this was a blood-borne disease, transmitted by mosquitoes. The virus had to incubate and replicate in the mosquito, and thus had a unique life cycle [12–14] (An excellent description of the work of the commission can also be found in [1]).

Once the mosquito was discovered as the source of yellow fever, Major William C. Gorgas, who had been appointed at the time of the Spanish-American War to oversee sanitation in Havana, set out to eradicate the *Aedes aegypti* mosquito from Havana. Gorgas's campaign was one of the most successful sanitation campaigns in history. He destroyed all sources of stagnant water, covered barrels of drinking water, removed vases, etc., from houses, and every possible breeding ground for mosquito larvae was removed. Gorgas and his men went from house to house destroying possible breeding grounds, and putting oil on top of water sources to suffocate the larvae. Within a few months, yellow fever was almost completely wiped out in Havana.

10.9 Panama Canal

The economic importance of the Panama Canal cannot be underestimated. It is today the main shipping route between the Pacific ports and European and North American East Coast ports. The idea of building a canal across the narrow Isthmus of Panama is as old as the discovery of America. However, actual attempts to build this canal were not made until 1881 when the French decided to build the canal, under the leadership of Ferdinand De Lesseps, at that time 74 years old. De Lesseps had been instrumental in the building of the Suez Canal, and felt that it would not be difficult to build a canal across Panama. This was a huge undertaking, and started off with great pomp. It was called "La Grande Entreprise" and was one of the largest financial ventures ever attempted at the time, launched with a great deal of French pride and nationalism. About 200 engineers and 800 laborers began building the infrastructure necessary for a project of this size. By the end of 1881, 2,000 men were employed in building the canal. In 1882, approximately 400 deaths were reported from disease and the next year this number climbed to 1,500, mostly from yellow fever. Over 20,000 workers, mostly French, died of malaria or yellow fever, making it difficult to recruit engineers and other workers. After a few years, the Canal Company collapsed and De Lesseps and his son returned to France to face trial for embezzlement of funds; they were found guilty, but did not serve time in jail.

A U.S.-owned company built the canal on territory leased from Panama. Building began in 1904, after the area had been cleaned of mosquitoes (now known to spread both yellow fever and malaria), and was successfully completed in 1914. The hero of this venture was again Colonel George William C. Gorgas, who carried out a policy of eradicating mosquitoes in the isthmus area of Panama, at a time when the mosquito theory of transmission of the disease was not readily accepted. Although his approach to dealing with yellow fever faced much skepticism, in 1908 he was appointed president of the American Medical Association, and then named surgeon-general of the U.S. Army, a position he held during the 1918 influenza epidemic. He was still in charge when the first ships sailed through the canal in 1914. He suddenly died in 1920 after having been knighted in England, and he was given a hero's state funeral in Washington and London.

10.10 The Virus

Yellow fever virus belongs to a class of viruses known as arboviruses, i.e., "ARthropod BOrne" viruses, since it is insect-transmitted. Within this general classification it belongs to the, flavivirus genus. It is an icosahedral-enveloped virus of about 50 nm diameters containing single-stranded RNA. It is related to the West Nile, St. Louis encephalitis, and Japanese encephalitis viruses (Fig. 10.9).

As discussed above, the infection is often innocuous in the young, who then acquire immunity, but can also lead to internal bleeding, black vomiting, high fever, and death in the susceptible individual.

The epidemiology of the yellow fever virus is complicated. It appears that there are three routes of transmission:

1. Jungle yellow fever (Sylvatic), where the virus is spread from a non-human primate to another non-human primate by mosquitoes that live on the canopy of the jungle;
2. Savannah, which is at the edge of the jungle and the virus is spread between primates and humans; and
3. Urban yellow fever, where there is human-to-human transmission through the mosquito vector (Fig. 10.10). One puzzling fact is that the virus does not appear in Asia, even in the tropics, which suggests some possible competing immunity, or the lack of a suitable vector in that environment. Even today, in the era of air travel and population mixing, there is no evidence of yellow fever in the Orient.

Although the virus can be transmitted by a number of strains of mosquitoes, the primary vector is *Aedes aegypti*, which prefers humans, and lives close to human habitats in vessels containing stagnant water (Fig. 10.11). The mosquito can be found in old tires, and in uncovered water sources; it can live for a long time, up to 70–160 days and has less than a 300 m flight range. Unlike the *anopheles* that spreads malaria, this insect bites during the day rather than at night, and lives in different habitats. It is the female who imbibes the blood, incubates the virus and transmits the disease. There is an incubation period of 10–12 days in the insect, and after being digested by the insect, the virus remains in the gut where it replicates and is then transmitted to the salivary gland, a time known as the extrinsic period. Many insects transmit the virus to the ovary (transovarial transmission), but this does not seem common in *Aedes aegypti*, although it does occur in other mosquitoes that transmit dengue, and in tick viruses.

An interesting social factor is related to the mosquito involved in the transfer of yellow fever. In the outbreaks in Philadelphia and other cities, it was noted that the wealthy and those who lived in sanitary conditions became sick at the same rate as the poor, who lived in overcrowded and unhygienic conditions. Thus the mosquito was able to breed in water closets outside of wealthy homes or in barrels collecting water for household use; mosquito eradication was essential for the elimination of the disease.

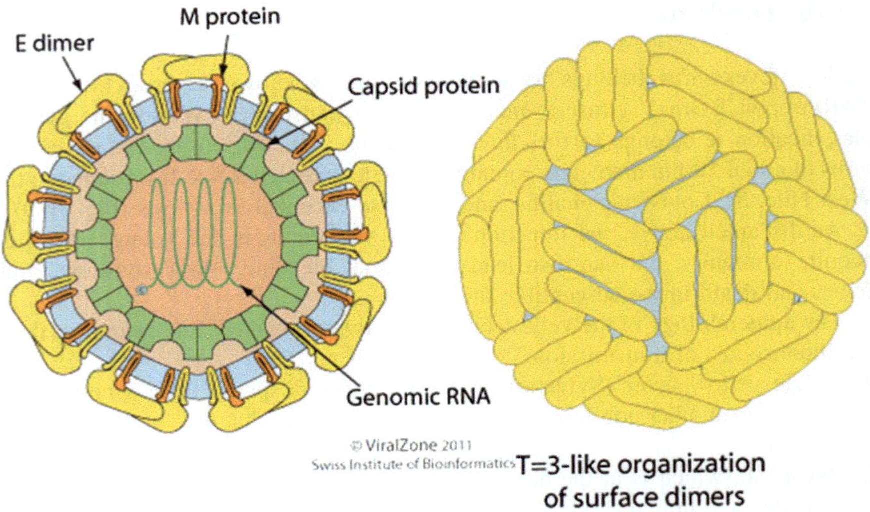

Fig. 10.9 Yellow fever structure (ViralZone: www.expasy.org/viralzone, Swiss Institute of Bioinformatics)

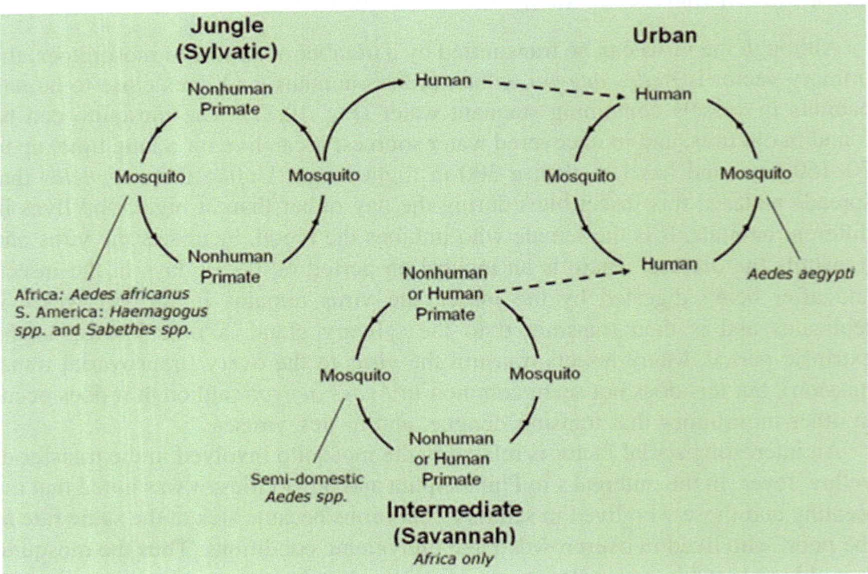

Fig. 10.10 Cycle of yellow fever virus and hosts (courtesy of the CDC)

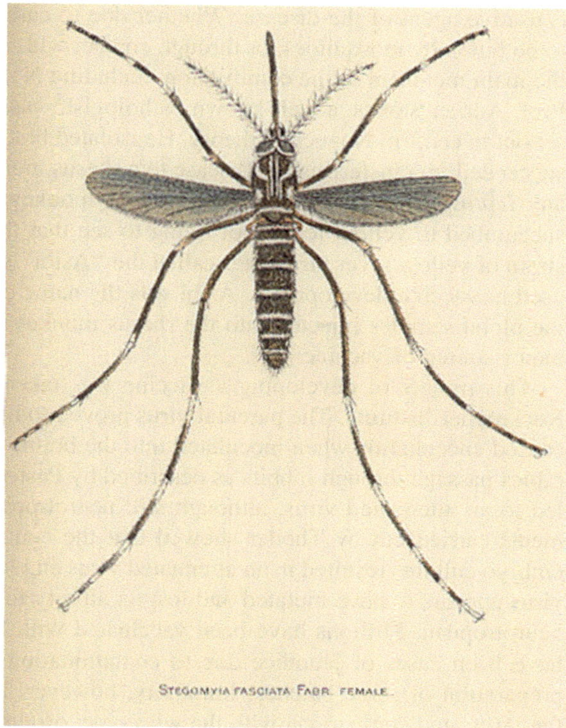

Fig. 10.11 Aedes aegypti

10.11 Vaccine Development

Despite the identification of a virus as the agent of yellow fever, initial attempts at production of a vaccine were based on the finding of a bacterial spirochete (mobile spiral-like organism) in the livers of many victims of the disease—which led to the hypothesis of a bacterial origin for the disease. Hideyo Noguchi, a scientist at the Rockefeller Institute, isolated this organism and named the novel spirochete "*Leptospira icteroides*." Although previously proven by the Reed Commission not to be the causative agent of yellow fever, he pursued his idea and developed a vaccine against it. Noguchi published successful results for over 7,000 doses of the vaccine. However, others could not repeat his experiments and by 1926 the use of this vaccine slowly disappeared, but this was not the end of the bacterial origin of yellow fever. The second West African Yellow Fever Commission was formed in 1925 (a prior commission in 1920 had accomplished little) to investigate whether South American yellow fever was similar to that of African yellow fever, and to explore the root of the disease. The commission was made up primarily of bacteriologists who had previously studied *Leptospira*, and they were also joined by Noguchi, who still had not abandoned the idea that *Leptospira icteroidase* was the causative agent of the disease. Whether due to carelessness, allowing themselves to be bitten by mosquitoes, or through contact with infected blood of monkeys, all the main members of the commission, including Noguchi, died from yellow fever. Prof. Adrian Stokes, a well-known pathologist, was the head of the group, and he set out to confirm Noguchi's theory. He isolated blood from an infected person and succeeded in transferring the disease to a rhesus monkey. He was unable to isolate any *leptospira* from the large number of monkeys he infected and he himself succumbed to yellow fever, not living to see that the infected monkeys carried a strain of yellow fever virus later called the "Asibi" strain that would eventually be used in vaccine development. Asibi was the name of the individual who supplied the blood samples injected into the rhesus monkey, and this virus later became a major source of vaccine [15].

This project of developing a vaccine was taken over by Max Theiler of the Rockefeller Institute. The parental virus proved to be fatal to rhesus monkeys, and caused encephalitis when inoculated into the brains of mice. But, as in the case of rabies passage through rabbits as described by Pasteur, passage through the mouse led to an attenuated virus, although still neurotropic in man. Subsequent experiments carried out by Theiler showed that the continuous passage through chick embryo cultures resulted in an attenuated virus and loss of neurotropic effects. This virus appears to have mutated and lost its affinity for internal organs as well as its neurotropism. Millions have been vaccinated with little or no side effects. There have been cases of jaundice due to contamination of human serum used in the preparation of some batches. Immunity, however, is long-lasting. Sequencing of the virus, and comparison with the wild type, original Asibi strain, have shown as many as 69 substitutions in nucleic acid between the parental Asibi strain and the strain used for the vaccine. Many of these mutations are in the envelope of the virus and may have altered its neurotropism [15, 16]. Since this is an RNA virus, this high rate of mutation is not unexpected, and the recognition that the virus continued to mutate at a high frequency led to the development of seed lots that were frozen and used as required (For more details, see Chap. 19 on vaccine development).

10.12 Other Hemorrhagic Fever Viruses

There are many other viruses that cause symptoms similar to yellow fever and are transmitted by insects. In all cases, the symptoms are similar, although not as severe as in yellow fever. The virus attacks the blood vessel walls, causing internal bleeding. This is accompanied by high fever, nausea, headaches, and jaundice. Among these viruses are dengue, equine encephalitis, West Nile virus, Ebola, Lassa fever and Colorado tick fever. Many of these viruses have an avian or primate reservoir and are transmitted by mosquitoes, ticks, sand flies or gnats or other insects. Not all are flaviviruses, but belong to a different genus. Vaccines are not currently available for these infections.

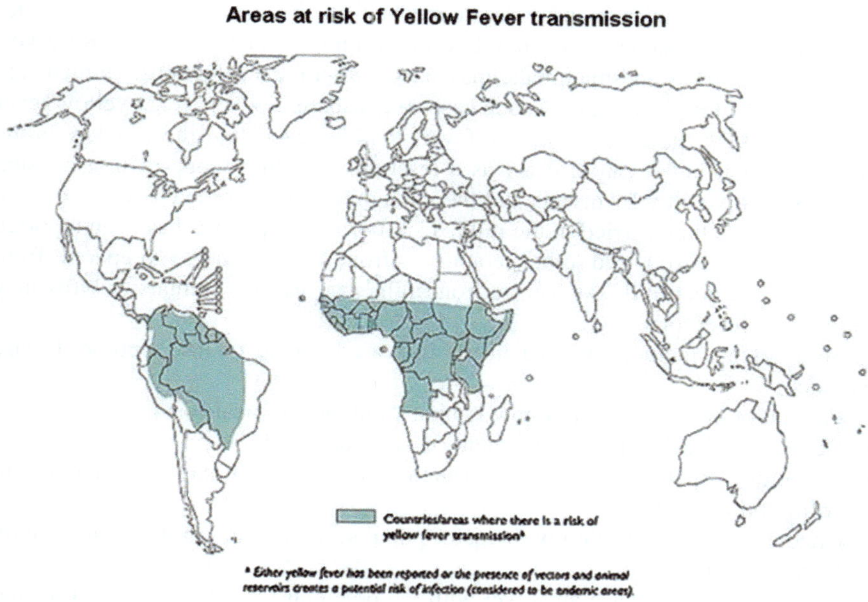

Fig. 10.12 Endemic areas of yellow fever in 2014

10.13 Yellow Fever Today (2014)

There are an estimated 200,000 cases of yellow fever, causing 30,000 deaths worldwide every year, according to the WHO, and the virus is still endemic in tropical areas of Latin America and Africa. There has been an increase in the incidence the last few years due to declining numbers of people being vaccinated, movement of populations into "virgin" areas, deforestation, urbanization. and possibly climate change. Yellow fever can be totally prevented by vaccination, and one inoculation appears to last for life. In order to stamp out the epidemic, 80 % of the population of any one area would have to be vaccinated.

The use of insecticides can eliminate mosquitoes but the campaigns against the mosquitoes have lapsed in many countries due to government inefficiency. The WHO recommends total vaccination of populations above nine months of age, as well as maintaining surveillance to identify local outbreaks. There appears to be enough vaccine for everyone in endemic regions. Between 2007 and 2012, 12 countries completed preventative yellow fever vaccination campaigns: Benin, Burkina Faso, Cameroon, Central African Republic, Côte d'Ivoire, Ghana, Guinea, Liberia, Mali, Senegal, Sierra Leone and Togo (WHO) (Fig. 10.12).

Arthropod transmission provides a mechanism for a virus to cross species barriers, since the same insect may bite birds, reptiles, and mammals. Reservoir hosts, such as birds and mammals, may have sub-clinical infections. Humans are often "accidental" hosts, and in such cases the disease is severe (e.g., yellow fever and dengue). In the case of yellow fever, it is believed that the monkey is the primary target, but it can also act as a reservoir. Another mode of transmission does not involve bites by insects, but the virus may be transmitted to a cow or goat by a tick and is then carried in the milk. Tick-borne encephalitis has been reported in man (in Austria) and other countries from eating cheese prepared from unpasteurized goat milk. As we see from influenza, various animals or birds may act as viral reservoirs.

Some of the factors involved in the increased spread of insect-borne disease include:

- Population movements into new habitats, notably tropical forests
- Deforestation with the development of new farmlands
- Irrigation, especially primitive irrigation systems that pay no attention to arthropod control
- Uncontrolled urbanization with vector population breeding in the accumulation of water and sewage
- Increased long-distance air travel with the potential for carriage of vectors and persons incubating virus. This is the greatest danger today
- New routings of long-distance bird migrations brought about by man-made water projects.

References

1. Oldstone MBA (2010). *Viruses, plagues and history*, 2nd edn. Oxford: Oxford University Press.
2. Watts, S. (2001). Yellow fever immunities in West Africa and the Americas in the age of slavery and beyond: A reappraisal. *Journal of Social History, 34*(4), 955–967.
3. Bray, R.S. (1996). *The impact of disease on history.*Cambridge: James Clarke Co.
4. Carter, H. R. (1931). *Yellow fever, an epidemiological and historical study of its place of origin*. Baltimore: Williams and Wilkins Company.
5. Lepiniec, L., Dalgarno, L., Huong, V. T., Monath, T. P., Digoutte, J. P., & Deubel, V. (1994). Geographic distribution and evolution of yellow fever viruses based on direct sequencing of genomic cDNA fragments. *Journal of General Virology, 75*(Pt 2), 417–423.
6. Watts, S. (1997). Epidemics and history: Disease, power and imperialism. New Haven: Yale University Press.
7. Crosby, M. C. (2006). *The American plague*. New York: Berkeley Publishing Group.
8. Marr, J. S., & Cathey, J. T. (2013). The 1802 Saint-Domingue yellow fever epidemic and the Louisiana Purchase. *Journal of Public Health Management and Practice, 19*(1), 77–82.
9. Kohn, G.C. (2007). Encyclopedia of plagues and pestilence—from ancient times to the present. In G.C. Kohn (Ed.) New York: Facts On File.
10. Chaves-Carballo, E. (2005). Carlos Finlay and yellow fever: Triumph over adversity. *Military Medicine, 170*(10), 881–885.
11. Delaporte, F. (1991). *History of yellow fever. An essay on the birth of tropical medicine*. Cambridge: MIT press.

12. Reed, W. (1902). Recent researches concerning the etiology, propagation, and prevention of yellow fever, by the United States Army Commission. *The Journal of Hygiene, 2*(2), 101–119.
13. Reed, W., & Carroll, J. (1901). The prevention of yellow fever. *Public Health Papers and Reports, 27*, 113–129.
14. Reed, W., Carroll, J., Agramonte, A., & Lazear, J. W. (1900). The etiology of yellow fever—a preliminary note. *Public Health Papers and Reports, 26*, 37–53.
15. Frierson, J. G. (2010). The yellow fever vaccine: A history. *Yale J Biol Med, 83*(2), 77–85.
16. Hahn, C. S., Dalrymple, J. M., Strauss, J. H., & Rice, C. M. (1987). Comparison of the virulent Asibi strain of yellow fever virus with the 17D vaccine strain derived from it. *Proceedings of the National Academy of Sciences of the United States of America, 84*(7), 2019–2023.

11 Influenza

Abstract

Every few years the world faces the risk of a new influenza pandemic. Novel strains of influenza are found in chickens, ducks, or other aquatic birds and result in speculation and occasionally panic that a new influenza epidemic is imminent. These are unique recombinant classes of influenza, an "old" virus that undergoes rapid shifts in its ability to spread from birds or animals to humans, and then from human to human. Influenza changes its antigenicity by two methods: antigenic drift, which is the result of a mutation in a viral gene, usually the hemagglutinin or neuraminidase, and antigenic shift, which results from the reassortment of the eight segments of the influenza genome. More than 50 million people died as a result of the 1918 pandemic, a novel virus infecting humans (H1N1 influenza). The 1918 pandemic flu virus has been reconstructed from samples stored in hospitals and dug up from permafrost. This virus is unique and appears to have been a novel virus not related to swine flu but possibly to an avian source. Other pandemics, such as Asian flu or Hong Kong flu, have followed periodically. The last few years have seen massive culling of poultry due to outbreaks of other recombinant influenza viruses such as H5N1 and H7N9. Vaccines against influenza are rapidly changing by growing the virus in cell culture, and the development of recombinant techniques.

11.1 The Virus

Every few years the world faces the prospect of a new influenza pandemic. Novel strains of influenza are found in domestic poultry, ducks, or other aquatic birds and result in speculation and occasional panic that a new influenza epidemic is imminent. These are novel recombinant classes of influenza A virus that undergo rapid shifts in its ability to spread from birds to man and then from man to man.

M. W. Taylor, *Viruses and Man: A History of Interactions*,
DOI: 10.1007/978-3-319-07758-1_11,

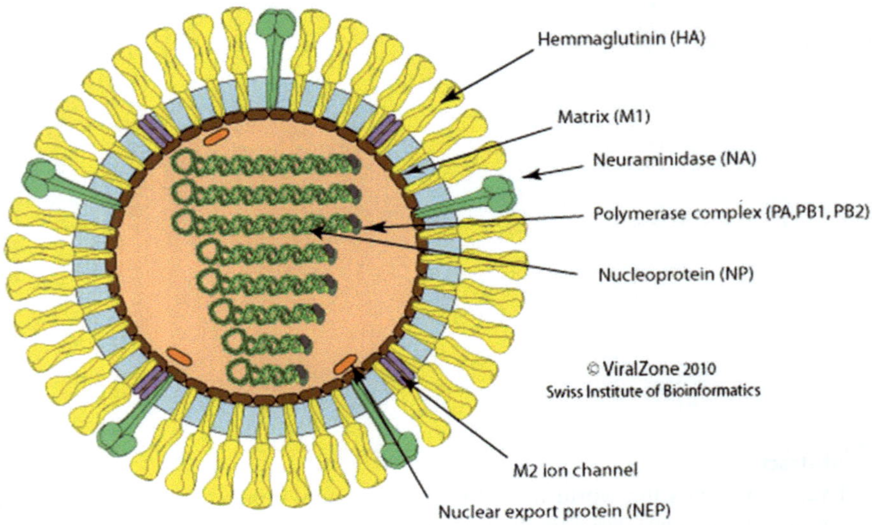

Fig. 11.1 Schematic of influenza virus. *Note* eight genomic segments and spikes with neuraminidase and hemagglutinin activity (ViralZone, SIB Swiss Institute of Bioinformatics)

The influenza virus belongs to the orthomyxovirus family, of which there are three main types: influenza A, the one responsible for pandemics; influenza B, which causes flu-like symptoms and is connected to local epidemics; and influenza C, which is the least pathogenic. Whereas influenza B and C are confined to humans, influenza A infects a wide variety of animals, including birds, swine, and horses. Aquatic birds serve as the major reservoir for all known subtypes of influenza A; the virus replicates in the bird's gut, in many cases without causing symptoms.

Influenza is a negative-stranded RNA virus with eight unique segments of RNA, each containing information for one or more viral genes (Fig. 11.1). Each segment of RNA can be considered a mini-chromosome. The expression "negative-stranded" denotes viruses in which the RNA is copied into a complementary strand that is then used for the manufacturing of viral proteins. The virus is enveloped and its shape can be pleomorphic. Virions are 80–120 nm in diameter.

The major genes involved in the immune response to influenza are the hemagglutinin (H) and neuraminidase (N) genes. The combination of these two genes gives rise to the different genotypes of the virus. The hemagglutinin is involved in the attachment of the virus to the cell, and the neuraminidase in the release of the virus. Modifications in the polymerase gene may affect the pathogenicity of the virus.

11.2 Antigenic Drift and Antigenic Shift

As is common with all RNA viruses, influenza undergoes a high frequency of single-base mutations, giving rise to new "antigenic strains." This process is known as "antigenic drift" and results in an altered amino acid in proteins, such as the hemagglutinin (H) and neuraminidase, (N), thus changing its antigenicity. This is the process by which RNA viruses form quasi-species and evade antibodies.

In the event that two separate influenza viruses infect the same cell, there is likely to be shuffling of the viral RNA segments resulting in re-assortment and a recombinant virus. Thus different chromosomes from different viruses may be packaged in the same coat. This event is known as "antigenic shift" and is the basis of new strains of influenza. Antigenic shift can involve two specific RNA segments containing genes coding for proteins involved in receptor binding and penetration of the cell. These are the hemagglutinin, which are receptors binding protein that attach to specific sites on the cell surface (particularly of the lung) and the neuraminidase, which is involved in the penetration of the virus into the cell; thus, H7N9 refers to hemagglutinin of type 7 and neuraminidase of type 9. The pandemic influenza virus of 1918 (discussed below) was called H1N1 and is the reference strain for all other epidemics. Following infection, the infected individual produces antibodies to these two proteins. Changes in the antigenicity of the virus have occurred since the 1918 pandemic.

The first human influenza virus A subtype, H1N1, was isolated in the laboratory in 1933. It was responsible for all recorded influenza infections until 1957. In sera (blood) from persons born during this period—regardless of their age—the highest influenza antibody titers interact with the H1N1 subtypes. In 1947 another H1N1 subtype (variant) emerged that could not be neutralized efficiently with antibodies to the 1933 variant. Antibodies to this 1947 isolate supplanted all prior strains in the human population. In 1957 a new recombinant, the "Asian" influenza virus H2N2 took over, a strain that had acquired three new segments of avian origin. This virus resulted in a serious pandemic with high numbers of infections and

Table 11.1 Time line of human influenza pandemics

Year	Type	Description
1918	H1N1	Most devastating pandemic, killing over 50 million people
1957–1958	H2N2 "Asian" flu	About 70,000 deaths in U.S. Fatality predominantly in the elderly
1968–1969	H3N2 "Hong Kong" flu	Approximately 34,000 deaths in U.S.
1977	H1N1 "Russian" flu	Similar to previous non-pandemic strain
1997–2013	H5N1	Avian flu that on occasion has jumped to humans
2009	H1N1	Mild epidemic, started in Mexico and spread to U.S. and rest of world

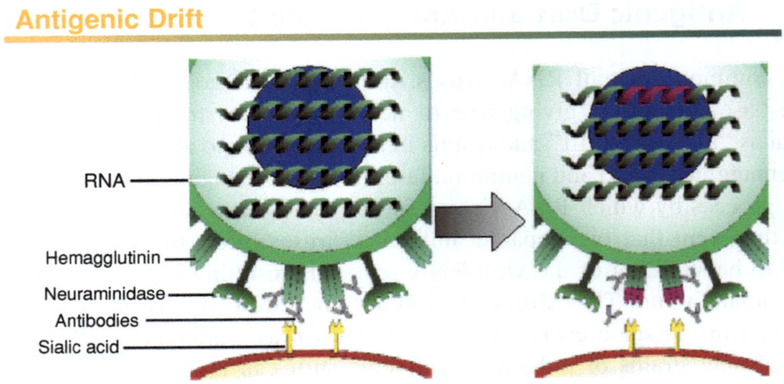

Fig. 11.2 Antigenic drift results in a mutated gene. In the diagram, a mutated hemagglutinin will no longer be recognized by antibodies that inactivate the virus

deaths. In 1968 another recombinant variant, the "Hong Kong flu" (H3N2) emerged. In this case, the H gene and another gene, PB1, replaced the previous genes. In 1977 a novel H1N1 strain arose, similar to the 1950 variant. (These pandemics are summarized in Table 11.1.) Thus, individuals born at different times may have serum antibodies to different subtypes. These different subtypes arose by antigenic shifts, probably by recombination events in different animals. Human type 1 influenza can be isolated from horses, pigs, and aquatic birds. This antigenic shift, as well as antigenic drift, make it difficult to develop an efficient vaccine since it is impossible to predict what strain of virus will infect humans at a future time. The differences between antigenic drift and antigenic shift are explained in Figs. 11.2 and 11.3 from the National Institute of Allergy and Infectious Diseases (USPHS).

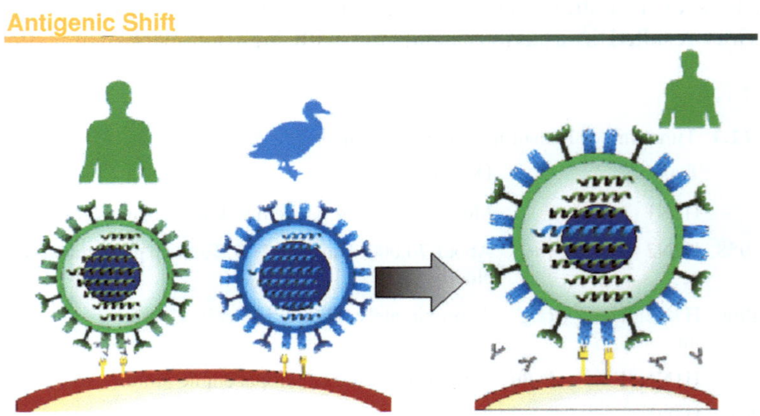

Fig. 11.3 Mechanism of antigenic shift. Reassortment of viral genes in the same host or a new host. Antigenic shift occurs as a result of genetic mixing of influenza "gene segments," which can spread from bird to human to pig or any other animal, and human-to-human transmission

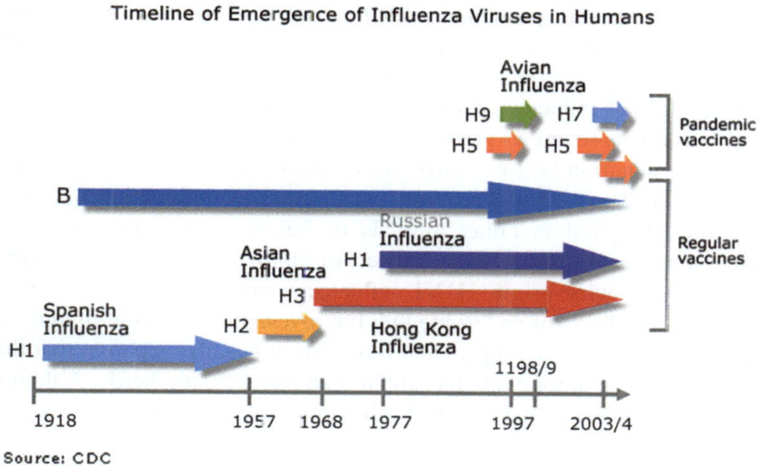

Fig. 11.4 Timeline of emergence of influenza virus in humans (CDC)

11.3 History of Influenza Epidemics

Influenza epidemics have occurred periodically since the Middle Ages. Since 1500, there appear to have been 14 or more influenza pandemics; the epidemics appear to begin in Central Asia (China) and move from east to west. This may be related to the proximity of humans to domestic fowl, or to the migratory patterns of birds.

Since the beginning of the "germ theory" era, pandemics have been reported to have occurred in 1847–1848, 1889, 1918, 1957, 1968, 1977, and 2009. The most serious of these was in 1889–1890 and 1918. The 1889 virus spread rapidly through railway connections and steam ships throughout the world. The virus killed millions, mostly elderly people, but without greatly affecting society [1]. The pandemic of 2009 was not as severe as the others. Table 11.1 and Fig. 11.4 present a timeline of the pandemics since 1918.

The 1957 pandemic (Asian flu) was caused by the H2N2 virus, and the 1968 pandemic (Hong Kong flu) by H3N2. The Asian flu began in China in 1956 and lasted until 1958. This flu outbreak—with an estimated 69,800 deaths in the U.S. alone—appeared to be the result of the re-assortment of virus segments between avian and human sources, The outbreak was accompanied by very high fever, general pain and malaise. A vaccine was developed against this strain in 1957. The Asian flu caused an estimated one to two million deaths worldwide.

After 10 years of evolution, the Asian flu virus disappeared, having been replaced through antigenic shift by a new influenza A subtype, H3N2, which gave rise to the Hong Kong flu pandemic, which began in the spring of 1968, and by the fall had spread to the U.S. There were about 34,000 deaths in the U.S., and a million worldwide; this pandemic ended in 1970. Currently, the predominant antibodies in the human population are to this particular flu. Endemic influenza kills about 30,000 people in the U.S. each year, mostly the aged.

Another strain of influenza, a variant of H1N1, occurred in 2009–2010, but the epidemic was not as severe as predicted. This virus appears to have developed as a result of recombination between two swine variants and was first detected in the United States in April 2009. The virus was a unique combination of viral genes never before identified in either animals or humans. The virus genes were a combination of genes most closely related to North American swine H1N1 and Eurasian lineage swine-origin H1N1 influenza viruses. Because of this, initial reports referred to the virus as "swine flu" but investigations of initial human cases did not show any contact with pigs and it quickly it became apparent that this new virus was circulating among humans and not among U.S. pig herds. Although dire predictions were made of the effect of this virus, it was not as severe as expected and resulted in a moderate pandemic [2].

11.4 The H5N1 Influenza Epidemic

At the time of this writing, another flu epidemic, milder than predicted, is ongoing, even though some virologists are predicting a worst-case scenario of subsequent epidemics. The H5N1 virus (avian flu) has been around for quite a few years and its potential to infect man is still unpredictable. Its appearance has led to the destruction of millions of heads of poultry and ducks. The human toll has not been so great—about 350 fatalities out of 650 cases—mostly in those in close contact with infected poultry, either by direct contact or handling uncooked meat. The fatality rate is about 60 %. Table 11.2 presents the total number of cases and deaths from the beginning of the outbreak until 2013.

The financial losses from H5N1 has been staggering. Billions of dollars have been lost in the poultry industry due to the culling of chickens and ducks, which has meant large losses in the Far East to poultry farmers, most of whom are poor. According to the *New York Times*, due to the H5N1 threat, as of March 2006, "Governments worldwide have spent billions planning for a potential influenza pandemic: buying medicines, running disaster drills, [and] developing strategies for tighter border controls." The major Southeast Asian economies have seen direct costs, approximately 140 million birds culled and the stated costs of containing the epidemic are approximately U.S. $10 billion (World Bank 2005). In November of 2005, President Bush announced a "National Strategy to Safeguard against the Danger of Pandemic Influenza," at a cost of $7.1 billion, while other countries have carried out similar but less expensive plans to curb the outbreak. The precautions taken included the culling of flocks; mass vaccinations; closing of borders; and banning of tourism in certain areas of Southeast Asia. The last two large poultry culling exercises in Hong Kong took place in June 2008, after inspection uncovered H5N1 in chicken feces samples, and in December 2008,

Table 11.2 Number of cases and deaths from H5N1 influenza A from 2003 to 2013 (WHO)

Country	No. of cases/deaths
Azerbaijan	8/5
Bangladesh	7/1
Cambodia	39/29
China	45/30
Egypt	173/63
Indonesia	193/161
Iraq	3/2
Laos	2/2
Myanmar	1/0
Nigeria	1/1
Pakistan	3/1
Thailand	25/17
Turkey	12/4
Vietnam	125/82

when the virus killed scores of chickens at a chicken farm. In 1997, six people died from the H5N1 virus in Hong-Kong and 1.3 million chickens were culled. Hong Kong's last recorded human case of H5N1 was in November 2010—a 59-year-old woman who came down with the illness after returning from a trip to Mainland China, but survived. There have been reports of farmers committing suicide as a result of financial ruin, and others hiding their chickens to avoid slaughter. The cost to the poultry sector in Cambodia, Thailand and Vietnam has been estimated at U.S. $560 million [3]. Vaccination is an especially labor-intensive task, costing huge amounts. Even then, there are doubts as to whether a vaccine will work, given the viruses' propensity to undergo antigenic drift; in fact, vaccines against H5N1 have been prepared and do not seem to be very effective.

Has the effect of H5N1 been overstated and led to panic? Luckily it has not mutated so that human-to-human transmission occurs only on a very limited scale. The hemagglutinin of this specific virus binds to a sialic-acid receptor that is found deep in the lower respiratory tract, whereas other influenza viruses, with H1, H2, and H3, bind to a receptor in the upper respiratory tract. This has been suggested as a possible reason for the lack of human-to-human spread on a pandemic scale [4, 5]. Many countries have set up rapid response teams to cull poultry as soon as a case of H5N1 is reported. On the positive side, this outbreak has increased the efficiency of poultry husbandry by introduction of new methods, as well as better hygiene in chicken coops. Since the virus is carried in wild aquatic birds and other species of birds, it is very difficult to control. A schematic of the spread of the different strains of influenza virus is presented in Fig. 11.5.

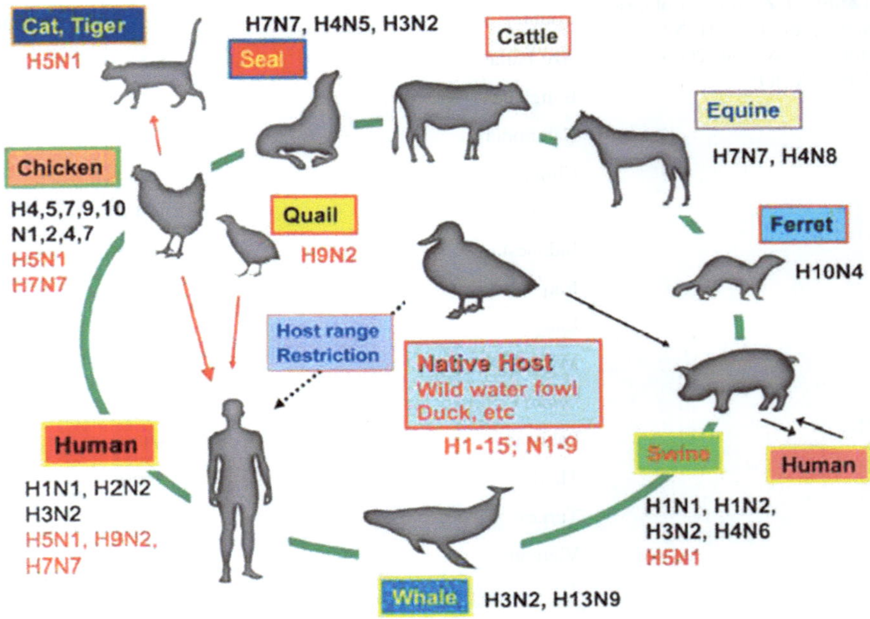

Fig. 11.5 Schematic of the spread of different strains of influenza A CDC Health Advisory Swine Flu H3N2v

Recent experiments have been performed by scientists in Holland and the U.S. to alter the H5N1 virus host range. These experiments are dangerous and have been met with consternation and controversy and are discussed in the chapter on creating viruses in vitro (Chap. 21).

There are as many as 16 known variants of the hemagglutinin (H) gene, and 9 known variants of the Neuraminidase (N) gene, [6] but only the H1, H2 and H3 and N1 and N2 subtypes appear to infect humans, although recently (in 2013) there was an outbreak of H7N9—a totally novel strain of influenza pathogenic to humans.

11.5 H7N9 Influenza Epidemic

Another recombinant virus that has recently come on the scene is the H7N9 influenza virus. As in the case of H5N1, this is an avian virus with low transmission rates to humans. H7 is not usually associated with human viruses. Over the past ~135 years, H7 influenza viruses have neither caused pandemics nor undergone human adaptation [7]. However, as in the case of H5N1, there have been some human fatalities among poultry workers and there is fear of a pandemic.

It has been associated with 132 cases and 28 related deaths in China as of June 2013. Although not pathogenic to humans, as discussed above, there is danger of the virus infecting other mammals, mixing with other influenza subtypes. As pointed out by David Morens and Anthony Fauci [7] no poultry influenza has infected humans on a large scale since records have been kept after 1918. As in the case of H5N1, there was some degree of panic when this virus was discovered, with the expectation, at least in the popular press, that this virus would lead to another pandemic. Some changes have occurred in the virus since May 2013 that possibly make the virus more pathogenic and enhance human transmission. However, so far the virus has been confined to specific areas of China, with the human infections resulting in clusters of families who live close to their poultry. One case was recently found outside of China, in Malaysia, in a traveller who arrived from China.

11.6 Viral Reservoirs

These periodic outbreaks bring up some interesting questions: Where is the virus in between pandemics? It is thought that type A influenza is maintained in aquatic birds. There is no evidence of viral latency in humans, as there is for herpes viruses. Any discussion of influenza should consider the work of Robert Webster and his group at St. Jude's Hospital in Memphis. Webster has spent a lifetime tracking influenza spread and evolution, isolating the virus from bird droppings throughout the world. He concludes that studies on the ecology of influenza viruses have established that all known influenza A viruses are maintained in aquatic birds, and he suggested that aquatic birds are the primordial source of all influenza viruses in other species [8]. These birds are adapted to the virus and are not made sick by it, indicating an old evolutionary relationship. By following the migratory patterns of birds, it was concluded that ducks harbor the virus and spread the virus to other birds [6] (Fig. 11.6).

What triggers the pandemics? There is no answer to this question, since it appears to be a random event in which a subtle change in the virus may lead to increased infectivity to humans through human-human transmission or increased pathogenicity. Two strains of influenza may infect the same animal (duck, chicken, pig), undergo recombination by gene reassortment, (antigenic shift), and result in a novel virus with increased infectivity to humans and human-human transmission. That aquatic birds and ducks may be the major reservoir of influenza A has recently been challenged by new research that has shown that influenza may have originated in horses in the nineteenth century [9].

Owing to an error-prone RNA polymerase, the rate of mutation is very high. In the hemagglutinin gene, it is estimated at about 2×10^{-3} base substitutions per position per virus generation, or about one base substitution in the HA gene per virus generation. Selection for amino acid substitutions is driven at least in part by immune pressure, as the HA is the major target of the host immune response.

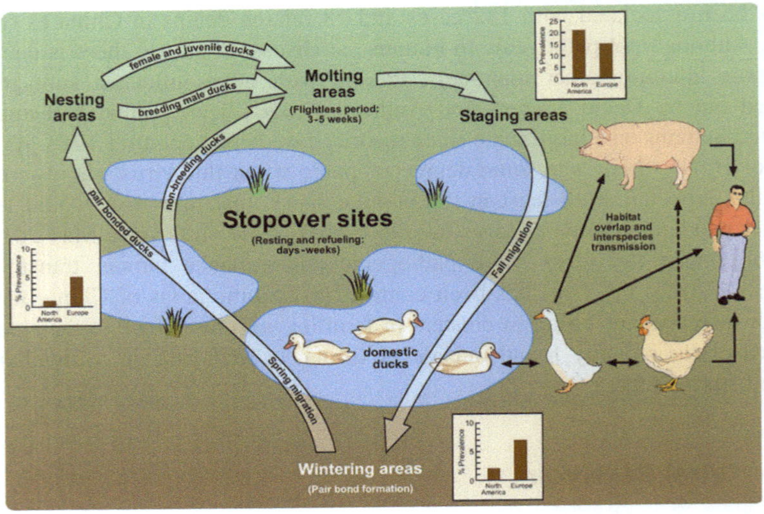

Fig. 11.6 Central role of ducks in the spread of influenza A (http://www.ncbi.nlm.nih.gov/core/lw/2.0/html/tileshop_pmc/tileshop_pmc_inline.html?title=Click%20on%20image%20to%20zoom&p=PMC3&id=2749972_nihms144278f1.jpg) [6]

The clinical outcome of infection depends on the host and virus strain. Domestic poultry are very susceptible to type A, particularly with H5 and H7 combinations. These strains are not highly pathogenic in wild birds. However, the combination of H5N1 does appear to be pathogenic to many species. This particular strain has caused pathogenic effects in wild aquatic birds, and has been deadly to the poultry industry in Southeast Asia and in parts of Europe. In some cases, ducks have become quite sick, but in other cases they have acted like Trojan horses and infected poultry [10].

11.7 The Great Pandemic of 1918

The current fear of influenza ("flu") is a reflection of the devastating outbreak the world faced in 1918, at the end of World War I. That particular strain of influenza killed at least 50 million people worldwide—675,000 in the U.S. alone. It was the most serious recorded plague to hit mankind, more deadly than the Black Death. It was called "Spanish flu" because the initial epidemic appeared to have begun in Spain, but this is probably incorrect; recent research indicates that it originated in the Western United States [9]. The epidemic killed an estimated 43,000 U.S. servicemen—40 % of the U.S. casualties reported for WWI. In 1 month alone—May 1918—it killed 8 million people worldwide. It spread to all corners of the earth, and occurred as three separate recurring epidemics, targeting in particular

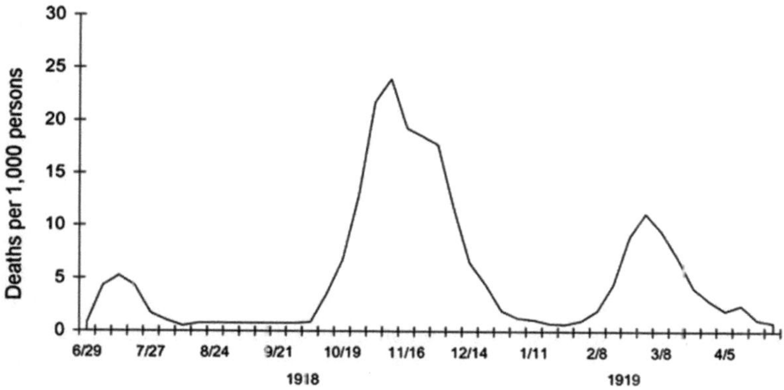

Fig. 11.7 Three pandemic waves: weekly combined influenza and pneumonia mortality, United Kingdom, 1918–1919 [11]

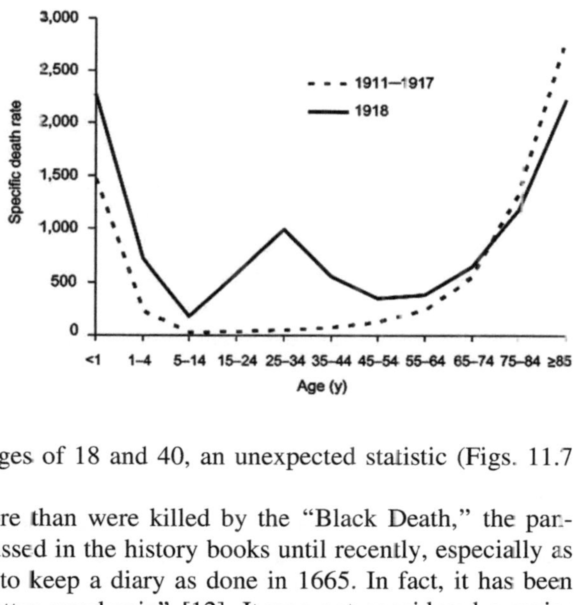

Fig. 11.8 "U-" and "W-" shaped combined influenza and pneumonia mortality, by age at death, per 100,000 persons in each age group, United States, 1911–1918. Influenza- and pneumonia-specific death rates are plotted for the interpandemic years 1911–1917 (*dotted line*) and for the pandemic year 1918 (*solid line*) [11]

young people between the ages of 18 and 40, an unexpected statistic (Figs. 11.7 and 11.8).

Despite having killed more than were killed by the "Black Death," the pandemic of 1918 was not discussed in the history books until recently, especially as there was no Samuel Pepys to keep a diary as done in 1665. In fact, it has been referred to as the "the forgotten pandemic" [12]. It was not considered a major plague until the last 20 or 30 years, and there is not much discussion about its economic or social impact in the scientific literature. The 1918 influenza pandemic charged across America in 7 days and across the world in 3 months.

The initial phase of the epidemic began in the spring of 1918. There are many theories as to where the flu originated; these include among U.S. troops in France, possibly in China or with an outbreak among soldiers in Kansas spreading eastwards across the U.S. In the initial phase, there were not many deaths. The virus spread, probably by troop transportation, and by May of 1918 it had reached

France, where American troops were fighting the Germans; from there it spread to Spain, Egypt, Palestine, and the Far East (China). By the summer (July) of 1918, the pandemic was very serious in France, Germany, Spain, Portugal, and parts of the South American continent; this is referred to as the "first wave."

The second wave was much more extensive. It appeared to have begun both in France and India (possibly spread by troops) in August of 1918, quickly spreading through all of western Europe, most of the United States and South America, Canada, India and China. The highest death toll was in India, about 50 per 1,000 inhabitants, and there appeared to be a correlation between poverty and high death rate. The second wave arrived in the U.S. in Boston in September of 1918 through the port, busy with war shipments of machinery and supplies. The war also enabled the virus to spread and diffuse through the mobilization of civilians into the army. As they came together, they brought the virus with them and to those with whom they had contact. The virus killed almost 20,000 in October of 1918 alone. On November 11, 1918, when Armistice Day was celebrated at the end of the war with parades and large gatherings (a complete disaster from the public health standpoint), a rebirth of the epidemic occurred in some cities. The flu that winter was beyond imagination as millions were infected and thousands died [13].

The incubation period and the onset of symptoms were so short that apparently healthy people in the prime of their lives (ages 15–35) were suddenly overcome, and within an hour could become helpless with fever, delirium and chills. Additional symptoms were severe headache, pain in muscles and joints, hair loss, acute congestion, and temperatures of 101–105 °F. The most unusual pathologic finding was massive pulmonary edema and/or hemorrhage. This was a unique viral pneumonia—a patient could be convalescing one day and dead the next. Those who did not die of the viral infection often died of secondary bacterial pneumonia. One anecdote shared in 1918 was about four women who played bridge together late into the night—and overnight three of them died from influenza. Others told stories of people on their way to work suddenly developing the flu and dying within hours. One physician writes that patients with seemingly ordinary influenza would rapidly "develop the most viscous type of pneumonia that has ever been seen," and later, when cyanosis appeared in the patients, "it is simply a struggle for air until they suffocate." Another physician recalls that the influenza patients "died struggling to clear their airways of a blood-tinged froth that sometimes gushed from their noses and mouths." The physicians of the time were helpless against this powerful virus [13].

Sir Martin Gilbert, the renowned British historian, writes, [14] "That October, 225 Londoners died of Spanish 'Flu. This was more that all the deaths from 4 years of Zeppelin and bomber raids over Britain. In Vienna the expressionist painter Egon Schiele was among the influenza victims. On the Western Front the Canadian air ace, Captain Quigley, who had shot down thirty-four German planes in combat was another victim. Of the three thousand Eskimos living on the coast of Labrador, two thousand died of the disease. In South Africa, 20 % of the population of Kimberley died. The disease swept the country 'like a grass fire', a South African reader reported to the Times, 'with leaps of hundreds of miles'. In some places within a few days, doctors, chemists, nurses, butchers, bakers, and railway staff were struck down and the complete paralysis of communities resulted as from a stroke.

"In Bombay, a thousand Indians were reported to have died of influenza in 1 month, and with each month the epidemic intensified throughout the sub-continent. ANZAC soldiers returning from Europe carried the epidemic to Australia and New Zealand; more than 12,000 Australians died of influenza, despite two compulsory preventative measures, the wearing of face masks and the closing of theatres and cinemas. On November 9, as negotiations for the armistice had been taking place in Compiegne the French Poet, Guillaume Apollinaire, died of influenza. In Toronto on the day chosen by the city fathers to mark the Allied victory, the victory parade itself had to come to a halt, during its triumphant march through the city, to allow a funeral cortege of two victims of influenza to pass…The prisoners of war held by the Allies were sent home as quickly as the Allied prisoners. Among 300,000 Austrian prisoners of war being held in camps in Italy, 30,000 died of influenza while still in captivity: A British army officer, Captain Colin Coote, who hoped to be elected to Parliament in the election of November 1918, recalled reaching a village in his would be constituency, intent on encouraging the voters to vote for him, and finding that every villager was dead…..Six million died in India, the country where plague had killed a similar number at the turn of the century."

That this epidemic is not much discussed in history books, or was not much discussed until recently, might be a reflection of the fact that it occurred at the end of a terrible war. In some texts it is treated as part of the history of World War I. We must remember that when the Black Death or the ravages of smallpox occurred, the population of the world could be measured in millions and not billions. In 1918, other infectious diseases, such as tuberculosis, were rampant and killed millions. However there is no doubt that a combination of WWI and influenza wiped out a whole generation of young men in Europe and even in North America. This reflected the ease of spread in the army where soldiers were billeted in crowded conditions, and were packed in terrible conditions in the trenches. Initially, historians paid most of their attention to the deaths in the military, but more modern histories have emphasized the number of people killed in the cities of the world by the pandemic.

11.8 Molecular Characteristics of the 1918 Pandemic Virus

Why was the pandemic of 1918 so severe? Jeffrey Taubenberger and colleagues [15] recently sequenced the entire 8-segment genome of the 1918 influenza virus. RNA from a victim of the 1918 pandemic was isolated from a formalin-fixed, paraffin-embedded, lung tissue sample [16]. Segments of RNA were copied into DNA sequences, and overlapping sequences reconstructed to form the complete genome. All 8 segments were thus copied into DNA to complete the genome.

Since regulatory regions were missing, these were substituted from other strains of human influenza and the virus was then tested for infectivity in influenza-sensitive mice. The virus was undoubtedly an H1N1 virus. Surprisingly, this virus does not appear to be the result of reassortment between two influenza strains, nor is it the result of avian-pig hybridization, as originally hypothesized. Rather, it appears to be a "novel" virus, probably of avian (water-fowl) origin that mutated and became pathogenic and infectious to man. All eight of the genome segments from the 1918 virus differ in significant ways from other human flu sequences, suggesting that none of the genome came from a strain that had previously infected people. "It is the most bird-like of all mammalian flu viruses," says Taubenberger [15, 16].

The reconstructed virus was much more pathogenic in mouse strains than the control virus; in 4 days it had generated 39,000 more virus per mouse than the control strain. In contrast to the H1N1 virus (the standard control strain), the 1918 virus could grow in cells without treating it with trypsin, was lethal to mice and embryonated chick eggs, and grew to a higher titer in cells in culture than the standard virus (Fig. 11.9). The virus differed from the standard strains in all eight segments of RNA. Replacing the hemagglutinin gene with that of H1N1 reduced pathogenicity, and replacing the polymerase genes, the genes responsible for viral replication, also reduced virulence. Thus the 1918 strain had many differences from the control H1N1 strain, making the combination much more lethal. A total of 10 amino acid changes in the polymerase proteins differentiate the 1918 virus from subsequent influenza viruses. Similar changes have been noted in the H5N1 virus, which might explain the enhanced virulence of this strain and its adaptation to humans [17]. Phylogenetic analysis of the 1918 virus indicates that it is the common ancestor of the human and swine H1N1 virus.

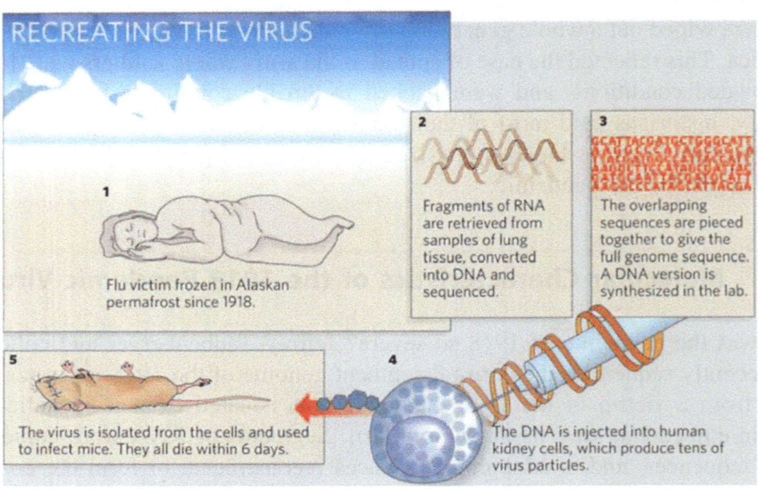

Fig. 11.9 Reconstructing the 1918 virus (*Nature* with permission)

The 1918 flu was possibly transmitted from humans to pigs and was therefore mistakenly called swine flu, so for many years it was believed that the transmission was in the other direction, i.e., from pigs to humans. It is not known why the primary target of the infection was in the 20–40 age group. The alterations in the polymerase may explain the high titers of the virus. The virus infection was also accompanied by a very virulent bacterial pneumonia. This appeared to be the result of a "cytokine storm," in which acute infection with influenza virus results in an unusually strong immune response with hundreds of cytokines being activated and damaging the lungs.

The history of the attempts to isolate this 1918 virus is fascinating. In 1951, scientists from the University of Iowa, including a Swedish graduate student, Johan Hultin, went as far as the Seward Peninsula in Alaska seeking the 1918 strain. He dug up the graves (with permission) at the Brevig Mission, an Inuit fishing village. These bodies were buried in permafrost, and it was hoped to find some residual virus in the lungs of the flu victims. However, when dug up, it was found that the tissue and virus were completely degraded and virus could not be recovered. In 1995, a group at the Armed Forces Institute of Pathology (AFIP) in Rockville, Maryland, headed by Jeffrey Taubenberger, isolated virus sequences from autopsy specimens maintained at the institute. The army had preserved these lung specimens in formaldehyde and embedded in paraffin. It was exceedingly difficult to isolate virus material from such specimens, but eventually a virus sequence was isolated from the lung of a soldier who had died in September 1918 at Fort Jackson, South Carolina; subsequently, another sequence was isolated from another flu victim that confirmed the sequences. However, the amount of material was still very limiting. After all those years, Johan Hultin returned to Alaska in 1997 and obtained frozen lung tissue from four graves. From one of these the full sequence could be obtained [18]. Afterwards, samples stored at the Royal London Hospital were received and sequenced, confirming the other sequences.

11.9 Can We Avoid Pandemics?

Can we predict another pandemic? Not really. Today we are better prepared for this type of event; there are new vaccine strains; it is easier to develop vaccines from cell culture than from eggs, the standard method; and there is a supply of anti-flu medications. However, the change to producing the virus in cell culture is a very recent development, and the production of vaccine could be improved using recombinant DNA techniques and modern molecular biology. The 1918 flu and subsequent pandemics spread very quickly around the globe, and the same can be expected of the next pandemic. It is difficult or impossible to predict the consequences and pattern of the next one; worst-case scenarios predict 40,000,000 ill and about 200,000 deaths.

Most people retain antibodies to the original infection they have experienced. This, in part, could be the reason for the high death rate among young people in the 1918 pandemic, in that older people had antibodies to a previous exposure to a virus similar to the H1N1. Today, people born much later have antibodies to other strains, either H2N2 or H3N2. Attempts are being made to produce a vaccine that contains all possible combinations of these antigens—something that should be possible within the next few years. Thus there is a need for a universal vaccine. It takes a few months to develop a vaccine, and by the time a suitable one has been developed, it is possible that the epidemic will be over.

Each year, the seasonal influenza vaccine contains three influenza viruses—one influenza A (H3N2) virus, one seasonal influenza A (H1N1) virus, and one influenza B virus. The influenza viruses in the seasonal flu vaccine are selected each year based on surveillance-based forecasts about what viruses are most likely to cause illness in the coming season. Therefore, each year's seasonal flu vaccine is designed to protect against the influenza viruses expected to cause disease during the upcoming influenza season.

Why does virus pandemic always seem to begin in Asia? This is probably because the type of farming used in rural China involves poultry, ducks or other animals living in close proximity to humans and to each other, and also due to crowded markets where there is a chance of spread of the virus. These markets contain aquatic birds, chickens, and other types of birds, as well as pigs—all susceptible to the virus in close proximity to one another.

11.10 Molecular Biology of Influenza

Influenza subverts the host cell to produce more of itself by inhibiting host protein synthesis. Once bound to the cell receptor, a small vesicle is formed around the virus. In the acidic environment of the endosomes (small vacuoles filled with liquid), the virus changes shape and fuses its envelope with the endosomal membrane. This is followed by a signal to release the viral nucleocapsid into the host cytoplasm. From there, the nucleocapsid travels to the host nucleus, where the virus undergoes primary transcription to produce proteins necessary for replication. The primary transcription involves what is known as "cap-snatching." A viral endonuclease, an enzyme that digests nucleic acids internally, cuts the 5' methylguanosine cap as well as 10–13 nucleotides adjoining it from the host RNA. Most host mRNAs are capped at the end with 5-methylguanosine, which is the signal for initiation of transcription. This is then added to the replicated viral RNA for the transcription of a protein necessary for virus replication (Fig. 11.10).

Once the initial proteins are produced, eight complementary positive-sense RNA strands are transcribed from the eight negative-sense RNA segments (at least in influenza A and B influenza C has seven segments). From this, cRNA, a negative-sense RNA, is produced. Various proteins then help this negative-sense RNA exit the nucleus for the cytoplasm of the host, where it interacts with viral proteins

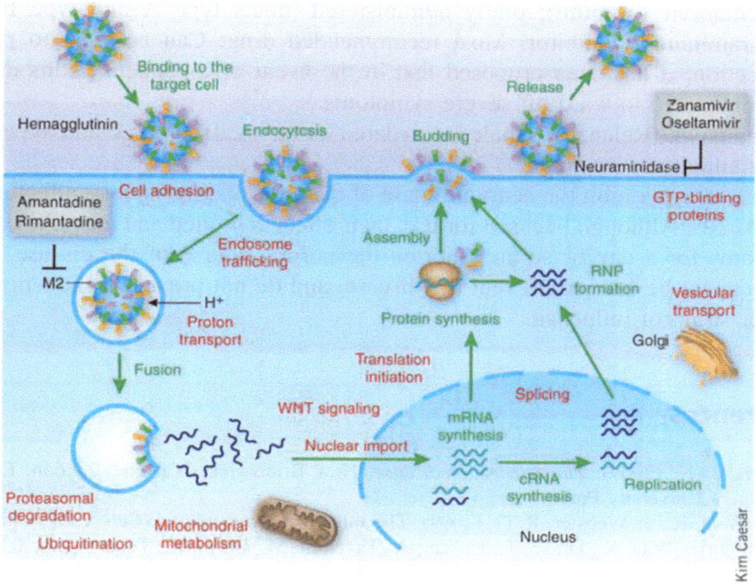

Fig. 11.10 Replication cycle of influenza virus and points at which anti-influenza drugs interact with the virus (permission from *Nature Biotechnology* [19])

to be packaged. In the cytoplasm, the hemagglutinin and neuraminidase proteins have undergone various modifications. The hemagglutinin, neuraminidase, and the two matrix proteins (M2) all travel together to the plasma membrane. There the proteins meet with the other matrix protein (M1) and begin the budding process. At least eight RNA segments come to the site (the mechanism is not completely known), and the virus buds. The neuraminidase finally destroys the sialic acid receptors on the membrane, thus allowing the virus to leave the cell. It is the neuraminidase and the hemagglutinin that give the type specificity to the virus.

11.11 Drugs Developed Against Influenza

A number of drugs have been developed to inhibit influenza, although one of the major problems in treating influenza is that the illness comes on very quickly and the drugs are taken after the virus has infected the individual and thus are not prophylactic (preventative) but decrease the symptoms. These drugs are:

- Rimantadine (Flumadine): orally administered, treats type A, only for adults. Mechanism unknown, maybe involved in un-coating. Also a related drug, amantadine. H1N1 strains of influenza have developed resistance to this class of drugs.

- Oseltamivir (Tamiflu); orally administered, treats type A and type B. Is a neuraminidase inhibitor. Most recommended drug. Can be used to prevent infection. It has been proposed that in the event of a pandemic, this drug be administered to ward off severe symptoms
- Zanamavir (Relenza): inhaled powdered drug, treats type A and type B. A neuraminidase inhibitor.

These drugs inhibit the neuraminidase of the virus and thus do not affect normal human cells. Although licensed for use, their effect is limited and may alleviate the symptoms for a day or so and shorten the normal course of the disease. These drugs cannot be used to prevent the disease, and do not prevent re-infection with another strain of influenza.

References

1. Hays, J. N. (2009). *The burden of disease*. New Brunswick, N.J. and London, England: Rutgers University Press.
2. Salomon, R., & Webster, R. G. (2009). The influenza virus enigma. *Cell, 136*(3), 402–410.
3. Humphries-Waa, K., Drake, T., Huszar, A., Liverani, M., Borin, K., Touch, S., et al. (2013). Human H5N1 influenza infections in Cambodia 2005–2011: Case series and cost-of-illness. *BMC Public Health, 13*(1), 549.
4. Shinya, K., Ebina, M., Yamada, S., Ono, M., Kasai, N., & Kawaoka, Y. (2006). Avian flu: Influenza virus receptors in the human airway. *Nature, 440*(7083), 435–436.
5. Oldstone, M. (2010). *Viruses plagues and history*. Oxford: Oxford Univeristy Press.
6. Kim, J. K., Negovetich, N. J., Forrest, H. L., & Webster, R. G. (2009). Ducks: The "Trojan horses" of H5N1 influenza. *Influenza Other Respiratory Viruses, 3*(4), 121–128.
7. Morens, D. M., Taubenberger, J. K., & Fauci, A. S. (2013). H7N9 avian influenza a virus and the perpetual challenge of potential human pandemicity. *MBiology, 4*(4), e00445–13.
8. Webster, R. G., Bean, W. J., Gorman, O. T., Chambers, T. M., & Kawaoka, Y. (1992). Evolution and ecology of influenza a viruses. *Microbiological Reviews, 56*(1), 152–179.
9. Worobey, M., Han, G. Z., & Rambaut, A. (2014). *A synchronized global sweep of the internal genes of modern avian influenza virus*. Nature.
10. Yen, H. L., & Webster, R. G. (2009). Pandemic influenza as a current threat. *Current Topics in Microbiology and Immunology, 333*, 3–24.
11. Taubenberger, J. K., & Morens, D. M. (2006). 1918 Influenza: The mother of all pandemics. *Emerging Infectious Diseases, 12*(1), 15–22.
12. Crosby, A. W. (1989). *America's forgotten pandemic: the influenza of 1918*. Cambridge: Cambridge University Press.
13. The 1918 Influenza pandemic. http://virus.stanford.edu/uda/.
14. Gilbert, M. (1997). *A history of the 20th century* (1st ed., Vol 1). New York: William Morrow and Co.
15. Tumpey, T. M., Basler, C. F., Aguilar, P. V., Zeng, H., Solorzano, A., Swayne, D. E., et al. (2005). Characterization of the reconstructed 1918 Spanish influenza pandemic virus. *Science, 310*(5745), 77–80.
16. Taubenberger, J. K., Reid, A. H., Krafft, A. E., Bijwaard, K. E., & Fanning, T. G. (1997). Initial genetic characterization of the 1918 "Spanish" influenza virus. *Science, 275*(5307), 1793–1796.

17. Taubenberger, J. K., Reid, A. H., Lourens, R. M., Wang, R., Jin, G., & Fanning, T. G. (2005). Characterization of the 1918 influenza virus polymerase genes. *Nature, 437*(7060), 889–893.
18. Taubenberger, J. K., Reid, A. H., & Fanning, T. G. (2005). Capturing a killer flu virus. *Scientific American, 292*(1), 48–57.
19. Min, J. Y., & Subbarao, K. (2010). Cellular targets for influenza drugs. *Nature Biotechnology, 28*(3), 239–240.

Poliovirus and Other Picornaviruses 12

Abstract

Poliovirus is an epidemic virus of the nineteenth and twentieth centuries. Although there were infections of polio in ancient times, the improvement in sanitation in the twentieth century led to epidemic outbreaks, particularly in western Europe and the U.S. Panic often accompanied outbreaks, particularly in the U.S., with demands for strict quarantine, prohibition of movement of people, and closing of places of recreation. Public interest in eradication of the virus led to fund-raising with the "March of Dimes," whose aim was to alleviate suffering and develop a vaccine. The fact that President Roosevelt was a victim of polio helped with the campaign. The treatment of patients with polio was controversial; Sister Kenny claimed that exercise was important, while other physicians recommended rest. Patients were often kept in iron lungs for a long time. In 1955 the USPHS licensed an inactive virus vaccine, the Salk vaccine, which was used in mass vaccination of children in the U.S. Albert Sabin and Hilary Koprowski independently developed two attenuated viral vaccines, which were taken orally and replaced the Salk vaccine in the 1960s. Poliovirus belongs to the family of viruses known as picornavirus; this includes foot and mouth disease, which is of great economic importance and outbreaks of which have plagued the British cattle industry, and rhinoviruses—the common cold virus.

12.1 Introduction

Growing up in Scotland in the 1940s, I was not allowed to go to the local swimming pool or cinema during the summer because of the threat of poliovirus. There was always fear of diseases such as diphtheria, scarlet fever, and polio. The condition caused by polio was known as "infantile paralysis," since one of the

M. W. Taylor, *Viruses and Man: A History of Interactions*,
DOI: 10.1007/978-3-319-07758-1_12,

outcomes (although quite rare, only in about 1 % of the cases) was paralysis of the lower limbs. In some cases death resulted, particularly among young children. The virus not only affected the legs but also the lungs, destroying the nerves involved in breathing.

The poliovirus has probably existed since the beginning of human history and is now recognized to be associated with raw sewage and fecal contamination. It was the result of "cleaning" the environment of human waste that resulted in polio epidemics, and it was no longer confined to young children. There was great panic in the U.S., where polio epidemics occurred between the 1930s and 1950s. The first major reported epidemic occurred in Sweden in 1887 [1].

Perhaps no epidemic, except for the Black Death in the Middle Ages, and AIDS these days, has resulted in such major social upheavals as poliomyelitis. Polio caused panic in most of the major cities of the U.S. and western Europe in the twentieth century. Though a large percentage of those afflicted by the disease survived, a large number were severely handicapped, resulting in pressure for new legislation dealing with disabilities and access to public, and later private, facilities. Moreover, the fact that a president of the U.S. (Franklin Roosevelt) had been stricken with polio had a tremendous impact on a search for its cause and treatment and the development of novel ways of funding research. The creation of the National Foundation for Infantile Paralysis and "The March of Dimes" involved a whole population in fundraising and medical research, and gave birth to many organizations aimed at supporting other specific diseases.

12.2 Early History of Poliomyelitis

A stele of an Egyptian priest with a withered leg—suggestive of polio—exists from the XVIII dynasty in Egypt (1580–1350 BCE). It shows a priest named Ruma at the sanctuary of the goddess Astarte in Memphis supporting himself with a crutch [1] (Fig. 12.1). The "foot drop" is similar to that found in cases of infantile paralysis, and this picture indicates that polio possibly existed in ancient times, although the withered leg could be due to other reasons. Speculation points to polio being a biblical disease; in II Samuel there is mention of a son of Jonathan, the son of King Saul, with a lame foot. The passage explains that this happened when the child was five years old and fell during the flight from the Philistines. Although mentioned in the classic by William Osler (1849–1928) in his *Principles and Practices of Medicine* as the first recorded case of polio, a rereading of the biblical text does not support this thesis, since the child was dropped and both legs were broken. There were occasional cases of polio throughout the ages, but epidemics did not take place until the late nineteenth century. Among those afflicted with polio was Sir Walter Scott (1771–1832), the author of *Ivanhoe*, and the Bride of Lammermoor.

"I showed every sign of health and strength until I was about eighteen months old. One night, I have been often told, I showed great reluctance to be caught and put to bed, and after being chased about the room, was apprehended and consigned

Fig. 12.1 An Egyptian stele of a priest with a withered leg, typical of poliomyelitis

to my dormitory with some difficulty. It was the last time I was to show much personal agility. In the morning I was discovered to be affected with the fever which often accompanies the cutting of large teeth. It held me three days. On the fourth, when they went to bathe me as usual, they discovered that I had lost the power in my right leg. My grandfather, an excellent anatomist as well as physician, the late worthy Alexander Wood, and many others of the most respectable of the faculty, were consulted. There appeared to be no dislocation or sprain; blisters and other topical remedies were applied in vain.

When the efforts of regular physicians had been exhausted, without the slightest success, my anxious parents, during the course of many years, eagerly grasped at every prospect of cure which was held out by the promise of empirics, or of ancient ladies and gentlemen who conceived themselves entitled to recommend various remedies, some of which were of a nature sufficiently singular.

The English physician Michael Underwood (1737–1820) in 1789 provided the first clinical description of polio, in his *Treatise on Diseases of Children*, with a chapter on "Debility of the Lower Extremities." In the chapter [2], Underwood mentions a form of paralysis that usually "attacks children previously reduced by fever; seldom those under one, or more than four or five years old." In references to the legs, he advocated metal braces similar to those used until fairly recently [1]. One can interpret Underwood's reference to the time of infection since the one-year-old is protected by maternal antibodies, but children past that age would pick up the virus while playing in the unsanitary conditions that existed everywhere at that time. Since it was predominantly a childhood disease, most adults were immune and the majority of children were infected without symptoms at a young age and acquired long-term immunity.

There are occasional mentions of polio in the medical literature prior to twentieth century. The first real clinical description was that of the German physician Jacob Heine (1800–1879) in 1840. The disease for some time was known as the Heine-Medin disease, so called after Heine and his Swedish colleague, Karl Medin (1847–1927), who was one of the first to describe the epidemic nature of polio.

12.3 The Polio Epidemics

At the end of the nineteenth century, the major emphasis in the public health movement was the eradication of diseases such as typhoid, smallpox, tuberculosis, and cholera, all thought to be associated—rightly or wrongly—with sanitary conditions. Prior to the nineteenth century, children were initially protected from polio by maternal antibodies and exposed to the virus at a very young age, about one year old, and in most cases acquired immunity. Thus, enhanced hygiene postponed exposure to the virus until later in life, but with grave consequences. The first "mini-epidemic" of polio was a report by the Swedish physician Nils August Bergenholtz (1816–1922), of 13 cases near the city of Umea in 1881 [3]. The first recorded major epidemic of polio was reported in Stockholm in 1887. The symptoms were mostly minor, with an outbreak of fever and headache, etc. Damage to the central nervous system occurred in later stages of the illness, and in only a few patients. Other major epidemics occurred in Norway and Sweden in 1905 and 1911.

During the 1905 epidemic, the Swedish physician Dr. Ivar Wickman (1872–1914), a student of Karl Medin, followed the course of the disease in the countryside, and reported that it was highly contagious and that even those who experienced only minor symptoms could spread the virus. Such cases were termed "abortive infections." He reported that patients with an abortive infection were just as likely to spread the disease as those with a paralytic infection and that the time between contact and transmission of the disease was only three days. Wickman visited over 300 homes where victims resided, and although he proposed the disease as contagious, he could not find a link between the various victims, who lived quite far apart. He was recognized worldwide for his work on poliovirus, and during the great epidemic of 1911–1913, Wickman's proposal that this was a contagious disease was accepted. Swedish scientists suggested that polio was a disease of the alimentary system caused by an infectious agent that entered through the mouth and exited in the feces (oral-fecal transfer). Although presented at international meetings, this model was rejected by the Rockefeller Institute's Simon Flexner, who insisted that poliovirus entered through the nose, advanced to the brain and spinal cord, causing paralysis, and exited in the same manner [4]. Wickman performed unique studies, tracking the course of polio epidemics in small towns, following the spread along railway lines and among school children living on the same street. Swedish physicians noted that in areas hit hard by the 1905 epidemic, there was no subsequent outbreak in 1911, even when polio occurred nearby, indicating that immunity could be easily established. Unfortunately, Wickman suddenly committed suicide in 1914 without explanation, although there were suspicions that he was depressed from not being offered the position of chief pediatrician at the Karolinska Institute, a position previously held by Karl Medin. He is now considered one of the most important pioneers in poliovirus research and in 1958 he was posthumously honored by being inducted into the Polio Hall of Fame in Warm Springs, Georgia, alongside Heine and Medin.

In the summer of 1894, paralysis descended on Rutland, a small town in Vermont. A local physician, Charles Caverly, tracked all 123 cases, noting that 50 remained paralyzed and 18 died. He thought it might have something to do with playing in the sun, and he also realized that many children were infected with minor symptoms with no lasting effect [5]. Caverly served as the president of the Vermont Board of Health from 1891 until 1924, and he continued to analyze polio outbreaks in the state of Vermont; following an outbreak in Montpelier in 1916, he advised that a quarantine be established to prevent movement into the town and that religious gatherings be cancelled.

As the number of cases increased in Montpelier, the city council took measures to establish a quarantine that restricted the movements of children under 16 years of age and prohibited them from attending all places of public gatherings such as theaters, schools, ball games, and churches. However, with more and more cases diagnosed daily, the local board of health and Montpelier City Council decided to extend the quarantine by restricting the movement of children beyond their homes (see Dorman Kent's diary entry for June 28, 1917 in [6]). In addition to keeping their children at home, parents were advised to disinfect their children by washing their noses, throats, and mouths with a saline solution to prevent the spread of the disease. This recommendation reflects the belief at the time that polio was spread through nasal mucus transmission rather than by oral-fecal routes, which were eventually shown to be the case [6]. This behavior became typical of the time.

Panic accompanied the outbreaks that occurred in the USA. The first major outbreak in a series occurred in New York City in 1907, with 2,500 cases and 125 deaths. The great U.S. epidemic occurred in 1916, with approximately 27,000 cases and 7,000 deaths. Even though the virus had by this time been isolated, there were still reports that the disease was spread by moldy flour, gooseberries, poisonous caterpillars, and even infected milk bottles. The City of New York clamped down with quarantines, and prevented the movement of people into the city during

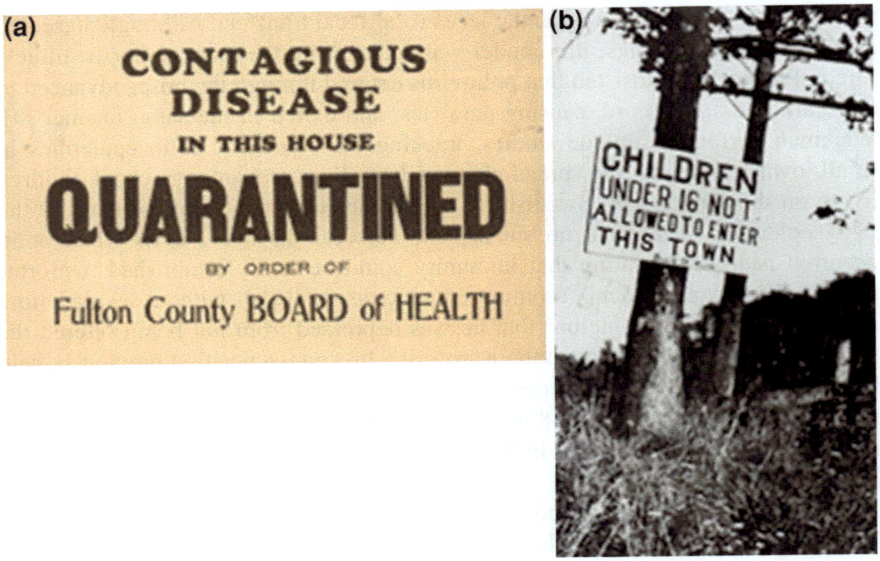

Fig. 12.2 **a** and **b** posters regarding quarantine

the summer months. The names of people with polio in their household were published daily in the press, and their houses were quarantined (Fig. 12.2).

The police prevented the movement of populations, and many cities refused to allow people to pass through them. Police were stationed along the Delaware River to prevent the entry of children into Pennsylvania; in many respects, this was like martial law. This might have been the first time in U.S. history that there was legal ban on freedom of movement since the rights of the community were placed before individual freedoms. Homes were quarantined and individuals hospitalized or often taken away against the wishes of their families.

A 1918 report indicated that quarantine had not worked, animals had not played any part in spreading the disease, the majority of cases were abortive, and the disease could be spread by such individuals. In 1931 there was another epidemic, in which 4,138 patients died, with even more paralyzed. An epidemic hit Los Angeles in the summer months of 1934, accompanied by a great deal of hysteria and probably exaggerated numbers, as many of the reported cases were most likely not polio. The worst outbreaks occurred between 1942 and 1953. In 1952 there were about 60,000 cases in the U.S. of them, 3,145 persons died and 21,269 were disabled. This was the worst year of the U.S. epidemic. Although paralysis was infrequent, polio caused great fear. Children were often separated from their parents against their wishes and they did not see each other for a week or 10 days; often the child would return home paralyzed and suffering from trauma. A large percentage of physicians and nurses also were infected. Due to improvements in sanitation, the disease afflicted both young and old, rich and poor. Outbreaks of the disease also occurred in the British Isles and European cities. In less developed countries, the disease occurred mostly among young children, since the adult population was immune.

One of the puzzles of this epidemic was the random nature of the victims. I remember being in a group where there was one case of paralytic polio in a young person of 20. Initially we were unaware of the fact that this was the diagnosis. Although many of us visited the sick person, sat with him and brought him food, no one else developed the illness—and this was in the days before the polio vaccine.

The general consensus in the USA at the time was that polio must be associated with dirt. Immigrants were blamed for bringing in the disease, since they were poor and lived in squalid conditions. However, improvements in waste disposal and sanitation in the early part of the twentieth century probably made the situation worse. Children living in poverty were most likely immune to the illness, having been infected asymptomatically in childhood.

12.4 Discovery of the Poliovirus

In 1908 the renowned Austrian physicians Karl Landsteiner and Erwin Popper isolated poliovirus from infected human tissue and from the upper respiratory tract and gut of infected monkeys. Landsteiner had graduated from medical school in Vienna in 1891 and returned to Vienna in 1908, at a time when Friedrich Loffler and Paul Frosch had discovered the foot and mouth disease virus (see Chap. 1). In 1908, during an epidemic of polio in Vienna, Landsteiner, working with Popper, obtained spinal cord material from a nine-year-old boy who had died of the disease. They injected this material into two "Old World" monkeys, the only ones available to them; luckily the virus replicated in the animals and developed the symptoms of polio [7]. They isolated poliovirus from infected tissue and passaged it through an animal system, providing a model for future studies. This was the same Karl Landsteiner of blood group fame who was to receive the Nobel Prize for this work in 1930.

Landsteiner then joined with Constantine Levaditi of the Pasteur Institute, and reported the successful filtration of polio through a Berkefeld filter, proving that polio was indeed a virus. The virus could be isolated from tonsils, the lining of the throat, from nasal fluid, and from lymphocytes in the gut; Simon Flexner and Paul Lewis at the Rockefeller Institute in New York confirmed these findings. The ability to grow the virus in an animal model (the rhesus monkey) should have resulted in a vaccine, since Pasteur had demonstrated with rabies virus that it was possible to passage a virus in animals and obtain an attenuated virus, so it was expected that the same could be done with poliovirus.

Flexner was director of the Rockefeller Institute for Medical Research. He went one step further than Landsteiner in establishing a monkey model for poliovirus, claiming that the portal of entrance of the virus in the monkey was through the nose. He concluded from his experiments that polio was a disease of the nervous system, passing directly through nasal nerves to the brain and spinal cord. From this observation he abandoned the idea of a preventative vaccine, reasoning that antibodies could not reach the neural system. Flexner's ideas dominated the polio field through the 1920s and 1930s. He encouraged the head of his virology department, Peter Olinsky, and his assistant, Albert Sabin, to work on the problem. His approach was to prevent the spread of polio by nasal sprays, but there was no indication that this was successful [5].

12.5 Roosevelt and Polio

In 1921 polio devastated the man who was later to become one of America's greatest presidents. Franklin Delano Roosevelt was 39 when he was infected with the virus and became paralyzed from the waist down. He had been Secretary of the Navy in Woodrow Wilson's cabinet during World War I and had unsuccessfully run for Governor of New York State. He was a very energetic man, involved in many sports and interactions with young people, particularly the Boy Scouts. During his tenure in the government he had organized physical education for members of the staff as well as for other cabinet members. He had been educated at Groton, a private Episcopalian college preparatory boarding school, and then went on to Harvard and Harvard Law School. He had grown up in very wealthy, pampered conditions, and was probably never exposed to viruses such as polio.

Roosevelt's name was well known nationally, being a cousin of former president Theodore Roosevelt. He was nominated as vice-presidential candidate to Governor James Cox of Indiana in an unsuccessful campaign against Warren Harding. During this period he was the picture of health. After losing the election, he returned to his law firm in New York to prepare for the next political battle, the governorship of the State of New York. He also volunteered to be president of the Boy Scout Foundation of New York. He visited and stayed at a Boy Scout camp at Bear Mountain on the Hudson River in July 27, 1921, and from there he sailed a few days later with a friend to the family's retreat at Campobello Island off the coast of Maine. During the first few days in Campobello he was very active, playing with his children and swimming, and again the picture of good health. On August 10, he suddenly felt too tired to dress for dinner and retired early, not feeling well. When he awoke in the morning he had a fever and had difficulty getting out of bed, his left leg dragging. As the day wore on, he lost all ability to move either leg and was in great pain. He may have picked up polio from his stay with the Boy Scouts, although this was two weeks after his stay there (the normal incubation period is 3–7 days).

The local physician was called and he diagnosed the condition as a cold, but he was worried about the situation and looked for some help in the area. He found Dr. William Williams Keen, who had performed surgery on President Cleveland in 1893. Keen had been the first to perform brain surgery in the U.S. However, on examining Roosevelt, he misdiagnosed the illness as a blood clot causing loss of power in the lower body and he recommended heavy massage, which increased the pain. After 10 days the family found Dr. Robert Lovette, an orthopedist from Rhode Island, who confirmed the diagnosis of polio.

For adults, polio was much more severe than in children. It was exceedingly difficult for Roosevelt to accept being an invalid, and he tried to hide it most of his life, by rarely showing himself standing (which he could not do without support). Most pictures show him sitting behind a desk, or in a wheelchair covered with a blanket since he was very embarrassed by his disability. Without his wheelchair, he had to drag himself along the floor or require assistance in moving. Very few saw the man as an invalid.

Despite this condition, Roosevelt decided to continue his political career. He enforced on himself a strict regimen of exercise, and found swimming a great help since in the water he did not feel as encumbered as he did on land (Fig. 12.3).

In 1924, a friend, George Foster Peabody, introduced Roosevelt to a spa he owned in Western Georgia where the mineral-laden water comes gushing out at 88° Roosevelt felt good bathing in the water, which relieved his pain, and allowed him to stand without aid. In 1926, Roosevelt bought Warm Springs, then a run-down thermal springs resort, and turned it into a polio rehabilitation center, formally incorporating it as the Georgia Warm Springs Foundation, a nonprofit company eligible for tax-free gifts [5]. This became a major rehabilitation center, later promoted by the March of Dimes. It was open to paying victims of polio, and later to all who could get outside support. Eleanor Roosevelt said that if her husband had not become involved in politics, he would have been manager of this

Fig. 12.3 President Roosevelt in wheel chair with another polio victim. Figure in Wikipdia and many other places in public domain

rehabilitation spa. Although turned down by the American Orthopedic Association, initially as a rehabilitation center, it was later approved. To help pay for those unable to afford treatment, Basil O'Connor, Roosevelt's former law partner, established a fund to support such individuals and their treatment at the spa. This fund developed into the National Foundation for Poliomyelitis, later known as the National Foundation, and gave rise to the March of Dimes (see below). Roosevelt visited Warm Springs every year for rehabilitation and built what became known as "The Little White House," where he would work and at the same time have access to the warm pool. Roosevelt died at the Little White House" in April of 1945; today it is a Georgia State Park.

From our perspective today, there was one major problem with Warm Springs: no blacks (African-Americans) were allowed because most of the medical facilities established in the Southern States of the USA banned African-Americans. In fact, there was a theory, proposed in part by members of the medical establishment, that African-Americans were not susceptible to polio. It is true that there were fewer African-American patients than white, Caucasian patients, but this reflected early exposure to the virus and state of hygiene in the neglected African-American community. There were many fewer cases of polio in general in the southern states than in the northern states, due to poorer sanitation. It was not until the 1930s Presidential election that the question of segregation in Georgia Warm Springs arose. Despite discussion the Georgia Warm Springs continued to be segregated. In 1941, a similar facility for African-American victims of polio was opened in Tuskegee, Alabama [8]. Today, the March of Dimes is still active and concentrates its efforts on birth defects, obviously without discrimination. The Georgia Warm Springs was sold to the state of Georgia for one dollar in 1974.

12.6 The March of Dimes

During this period we saw the establishment of the National Foundation for Infantile Paralysis, and an offshoot. The March of Dimes campaign, which was the model for subsequent charitable medical organizations. One of the driving forces behind the campaign to find a cure was Franklin D. Roosevelt. A unique aspect of this epidemic was the use of mass communication at the time—radio—for fund-raising and support of research.

From the March of Dimes Web site: http://www.marchofdimes.com/mission/eddie-cantor-and-the-origin-of-the-march-of-dimes.aspx

Comedian Eddie Cantor (1892–1964) holds a very special place in the history of poliovirus research for it was he who coined the phrase "March of Dimes" used to identify the Foundation National Foundation. Eddie Cantor was a multi-talented performer—vaudeville star, singer, actor, comedian, radio and television personality—whose rise to fame began in the New York City theatrical revue, the Ziegfeld Follies, in 1917. Beloved by the American public and known as "banjo eyes" for his wide-eyed visage, Eddie Cantor had a deep well of compassion to match his enormous talent, for he embraced a variety of humanitarian causes over the course of his long career in show business. One of these was the March of Dimes, and the story of how he created this name has special distinction.

Beginning in 1934, the fight against poliomyelitis (also known as infantile paralysis, or polio) was commonly associated with the annual Birthday Balls held each January 30th in honor of President Franklin D. Roosevelt's birthday. These lively fundraising parties were organized by Basil O'Connor, a Roosevelt advisor and lawyer. In cities throughout the United States just as the country emerged from the Great Depression and were unique in their appeal to ordinary citizens to join the campaign of finding a solution to this dreaded disease. After FDR issued his proclamation announcing the creation of the National Foundation for Infantile Paralysis on September 23, 1937, to carry on the battle against polio on a national basis, it was left to Eddie Cantor and other promoters to organize a fundraising strategy for the next Birthday Balls in California. On November 22, 1937, Cantor met with W. S. Van Dyke II and Harry Mazlish of Warner Brothers in the office of John Considine, Jr. in the studios of Metro-Goldwyn-Mayer to discuss their plans. In the meeting, Cantor recalled a successful 30-second radio appeal for relief funds after a catastrophic Mississippi River flood. Applying this idea to the National Foundation, Cantor said, 'I am sure that all of the national radio programs originating in Hollywood would devote 30 s to this great cause' He suggested that the money raised could be directed to the White House, pending the approval of the President. After another moment of reflection he suggested, 'We could call it the March of Dimes.' This idea brought the general approval of everyone in the meeting; it was a play on the 'March of Times,' the local newsreels shown in cinemas.

Eddie Cantor was not the only 'star' involved in the March of Dimes. Hollywood stars became very involved, including Mickey Rooney, Judy Garland, and many others. This later developed into a pattern where film stars and later television personalities were used in fund raising for many different charities, and of course in our times for election campaigns. An advert for the March of dimes is shown in Fig. 12.4.

The March of Dimes was a huge success, and brought in millions of dollars, which were used to develop the polio vaccine, first announced by Salk in 1955 and Sabin a few years later. The campaign was enhanced by the use of radio and later television. Between 1938 and 1962 the NFIP raised $630 million. The NFIP mission was threefold, to care for polio victims, research and training of medical personnel, and disseminate general information on the disease. Today the March of Dimes continues to support research into birth defects and genetic diseases of children.

12.7 Treatment for Polio

One of the symptoms of polio infection was the paralysis of the nerves controlling breathing and resulting suffocation. Philip Drinker and Louis Agassiz Shaw at Harvard University invented the iron lung in 1927 to alleviate breathing problems.

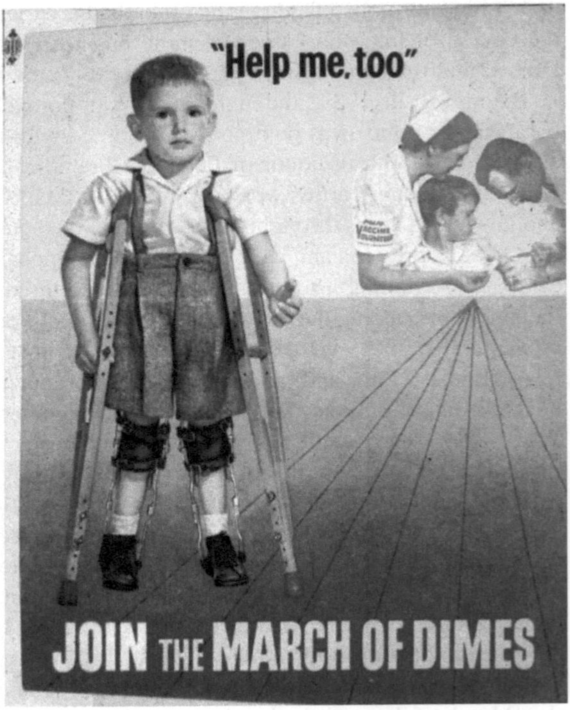

Fig. 12.4 Poster child for the March of Dimes

They devised a version of a tank respirator that could artificially maintain respiration until a person could breathe independently, usually after a week or two. The machine was powered by an electric motor with two vacuum cleaners; the pump changed the pressure inside a rectangular, airtight metal box, pulling air in and out of the lungs. These were large contraptions with "windows" so that nurses could look in on the patients. Beds were adjusted so that they could slide into the iron lung. In 1930 an iron lung cost about $1,500—the average price of a home. Money for iron lungs was collected and distributed by the National Foundation for Infantile Paralysis (two examples are shown in Fig. 12.5).

The method of treatment of withered limbs, such as the legs and arms, was also controversial, with Sister Elizabeth Kenny of Australia advocating physiotherapy as a treatment, in order to restore feeling in the muscles of diseased legs and arms. Others wanted to maintain the withered limbs in casts or plaster. Sister Kenny, from a small town in rural Australia, worked as a nurse during World War 1, and on her return to Australia treated children who had muscle injuries from accidents, and later treated children with polio. She was not aware of the standard treatment at the time, and developed her own methodology. Instead of immobilizing affected limbs in casts or stents, she used hot packs and massage to loosen afflicted muscles. She reported some success in getting children to walk who had previously been paralyzed. She then established clinics in Western Australia, but ended up at

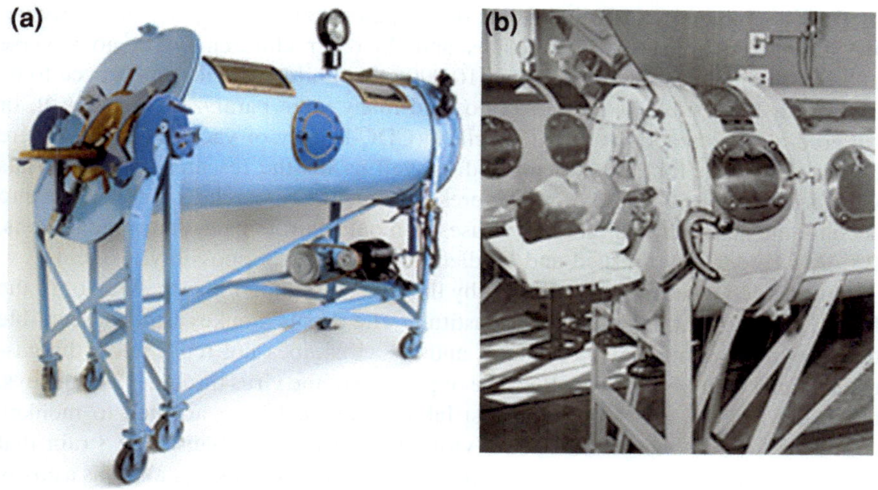

Fig. 12.5 **a** and **b** Iron lungs (Wikipedia)

odds with the medical establishment over her methods. She also established treatment centers all over the U.S. and had her headquarters in Minneapolis, Minnesota. She was one of the founders of the science of physical therapy, and her methodology is still quite acceptable today [9, 10].

In the U.S., the response to the polio epidemic was a philanthropic and public one with little interference from the government. Canada was also hard hit by outbreaks of polio but here the response differed because in Canada the provincial governments took over the care of victims of polio and paid for hospitalization and treatment. In fact, it could be looked at as the beginnings of the Canadian health insurance system.

12.8 Development of a Polio Vaccine

In 1935, two separate teams were at work developing and testing a polio vaccine. Both projects came to disastrous ends. At New York University, Maurice Brodie, MD (1903–1939), a young Canadian researcher, prepared a killed poliovirus vaccine, testing it on chimpanzees, on himself, and finally on children. He used polio grown in monkey spinal cords, and inactivated the virus with formalin. When injected into monkeys, the animals produced antibodies to the virus—indicating an active vaccine. Park, the department head, and Brodie self-injected the virus, with no adverse results. He enrolled and injected about 11,000 individuals (in both control and vaccine groups) in a vaccine trial. The results were inconclusive, and the vaccine produced an allergic response in a number of children, possibly due to the presence of monkey myelin basic protein [5]. Meanwhile, John Kolmer, MD, of Temple University in Philadelphia, developed an attenuated poliovirus vaccine. He vaccinated himself, his two sons and 23 other children—with no adverse effects. He then went on to vaccinate 10,000 other children. The tests proved to be a disaster. Nine subjects died of polio, and many were paralyzed, became ill, or suffered allergic reactions to the vaccines (CDC-history of vaccines). These were tragic cases both for those vaccinated as well as for the researchers. Both trials were done in haste because of public pressure, and the researchers did not take into account the possibility of monkey viruses or monkey proteins in their preparations. Brodie's career was finished and he died at the age of 36, possibly a suicide.

Progress on a vaccine was delayed by the ideas of Simon Flexner, director of the Polio Laboratory at the Rockefeller Institute. The rhesus monkey is not susceptible to polio in the same way as humans, and the virus does not replicate in the gastrointestinal tract of monkeys. Because of passage of the virus through the monkeys, the strain isolated and grown in most laboratories had been adapted to monkey neural tissue and was not the same virus that attacked humans. The strain that emerged from these passages was called the "MV" strain. This mistake was a costly one in lives and it delayed the development of the "real" vaccine by some 20 years.

Because of these failures, and the continuing toll of the disease, Basil O'Connor appointed Thomas Rivers, director of the Rockefeller Institute Hospital, to head the National Foundation Committee on Scientific Research. Rivers set down a list of priorities that had to be answered in order to continue research towards a vaccine. These included basic questions, such as how the virus gets into a person, how is it spread, and the characterization and life cycle of the virus. Groups of scientists received grants to investigate these questions. Another question that needed answering was how many species of polio virus are there?

In 1941 Albert Sabin performed autopsies on polio victims and found that the virus was plentiful in the digestive tract but not in nasal tissue. It was clear that this was an enterovirus, entered through the mouth and replicated in the gut. By 1948 there was a major breakthrough, (already discussed in Chap. 3 of this book), when John Franklin Enders and his colleagues discovered that poliovirus could be grown in cell culture, in embryonic skin, and in muscle cells. The virus could now be grown in large amounts in culture. This paved the way for the possible development of a vaccine. The true nature of the virus had been discovered and its mode of transmission, and it was now possible to grow the virus in large quantities.

The National Foundation approached Dr. Jonas Salk, a young researcher at the University of Pittsburgh with a grant to investigate whether there were two or three different species of polio affecting man. If there were multiple species, then a vaccine would be required against all of them. Salk's laboratory tested clinical samples from all over the U.S. In 1949 it was reported that there were three serotypes [11]. Meanwhile, in 1949 Dr. Isabel Morgan showed that formalin could inactivate the virus and when injected into monkeys protect them from polio [12, 13].

Independently, Lederle, a pharmaceutical company in New York State, in 1945 had hired Dr. Hilary Koprowski, a refugee from Warsaw University's Medical School. He did research on the Lansing strain of poliovirus (the most pathogenic strain) and passaged it through the brains of mice and cotton rats, and after 30 or so passages isolated a weakened virus that he fed to nine chimpanzees. All nine were protected from polio. He then drank the material himself, as did his colleague, Thomas Norton. Apparently it tasted like cod liver oil, something that those of us who were children during World War II remember well [14]. This was the first trial of an oral vaccine.

This vaccine was the first tested in children in the U.S. In early 1950 Koprowski administered the vaccine to children at a home for the "feeble-minded and epileptic" in New York State. Twenty-one children were vaccinated with no ill effect, and all developed neutralizing antibodies. This was kept secret until a conference of the National Foundation in 1951. Koprowski went on to produce other attenuated strains of virus, taking advantage of growing the virus in cell culture. A shake-up in the company in 1954 resulted in Koprowski leaving Lederle, by now American Cyanamid and he became director of the Wistar Institute in Philadelphia.

Koprowski and Sabin were both working on an attenuated vaccine. It was their belief, and that of Enders and others in the "polio establishment," that only an attenuated vaccine would work. Jonas Salk was a relative newcomer to the group and was not easily accepted by them. He graduated from NYU in 1939, and completed his internship at Mt. Sinai Hospital in New York. While at NYU, he had come to the attention of Thomas Francis, a professor of Microbiology; this later proved a useful contact. After completing his internship and not being able to find another position, he applied for a position at the Department of Epidemiology at the University of Michigan, where Francis was now chairman. Together they worked on inactivating influenza virus with formalin to produce a vaccine for the army. In 1947 he applied for an independent position, and found one at the University of Pittsburgh Medical School. He brought with him a grant to continue working on influenza, but he also received funding from the National Foundation to perform polio typing. By 1951, with the end of the typing experiments, proving that only three strains of polio existed in the wild, he received more grants than anyone else at the medical school.

By 1950 Salk had tested both live attenuated and killed poliovirus in monkeys, and he was now ready to test his killed vaccine in people. In December 1951 an expert committee was convened by the Director of the National Foundation to consider Salk's request to start treating children with the vaccine. The committee, among whom were Sabin and Enders, turned the idea down, considering it too dangerous, and they wanted to wait until Sabin had completed his work on an attenuated vaccine. However, there was great pressure to do something, so the administration of the National Foundation agreed to perform vaccination with Salk's vaccine in secret, just as Koprowski had done around the same time. The State of Pennsylvania approved the plan to test the vaccine on children in the Watson Home for crippled children—most of them crippled by polio—and in a school for the mentally retarded and feebleminded. Testing began in June 1952, when 42 children in the Watson Home were given the vaccine with no ill effects, and with a general rise in antibody titer. This was the most suitable group of children for a trial since they would all be immune to polio (at least one strain of it) and would already have antibodies. Salk then tested the vaccine in the home for retarded children, again with no ill effects. At a meeting of the National Foundation Immunization Committee, Salk presented his data: he had vaccinated 161 children with his vaccine. It caused a commotion with Sabin, Enders and others, believing that much more research was needed and that it was too premature. 1952 was one of the worst years for polio, and by the year's end more than 20,000 had been paralyzed and 3,000 had died [5].

In contrast to the Koprowski trials, Salk leaked his results to a newspaper correspondent and in February 1953, Time Magazine carried the news. Salk emphasized that the results were preliminary and that there was the need for a large trial before declaring the vaccine a complete success. As expected, Sabin, Enders, and others were shocked by the news. The National Foundation, however, convened a new committee to design a large-scale trial. There was, as expected, much discussion on the form of the trial. A large number of children would have to be vaccinated in order to test the efficacy of the vaccine, since paralysis (or death) affected only a small percentage of the infected. There was a proposal to perform a double-blind study, in which half the children would receive a placebo and the other half the actual vaccine. None of the administrators of the vaccine would know who was receiving the actual vaccine, and the data would be analyzed after the trial was finished, so this would avoid bias in the analysis. Sabin argued that the Salk vaccine was not ready for testing, and after strenuous arguments, the idea of the double-blind test was abandoned.

Rather than give up, Basil O'Connor and others on the committee of the National Foundation asked Thomas Francis, known to be a meticulous scientist, to oversee a large trial. He reinstated the idea of a double-blind trial, with a placebo given to half the children. Because the frequency of active infection—as opposed to abortive infection—was so low, half a million children would have to be vaccinated to make the trial statistically meaningful. Over 600,000 children between the ages of seven and eight would receive three injections over an eight-week period. To obtain enough vaccine, the National Foundation arranged commercial contracts with Connaught Laboratories of Canada, Parke Davis, and later Eli Lilly and Wyeth. Some smaller laboratories, such as Cutter, were also involved, with Eli Lilly and Park Davis suppling most of the vaccines. By the end of June 1954, vaccination of the children was complete and on April 12, 1955, Francis announced the results of the trial. Comparing vaccine- and placebo- injected groups, the Salk vaccine was 80–90 % effective. The excitement of that day, and the events following the announcement, were captured in an excellent book, *The End of Plagues* [5]. The vaccine was quickly licensed by the newly created Department of Health, Education and Welfare, and Jonas Salk was proclaimed a hero.

On April 24, a six-year-old girl developed the symptoms of polio six days after receiving the vaccine; she died three days later. Thus began a series of polio cases that eventually reached 200. They were all traced to one batch of vaccine, prepared by Cutter. After investigation, it was found that faulty filters were used, which allowed clumps of virus, some not inactive, to get through the preparation. Once this glitch had been found, the Salk vaccine proved safe. The number of annual cases of polio plummeted from 35,000 in 1953 to 5,600 by 1957. In 1961, only 161 cases were recorded in the U.S. Thus, despite the many misgivings, the vaccine worked.

Sabin continued to work on his vaccine, and clinical trials began in 1954 at Chillicothe Federal Penitentiary in Ohio. The trial was a success with high antibody titer being induced. Despite the Cold War, President Eisenhower had promised to cooperate with the Russians on defeating polio. In 1956 a Russian delegation arrived in the U.S. to assess the Salk and Sabin vaccines, and the Russians chose to use Sabin's. In 1959, 10 million Russian children were successfully vaccinated, proving that the Sabin vaccine strains were safe and effective. The Russian authorities then proposed the vaccination of every Russian under the age of 20, and a massive campaign was undertaken to vaccinate 70 million people; the vaccine worked.

In 1960, Lederle Laboratories revived the Koprowski vaccine, which could be given in one oral dose. A trial was launched in Florida, with 400,000 receiving the vaccine. A few cases of polio occurred; whether they were caused by or related to the vaccine is unknown, but this was the deciding factor in licensing the Sabin vaccine for use in the U.S. Although the Salk vaccine worked well, and by 1961 the occurrence of polio was considerably decreased, the American Medical Association demanded clarification on which vaccine to use. In September 1961 the USPHS licensed the Sabin vaccine for use in the U.S., and by 1979 polio had been eliminated from the U.S.

The Koprowski vaccine was eventually used in Africa. However, Koprowski never received the acclaim he deserved, perhaps owing to his work in industry or due to his larger-than-life personality. The National Foundation did not support his research (see http://www.nytimes.com/2013/04/21/us/hilary-koprowski-developed-live-virus-polio-vaccine-dies-at-96.html?pagewanted=2&_r=0).

Today polio is almost non-existent in the U.S. and almost completely eradicated world-wide, although there are still a few holdouts where complete vaccination has not occurred, such as India and certain regions of Africa.

The success in developing a polio vaccine was due to the ease of growing the virus in culture, particularly in non-neural tissue (although this took a long time of trial and error) and because the virus had no animal reservoir. The development of these vaccines allowed for mass vaccination. Utilizing the Sabin attenuated vaccine, herd immunity could be established in populations, since this attenuated virus could replicate and spread in a population without causing any of the symptoms of the disease.

Although it was assumed that by 2010 polio would be wiped out worldwide, this has not been the case. There have been sporadic outbreaks in India, particularly among undernourished children, who do not seem to respond to the vaccine, and in parts of Africa where there has been resistance to vaccination. However, this year it appears that polio is close to being eradicated in Nigeria, where a few cases have been reported. Other sources of polio are Pakistan and Afghanistan. The World Health Organization and the Indian government, in the hope that the virus has finally been eradicated or close to eradication, have carried out a massive campaign against polio. Recently, polio workers were murdered in Pakistan, and resistance to evacuation has increased in Nigeria due to Islamic Fundamentalists who see a "Western plot." Within the last few weeks of writing this chapter, there have been outbreaks of polio in Syria, due to the upheavals caused by the civil war; fighters may have brought the virus from Pakistan (Table 12.1).

Table 12.1 Polio history timeline (adapted from CDC)

1894	First U.S. polio epidemic
1905	Contagious nature of polio discovered
1908	Poliovirus identified as disease-causing agent
1910	Flexner investigates polio immunity
1916	New York City polio epidemic
1921	Polio strikes Franklin D. Roosevelt
1929	Invention of the iron lung
1934	First Roosevelt Ball organized by Basil O'Connor
1935	Early polio vaccine trials
1936	Virus grown in human nervous tissue in culture
1938	March of Dimes established
1941	Sabin and Ward showed that virus was in the digestive system
1948	Hilary Koprowski tests polio vaccine he developed on himself with no ill effect
1949	Three types of immunologically distinct types of poliovirus identified
1949	Enders and colleagues grow poliovirus in different tissues
1950	Koprowski tests his vaccine in a group of children, with no ill effect
1951	Salk develops methods of growing poliovirus in large amounts in monkey kidney cells
1952	Major epidemic of polio in the U.S.
1952	Salk begins testing inactive vaccine on children
1953	Salk injects himself and family with his vaccine
1954	Massive polio vaccine trial initiated with children in the U.S.
1955	Salk polio vaccine results announced: 90 % protection. Vaccine licensed
1955	Cutter incident, vaccination discontinued
1957	Koprowski begins vaccination in the Congo and other countries
1959	Sabin vaccine used in trials in the Soviet Union with success
1960	Sabin vaccine licensed in the U.S.
1968	Salk vaccine phased out in the U.S.
1985	Goal set for eradication of polio in the Americas
1994	Polio eradicated from the Americas, including South America
1997	Massive immunization efforts in India, Sudan and other countries
2000	99 % eradication of polio in the world

12.9 Picornaviruses

Picornaviruses have played an important role in the history of virology (see Chap. 1). One of the first diseases identified as viral was Foot and Mouth Disease virus of cattle by Friedrich Loeffler in 1897. In more recent times, picornaviruses have also been used as models for studying virus replication and virus structure, due to the ease in growing such viruses in culture and the high yields obtained. Poliovirus was among the first animal viruses to be shown to undergo self-assembly in the test tube and the first viruses produced from RNA in vitro [15, 16].

As the name suggests, these are small ("pico"), single-stranded RNA viruses. This family of virus contains 12 different genera: enterovirus, which includes polio and rhinovirus, the common cold virus; apthovirus, which includes foot and mouth disease virus of cattle; and hepatovirus, which includes hepatitis A. A major reason for lack of immunity to the common cold is that there are 100 or so unique antigenic strains (serotypes), strains that are constantly arising as a result of mutations in the RNA.

These viruses are small icosahedron structures (see Chap. 2), typically with four different coat polypeptides wrapped around single-stranded RNA (Fig. 12.6). This single-stranded RNA is termed "+polarity" since the RNA immediately acts a messenger RNA for the synthesis of enzymes used in viral replication. Thus viral replication and translation of the mRNA into viral proteins necessary for replication, and viral coat proteins, occur simultaneously. Picornaviruses have very limited information content, coding for only five or six proteins. When the virus attaches to the cell receptor, the virus is surrounded by the cell membrane to form a vesicle that is taken into the cell cytoplasm, and the pH of the vesicle changes, resulting in un-coating of the virus (Fig. 12.7). The viral RNA is translated into a long polyprotein that is then cleaved into individual proteins that will eventually form the four viral coat proteins and enzymes necessary for viral replication and processing of proteins. At the same time, the enzyme responsible for viral replication (replicase), is synthesized from the input strand of RNA and results in the copying of the RNA into a complementary RNA, which in turn is copied back into the + strand viral RNA. Part or all of these + and − strand RNAs may form double-stranded RNA structures through complementary binding. This double-stranded RNA plays a role in activating the immune and interferon response that occurs in vivo by binding to toll-like receptors. The protein coats accumulate in one part of the cell and the nucleic acid in another. By a mechanism still not completely understood, the assembly of the virus occurs either in unique structures in the cell known as Golgi bodies, or on internal membrane structures (endoplasmic reticulum), in which the viral proteins are wrapped around the folded RNA or empty capsids are formed into which the RNA is packaged, and hundreds

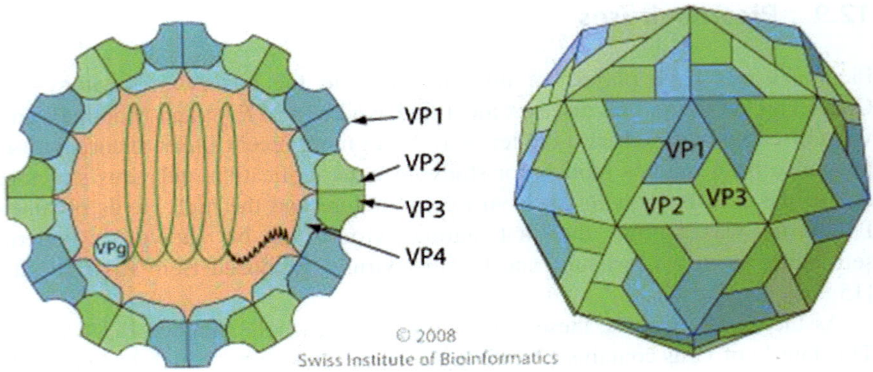

Fig. 12.6 Structure of typical picornavirus. Note that VP4 is internal to the virus and undergoes a conformational change (with permission: ViralZone, SIB Swiss Institute of Bioinformatics)

of thousands of viral particles are released from the cell. In the case of most picornaviruses, the cells are lysed (destroyed) by the large amounts of virus released. The structure of the virus and mechanism of viral replication are illustrated in Figs. 12.6 and 12.7.

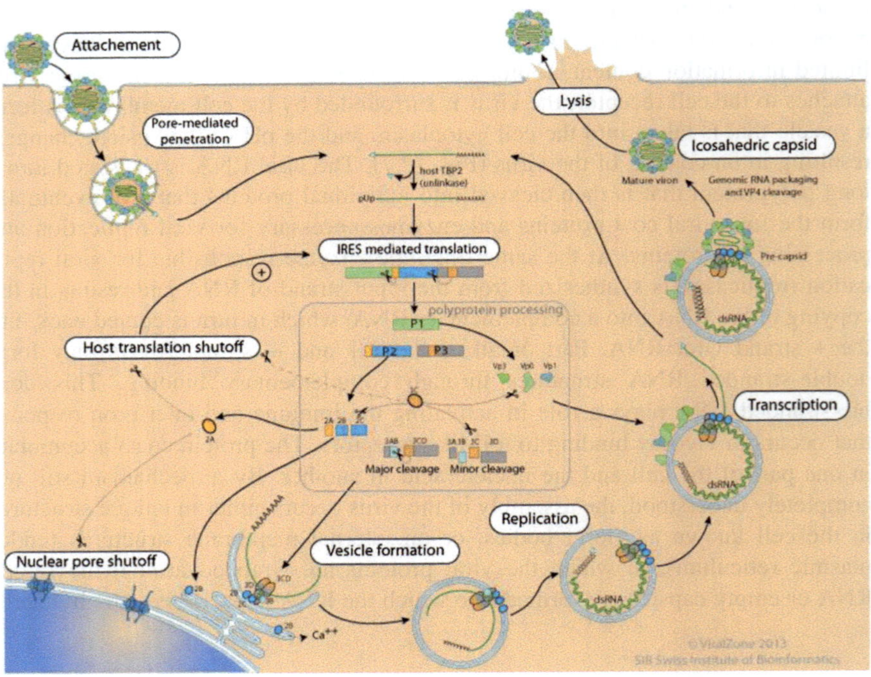

Fig. 12.7 Poliovirus replication cycle (with permission: ViralZone, SIB Swiss Institute of Bioinformatics)

12.10 Foot and Mouth Disease Virus

Although FMDV is not a human virus, it has had a major impact on the economies of many western European countries, particularly that of the U.K. This virus also belongs to the picornavirus family (apthovirus). It causes high fever, lesions in the mouth and on the legs (hoofs), and lameness in many domestic animals including cattle, pigs, goats and deer. It has low mortality but can affect milk production over a long period of time. However, mortality in young animals is high. The virus can be spread by contact between animals, is found in animal excretions, and can even be transmitted by the wind and carried in clothing. Many countries live with the disease and vaccinate their herds of cattle, while other countries have found it more economically efficient not to vaccinate the cattle, and attempt to cull them if necessary in order to be free of the disease. This allows for greater export of meat products; the U.S., for example, has such a policy. The U.K. has tried this policy but has had many outbreaks of FMDV, so rather than vaccinate against the disease, they have followed a policy of mass slaughter of infected herds, and nearby herds. Cattle and people have been strictly "quarantined" during periods of outbreaks. The mass slaughter of cattle has taken a heavy toll on British agriculture. Vaccination is banned in the EU in an attempt to eliminate the disease by culling. To me as a virologist, this seems short-sighted, since the cost of mass slaughter is prohibitive and has not been successful in eliminating the disease in the long run. The arguments against vaccination are that it would be difficult to identify which animals have had FMDV from those vaccinated. However, new vaccines that use recombinant adenovirus-FMDV combinations are being developed. These vaccines would only express some of the FMDV proteins and not others, thus allowing for differentiation between true infection and vaccinated animals. Taiwan and some countries of western Europe have adopted vaccination rather than wholesale slaughter, or a combination of both. This approach has also been supported by top-notch scientists working at Pirbright and Plum Island, the two major centers of FMDV research. The virus is endemic in parts of Asia, the Middle East and South America; it does undergo mutation at a high rate, making it difficult to develop a long-lasting vaccine. Luckily, most meat-producing countries have been FMDV-free for a considerable time, with the last major outbreak in the U.S. in 1927.

The last major foot and mouth epidemic in the United Kingdom occurred in 2001, with a smaller outbreak in 2007. The total costs arising from the 2001 outbreak have been put at no less than £9 billion, with at least £3 billion in direct costs to the public sector and about £5 billion in costs to tourism and the rural economy. Out of the total costs incurred during the outbreak, compensation to farmers for the slaughter of their animals and welfare reasons were placed at £1.34 billion. The cause of this outbreak may have been pigswill that was not completely boiled and processed according to law; it also contained contaminated meat. During this outbreak, over ten million sheep and cattle were killed. Tourist trade in the Lake District, a major tourist area, was closed off. (For a description of the conditions in the U.K., and a critique of the handling of the outbreaks, see. http://www.numberwatch.co.uk/lest_we_forget.htm).

I visited the U.K. in 2001 spending my vacation in the south of the country, not the hardest-hit area, although large segments of Dorset were off-limits. Many parts of Britain, including parkland, were closed to the public. One had to walk through troughs of disinfectant while hiking in the country. Not only was this a great cost to the farmers, it also cost the hotels and other business catering to tourists dearly. The picture is not a pleasant one, with hundreds of head of cattle being slaughtered and then burned and the smoke and odor spreading throughout the countryside.

References

1. Bollet, A. J. (2004). *Plagues and Poxes*. New York: Demos Medical Publishing, Inc.
2. Underwood, M. (1789). *A treatise on the diseases of children with general directions on the management of infants from birth*. London: John Churchill.
3. Axelsson, P. (2009). Do not eat those apples; they've been on the ground!: polio epidemics and preventive measures, Sweden 1880s–1940s. *Asclepio; archivo iberoamericano de historia de la medicina y antropologia medica, 61*(1), 23–38.
4. Grimshaw, M. L. (1995). Scientific specialization and the poliovirus controversy in the years before World War II. *Bulletin of the History of Medicine, 69*(1), 44–65.
5. Rhodes, J. (2013). *The End of Plagues*. United States: Palgrave, MacMillan.
6. Renne, E. P. (2011). The 1917 Polio Outbreak in Montpelier. *Vermont. Vermont Historical Society Journal, 79*, 162–181.
7. Oldstone M. B. A. (2010). Viruses, plagues and history. 2nd edn, Oxford: Oxford University Press.
8. Rogers, N. (2007). Race and the politics of polio: Warm springs, Tuskegee, and the March of Dimes. *American Journal of Public Health, 97*(5), 784–795.
9. Oppewal, S. R. (1997). Sister Elizabeth Kenny, an Australian nurse, and treatment of poliomyelitis victims. *Image–the Journal of Nursing Scholarship, 29*(1), 83–87.
10. Golden, J., & Rogers, N. (2010). Nurse Irene Shea studies the "Kenny method" of treatment of infantile paralysis, 1942–1943. *Nursing history review: official journal of the American Association for the History of Nursing, 18*, 189–203.
11. Bodian, D., Morgan, I. M., & Howe, H. A. (1949). Differentiation of types of poliomyelitis viruses; the grouping of 14 strains into three basic immunological types. *American Journal of Hygiene, 49*(2), 234–245.
12. Morgan, I. M. (1948). Immunization of monkeys with formalin-inactivated poliomyelitis viruses. *American Journal of Hygiene, 48*(3), 394–406.
13. Morgan, I. M. (1950). Persistence of neutralizing antibody for a year following vaccination of monkeys with Lansing poliomyelitis virus. *Proceedings of the Society for Experimental Biology and Medicine, 75*(2), 305–308.
14. Koprowski, H. (1960). Historical aspects of the development of live virus vaccine in poliomyelitis. *British Medical Journal, 2*(5192), 85–91.
15. Cello, J., Paul, A. V., & Wimmer, E. (2002). Chemical synthesis of poliovirus cDNA: generation of infectious virus in the absence of natural template. *Science, 297*(5583), 1016–1018.
16. Franco, D., Pathak, H. B., Cameron, C. E., Rombaut, B., Wimmer, E., & Paul, A. V. (2005). Stimulation of poliovirus synthesis in a HeLa cell-free in vitro translation-RNA replication system by viral protein 3CDpro. *Journal of Virology, 79*(10), 6358–6367.

Hepatitis 13

Abstract

There are many viruses that cause hepatitis. In this chapter we discuss three of them: hepatitis A (jaundice), hepatitis B (serum hepatitis), and hepatitis C. Hepatitis A is an acute infection, whereas hepatitis B and C are blood-borne diseases and cause chronic infections and eventually cirrhosis of the liver and liver cancer. Outbreaks of hepatitis A were once very common in most of the world, since it was spread by unsanitary conditions Hepatitis A disrupted military campaigns throughout the ages. A vaccine is now available, and hepatitis A has almost disappeared from the developed world. A recombinant vaccine has also been developed against hepatitis B, resulting in a decline in cases. A vaccine is not yet available for hepatitis C, although there is active research in this area. Until recently Hepatitis C was routinely treated with a combination of interferon and ribavirin, which had severe side effects and low cure rates. This treatment is being replaced by nucleotides/nucleosides analogues that inhibit viral protein processing and replication enzymes. The large numbers of hepatitis B cases in the world are due to sexual transmission and mother-to-child transfer. The sharing of needles when using recreational drugs, blood transfusions, and unsanitary tattoo parlors—all these spread hepatitis C. All three viruses infect millions of people worldwide.

13.1 What Is Hepatitis?

The term "hepatitis" refers to an inflammation of the liver usually accompanied by increases in liver-specific enzymes; it is also known as "jaundice." Alcohol use, toxins, specific prescription drugs, or a virus can cause this condition. The symptoms of hepatitis include yellowing of the eyes and skin due to an increase in bilirubin (a breakdown product of heme in the blood or from the bile duct),

M. W. Taylor, *Viruses and Man: A History of Interactions*,
DOI: 10.1007/978-3-319-07758-1_13,

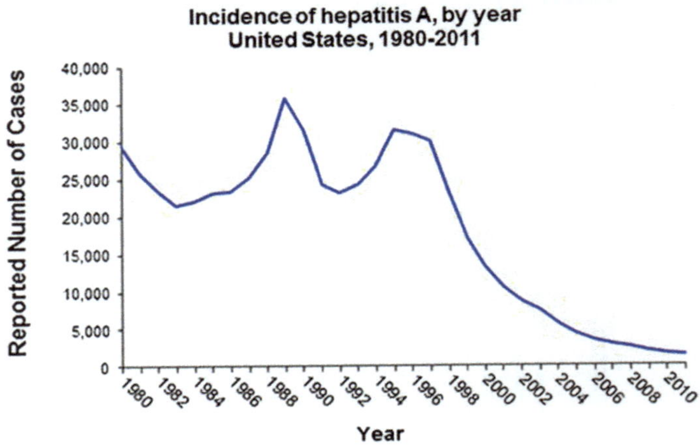

Fig. 13.1 Number of acute hepatitis A cases in the U.S. from 1980 to 2011. Note effect of the vaccine (courtesy of the Centers for Disease Control, Atlanta, USA)

clay-colored feces, fatigue, vomiting, and nausea. The yellowing of the eyes and skin are the most obvious symptoms. Three major viruses known as hepatitis A, B, and C, which reflect the order in which they were discovered, cause hepatitis. The above-mentioned symptoms are common to all three viral infections. In the last few years, there were about 1,300 cases of hepatitis A each year in the U.S., in contrast to the hundreds of thousands reported before the introduction of a vaccine (see Fig. 13.1). All children at age 1, and adults when they travel to countries with a high occurrence of the disease, are currently vaccinated. Hepatitis A infections result from contaminated food and unsanitary conditions and is most common in countries around the Mediterranean and the Far East. In the U.S. in the past, outbreaks have been associated with contaminated meat, vegetables and seafood.

13.2 History of Jaundice

Jaundice has a long history. Hippocrates mentions it in his work *Epidemics*, and the Babylonian Talmud of the fifth century A.D. discusses "*yerakon*" (greenish), an allusion in the Bible to a greenish pallor of people, or yellowing of plants (a controversial translation of the Hebrew). In the Talmud, yerakon is interpreted as a disease similar to modern-day jaundice. Pope Zacharias, who is credited with recommending quarantine for the disease, recognized it as a communicable disease in the eighth century [1]. It reached epidemic proportions with the increased populations of European cities in the seventeenth–nineteenth centuries, and was particularly rampant during military campaigns, where there was overcrowding of troops in unsanitary conditions. From the seventeenth century onwards, jaundice among soldiers increasingly became a hazard of military life and was reported from many theaters of war—in the Netherlands in the War of the Austrian Succession (1740–1748), in Egypt during Napoleon's brief incursion there (1798), and in Paris during the Franco-Prussian War (1870) [1].

There were more than 40,000 cases among 2,000,000 soldiers on the Union side in the American Civil War (1861–1865), 6,000 cases among almost half a million British and Dominion soldiers in the Boer War in South Africa (1899–1902), and huge numbers of Japanese sailors in the war against Russia (1904–1905). In the Middle East campaign of World War I, it affected up to 25 % of some units. During World War II, over 5,000,000 soldiers and civilians on the German side, and more than 150,000 U.S. army soldiers suffered from hepatitis [2]. In the battle of El Alamein in the Libyan desert, there was an outbreak of hepatitis among Australian and German troops at the frontline; this was possibly due to the large numbers of dead bodies and feces in the field of action. During the Italian campaign, American troops suffered greatly from jaundice at a rate of 37 cases per 1,000 soldiers. Of about 16,000 cases, each one spent an average of six weeks in hospital. A delay in D-day by one month may have resulted from 22,000 cases reported before June 1944 [3]. Thus, this disease had a profound effect on military history, as we have also seen for smallpox and influenza. Most of this was due to hepatitis A.

13.3 Hepatitis A

Hepatitis A is a self-limiting acute infection, which may last from a few weeks to a few months. The virus is a picornavirus, from the same family as polio and other enteroviruses and is spread by oral-fecal transmission. Hepatitis A appears cyclically in the U.S. with peaks every 15 or so years. Since the introduction of the vaccine in 1996, the number of cases has decreased considerably (Fig. 13.1) and it is no longer considered a hazard in the U.S. or Western Europe. Hepatitis A infections still occur frequently in Africa, South America and parts of Asia (Fig. 13.2). Hepatitis A vaccine was introduced incrementally in 1996, first for children dwelling in communities with the highest rates of disease, and expanded in 1999 to children living in states/communities with consistently elevated rates of infection (http://www.immunizationinfo.org/vaccines/hepatitis). The level of hepatitis A infection is now the same in most areas of the U.S. There are a number of different vaccines available, both attenuated virus and inactive (dead) virus. A new vaccine containing hepatitis A and hepatitis B antigens was introduced in the U.S. in 2001 for persons age 18 or older.

An outbreak of hepatitis A involving 601 patients and three fatalities occurred in November 2003, associated with a restaurant in Pennsylvania; [4] the cause of the infection was salsa that contained green onions grown in northern Mexico. Another outbreak of hepatitis A occurred in the U.S. in the spring-summer of 2013, making 168 people sick; it was tracked down to a frozen berry mixture that contained pomegranate seeds imported from Turkey. A similar eruption occurred

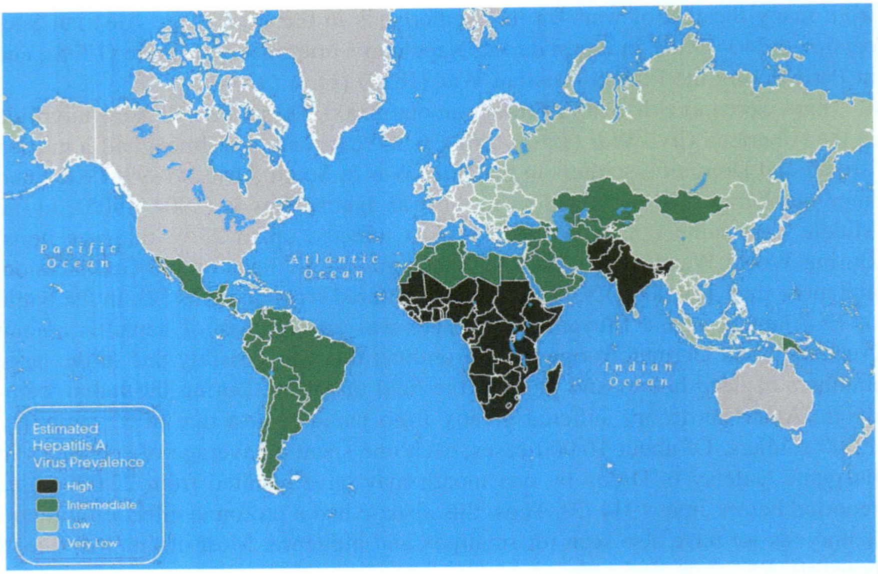

Fig. 13.2 Prevalence of hepatitis A in the world (courtesy of the Centers for Disease Control, Atlanta, USA)

in 2012 in British Columbia, and also in Europe, from pomegranate seeds imported from Egypt; none of those who became ill had been vaccinated.

An unusual feature of the virus has only recently been discovered. Although hepatitis A is a picornavirus, a non-enveloped virus, some particles isolated from the blood of infected individuals have a membranous envelope around them, [5] which may be a mechanism whereby the virus avoids neutralization by antibodies.

13.4 Hepatitis B

The detection of the other hepatitis viruses, now known as hepatitis B and hepatitis C, called for a time non-A, non-B, is a fascinating story. Hepatitis B virus, at one time known as "serum hepatitis," was found by accident although viral causes of liver disease other than hepatitis A was suspected for some time.

Dr. Baruch Blumberg travelled throughout the world in the 1950s collecting blood samples in order to analyze serum proteins, particularly those from hemophiliacs, receiving blood from many different sources. His purpose was to examine whether such blood contained antibodies to "foreign" antigens that originated from the transfused blood. He discovered a unique antigen, termed the "Australia antigen"in the serum of an Australian aborigine that reacted with an antibody found in leukemia patients and in the blood of inmates of institutions such as prisons and old age homes. It was not present in the normal serum of adults in the U.S. Blumberg proposed that this antigen was connected with either the onset or presence of an agent responsible for leukemia. This antigen was also present in a relatively high percentage of Taiwanese [6] and later found in serum of individuals in mental institutions, e.g., Down's syndrome patients, and also occasionally in persons diagnosed with hepatitis or hemophilia. A lab technician working in the Blumberg lab with Australia antigen, serum converted—that is, her blood now contained antibodies to the Australia antigen and she developed mild symptoms of hepatitis. Another lab associate whose job it was to inject the antigen into rabbits also sero-converted. The antigen was later identified in some patients who underwent blood transfusions, and also in a group of patients with symptoms of hepatitis but not those with other liver diseases such as alcohol induced cirrhosis. Blumberg and colleagues [7] initially proposed that this factor was a genetic autosomal inherited factor, and family studies revealed that the antigen could be transmitted from mother to child.

These data, and the two accidental cases described above, indicated that the antigen was an infectious agent and not a genetic condition [8]. It was hypothesized that the Australia antigen was linked to a virus related to hepatitis. Electron micrographs indicated that the antigen had characteristics of a virus-like particle, about the size of a picornavirus (Fig. 13.3). This antigen, initially termed "Australia antigen" was later called "hepatitis B surface antigen" (HBsAg). As shown in Fig. 13.4, it is present during the acute phase of the disease.

Baruch Blumberg, in his Nobel Prize acceptance speech, made a point of the fact that his research on antigen polymorphisms accidently led to the discovery of the virus. By 1966 Blumberg and his colleagues were convinced that Australia antigen was associated with serum hepatitis, that it might be a virus, and that it was transmitted during blood transfusions. The paper claiming this was initially rejected, since many different viruses had been proposed as the cause of serum hepatitis. However, the research results were confirmed by a group of Japanese researchers [9] and others [10], and the results were immediately translated into practical use. Since serum hepatitis was very common after blood transfusion (about 18 %), blood was tested for the Australia antigen, and, if present, discarded. The exclusion of such blood led to a decrease in the number of new cases of hepatitis after blood transfusion, and within a few years this had become a routine test. Despite this, there were still many cases of hepatitis after surgery, indicating the possible presence of another virus.

In parallel work on the transmission of the serum antigens, the same group initiated research to visualize the antigens using electron microscopy. A number of different-size particles could be found in antigen-positive serum, representing different stages of the virus development. These included particles about 20 nm in diameter, which were aggregated by anti-Australia antigen antiserum. There were also elongated-like particles of the same diameter. Subsequently Dane, Cameron and Briggs identified a larger particle, about 42 nm in diameter, with an electron dense core of about 27 nm [11]. This was called a "Dane particle" and it represents the whole virus particle. Both the 20 nm and 42 nm particles contain Australia antigen on their surfaces; this is now termed "hepatitis B surface antigen" (HBsAg). The surface antigen can be removed from Dane particles by the action of

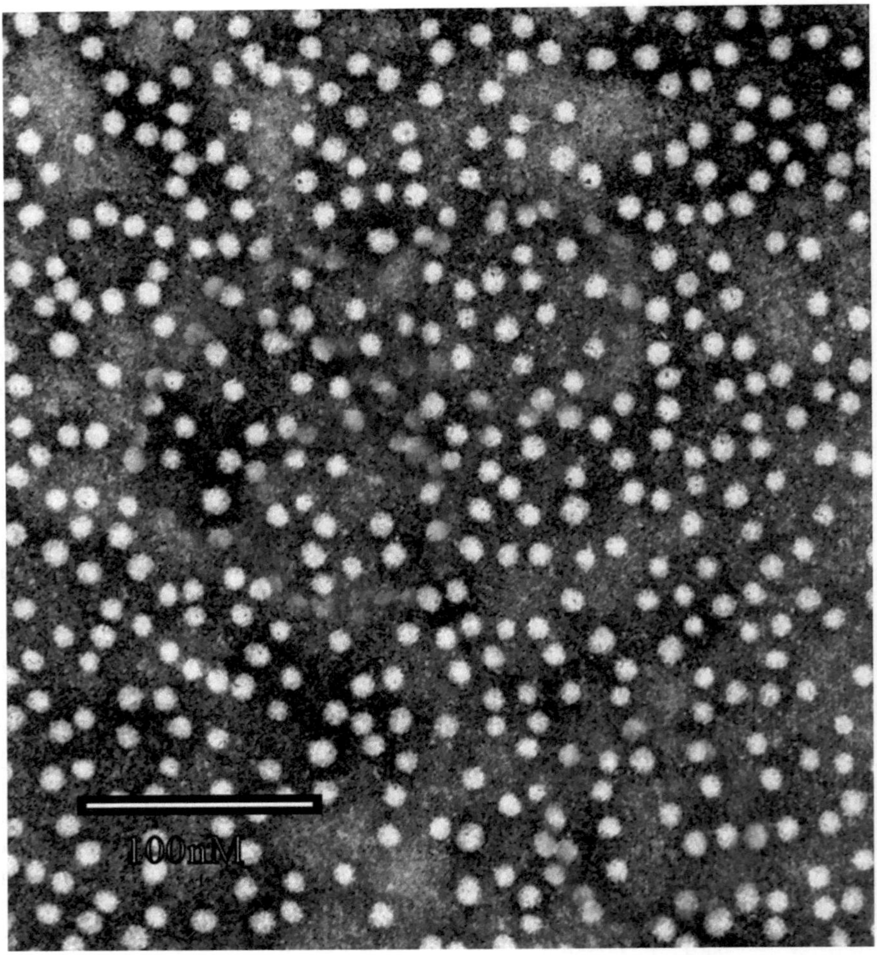

Fig. 13.3 Australia antigen (picture from Wikipedia)

detergents to reveal the core, which produces its own antigen, hepatitis B core antigen (HBcAg). All of these antigens can be found in the blood of hepatitis B patients at different times during infection. HBsAg and HBcAg can be found in the liver of infected patients (see Fig. 13.4).

Hepatitis B virus has been extensively studied and characterized [12–14] (Fig. 13.5). The virus contains double-stranded circular DNA with a single-stranded DNA region. The full-length strand is 3.2 kb and the single-stranded region 1.7 kb in length. The virus contains a reverse transcriptase, the enzyme that converts RNA to DNA, which is unusual for a DNA virus. This virus, and other related viruses found in other animals such as the woodchuck, Peking duck and ground squirrel, were called hepadnaviruses. They form a separate family in viral

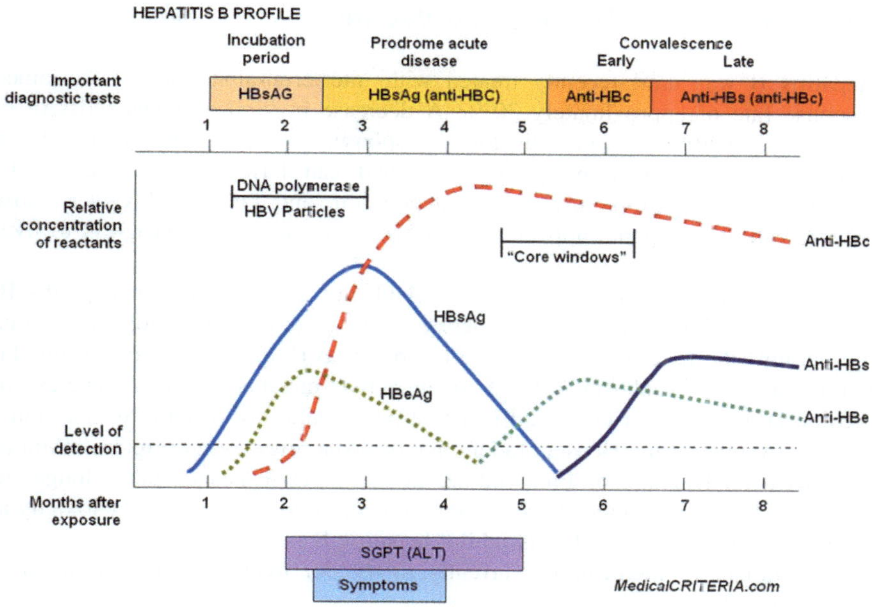

Fig. 13.4 The appearance of various antigens during the course of hepatitis B infection and increase in liver enzymes (ALT) (from the NIH Web site) http://labtestsonline.org/understanding/analytes/hepatitis

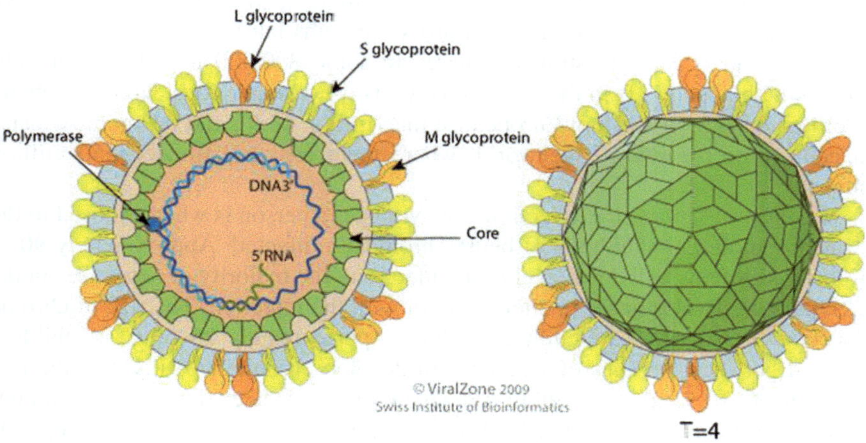

Fig. 13.5 Schematic of hepatitis B (courtesy of ViralZone, Swiss institute of Bioinformatics)

classification. The virus can be propagated in culture of human liver cell lines following transfection with DNA (personal communication). A schematic of the virus is presented in Fig. 13.5, which illustrates the partial double-stranded DNA.

13.5 Hepatitis B Therapy and Vaccine

Hepatitis B was until recently treated with interferon-alpha, with a sustained response rate of approximately 30 %. A decrease in viral antigens (HBsAg or HBeAg) is measured as an indicator of response after six months of interferon therapy. Interferon therapy is expensive and can have serious side effects, including flu-like symptoms, nausea, depression, and fatigue. However, those patients who do respond, appear to clear the virus and do not progress to liver cancer [15].

A series of newer drugs are now available for the treatment of hepatitis B. These include Lamivudine (2′,3′-dideoxy-3′-thiacytidine), an analog of cytidine that inhibits the reverse transcriptase of hepatitis B (http://en.wikipedia.org/wiki/Lamivudine). It is phosphorylated to an active metabolite that competes for incorporation into viral DNA and is also a potent inhibitor of HIV reverse transcriptase. Another drug, Hepsera (adefovir), is also a reverse transcriptase inhibitor and has the advantage of Lamivudine, in that viral resistance takes longer to develop. The FDA rejected Hepsera for treatment of HIV due to kidney toxicity at the level indicated. Lower levels inhibit hepatitis B.

A summary of medications currently in use for treatment of hepatitis B is presented in Table 13.1.

The hepatitis B vaccine was introduced in 1982, and is 99 % effective in preventing infection. It is the first vaccine to prevent a specific type of cancer, and is one of the first recombinant vaccines produced in yeast. The vaccine contains the viral surface antigen cloned into yeast DNA and thus the vaccine does not contain virus. Full immunity requires three rounds of vaccination, the first two administered a month apart, and the third six months later.

According to current estimates, there are some 800,000 to 1.4 million individuals with hepatitis B in the U.S. Hepatitis B is no longer considered a problem in either the U.S. or western Europe (see map in Fig. 13.6), and this number pales in contrast to the rest of the world, where there are an estimated 350 million infected.

In contrast to most viral infections, the younger a person is when exposed to the virus, the more his likelihood of being chronically infected. Approximately 90 % of infected infants will develop chronic infection. The majority of new cases in the U.S. (and the world) occur in children born with the infection. The risk of chronic infection decreases as the child gets older. Approximately 25–50 % of children infected between the ages of 1 and 5 will develop chronic hepatitis, and the risk drops to 6–10 % if the child is over 5 years of age (http://www.who.int/mediacentre/factsheets/fs204/en/). Worldwide, most people with chronic hepatitis B were infected at birth or during early childhood. Hepatitis B is currently predominantly spread by sexual transmission. Other modes of spread include intravenous drug injection (sharing of needles), hemodialysis, and accidents among health care workers. The virus is 50–100 times more infectious than HIV and can be transmitted in semen and vaginal fluids as well as blood. The virus is more

Table 13.1 http://www.fda.gov/ForConsumers/ByAudience/ForPatientAdvocates/ucm151494.htm

Brand name	Generic name	Manufacturer's name	Indication
Baraclude	Entecavir	Bristol-Myers Squibb	Chronic hepatitis B virus infection with evidence of active viral replication
Epivir-HBV	Lamivudine	GlaxoSmithKline	Chronic hepatitis B associated with hepatitis B viral replication and active liver inflammation
Hepsera	Adefovir dipivoxil	Gilead sciences	Chronic hepatitis B in patients ≥ 12 years of age
Intron A	Interferon alpha-2b	Schering	Chronic hepatitis B in patients 1 year of age or older with compensated liver disease
Pegasys	Pegylated interferon	Roche	Treatment of adult patients with HBeAg positive and HBeAg negative chronic hepatitis B who have compensated liver disease and evidence of viral replication and liver inflammation
Tyzeka	Telbivudine	Novartis	Chronic hepatitis B in adult patients with evidence of viral replication and either evidence of persistent elevations in serum aminotransferases (ALT or AST) or histologically active disease
Viread	Tenofovir	Gilead sciences	Chronic hepatitis B in adults

Fig. 13.6 Worldwide incidence of hepatitis B (from WHO)

robust than HIV and can exist outside the body for as long as seven days. The conditions for infection are similar to those for HIV, and the frequency increases with the number of sexual partners, whether hetero- or homosexual.

Hepatitis B's occurrence is highest in Sub-Saharan Africa and East Asia, including China (Fig. 13.6). Most people in these regions become infected with the hepatitis B virus during childhood and between 5 and 10 % of the adult population is infected. High rates of infections are also found in the Amazon and the southern parts of Eastern and Central Europe. In the Middle East and the Indian subcontinent, an estimated 2–5 % of the general population is chronically infected. Death from liver disease occurs in 15–25 % of chronically infected individuals (http://www.who.int/mediacentre/factsheets/fs204/en/).

The major consequences of HBV infection are loss of work and the psychological damage to the individual when he or she becomes aware that he or she is a carrier; there is also the danger of transmitting the infection to sexual partners and children. There has also been a stigma attached to those who are carriers of the virus, similar to that found in cases of AIDS: loss of jobs, fear of loss of insurance and fear of general discrimination.

For those who reach end-stage cirrhosis of the liver, transplant is the only alternative. This is a very expensive procedure; it is estimated that a liver transplant can cost as much as $577,000 in the U.S. (http://www.transplantliving.org/).

As described above for hepatitis A, there have been outbreaks of hepatitis B in the military, often of unknown cause. One of the worst cases was an outbreak of hepatitis following vaccination against yellow fever in a group of American servicemen in WW II; 50,000 cases of jaundice were observed and 62 deaths reported. It is suspected that as many as 300,000 men were infected with hepatitis B from the serum used to manufacture the yellow fever vaccine [3]. Hepatitis B increased among military personal with the increase of drug abuse among soldiers during and after the Vietnam War. Drug abuse has been a serious problem in the army since the 1960s. Since hepatitis B is sexually transmitted, there was also a high frequency among army personnel posted to the Far East, where hepatitis B is endemic in the local population. Hepatitis B vaccine is now routinely given to all military personnel.

The rates of chronic infection vary greatly in different parts of the world, with about 10–25 % of long-term carriers developing liver cancer. About 1 million people die of this disease annually. The mechanism of oncogenesis is unknown, although the immune system probably constantly attacks hepatocytes (liver cells) infected with the virus, causing extensive liver damage due to inflammation, as well as the production of cytokines. In particular, CD8 cytotoxic T-cells attack liver cells and the resulting cytokines cause damage to the liver. Viral DNA can be found in the chromosomes of liver cells, but it is not known whether it activates a specific oncogene. Since it takes a long time to develop cancer, there may be many steps along the oncogenic pathway that finally lead to tumor formation.

13.6 The Discovery of Hepatitis C

Despite the discovery of hepatitis A and hepatitis B, a large population still had the symptoms of hepatitis—that is, jaundice-like symptoms and elevated liver enzymes—but with no trace of either of these two viruses. Such patients were classified as non-A, non-B hepatitis infected, and the virus could not be identified directly in the serum or liver of such individuals. A similar situation with similar symptoms existed in chimpanzees. A group of scientists at the Chiron Corporation between 1982 and 1988 devised a novel approach to determine whether patients with non-A, non-B were infected with another virus [16–19]. The approach was based on purely recombinant-molecular biology, without knowing whether or not a virus existed, and certainly not knowing anything about the presumptive virus. Total complementary DNA (cDNA) from the plasma of chimpanzees suspected of having a high titer of virus was cloned into bacteriophage lambda, λg11, which expresses cDNA-encoded polypeptides at a high level. The phage library was then screened with serum from a non-A, non-B patient to detect possible viral epitopes that might react with the patients' antibodies. On screening millions of phage with serum from the infected individual, one complementary clone (5–11) was detected. The DNA from this clone did not hybridize to DNA isolated from either human or chimpanzee uninfected tissue. Single-stranded RNA, of between 5,000–10,000 nucleotides isolated from the liver of infected chimpanzees, hybridized to this clone. The RNA was further purified and sequenced, and contained sequences coding for a long polypeptide typical of a virus. The non-A non-B patients were infected with a single-stranded RNA virus belonging to the flavivirus family, and it was found that the majority of hepatitis cases not identified as hepatitis A or B were infected with what was to be called "hepatitis C".

13.7 Risk Factors for Hepatitis C; Testing

Hepatitis C may occur as an acute infection, or, more likely, develop into a chronic infection. In most cases, the disease can go unrecognized for years since symptoms develop slowly, with an elevation of liver enzymes, general fatigue and nausea, and, in relatively rare cases, it may develop into hepatocellular carcinoma (liver cancer). Hepatitis C is the leading cause of liver cancer and cirrhosis of the liver. It is important to test for the virus if there is any suspicion of infection due to past behavior, such as sharing of needles, intravenous injection of drugs, or even tattooing. Figure 13.7 illustrates the risk factors for hepatitis C and Fig. 13.8 the outcome of the infection. Note about 1 % will develop liver cancer.

According to the World Health Organization, there are 170 million people infected worldwide with chronic hepatitis C, of whom approximately 350,000 die each year. The CDC estimates that there are 3.5 million cases in the U.S., with about 16,000 new cases per year and an equal number of deaths. It is the leading cause of cirrhosis of the liver. (Note in Fig. 13.9 the extremely high numbers of infected individuals in Egypt. Pakistan, Romania and Italy).

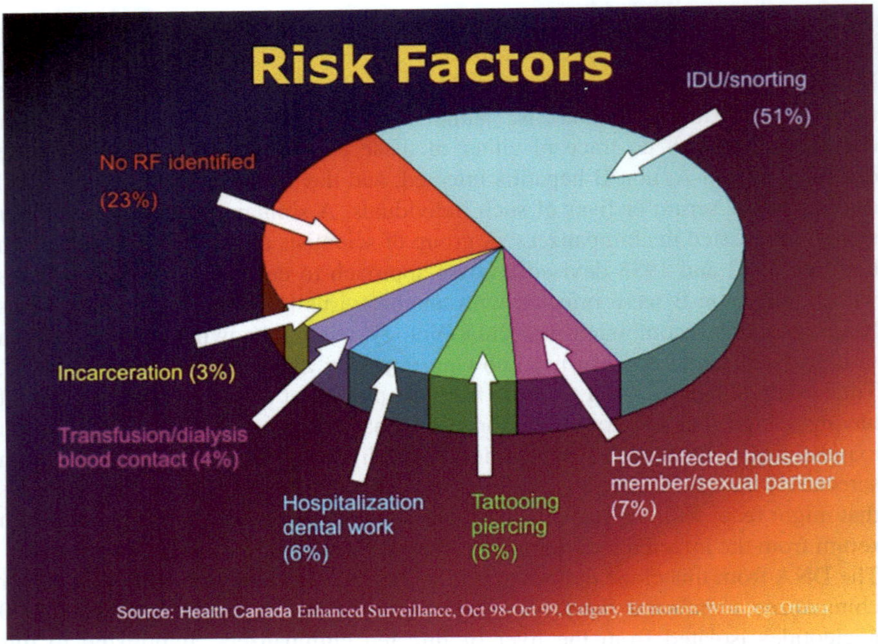

Fig. 13.7 Risk factors for hepatitis C

There is no strong evidence of sexual transmission of HCV, unlike the case of hepatitis B, although it may be sexually transmitted along with HIV. It can be transmitted from mother to child, but the rate of such transmission is low, about 4 %. Until 1992 blood was not screened for HCV, thus blood transfusions or blood products infected many people accidentally before that date.

Initial testing for the presence of antibodies in the blood, indicates whether a person had been exposed to the virus. A second test the polymerase chain reaction (PCR) measures the number of copies of viral RNA in the blood. During chronic infection, this may be as high as 10^9 particles per ml of blood. A third test will look for the strain of hepatitis C, since that will help to predict treatment response. Types 1a and 1b are the most common in the U.S., and the most difficult to treat with interferon.

The prevalence of HCV infection among prison inmates is substantially higher than that in the general U.S. population. Among prison inmates, about 12–35 % are chronically infected, compared to 1–1.5 % in the overall U.S. population. Since HCV infection is primarily associated with a history of intravenous drug use, the CDC recommends that correctional facilities personnel ask inmates about their risk factors for HCV infection during their entry medical evaluations. Inmates reporting risk factors should be tested for HCV infection and those who test positive for it should receive further medical evaluation to determine if they have chronic infection and/or liver disease (http://www.cdc.gov/hepatitis/Settings/corrections.htm).

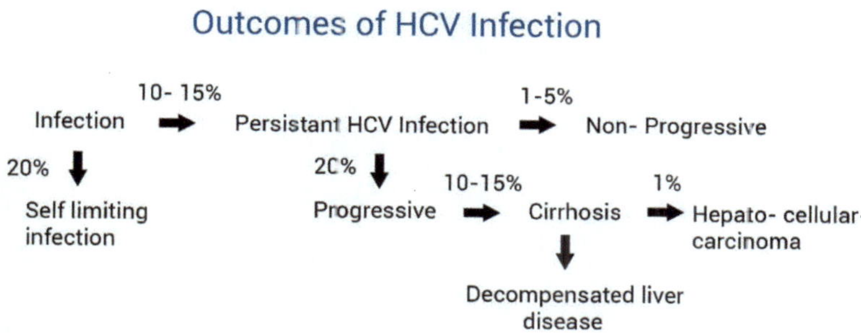

Fig. 13.8 Possible outcomes of HCV infections

While working as a volunteer in a medical facility in Indiana, most of those patients I met having HCV had recently been released from prison. The proportion of prisoners with hepatitis B is also high, and prisoners now receive the vaccine.

Large numbers of veterans of the Vietnam War and, more recently, the Iraq and Afghanistan conflicts, return with hepatitis C due to the increase in drug addiction and tattooing among military personnel. Approximately 8–9 % of Veterans Affairs medical center patients are positive for HCV antibodies, and 1 in 10 veterans returning from Vietnam are HCV positive.

13.8 VIII Vaccine Development

Since this is an RNA virus, it undergoes mutation at a very high rate because of an error-prone replicating enzyme. This has resulted in at least six genotypes of the virus, the most common in the USA being antigen type 1a and 1b, and within each genotype there develops a large number of quasi-species. The viral mutants that accumulate avoid being neutralized by antibodies or T-cells, making it extremely difficult to develop a vaccine against the virus.

However, some new studies indicate that a vaccine is on the way. A phase 1 trial has used recombinant human adenoviruses Ad6 and a chimpanzee Ad3 containing the non-structural genes of HCV. These constructs elicited an HCV T-cell immune response against the virus, including the induction of interferon-gamma and other interleukins. This response has been sustained for one year with a boosting of the vector viruses [20]. Similar studies are progressing in Korea with a DNA vaccine containing many of the HCV structural genes. This is also an ongoing phase 1 clinical trial. Inovio's SynCon® DNA vaccine, INO-8000, carries the genes to produce the antigens NS3/4A, NS4B, and NS5A of HCV genotypes 1a and 1b, the most difficult-to-treat genotypes (Inovio Pharmaceuticals, Inc.). These are all non-structural genes of the virus that code for proteases necessary for processing viral polyprotein (the first translational product of the virus) and the replicase. Since these are not viral coat proteins, it is unclear whether neutralizing antibody will form.

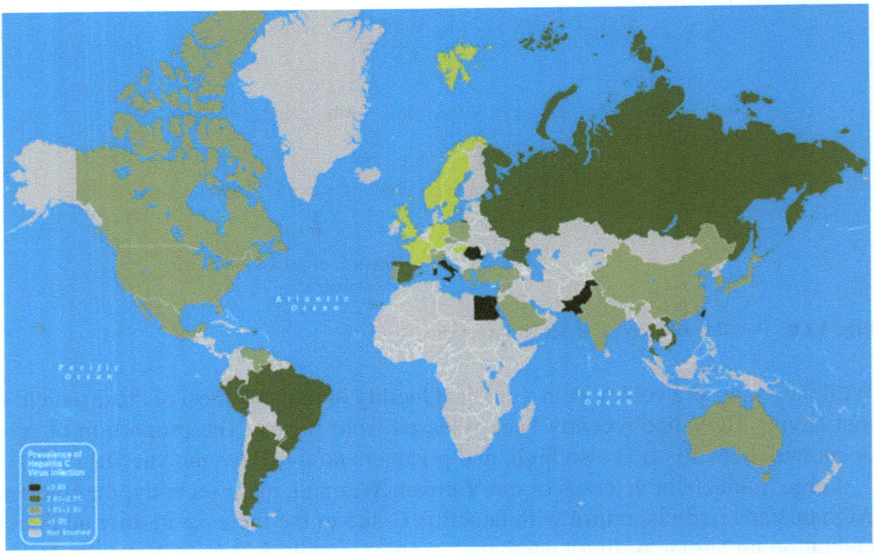

Fig. 13.9 Frequency of HCV infections worldwide. Note the absence in Africa. This is due to lack of information and reporting from this area

13.9 Treatment

Until recently, the treatment for HCV was a regimen of interferon (usually of the alpha type) plus ribavirin, an anti-viral, However, the response rate depends on the antigenic type of the virus, the race of the patients, and the response to interferon [21]. The sustained viral response rate, i.e., complete clearing of the virus among Caucasians and Oriental patients with this regimen, is usually slightly under 50 %, and among African Americans under 30 % [22]. These drugs are difficult to tolerate, and treatment continues for 48 weeks. Interferon/ribavirin treatment is very expensive, (about $1,000 per course of treatment) and interferon is not readily available in many Third World countries. New drugs, based on inhibitors of viral replication or specific viral enzymes, are now appearing on the market. Two new therapeutic agents, telaprevir and boceprevir, have recently been licensed in some countries, including the U.S. The Food and Drug Administration has approved newer drugs including semaprevir and danoprevir, all protease inhibitors. Sofosbuvir, an inhibitor of the viral replicase, was tested in Europe, and inhibits the virus without the addition of interferon. Conclusions from the clinical trial showed that of the 40 patients who were in the randomized group and who received sofosbuvir plus ribavirin there was a sustained viral response, measured as six months without HCV RNA in the blood. However, these were not HCV type 1 patients. Further studies have shown that a combination of protease inhibitors and replicase inhibitors can "cure" HCV without the use of interferon [23]. It is a widely held opinion among hepatologists that interferon treatment eventually will be discontinued.

One of the major difficulties in the development of anti-virals was the specificity of the antiviral agent. Could one target the virus without hurting the host cell? With the development of small molecules that can target specific viral enzymes, or RNA virus replicase, this specific difficulty has been overcome. However, the new drugs are very expensive; Gilead, the makers of sofosbuvir, plans to charge 1,000 per pill. The medication has to be taken daily for 12 weeks, at a total cost of $84,000. It is true that this regimen will cure the patient, and may cost less than liver transplant and accompanying hospitalization, but it still seems excessive.

During my scientific career, I devoted a large part of my time to examining the response to interferon/ribavirin treatment in HCV patients. I was a principal investigator in a large clinical trial examining whether there were racial differences in the level of response to interferon treatment. We also investigated the differential response at the genome level to interferon/ribavirin treatment between responders and non-responders. The biochemical and genetic response to the interferon treatment was analyzed using DNA-microarrays, a technique that measures messenger RNA induction and, assumedly, proteins following drug administration. Major differences could not be detected at the level of mRNA to explain racial differences or lack of response. Shortly after the completion of this study, a genetic polymorphism was discovered that might explain racial differences [24, 25]. This genetic polymorphism is within a few nucleotides of the IL28b gene, encoding interferon-λ3, a class of interferon not previously known. This polymorphism enhanced the rate of response to treatment twofold, and was more common among Caucasians that Africans, thus in part explaining the difference in racial response rates.

No vaccine currently exists for HCV. The newer medications are very expensive, and will be so for many years to come. The population diagnosed with HCV will increase as more young people are tested and if the use of recreational drugs and tattooing increases. This is a societal disease in which changes in behavior can decrease the incidence of HCV. Most patients realize that their disease was a result of risky behavior. Unlike the case of HIV/AIDS, there is little done by way of education to prevent the spread of HCV.

References

1. Reuben, A. (2002). The thin red line. *Hepatology, 36*(3), 770–773.
2. Zuckerman, A.J. (1979): The history of viral hepatitis. London: Academic Press.
3. Dooley, D.P. (2005). History of U.S. military contributions to the study of viral hepatitis. *Military Medicine* , *170*(4 Suppl), 71–76.
4. Wheeler, C., Vogt, T. M., Armstrong, G. L., Vaughan, G., Weltman, A., Nainan, O. V., et al. (2005). An outbreak of hepatitis A associated with green onions. *New England Journal of Medicine, 353*(9), 890–897.
5. Feng, Z., Lemon, S.M. (2013). Peek-a-boo: membrane hijacking and the pathogenesis of viral hepatitis. *Trends Microbiology.*

6. Blumberg, B. S., Alter, H. J., & Visnich, S. (1965). A "new" antigen in leukemia sera. *JAMA, 191*, 541–546.
7. Blumberg, B. S., Friedlaender, J. S., Woodside, A., Sutnick, A. I., & London, W. T. (1969). Hepatitis and Australia antigen: autosomal recessive inheritance of susceptibility to infection in humans. *Proceedings of the National Academy of Sciences USA, 62*(4), 1108–1115.
8. Blumberg, B. S., Sutnick, A. I., & London, W. T. (1968). Hepatitis and leukemia: their relation to Australia antigen. *Bulletin of the New York Academy of Medicine, 44*(12), 1566–1586.
9. Okochi, K., & Murakami, S. (1968). Observations on Australia antigen in Japanese. *Vox Sanguinis, 15*(5), 374–385.
10. Prince, A. M. (1968). An antigen detected in the blood during the incubation period of serum hepatitis. *Proceedings of the National Academy of Sciences USA, 60*(3), 814–821.
11. Dane, D. S., Cameron, C. H., & Briggs, M. (1970). Virus-like particles in serum of patients with Australia-antigen-associated hepatitis. *Lancet, 1*(7649), 695–698.
12. Summers, J., O'Connell, A., & Millman, I. (1975). Genome of hepatitis B virus: restriction enzyme cleavage and structure of DNA extracted from Dane particles. *Proceedings of the National Academy of Sciences USA, 72*(11), 4597–4601.
13. Robinson, W. S., & Lutwick, L. I. (1976). The virus of hepatitis, type B. (second of two parts). *New England Journal of Medicine, 295*(22), 1232–1236.
14. Robinson, W. S., & Lutwick, L. I. (1976). The virus of hepatitis, type B (first of two parts). *New England Journal of Medicine, 295*(21), 1168–1175.
15. van Zonneveld, M., Honkoop, P., Hansen, B. E., Niesters, H. G., Darwish Murad, S., de Man, R. A., et al. (2004). Long-term follow-up of alpha-interferon treatment of patients with chronic hepatitis B. *Hepatology, 39*(3), 804–810.
16. Choo, Q. L., Kuo, G., Weiner, A. J., Overby, L. R., Bradley, D. W., & Houghton, M. (1989). Isolation of a cDNA clone derived from a blood-borne non-A, non-B viral hepatitis genome. *Science, 244*(4902), 359–362.
17. Krawczynski, K., & Bradley, D. W. (1989). Enterically transmitted non-A, non-B hepatitis: identification of virus-associated antigen in experimentally infected cynomolgus macaques. *Journal of Infectious Diseases, 159*(6), 1042–1049.
18. Kubo, Y., Takeuchi, K., Boonmar, S., Katayama, T., Choo, Q. L., Kuo, G., et al. (1989). A cDNA fragment of hepatitis C virus isolated from an implicated donor of post-transfusion non-A, non-B hepatitis in Japan. *Nucleic Acids Research, 17*(24), 10367–10372.
19. Houghton, M. (2009). Discovery of the hepatitis C virus. *Liver international: official journal of the International Association for the Study of the Liver, 29*(Suppl 1), 82–88.
20. Barnes, E., Folgori, A., Capone, S., Swadling, L., Aston, S., Kurioka, A., Meyer, J., Huddart, R., Smith, K., Townsend, R et al. (2012). Novel adenovirus-based vaccines induce broad and sustained T cell responses to HCV in man. *Science Translational Medicine, 4*(115), 115ra111.
21. Howell, C. D., Dowling, T. C., Paul, M., Wahed, A. S., Terrault, N. A., Taylor, M., et al. (2008). Peginterferon pharmacokinetics in African American and Caucasian American patients with hepatitis C virus genotype 1 infection. *Clinical Gastroenterology and Hepatology: The Official Clinical Practice Journal of the American Gastroenterological Association, 6*(5), 575–583.
22. Conjeevaram, H. S., Fried, M. W., Jeffers, L. J., Terrault, N. A., Wiley-Lucas, T. E., Afdhal, N., et al. (2006). Peginterferon and ribavirin treatment in African American and Caucasian American patients with hepatitis C genotype 1. *Gastroenterology, 131*(2), 470–477.
23. Gane, E. J., Stedman, C. A., Hyland, R. H., Ding, X., Svarovskaia, E., Symonds, W. T., et al. (2013). Nucleotide polymerase inhibitor sofosbuvir plus ribavirin for hepatitis C. *New England Journal of Medicine, 368*(1), 34–44.
24. Ge, D., Fellay, J., Thompson, A. J., Simon, J. S., Shianna, K. V., Urban, T. J., et al. (2009). Genetic variation in IL28B predicts hepatitis C treatment-induced viral clearance. *Nature, 461*(7262), 399–401.
25. Thomas, D. L., Thio, C. L., Martin, M. P., Qi, Y., Ge, D., O'Huigin, C., et al. (2009). Genetic variation in IL28B and spontaneous clearance of hepatitis C virus. *Nature, 461*(7265), 798–801.

Herpesvirus

14

Abstract

The herpes family of viruses is associated with a large number of infectious diseases, including some types of cancer, such as Burkitt's lymphoma, and Kaposi sarcoma and with sexually transmitted diseases such as genital blisters. (HSV-2), and chickenpox. The virus can exist in an active phase or latent phase, and can manifest itself in different forms and pathological situations. Chickenpox can reappear many years later as shingles, in particular when the immune system is compromised. Epstein Barr-virus infection can lead to various cancers depending on environmental and genetic factors, and in the lytic form as mononucleosis. Burkitt's lymphoma is confined to areas of Africa with high malaria prevalence and is characterized at the genetic level by a gene translocation. HSV-8 infection results in Kaposi sarcoma and is found in individuals suffering from AIDS, although now recognized as a cancer in non-AIDS populations in Africa. There has been a dramatic increase in genital herpes during the last few years.

14.1 Burkitt's Lymphoma

One way (and perhaps the only way) to immortalize immune cells is to infect them with a herpes virus known as Epstein-Barr virus, so-called after two of the scientists, Antony Epstein and Yvonne Barr, who isolated it from a tumor derived from an African child with an unusual lymphoma, known as Burkitt's lymphoma. This is a tumor of the lymphatic system, particularly of B-cells, discovered by a surgeon, Dennis Burkitt, while working in Equatorial Africa in 1958. The cancer manifests itself in Africa mostly with deformation of the jawbones (Fig. 14.1). As Dennis Burkitt tells it [1], in 1957 while working as a surgeon in Kampala, Uganda, Dr. Hugh Trowell, who was in charge of the pediatrics ward at Mulago Hospital, called him for a consultation on a child who had a massive swelling on

M. W. Taylor, *Viruses and Man: A History of Interactions*,
DOI: 10.1007/978-3-319-07758-1_14,

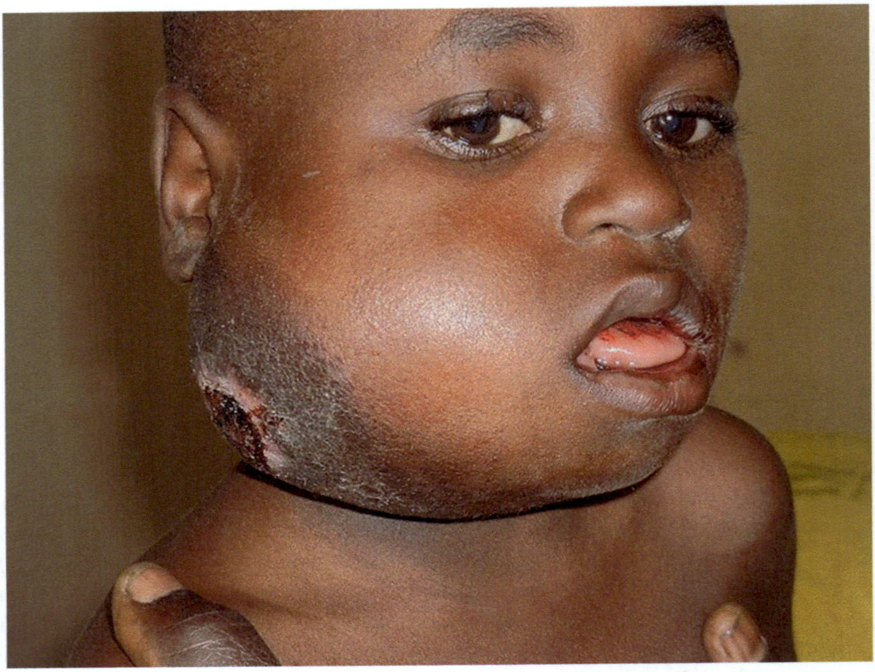

Fig. 14.1 This seven-year-old Nigerian boy presented with a several-months' history of jaw swelling which had been treated with antibiotics. The tumor was ulcerated and draining (courtesy of Mike Blyth, Wikipedia)

both sides of his upper and lower jaws. A biopsy suggested a type of granuloma, a tumor mass containing macrophages, but no conclusions were reached. While visiting another hospital a few weeks later, he noticed another child sitting outside with similar features. He immediately took the mother and child back to Kampala. On clinical examination, he detected a mass in the abdomen. How could a growth (if that was what it was) in the jaw region be linked to a tumor in the abdomen? A short time later, a number of children with this condition had been recognized, and in most cases there were tumors in the abdomen or other organs as well as in the jaw region. This did not seem to arise from an infection but was some unusual type of cancer. Burkitt then examined the records of other children admitted to the hospital over the last few years and found that many had similar symptoms. By looking at histological slides from these children, Burkitt and his colleagues came to the conclusion that this was some type of lymphoma, a cancer where cells of the immune system, either B- or T-cells, divide faster than normal, or live longer than normal, thus accumulating as a mass. Whether this was a novel phenomenon or not was investigated by examining records kept at the nearby Mengo Mission Hospital, the oldest hospital in East Africa. There were descriptions by Sir Albert Cook (1870–1951), a missionary doctor at the turn of the nineteenth century, of the same tumors of the jaw among children.

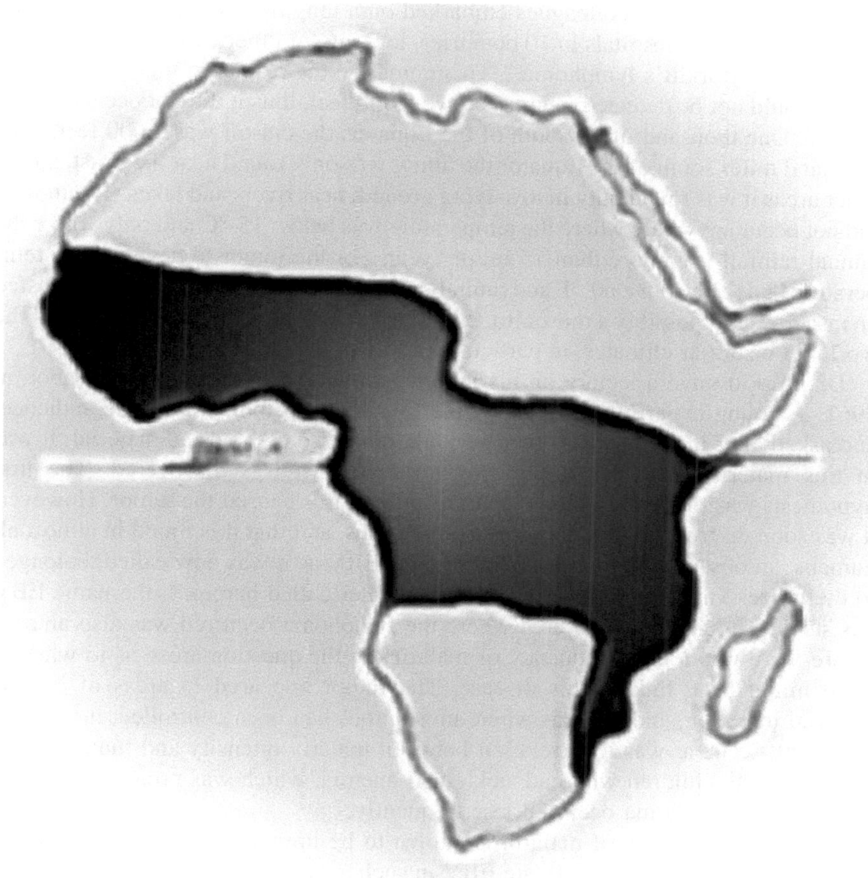

Fig. 14.2 Area of Africa where Burkitt's lymphoma occurs (*in black*)

Sir Albert Cook, who had founded the Mengo Missions Hospital in 1897, contributed greatly to the training of medical personnel in East Africa. He was outstanding among medical missionaries for his efforts to train Africans to become skilled medical workers. He and his wife opened a school for midwives at Mengo, Uganda. Cook began training African medical assistants at Mulago during World War I, and in the 1920s encouraged the opening of a medical college that initially trained Africans to the level defined by the colonial government as "Asian sub-assistant surgeon." The school grew to become a full-fledged medical school in his lifetime (http://en.wikipedia.org/wiki/Albert_Ruskin_Cook). Further investigation by Burkitt indicated that the tumors only occurred in specific areas of Uganda, and only in specific areas of Africa. These cancers were not found in North Africa above the Sahara or South Africa. They appeared in a belt 10 degrees on either side of the Equator, and along the South-Eastern east coast of the African continent (Fig. 14.2).

Dr. Burkitt and two colleagues embarked on a long trip of 10,000 miles across Africa, visiting 60 hospitals in 10 countries, looking for other cases of what was by now called "Burkitt's lymphoma." To summarize the results of this "safari," the tumor could not be detected in areas above 5,000 feet, thus it did not occur on high ground. One thousand miles south of the Equator, the cut-off was 3,000 feet. Two thousand miles south of the Equator the tumor was only found near the coast, and in other areas it was found only in low-lying ground, near rivers and lakes. The tumors did not occur anywhere where the temperature was below 15 °C and only where the annual rainfall was more than 15 in. per year. For the tumor to develop, the temperature had to be above 60 °F and rainfall more than 20 in. These data all suggested an insect vector, possibly a mosquito. A similar tumor was found in other parts of the world with similar climates, in particular in Papua, New Guinea.

Dr. Burkitt gave a lecture on his findings at Middlesex Hospital in London in 1961. A young experimental pathologist, Dr. Tony Epstein, was in the audience, and asked after the lecture whether he could obtain some biopsied material. It was in this material that the Epstein-Barr virus (EBV) was discovered. The first hypothesis was that an insect vector-transmitted virus caused the tumor. However, it was soon discovered that the virus is ubiquitous, and that it is found in almost all humans, an observation discussed below. The EBV as it was now called, belonged to the herpes viruses genus. Although it was later called herpes-4, the name EBV has stuck. The geographical area where the lymphoma occurred was also an area where there was a high frequency of malaria, so the question arose as to whether there might be a link to this disease. The tumor appeared in areas of intense malarial infection, and in areas where mosquitoes had been controlled, it was less prevalent, so there was a correlation between malaria intensity and tumor occurrence. Among children who had sickle-cell anemia, which was protective against malaria, the lymphoma occurred less frequently.

An intense outbreak of malaria is known to be immune-depressive over time. Could immune depression activate EBV in such a way as to cause the lymphoma? This hypothesis was supported by the fact that EBV existed in a "latent" state in most individuals, and yet was the causative agent of infectious mononucleosis in individuals not exposed to the virus early in life [2]. This was discovered by Werner and Gertrude Henle of Children's Hospital in Philadelphia when a lab technician working in their lab with Epstein-Barr virus became sick with mononucleosis. Infectious mononucleosis results in an abnormal increase in lymphoid cells. The Henles also showed that B-lymphocytes, when infected with EBV, became immortal in culture. This was an important observation and is used to this day to create human cell lines.

14.2 Epstein–Barr Virus and c-myc Translocation

There is another twist to the EBV and Burkitt's lymphoma story. Burkitt's lymphomas contain chromosomal translocations that result in c-*myc* up regulation. The myc gene is one of the best characterized oncogenes—that is, genes causing cancer in humans. This gene product is a regulatory protein that controls a large number of transcription factors and thus enhances cell replication. The mutant form is found in many cancers, including cervical, breast, and colon cancer. The most common chromosomal translocation, t(8;14), places a portion of the c-*myc* oncogene adjacent to an immunoglobulin heavy chain gene (a gene important in making antibodies, and many other functions in the cell). Less common translocations involve the c-*myc* oncogene and the kappa or lambda immunoglobulin light chain genes t(2;8) and t(8;22), respectively. These translocations result in high constitutive expression of c-*myc.* Abnormal expression of c-*myc* in EBV-immortalized B-cell lines results in highly transformed cells that form tumors when injected into immune-deficient mice. Transgenic mice that express a mutated human c-*myc* that is derived from a Burkitt's lymphoma and is fused to the Igλ locus, develop tumors that have histologic and phenotypic features of Burkitt's lymphomas.

EBV-associated endemic Burkitt's lymphoma is thought to develop in steps. First, EBV infection may expand the pool of differentiating and proliferating B-cells; second, chronic high-level malaria infection may cause T-cell suppression and B-cell proliferation. Third, enhanced proliferation of differentiating B-cells may favor the chance occurrence of a reciprocal c-*myc* (t[8;14] or t[8;22]) translocation, placing c-*myc* partially under the control of immunoglobulin-related transcriptional enhancers, with development of a monoclonal (derived from a single cell) tumor [3] (Fig. 14.3). Thus, Burkitt's lymphoma is an example of an "environmental" cancer that results from interaction of temperature, insects (mosquito), virus, and perhaps genetic propensity for a gene translocation.

Burkitt's lymphoma is no longer thought of as an "African disease," since there have been cases identified in the U.S. and in Brazil. The rate of Burkitt's in the U.S. is about 0.4 per 100,000—a very rare cancer. Recent evidence supports the concept that not only is the *myc* gene mutated in Burkitt's lymphoma, but that many other genes, important in B-cell biology and cell cycle regulation, are mutated.

14.3 Naso-Pharyngial Carcinoma and Other Tumors

Another cancer associated with EBV is naso-pharyngial carcinoma, which occurs predominantly in Southern China. The rate is about 20 cases per 100,000 population, although it may occur in other countries at a much lower rate. In the U.S., Chinese-Americans are the main victims. It has been associated with diet, and perhaps high salt intake, and vitamin C deficiency at a young age also may be a contributing factor. Finally, a study of HLA haplotypes, genes involved in the rejection process of the immune system, revealed a genetically distinct subpopulation in southern China, with an increased frequency of haplotype A-2/B-Sin-2, which may account for the higher disease incidence in the area. EBV has been shown more recently to be associated also with B-cell lymphomas such as Hodgkin's lymphoma, and other lympho-proliferative disorders. EBV-initiated tumors also occur in AIDS patients during immunosuppression. Burkitt's lymphoma responds to treatment with many chemotherapeutic agents.

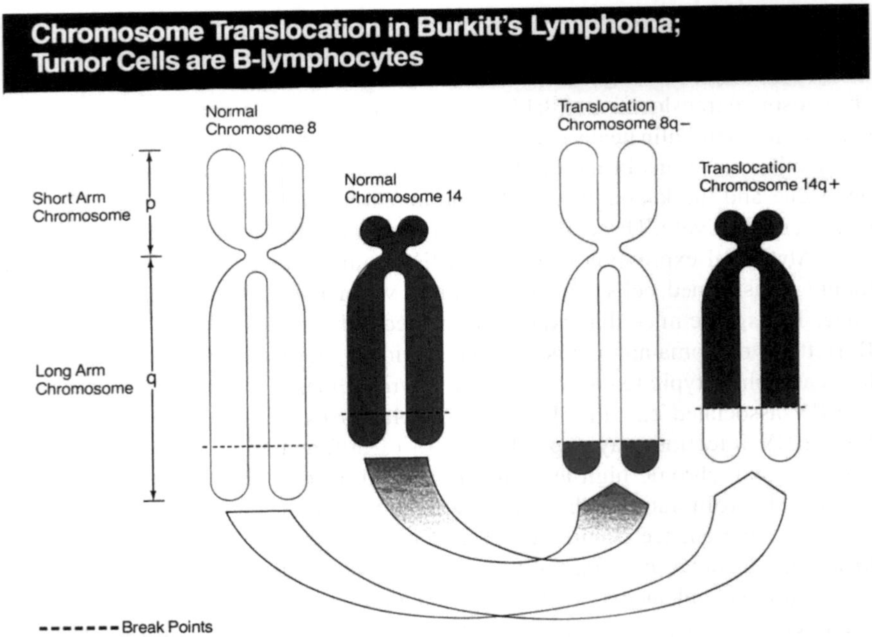

Fig. 14.3 Chromosomal translocation in Burkitt's lymphoma (NCI)

14.4 Epstein–Barr Virus and Mononucleosis

It turns out that EBV, as it is now called, is one of the most common viruses in man and almost everyone in the world carries it. Most human cells in culture are also EBV-positive. Infants become susceptible to it as soon as maternal antibodies disappear. When infection with EBV occurs among adolescents or young adults who have not been exposed to the virus, mononucleosis occurs about 50 % of the time. The virus is carried in saliva of infected individuals, and in some cases in people who have no symptoms but are EBV-positive. The symptoms of mononucleosis are swollen lymph nodes in the neck, sore throat, and fever. EBV remains dormant after the illness in a few cells for the rest of that person's life; as such, it is in the latent stage. There is no treatment for mononucleosis.

Epstein-Barr virus might also be associated with multiple sclerosis. There is circumstantial or correlative evidence that MS occurs 2–3 times more following mononucleosis than in the rest of the EBV-positive population, and that MS is very rare in EB-negative individuals, although as pointed out, such individuals must be very rare. Additional evidence supporting a role for EBV in MS pathogenesis includes the observations of elevated antibodies to EBV antigens (especially EBV nuclear antigen-1) prior to the onset of MS [4].

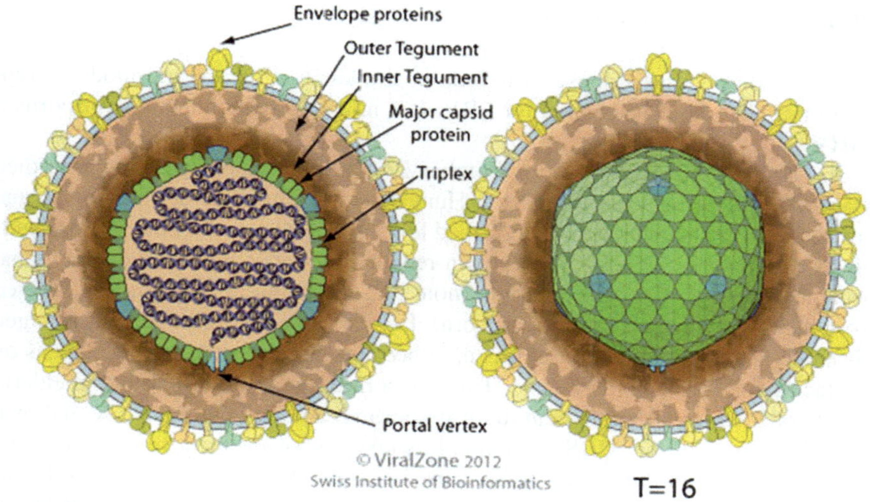

Fig. 14.4 Diagrammatic presentation of gammaherpes viruses. Capsid consists of 162 capsomers and is surrounded by an amorphous tegument. Glycoprotein complexes are embedded in the lipid envelope (Swiss Institute of Bioinformatics)

14.5 The E-B Virus

Epstein-Barr virus belongs to the gammaherpes virus family. It is a large virus and may code for as many as 85 genes. The genome consists of linear double-stranded DNA (Fig. 14.4).

The virus has both a lytic cycle in which replication takes place in the nucleus of the host cell, and a latent stage where it exists as an episome, an independent genetic entity, and replicates with host DNA. Like most herpes viruses, the transcription of the viral message is temporal, with different areas of the genome being transcribed at different times during infection. There are early genes, intermediate early genes and late genes expression. Transcription of immediate early genes promotes translation of early genes and protects the virus against innate host immunity, including interferon. Transcription of early viral mRNA is by host polymerase II, encoding proteins involved in replication of the viral DNA. DNA replication is by what is called the "circular model," similar to phage DNA replication, and transcription of late messages also uses host enzymes. The virus is assembled in the nucleus and buds through the inner layers of the nuclear membrane, which has been modified by the insertion of herpes glycoproteins, and final release at the plasma membrane. This virus is one of the largest human viruses and it contains genetic material that mimics many host genes that act as "decoys" and inhibit the immune system.

14.6 Kaposi's Sarcoma

The virus that causes Kaposi's sarcoma is similar in structure and mode of replication to that described above for EBV. It is now known as human herpesvirus-8 (HHV8).

In 1872, Moritz Kaposi (born October 23, 1837, in Kaposvár, Hungary, died March 6, 1902, in Vienna, Austria-Hungary), a Hungarian dermatologist and professor at Vienna University, described lesions that occurred on the feet or hands of six elderly men. These were brown-red or blue-red nodules, cancerous, and became known as Kaposi's sarcoma (note that the name of the tumor is derived from the town in which Kaposi was born). He was born Moritz Cohen but changed his name either to avoid anti-Semitism, or because there were too many Cohens on the faculty of Medicine in Vienna. This was a rare cancer usually found in elderly people of Mediterranean descent, until it appeared among young homosexual men at the beginning of the AIDS epidemic. It is now one of the most common cancers of our time, the fourth most common cause of cancer by an infectious agent. It was the observation of a higher-than-usual incidence of Kaposi's sarcoma that led to the first diagnosis of an immunodeficiency among young homosexual men in 1982. This was an unusual occurrence. Since it was present among gay men who were HIV-positive but not among others who had blood-borne diseases, such as hemophiliacs, it was suspected that there must be an infectious agent involved (Fig. 14.5).

The virus was first recognized by comparing the DNA from individuals who had KS and those who did not, and a DNA resembling a gamma herpes virus was recognized in 1994. As stated above, this was called HHV-8. A series of biochemical tests, comparing the presence of antibodies to the virus, viral DNA, and the viruses' presence in tumor cells, left no doubt that this was the causative agent. The major antigen appears to be a protein related to viral latency. The frequency is high among the HIV-positive homosexual population: 25–35 %. It is also high among children in sub-Saharan Africa, which may result from the high frequency of infection among pregnant women. In one study, approximately 40 % of pregnant women in Zambia had anti-latency-associated nuclear antigen (LANA) antibodies, including all the mothers of children with KS [5]. A role for sexual transmission has been convincingly demonstrated in a longitudinal study of men in San Francisco followed over a 10-year period. Among exclusively heterosexual men, no Kaposi's sarcoma sero-positivity was detected; however, among homosexual or bisexual men, the incidence of sero-conversion was linearly related to the number of male sexual-intercourse contacts. Men who had had more than 250 sexual partners in the preceding 2 years had a sero-positivity rate of 65 % [5] (Fig. 14.5).

The basis of the oncogenic affect of KS is still unknown. Preliminary data indicates an affect, possibly by a viral gene on T-cell cycle regulation.

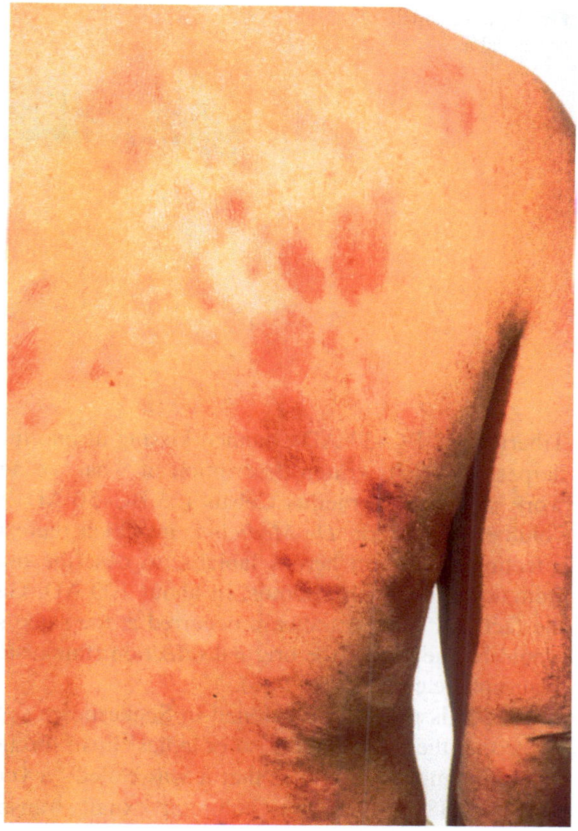

Fig. 14.5 Typical Kaposi's sarcoma rash (courtesy of NCI)

14.7 Herpes Simplex

I have discussed the "newer" herpes viruses first, since they are rather unique, and have a fascinating history. Human herpes viruses are divided into three families. "Alpha" herpes viruses include herpes simplex 1, and 2, and varicella-zoster (chicken pox-shingles); "beta" herpes viruses include herpes 6 and 7; and "gamma" herpes viruses are those discussed above (Epstein-Barr virus and Kaposi's sarcoma, HHV8).

Herpes simplex virus 1 (HSV1) and 2 (HSV2) are very common in the human population and we know that HSV-1 and 2 have been around in the human population for a long time. The symptoms we associate with herpes simplex 1—cold sores on the lips or in the mouth—were described as far back as Hippocrates (460–377 BC) in Greece. He used the term "herpes," which in Greek means "creeping" to describe a number of different conditions, including what we believe today is oral herpes caused by HSV1. It is said that the emperor of Rome,

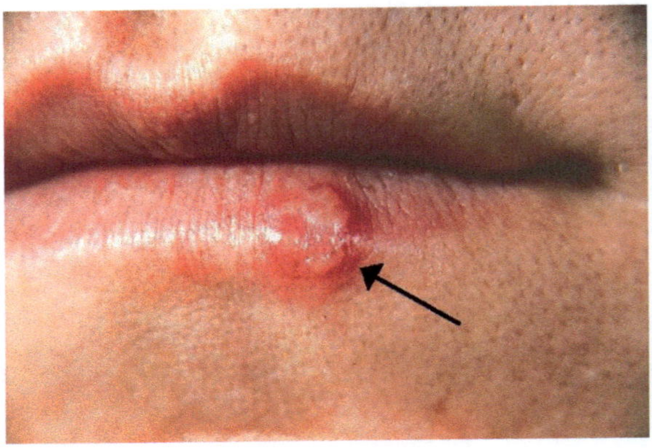

Fig. 14.6 Example of herpes labialis

Tiberius (42 BC–37 AD), banned kissing at one time because of an epidemic of herpes. The first written description of what we now call herpes appears in the work of Aulus Cornelius Celsus (25 BC–50 AD) who in 30 AD published six volumes of a work, *De Medicina*, in which there is a vague description of the infection. The term herpes labialis, to describe oral herpes, was coined by the Roman physician, Herodotus, in about 100 AD. Galen (130–201 AD) described the fact that the sores always occurred at the same site, which we now recognize would be the site of the original infection (this history adapted from http://www.livingsphere.com/) (Fig. 14.6).

Herpes is possibly mentioned in Shakespeare's "Romeo and Juliet," when he wrote that there are blisters "o'er ladies' lips." In the eighteenth century it was so common among prostitutes that it was called "a vocational disease of women."

Today HSV-2 is the most common sexually transmitted diseases in the U.S. The major mode of HSV-2 transmission is by sexual intercourse or oral sex. The virus can also be transmitted by touching lesions and then touching other locations in the body, in particular the eyes. The use of condoms during sex can decrease the frequency and chances of getting herpes. HSV-2 is a lifelong and incurable infection that can cause recurrent and painful genital sores and can make those infected with the virus two to three times more likely to acquire HIV, the virus that causes AIDS. It results in painful blisters on the penis and genital areas and is ten times more common than HSV-1. At least 50 million persons in the United States have a genital HSV infection, and an estimated 766,000 new cases of symptomatic, first-episode genital HSV-2 infections occur each year. About one in six people in the U.S. between the ages of 14 and 49 have genital herpes, more women than men, since it is spread more easily by men than by women. In many cases it is asymptomatic, but when symptoms do occur, they typically appear as one or more blisters on or around the genitals, rectum or mouth. The blisters break and leave painful sores that may take 2–4 weeks to heal. Outbreaks of the disease occur sporadically, more at the beginning of the infection and less as time goes on. The virus stays in the body indefinitely, and between outbreaks it is said to be in a latent state. It can be activated by stress and hormonal changes.

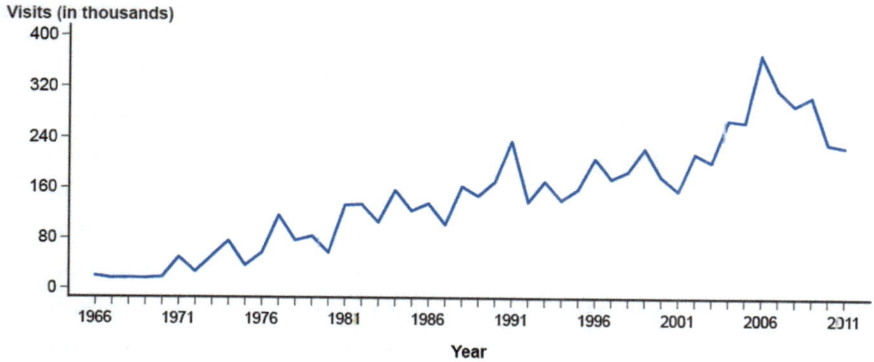

Fig. 14.7 Number of initial visits to physicians with HSV-2 (CDC)

Pregnant women should be very careful, since the virus can be transmitted to the unborn baby, resulting in death of the embryo, or transmission to the child. Eighty percent of babies born with herpes die at birth or shortly thereafter, and survivors are often brain-damaged. There has been a large increase in HSV-2 infections in the last few years, as shown in Fig. 14.7.

There is no cure for HSV-2. Symptoms can be alleviated with anti-viral drugs such as acyclovir. This compound can be used to treat both HSV-1 and HSV-2 infections as well as shingles, which results from another herpes virus (varicella-zoster or chicken pox). It alleviates pain and lengthens the time between outbreaks. A National Institute of Health-supported clinical trial showed that once-daily suppressive therapy using valacyclovir significantly reduces the risk of transmitting genital herpes to an uninfected partner. Valacyclovir is a pro-drug, that is, it converts to acyclovir in the body and is more efficient than acyclovir administration. This was the first time an antiviral medication had been shown to reduce the risk of transmission of a sexually transmitted disease. Topical microbicides, preparations containing microbe-killing compounds, are also in various stages of development and testing. These include gels, creams, or lotions that a woman can insert into her vagina prior to sexual intercourse.

14.8 Worldwide Prevalence of HSV-2

The estimated total number of people around the world aged 15–49 who were living with HSV-2 in 2003 was 536 million (Table 14.1). More women than men were infected, with an estimated 315 million infected women compared to 221 million infected men. The number infected increased with age, most markedly in the younger people, until it peaked in the 35–39 age group, after which it declined slightly. The number infected per age group is a combination of the size of the population in the group multiplied by the prevalence of infection and, as the pool of susceptible individuals is used up with increasing age, the rate of increase in

Table 1 Global estimates of the prevalence of the herpes simplex virus type 2 infection, in 2003 [6]

Age in years	Global prevalence in millions (percentage per population)		
	Females	Males	Both
15–19	25.8 (9.0)	14.6 (4.8)	40.4 (6.9)
20–24	39.4 (15.1)	24.1 (8.8)	63.5 (11.9)
25–29	46.5 (19.0)	30.5 (12.0)	77.1 (15.4)
30–34	51.5 (21.4)	36.1 (14.6)	87.6 (18.0)
35–39	52.9 (23.8)	38.8 (17.1)	91.8 (20.3)
40–44	50.8 (25.9)	38.8 (19.4)	89.6 (22.6)
45–49	47.9 (27.7)	37.8 (21.5)	85.6 (24.6)
Total	314.8 (19.4)	220.7 (13.1)	535.5 (16.2)

prevalence slows. In the model, prevalence itself does not decline with age, but because there were fewer older people than younger people in total, the actual number with current infections slightly decreased. This is shown in Table 14.1.

The regional incidence of HSV-2 is shown in Tables 14.2 and 14.3, one for females and the other for males.

14.9 Chicken Pox: Varicella–Zoster

Before the advent of a chicken pox vaccine, about 4,000,000 people would get chicken pox each year in the U.S.; about 10,600 of them would be hospitalized and 100–150 would die. Chicken pox is a very contagious disease causing a blister-like rash, itching, tiredness, and fever. Chicken pox can be serious, especially in babies, and in adults with weakened immune systems. The virus spreads by coughing or sneezing, or by touching or breathing in the virus particles that come from chicken pox blisters. (An example of the rash is shown in Fig. 14.8). The classic symptoms of chicken pox are a rash that turns into itchy, fluid-filled blisters that eventually form scabs. The rash may first show up on the face, chest, and back, then spread to the rest of the body, including inside the mouth, eyelids, and genital area. It usually takes about 1 week for all the blisters to become scabs. Giovanni Fillipo Ingrassia (1510–1580) was the first to describe chicken pox as a separate disease in the 1500s in Italy. Ingrassia was a Sicilian physician and anatomist, and professor at the University of Naples, and was the first to describe various infections. In the 1600s, an English physician named Richard Morton (1637–1698) gave the name chicken pox to what he thought was a milder form of smallpox. It is believed that in the 1700s, William Heberden (1710–1801) (another English physician), was the first physician to prove that chicken pox was actually different from smallpox. He wrote extensively on this disease in 1776, in *Medical Transactions of the Royal Society* (Fig. 14.8).

Table 2 Regional estimates of the prevalence of the herpes simplex virus type 2 infection among females, in 2003 [6]

Region	Regional prevalence in millions, by age							
	15–19 years	20–24 years	25–29 years	30–34 years	35–39 years	40–44 years	45–49 years	Total
North America	0.9	1.5	2.0	2.6	3.2	3.8	3.9	17.9
Latin America and the Caribbean	2.6	4.5	5.8	6.4	6.7	6.6	6.0	38.6
North Africa and the Middle East	1.0	1.5	1.6	1.5	1.4	1.3	1.1	9.6
Sub-Saharan Africa	9.0	13.1	13.6	12.5	11.2	10.0	8.8	78.2
Western Europe	0.7	1.3	1.8	2.2	2.6	2.6	2.5	13.7
Eastern Europe and central Asia	2.7	3.9	4.3	4.3	4.3	4.7	4.7	28.9
Eastern Asia	2.6	4.4	7.1	11.1	12.8	11.9	12.0	61.8
Japan	0.4	0.6	0.7	0.7	0.6	0.6	0.6	4.1
Pacific	0.03	0.04	0.05	0.06	0.06	0.06	0.05	0.3
South Asia	4.1	5.4	5.5	5.4	4.9	4.3	3.7	33.2
Southeast Asia	1.7	3.1	4.0	4.6	4.9	4.8	4.4	27.6
Australia and New Zealand	0.03	0.06	0.09	0.1	0.2	0.2	0.2	0.9
Total	25.8	39.4	46.5	51.5	52.9	50.8	47.9	314.8

Table 3 Regional estimates of the prevalence of the herpes simplex virus type 2 infection among males, in 2003 [6]

Region	Regional prevalence in millions, by age							
	15–19 years	20–24 years	25–29 years	30–34 years	35–39 years	40–44 years	45–49 years	Total
North America	0.6	1.0	1.4	1.7	2.2	2.5	2.6	11.9
Latin America and the Caribbean	0.9	1.6	2.1	2.4	2.7	2.8	2.7	15.1
North Africa and the Middle East	1.4	1.6	1.5	1.3	1.1	0.9	0.8	8.6
Sub-Saharan Africa	4.1	6.5	7.5	7.5	7.1	6.7	6.2	45.5
Western Europe	0.2	0.5	0.7	1.1	1.4	1.6	1.7	7.2
Eastern Europe and central Asia	0.6	1.1	1.5	1.8	2.1	2.6	2.8	12.3
Eastern Asia	2.0	3.4	5.4	8.4	9.8	9.3	9.5	47.8
Japan	0.02	0.05	0.08	0.1	0.1	0.1	0.2	0.7
Pacific	0.05	0.08	0.09	0.09	0.09	0.08	0.06	0.5
South Asia	1.8	3.1	4.0	4.8	5.2	5.4	5.2	29.4
Southeast Asia	3.1	5.2	6.3	6.9	7.0	6.6	6.0	41.2
Australia and New Zealand	0.02	0.03	0.05	0.06	0.08	0.1	0.1	0.4
Total	14.6	24.1	30.5	36.1	38.8	38.8	37.8	220.7

Note the high rates of infection in Sub-Saharan Africa and in South East Asia

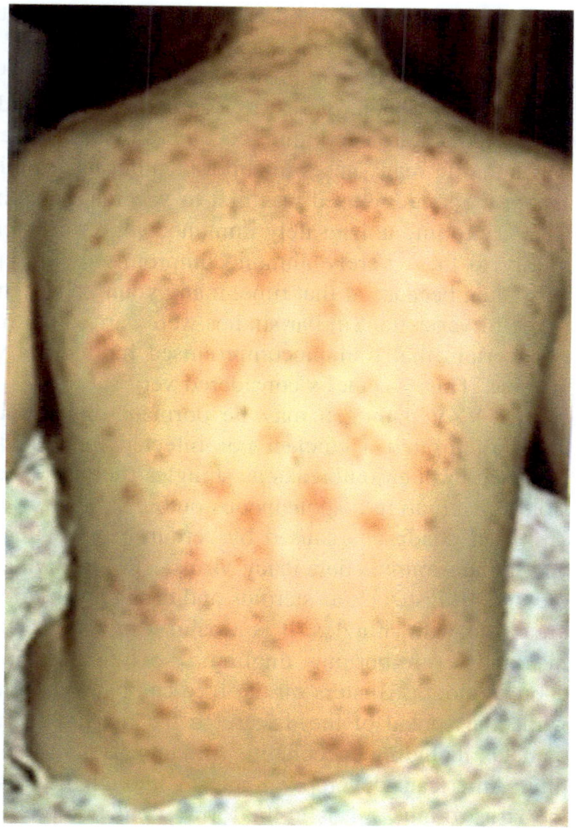

Fig. 14.8 Unvaccinated boy with chicken pox

Herpes zoster (shingles) has been recognized since ancient times and was described in the early medical literature. Clinical observations of the relationship between varicella and herpes zoster were made in 1888 by Von Bokay, when susceptible children acquired varicella (chicken pox) after contact with shingles. Varicella virus (VZV) was isolated from the vesicular fluid of both chicken pox and zoster lesions in cell culture by Thomas Huckle Weller in 1954.

Subsequent laboratory studies of the virus led to the development of a live attenuated varicella vaccine, the Oka strain, in Japan in 1974, by Michiaki Takahashi [7]. The original virus was isolated from a three-year-old, otherwise healthy Japanese boy with varicella. To prepare the seed lot, the virus was passaged 11 times at 34 °C in human embryonic fibroblasts, 12 times at 37 °C in guinea pig fibroblasts, and 5 times in human diploid fibroblasts (WI-38 and MRC-5 cells) at 37 °C [7]. Additional passages were carried out by the manufacturers to prepare the vaccine to be marketed. Because VZV is so strongly cell-associated, the final product had to be sonicated and clarified by centrifugation to produce live cell-free virus. The vaccine has been shown to be safe and effective in healthy and immunocompromised children, and in healthy adults; a booster injection of the vaccine is required 5 years later.

The vaccine has been licensed in the U.S. since 1995. When the virus was first introduced in Japan, there was controversy surrounding its use. It was tested on uninfected children and immunocompromised children, since chicken pox by itself was lethal in many cases in children suffering from leukemia or other cancers. In Japan, the strategy used was first to vaccinate healthy children and then to gradually try to immunize safely immunocompromised children, progressing from the mildly to the severely immunocompromised. This possibly protective approach was taken because at that time, many young children were surviving cancer, only to die of varicella. Similar studies were carried out in the U.S. over a long period, to ascertain that immunocompromised individuals were not in danger from the vaccine. This vaccine is considered very safe [8].

The VZV pathogen may lie dormant in the spinal nerve roots (dorsal root ganglia) through a chicken pox-infected individual's life, only manifesting its presence through outbreaks as shingles (herpes zoster). Many factors are responsible for initiating a spontaneous outbreak, such as age, stress, and immunocompromised conditions caused by certain cancers, including leukemia, lymphoma, and human immunodeficiency virus (HIV). Individuals who receive immunosuppressive drugs, such as steroids and drugs given after organ transplantation, often break out with shingles. It is a painful infection, which may include a blistering rash and severe burning, tingling, or extreme sensitivity to the skin and may last about a month. Blisters on the face can cause problems with vision and hearing, while the spread of the disease to the body's internal organs can cause serious damage, including encephalitis (inflammation of the brain). Approximately one in three Americans develops shingles. Although the person with shingles can transmit the virus to others, it is in the form of chicken pox, and not shingles and it is less infectious that an initial outbreak of chicken pox. A vaccine for shingles is available in the U.S. for persons over the age of 60. The vaccine was approved by the FDA in 2006 and appears to be 50 % effective and safe. The shingles vaccine contains a live attenuated virus and is marketed by Merck as Zostvax. The zoster vaccine is, essentially, a larger-than-normal dose of the chicken pox vaccine (Varivax, Varilrix).

14.10 Viral Latency

One of the unusual characteristics of the herpes virus family is their ability to "lie low" for long periods of time. The virus migrates from its original site in epithelial cells along nerve axons to nerve cells and remains there for long periods. During that period, specific genes associated with latency may be active, suppressing viral replication. A variety of factors, such as stress, hormonal changes, or immunosuppression associated with cancer treatment, all can trigger the virus. In the case of HSV-1 and HSV-2 there can be blisters at the original site of infection, but in the case of varicella-zoster, shingles may erupt in a section of the body where the virus is in hiding (latent). The herpes viruses remain partially or completely hidden from the immune system during latency, since they encode a variety of gene products—some of which are homologous to those of the host—that have the potential to evade the host immune responses. Co-evolution over millennia has resulted in an interaction between host and pathogen leading to the successful establishment of a lifelong latent infection with typically minimal impact upon the host, and subsequent maintenance and control of latently infected cells by the host immune system. Latency works to the benefit of the virus, since it allows for reappearance and spread of virus throughout the life of the host. It is a form of evolutionary adaptation.

References

1. Burkitt, D. P. (1983). The discovery of Burkitt's lymphoma. *Cancer, 51*(10), 1777–1786.
2. Henle, W., & Henle, G. (1969). The relation between the Epstein-Barr virus and infectious mononucleosis, Burkitt's lymphoma and cancer of the postnasal space. *East African Medical Journal, 46*(7), 402–406.
3. Cohen, J. (2003). *Epstein-Barr virus: An oncogenic human herpes virus* (6th ed.). Hamilton, Ontario, Canada.
4. Ascherio, A., & Munger, K. L. (2010). Epstein-Barr virus infection and multiple sclerosis: A review. *Journal of Neuroimmune {Harmacology: The Official Journal of the Society on NeuroImmune Pharmacology, 5*(3), 271–277.
5. Scadden, D. T. (2003). *Kaposi's sarcoma* (6th ed.). Hamilton, Ontario, Canada.
6. Looker, K. J., Garnett, G. P., & Schmid, G. P. (2008). An estimate of the global prevalence and incidence of herpes simplex virus type 2 infection. *Bulletin of the World Health Organization, 86*(10), 805–812A.
7. Takahashi, M., Otsuka, T., Okuno, Y., Asano, Y., & Yazaki, T. (1974). Live vaccine used to prevent the spread of varicella in children in hospital. *Lancet, 2*(7892), 1288–1290.
8. Gershon, A. A. (2007). Varicella-zoster vaccine. In A. Arvin, G. Campadelli-Fiume, E. Mocarski, P. S. Moore, B. Roizman, R. Whitley, & K. Yamanishi (Eds.), *Human herpesviruses: Biology, therapy, and immunoprophylaxis*. Cambridge: Cambridge University Press.

HIV and AIDS

15

Abstract

The first cases of AIDS were reported in 1981 with the occurrence of *Pneumocystis carinii* in a cohort of young homosexual men in the U.S. This "new" disease affected the immune system, destroying CD4 T-cells. The disease spread rapidly, initially among "gay" men, but later into the heterosexual community. At the same time, a "novel" cancer, Kaposi's sarcoma (KS), appeared in infected individuals; it was later found to be associated with a herpes virus. By 1982 it was obvious that AIDS was associated with an infectious agent, which was sexually transmitted or spread by intravenous drug use. The blood supply was contaminated since hemophiliacs became infected with the virus. Within 3 years of the initial outbreak, a virus was isolated. Initially called HTLV-3 by Robert Gallo (NIH), it was shown to be the same virus as LAV that was isolated by Montagnier at the Pasteur Institute a few months earlier. Because of a controversy as to who had isolated the virus first, an international committee decided to name the virus human immunodeficiency virus (HIV). Within a short time, HIV had spread all over the world, and by 1995 the number of infected people reached 30–40 million, at the peak of the epidemic. The virus actually crossed into the human population in the 1950s or earlier, and had crossed species from a chimpanzee. The disease began in Central Africa and the Congo and spread via Haiti to the U.S. and other countries. Today there is a regimen of drugs that controls the progression of the disease, but does not cure the patient, and attempts to develop a vaccine have been unsuccessful. There have been many scandals related to tainted blood, discrimination against HIV-positive individuals, and dubious government policies.

M. W. Taylor, *Viruses and Man: A History of Interactions*,
DOI: 10.1007/978-3-319-07758-1_15,

15.1 The Beginnings of a Pandemic

In 1981 the world was surprised to learn about the arrival of a completely new disease that has by now affected tens of millions of people worldwide. This disease can be called the "plague" of the twentieth and twenty-first centuries. It is known as AIDS—Acquired Immune Deficiency Syndrome—because the virus that causes the disease (HIV-1, human immunodeficiency virus-1) predominantly attacks the CD4 helper cells of the immune system and thus inhibits the immune system from functioning normally.

In 1981 a number of young men were treated at UCLA Hospital with a diagnosis of cytomegalovirus infection (normally not pathogenic in healthy individuals) including symptoms of diarrhea, and oral and anal thrush, indicating a collapse of the immune system. By May 1981 the number of such patients at UCLA Hospital had increased to five and the first died in March 1981. All were young homosexual men, and all suffered from pneumonia brought on by a fungus, *Pneumocystis carinii* (until 1988 classified as a protozoa) that is a normal component of the flora of the human lung, and only becomes pathogenic under conditions of suppressed immunity (he fungus is now known as *Pneumocystis jirovecii*). All the infected individuals clustered in one specific area of the country—Los Angeles—and all were young gay men. This type of pneumonia is usually cured with a single treatment of antibiotics, but in these cases it persisted even after continuous treatment. To quote from the Morbidity and Mortality Weekly Report of June 5, 1981: "In the period October 1980–May 1981, 5 young men, all active homosexuals, were treated for biopsy-confirmed *P. carinii* pneumonia at three different hospitals in Los Angeles, California. Two of the patients died. All five patients had laboratory-confirmed previous or current cytomegalovirus (CMV) infection and candida mucosal infection."

A year before, a number of cases of an unusual form of immune deficiency was observed in New York City. No one initially recognized the relationship to similar events in California. In March 1980, a young gay man ("N") had an illness that left his doctors perplexed; he had weight loss, spiking fevers and severe fatigue. An infectious agent was suspected when one of his friends was diagnosed with similar symptoms. N collapsed, and a brain scan showed cerebral lesions containing *Toxoplasma gondii*, which is normally benign. He died in January 1981. Beginning in 1980, and even earlier, a series of severe cases of *Pneumocystis carnii* was noted in New York City, and this led to an increase in the demand for a particular drug, pentamidine, used to treat antibiotic-resistant cases of *Pneumocystis*. In April 1981, Sandra Ford at the Centers for Disease Control brought the attention of her supervisor to the increase in orders for this drug [1].

At the same time, a rare type of cancer, Kaposi's sarcoma (KS) usually found only in elderly patients of Mediterranean ancestry, was reported in other clusters of gay men [2]. Initially it was suggested that there was some relationship to cytomegalovirus, a herpes virus associated with the defect in the immune system. The CDC formed a task force to look into the occurrence of these and other opportunistic infections, as diseases such as PCP and KS became known. It was initially thought that this was solely a disease of the "gay" community, and was termed as such: GRID, gay-related immunodeficiency [3]. Its mode of spread and its cause were unknown. By late 1981, the disease was also identified in drug addicts, likely spread through needle-sharing, but it was still considered predominantly a disease of the gay population. By the end of 1981 there had been a cumulative total of 270 reported cases, of whom 121 had died [4]. By July of 1982, it was clear that the disease was caused by an infectious agent, since it occurred among a group of individuals in various states who may have had contact with individuals in New York, Los Angeles, and San Francisco, and there seemed to be no doubt that the disease was sexually transmitted.

During 1982, a number of hemophiliacs were reported to have the symptoms of what was now called AIDS, resulting in the fear (warranted at the time) that the blood supply might be contaminated with this infectious agent. Hemophilia is a genetic disease in which factors that control blood clotting are missing. This requires the hemophiliac to have blood transfusions or injections of the missing factor after an injury of any type to stop the bleeding. The factors used by hemophiliacs (factors VIII or IX) were isolated from pools of blood that were bought commercially. These hemophiliacs were not "gay," and in many cases were children. Thus, not only was the disease sexually transmitted, but the infectious agent was also in the blood supply [5]. By the end of 1983, 21 hemophiliac AIDS patients had been identified in the U.S. and 8 in Europe. There was no doubt that the blood supply was contaminated, a fact not easily accepted by the medical profession nor by suppliers of blood. In 1980 a medical team at the Mayo Clinic in Rochester, Minnesota, had administered six units of whole blood, a unit of packed red cells, and four units of fresh frozen plasma to a 53-year-old man during open-heart surgery. Twenty-nine months later the patient had AIDS. He was neither homosexual nor a drug addict, nor did he belong to any group suspected of having AIDS. In 1983, the CDC reported 39 cases where the only factor related to AIDS was a blood transfusion.

Another group identified as carrying the disease were immigrants from Haiti, leading to discrimination against Haitians in the U.S. for a time [6, 7]; none of the Haitians was homosexual. Recent studies have indicated that the virus spread from Africa to Haiti, and was probably introduced into the U.S. from Haiti many years before the first cases were identified in either country. Later, Haiti would be recognized as one of the countries with the highest percentage of AIDS in the population, running second only to sub-Saharan Africa [1].

The disease is usually initiated by a short period of flu-like illness, after which there may be a period of normal health for months or years. Fever, diarrhea, thrush on the tongue, and the beginning of what has been called "opportunistic infections" begin anywhere from a few weeks after the initial infection (the one accompanied by flu-like symptoms) to anywhere up to 10 years. Opportunistic infections result from what might be termed "normal flora" that our immune system keeps in check. As we will discuss later, HIV destroys CD4 helper T-cells, thus impairing antibody production and the function of CD8 killer cells. When the

CD4 cells fall below 200/μl of blood, opportunistic infections occur and death usually results.

The first U.S. Congressional hearing on AIDS was in April of 1982, at which time the CDC announced that tens of thousands might be affected by the disease. Congress allocated some funds ($2 million) to explore the causes of the illness—not very much compared to large sums for other illnesses, such as Legionnaire's disease. President Reagan did not mention the word "AIDS" until 1987.

There was much speculation as to the origins of the infectious agent, whether it was a "new" disease of man, whether it was a disease that had jumped species, or an old disease that had been reactivated. As usually happens in the panic surrounding a plague, attempts were made in the media, as well as by demagogues, to blame the disease on "foreign" sources, such as Haiti, or in Africa on Europeans or "colonialists." There were also others with conspiracy theories blaming the CIA, "mad" scientists, and the deliberate contamination of the blood supply. By the end of 1982 it was obvious that this was an infectious disease, not limited to the U.S. (cases were also reported in France and the U.K.), and not confined to gay men. It occurred in people receiving blood transfusions, and in one case vertical transmission (mother to child). Later, it was shown that large numbers of children were HIV-positive from infected mothers, particularly in African countries.

By 1983 cases of AIDS-like diseases were identified in Zaire, the Congo, Belgium, Canada, Brazil and many European countries. The disease in most of these cases was transmitted not homosexually, but heterosexually. In 1985 another AIDS-like disease was diagnosed (not as severe as AIDS in Europe and the Americas) in a West African prostitute. It was called HIV-2 and was caused by a different infectious agent. HIV-2 has probably been around for some time in the human population but was not diagnosed as a separate illness because of general poor health due to malnutrition, in parts of Africa, and the disease was recognized in 1981 among refugees seeking medical care in Portugal and France. The virus was isolated in 1985 and shown to be different from HIV-1. According to the director of the Public Health services in Guinea-Bissau in West Africa, this is an old disease with a new name. For centuries people in the area died of diarrhea and tuberculosis. The disease is spread heterosexually and by blood transfusions. Infection evolves more slowly than with HIV-1, with longer survival and greater incidence of non-Hodgkin's lymphoma. There is no cross-protection by both viruses, and simultaneous infection with both viruses can occur. According to a 1985 survey of prostitutes in Dakar, most were infected with HIV-2. The first confirmed case of HIV-2 in the U.S. was reported in 1987 in a West African woman with symptoms similar to AIDS. By 2003 the number of cases of HIV-2 in the U.S. had reached 94, the majority of them among people of African origin and 66 of them born in West Africa. This virus was probably derived from a related virus, SIVsm (simian immunodeficiency virus) infecting sooty mangabeys (a type of primate) in the early part of the twentieth century. This virus never reaches the viral load detected in HIV1 and does not attack the immune system to the same extent as HIV [8].

15.2 Isolation of the Virus

The story of the discovery and identification of the causative agent of AIDS is one of rivalry between two groups of scientists, each claiming priority and a certain amount of misinformation and arrogance on the part of both groups. During the early stages of the epidemic, scientists in the U.S. and Europe endeavored to isolate the infectious agent. It was not known whether this was a viral, bacterial, or fungal infection, or even due to an environmental contaminant—or due to a specific lifestyle, as claimed by some. Robert Gallo of the National Cancer Institute in Washington D.C. had previously identified a human tumor virus causing symptoms of leukemia. He called this virus HTLV-1, since it was associated with human T-cell leukemia. This was followed by the isolation of HTLV-2, not directly associated with any human disease. These isolates were from blood lymphocytes cultured in a novel interleukin, IL-2 (cell producing growth factor) and could be characterized using a newly developed assay by the Gallo lab for the presence of the reverse transcriptase enzyme, unique to retroviruses and the hepatitis B virus. These observations were of immense importance in medical virology. Gallo and his colleagues were considered the world experts on retroviruses, and so Gallo threw himself and his group into the effort to find the causative agent of AIDS. Assuming it was related to HTLV.

By September of 1982, Jacques Leibowitch, a physician working in Paris, had been convinced that the virus causing AIDS was a retrovirus that specifically infected CD4 T-cells and was transmitted by blood. He suggested that it might be similar to HTLV-1, or at least a retrovirus. He discussed this with Gallo and with the Pasteur Institute group, headed by Luc Montagnier and François Barr-Sinnousi. Rather than looking for the virus in blood, the Pasteur group decided to look for the virus in the lymph node removed from the neck of a 33-year-old patient (RUB) who had early symptoms of immunodeficiency. Cells were grown in culture, in the presence of anti-interferon antibodies, in order to neutralize any interferon that might inhibit viral growth. After 15 days in culture, reverse transcriptase was detected and associated with viral particles, having the same density as a retrovirus. The virus grew slowly, and the culture seemed to die out, until fresh blood was added to it; the fresh blood stimulated growth of the virus. It was initially thought to be like HTLV-1, but this new virus killed T-cells in culture, whereas HTLV-1 promoted T-cell growth. The virus appeared to be a retrovirus, but not identical to HTLV-1, as initially claimed by the Gallo group and was later called "lymph-adenopathy-associated virus" (LAV) by the Pasteur Institute group. These results indicated to the French group that this virus was different from HTLV. Montagnier informed Gallo of his discovery, and Gallo sent antibodies capable of recognizing HTLV-1, but they did not react with the isolated virus. Gallo found it difficult to accept the results, since he was convinced that the "new" virus must be related to HTLV [1]. A paper announcing the isolation of a new virus appeared in *Science* on May 20, 1983, in the same issue that Gallo and Max Essex of Harvard separately announced that HTLV-1 was implicated in AIDS. In an editorial in the

"News and Views" section on this subject, the paper by Montagnier's group is discussed briefly after a review of five papers reporting that HTLV-1 is associated with AIDS. It describes the French virus as being slightly different from HTLV-1 [9]. In the paper itself, the French authors claimed that the new virus contained proteins not usually found in HTLV-1, and the electron micrograph of the virus was not identical to HTLV-1. However, at this stage no one was willing to claim that he or she had identified the causative agent of AIDS.

The group at the Pasteur Institute continued to isolate viruses from multiple cases of AIDS from patients of various ethnic and social classes, and in all cases the virus was the same as their primary isolate, LAV—not even related to the HTLV family. Despite presentations at Cold Spring Harbor, Gallo and the NCI establishment rejected the idea that LAV could be the causative agent of AIDS, and stuck with HTLV-1 as the causative agent, later calling it HTLV-3.

In early 1984 the Pasteur group began to compare the LAV with HTLV-1 and confirmed the differences previously noted, as well as other differences. In December 1983, a paper appeared showing an electron micrograph of a virus isolated from a young homosexual male with AIDS. This virus also differed in many respects from HTLV-1, and these antibodies could not be detected in AIDS patients [10, 11]. As it turned out, this newly discovered virus was the one that would finally be identified with AIDS, and it predated the discovery of the "correct" virus by Gallo and his group.

On April 24, 1984, the Secretary of the U.S. Department of Health and Human Services announced at a press conference that Robert Gallo and his colleagues had isolated a new virus, HTLV-3, had proved that it was the cause of AIDS, and that they were developing a test for diagnosis. The virus was described in detail in a series of papers in *Science* in the May 4, 1984, edition [12–15]. There is no doubt that these papers contributed greatly to the knowledge of the AIDS virus, even if the name, and thus the classification of the virus, were wrong. However, there are doubts as to whether the data presented in these papers accurate [16].

Before he isolated HTLV-3, Gallo had twice received samples of LAV, the first on July 17, 1983, the second on September 22. This virus sample was to be used for comparative purposes, and an agreement had been signed that the virus would not be used for any commercial purpose without permission from the French group. Were LAV and HTLV-3 the same virus or similar viruses, and how could one explain the discrepancy in the earlier papers between the two groups? Serology tests, cloning, and sequencing eliminated any doubt that the viruses were the same. Gallo refused to acknowledge that this was possible and claimed that he could not successfully produce the French virus in culture. In fact, one of his associates did grow the virus, and knew that Gallo misrepresented the data when he submitted his papers.

This led to an international controversy as to who had first discovered the virus causing what would later be called AIDS. It was not only a matter of prestige, but also of finances, since the discoverer's institute could patent tests to identify the virus, bringing considerable profit to the patent owner and institute. Although the French group was the first to apply for a patent, the U.S. patent office gave priority to Gallo's claim, resulting in a law suit later brought by the Pasteur Institute, claiming priority. Eventually an agreement was reached between the NIH and Pasteur Institute on the division of funds generated through the patents for HIV. In 1986 the International Committee for Viral Nomenclature decided that the new virus should be called HIV rather than LAV (Montagnier's nomenclature) or HTLV-3 (Gallo's nomenclature). Montagnier and his colleague François Barre-Sinoussi received the Nobel Prize for the discovery of HIV-1, but Gallo, who did much of the research, was not awarded the prize. The monograph "Science Fictions" by Crewdson [16] implies a "conspiracy" by Gallo and colleagues at the NCI to demean and ignore the French group, and to claim priority both for national prestige as well as financial gain.

It is now apparent that the Gallo isolate was the LAV sent to him by Montagnier. When the original isolates of HTLV-3 and LAV were compared at the nucleic acid level, the two viruses were 98 % identical. Since each isolate of what is now called HIV differs, there was no way that these could be separate isolates from different patients. The virus mutates at such a high rate that even isolates from the same person after a few months will differ from the original isolate. Gallo claimed a mix-up in his lab with the different virus strains. In fact, the electron micrograph published in the May 4th issue of *Science* describing the isolation of a new virus was actually a picture of the French isolate of LAV. This virus was grown in the Gallo lab by a visiting scientist, and the new virus he claimed to have isolated was really the one he received from France. There appeared to be a manipulation of facts, dates and actual data concerning the isolation of virus from differing AIDS patients. The scandal has been described in more detail in a series of articles in the Chicago Tribune and in John Crewdson's book [16].

15.3 Social Implications of AIDS

By 1983, the societal ramifications of the disease were being felt; there was fear, particularly in the U.S.—not that different from the fear that had occurred with previous plagues such as yellow fever and polio. There was talk of quarantine, or even isolation and segregation of AIDS patients. Actual quarantine facilities (sanatoriums) for infected patients, particularly Cuban soldiers returning from Angola, were set up. However, after ten years these were closed, but patients wanting to live in them could continue to do so. Cuba today has a very low incidence of AIDS, which might be related to its early quarantine step, although from a civil rights perspective this seemed cruel and certainly infringed on the right of movement and association. The U.S. and many other countries demanded HIV testing for immigrants or people traveling to these countries, and those who tested HIV-positive were denied entry. This, in fact, led to a boycott of AIDS-related conventions in the U.S. and the law was only rescinded (in 2011).

In the U.S., patients with AIDS were shunned in their communities: some lost jobs, others lost housing, and children with AIDS were even expelled from school. One source of AIDS among the gay community was the bathhouses. Promiscuity was the norm, men had random sex with other men and only after a struggle, particularly in San Francisco, were the bathhouses closed in 1984. During the next few years many conspiracy theories were proposed to explain the AIDS epidemic: that AIDS was a CIA plot; that it was targeting the African-American community; and that it was a contaminant of the polio vaccine used in Africa in the late 1950s. These unwarranted speculations remained as long as the source of AIDS remained a puzzle. The CDC attempted to calm the public explaining that the virus could only be transmitted by exposure to blood products, by sharing needles while taking drugs, or through accidents where blood was exchanged. It was a condition primarily spread through sexual intercourse, and not by sharing cups, hugging, kissing, or casual contact. Despite these assurances from the CDC, the general attitude in the country and in the press was one of fear and panic and, as described below, there were people who did not believe that AIDS was caused by the HIV virus; indeed, there are probably people who even now who do not accept it.

Infection with HIV resulted in a horrible disease, especially in the early years of the epidemic, when the only treatments were antibiotics or anti-fungal agents to inhibit opportunistic bacterial and fungal infections. The disease state was termed AIDS since one could be HIV-positive without progressing to full-blown AIDS. The film (and play) "Angels in America" by Tony Kushner (2003) presents a realistic depiction of the suffering of AIDS victims in the early years.

15.4 Why Were Homosexual Males More Likely to be Infected?

At the beginning of the epidemic, in 1982, HIV/AIDS was known as "GRID"—gay-related immunodeficiency disease. The first cluster of what we now call opportunistic infections occurred in gay men. The sub-culture of "gay" men was a very promiscuous one; oral or anal sex was performed with many different partners at "bath-houses" or during casual meetings in parks or bars. This resulted in the rapid spread of the virus throughout the gay community. In the days before AIDS, sex between men was performed without any protection, since condoms were only used during heterosexual intercourse to prevent pregnancy. The cause of the epidemic among gay men was initially thought to be due to the use of "poppers"—vials of alkyl nitrites used as an aphrodisiac and to enhance sexual pleasure. Poppers accompanied the use of drugs in the 1970s at discos and parties, but this idea was invalidated when AIDS was discovered in "straight" Haitian men. One reason for the higher frequency of HIV in gay men was that anal sex was the preferential mode of sexual behavior. Anal sex may damage the thin lining of the rectum, allowing the virus to invade the body and enter the systemic circulation.

HIV can be found in the blood, semen, and pre-seminal fluid in persons infected with the virus. In general, the person receiving the semen is at greater risk of getting HIV than the inserter. However, a person who inserts his penis into an infected partner is also at risk since HIV can enter through the urethra or small cuts, abrasions, or open sores on the penis, so it is possible for either sex partner to become infected with HIV during anal sex. (http://www.cdc.gov/hiv/resources/qa/transmission.htm)

A study in 2010 found that the risk of HIV transmission is 20 times greater with anal intercourse than with vaginal intercourse [17] http://www.sciencedaily.com/releases/2010/05/100525094900.htm. In a large study among homosexual men in 1987, there was a very high correlation between those who practiced receptive anal sex with more than one partner and seroconversion to HIV positive (seroconversion implies the presence of antibodies to HIV) than those who were not receptive [18].

In a study in San Francisco in 1987, [19, 20] among homosexual/bisexual men reporting no male sexual partners in the 2 years before inclusion in the study, seropositivity was 17.6 %. For those reporting more than 50 partners, seropositivity was 70.8 %. Thus, both anal intercourse and promiscuity played a large part in the initial epidemic. Similar results were reported later in other studies from many countries. The number of sexual partners that individual gay men had was in some cases startling; there are even quotes in the literature of 250 partners per year! (Whether or not this is true is difficult to say, as it could simply be boasting about sexual prowess.)

To quote the French physician Jean Paul Escande: "AIDS is not a divine lightning bolt, but its advent shows that when a community profoundly changes its lifestyle, a certain number of diseases will crop up without fail; the excess of sexual liberalization among homosexuals is probably responsible for the biological modification that fostered the rise of AIDS … AIDS mushroomed not because they transgressed certain sexual taboos, but rather certain rules for living which, until then, had concurred to maintain a relative biological equilibrium." [1].

15.5 AIDS in the USA in the 1980s

AIDS cases continued to be reported throughout the early 1980s, both in the U.S. and other countries. Two unique cases in 1985 brought interest in the AIDS pandemic to the wider American public through media exposure. What AIDS was and how it could be transmitted was unknown. What was known were its terrible effects. As usual in these conditions, the opinion of ignorant people, an opinion based essentially on senseless fear, becomes common and for a while even dominant. Popular opinion was reflected in the social stigma that was extended to the victims of the disease, irrespective of the causes of the infections.

Some cases are illustrative and should be remembered; among them is the case of Ryan White, a 13-year-old hemophiliac, who was diagnosed with AIDS after surgery for partial lung removal (Fig. 15.1). He had been receiving factor VIII for

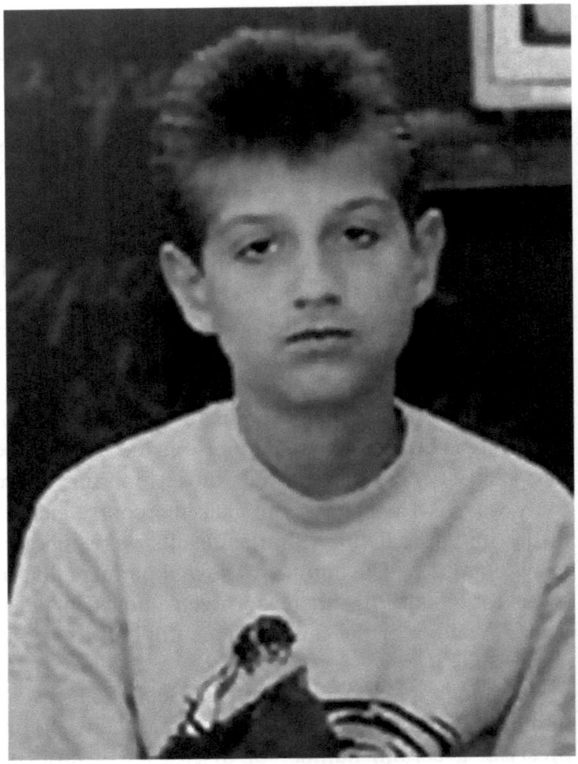

Fig. 15.1 Ryan White, 1985

treatment of his hemophilia and after it was found that he had contracted AIDS, he was banned from school in Kokomo, Indiana. He suffered from insults and indignities from his classmates and their parents. Ryan died of AIDS at age 18 in 1990. A similar case occurred in 1986 after Florida's Desoto County School Board refused to allow three HIV-positive brothers—Ricky, Robert, and Randy Ray—to attend school. A federal judge ordered the board to reinstate the three hemophiliacs, who had contracted HIV through transfusion of contaminated blood products. Outraged town residents refused to allow their children to attend school, and someone set fire to the Ray family's house on August 28, totally destroying it. This blind bias, so typical of ignorance, was also exemplified by the action taken by the U.S. Senate, which passed—96 to 2—the Helm's amendment banning federal assistance to groups that promoted AIDS educational material. An opinion piece by the Mayor of New York attacked the amendment as senseless and callous in a New York Times opinion piece (http://www.nytimes.com/1987/11/07/opinion/senator-helms-s-callousness-toward-aids-victims.html).

Another important media event, which increased awareness of how widespread the disease had become, was the October 1985 death from AIDS of Rock Hudson, the popular film star, and symbol of the "macho" male. Three months following his death, Randy Shilts published *And the Band Played On*, which presented an early history of the disease, particularly as seen in San Francisco, and the political struggles to close down the bath-houses, a major source of infection in the gay community. This book was later made into a film, which I would urge anyone interested in this early history of the epidemic to see. Randy Shilts himself died of AIDS in 1994 at the age of 42. Despite statements from the CDC that the disease was only spread by sexual relations or contact with blood, people were fired from their jobs because of the fear that they might transmit the virus through social contact, sharing the same cup, or being in the same room.

By the end of 1985, 15,400 individuals were diagnosed with AIDS, 12,500 deaths from AIDS were recorded, and the number of HIV-positive people was much greater. By the end of 1985, it became obvious that the epicenter of the epidemic was no longer the U.S. but Africa, where the disease had spread very rapidly because of sexual promiscuity, poverty, general ill health, and malnutrition. Patterns of transmission could be tracked by following truck routes and truck stops where the presence of sex workers (prostitutes) in both West and East Africa spread the disease. This was no longer a "gay" disease, but a heterosexually transmitted one. In February of 1987 the World Health Organization launched "The Global Program on AIDS" to raise awareness; formulate evidence-based policies; provide technical and financial support to countries; initiate relevant social, behavioral, and biomedical research; promote participation by non-governmental organizations; and champion the rights of those living with HIV.

In 1987 there was a turning point in the attitude of the U.S. government and other governments regarding the AIDS epidemic. By this time, the seriousness of the pandemic had been realized. A number of action groups had been formed, including AIDS coalition to Unleash Power (Act-Up) in the U.S. Their aim was to educate the public about the epidemic, influence medical research, and change political and social policy by pressuring the government and the NIH for more funding to study the disease. However, the social stigma of being a victim of HIV still existed despite an educational campaign by the CDC and other agencies.

By 1989 there were 117,508 reported cases of AIDS and 89,343 deaths (http://www.amfar.org/thirty-years-of-hiv/aids-snapshots-of-an-epidemic/).

15.6 Test for AIDS and Blood Supply

A number of pharmaceutical companies developed a test for HIV in 1985. This test, based on a method called "enzyme-linked immunosorbent assay" (ELISA), detects antibodies in the blood of infected individuals, who are termed "seropositive." A backup test using Western Blot technology[1] (antibody-antigen

[1] The story behind the nomenclature "Western blot" is amusing. The first type of "blot" that was used to compare DNA with other DNA samples following separation of DNA on a gel and the identification of spots was called "Southern blotting" after the creator of the technique, Ed Southern. This was followed by the development of a similar method to detect RNA samples on a

reaction on strips of paper following electrophoresis—a method that separates proteins and detected by a second antibody linked to a detector [dye]) was used as a backup. These tests were initially used to examine the safety of the blood supply. The licensing of the ELISA test was controversial, and a patent for testing for HIV (at that time called either LAV or HTLV-3) was filed first by the Pasteur Institute with the U.S. patent office, but was overlooked, and a second patent filed by NCI on behalf of Bob Gallo was approved. This second method was neither as efficient nor as reproducible as the Pasteur method, but with pressure from the NCI was the one adopted. This resulted in many HIV-positive samples testing as negative and the use of contaminated blood in transfusions in hospitals. It took 2–3 years and threats of lawsuits between the NCI and the Pasteur Institute before the Pasteur ELISA was licensed in the U.S. [16] and became the choice of the American Red Cross. Both patents were granted and both tests used in the U.S. Each test uses different antigens and may give different results [21].

The false-positive results caused much personal anguish, since many who believed themselves "clean" appeared to have acquired the virus, and required retesting. Some individuals did not retest and assumed they had AIDS, although they didn't. All positive tests have to be redone and confirmed by the Western blot. There is a period of approximately 2 weeks during which one can be infected but viral antibodies are not present in the blood, and thus one is sero-negative. The test should be redone at a later time if there is any suspicion of infection. Ethically, if one tests positive twice, one should stop having sexual relations or at least notify the sexual partner(s) of possible infection. Psychologically, it is difficult to live with the threat of AIDS, particularly during a period when no treatment is available. Also, in the U.S., Canada, Belgium, Italy and Germany, spreading the virus was often considered a criminal act and could lead to "isolation" in jail. The literature of this period contains many articles about the denying of the civil rights of HIV-positive individuals.

The tests, which became more reliable with time due to better purification of viral antigens, measured how many people were carriers of HIV and the extent of HIV-positive individuals in the population. In 1985 the French blood centers announced a rate of 1 per 1,000 of seropositivity, which was the same level as found in the French army. Similar results were found when the U.S. army tested applicants for military service; this was 100–1,000 times higher than those apparently suffering from AIDS.

Currently there are nine tests that can detect HIV antibodies, antigens (viral proteins) and viral nucleic acids. Recently, simple tests have been developed using urine, or saliva, and can be performed at home at very low cost. However, these are less reliable than the traditional Western blot. In the U.S., HIV testing is readily available at local hospitals and clinics for a nominal charge.

(Footnote 1 continued)
blot after electrophoresis and was called the "Northern Blot." This type of nomenclature was continued when protein blots were established, and called "Western blots." We now even have "South-western" and "Far-western" blots, which are variations on the Western blot.

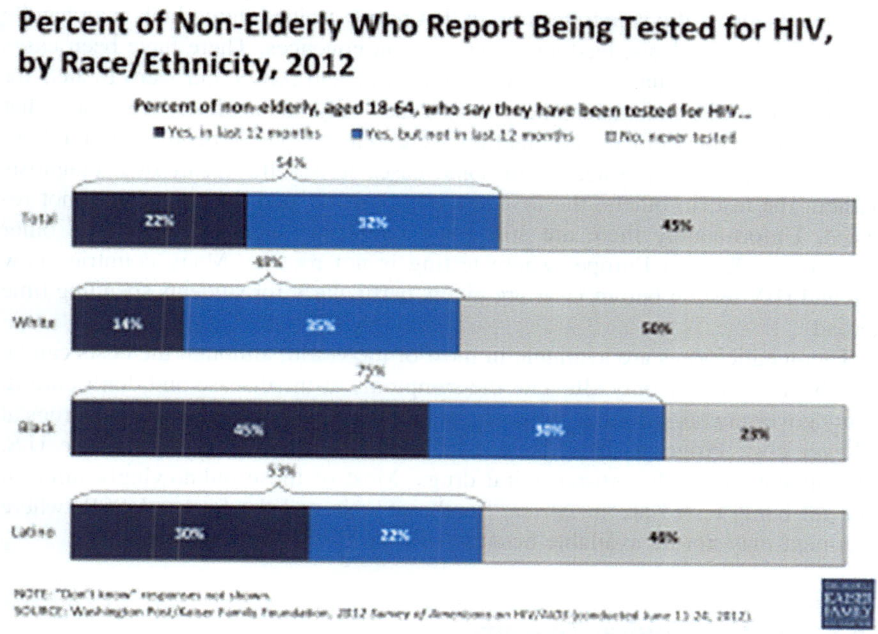

Fig. 15.2 Percent of population who report being tested for HIV (CDC)

15.7 Why Test?

To prevent an epidemic, it is necessary to know who are the carriers of the disease, and where these individuals live. Identifying who is positive for HIV may deter individuals from spreading the disease. Many people (probably the great majority) do not consider themselves candidates for HIV, and deny that there is such a possibility. This is particularly true among young people, since risky behavior is not considered by them to be risky. By the end of 2013, it is estimated that there will be 30 million people in the world with HIV. Of these, half will have been infected before the age of 25 and only about 15 % will know that they are HIV-positive. Since the symptoms of the disease are now treatable, particularly with drugs accessible worldwide, testing is crucial in preventing the spread. This assumes that individuals positive for HIV will either cease sexual activity or inform their partner(s) that they are infected, take precautions, and enter into a course of treatment. Determining one's HIV status is the first step in treating the disease, but unfortunately, large segments of the population shun testing. In the U.S. there are as many as 40,000 new cases of HIV per year. Figure 15.2 illustrates this point, where as many as 50 % of people of sexually active age are not tested for HIV.

There is also a flip side to testing, in that poorly trained individuals may handle the testing and analysis, leading to serious consequences. There have been cases [8] in which individuals have assumed that they had AIDS and were treated for AIDS with drugs that made them very sick—only to be discovered later that testing was not performed, and that they were "clean." These people suffered both psychologically and physically. In some cases it was the result of a diagnosis without the test; in others, the first test was positive, and the person was not re-tested. Unfortunately there are still some countries, such as Russia and other countries of Eastern Europe, where testing is not routine. Many countries now demand HIV testing before visas are given, particularly for students spending time abroad.

Therapeutic drugs are available in most of the world, although the costs can be prohibitive and they only alleviate the symptoms of the disease and don't cure it. Many governments subsidize the costs, or produce "generic" copies of the drugs at a lower cost. Programs have been established by the UN, WHO and the U.S. government to supply anti-retroviral drugs. Most of those unknowingly infected and not being tested are in Africa, Southeast Asia and the former USSR, where treatment may not be available because of cost.

15.8 Anti-retroviral Therapy

Although the epidemic began in 1981, the first drug to inhibit HIV replication was not identified until 1987 and received FDA approval that same year. For the first years of the epidemic, there was no reasonable medication. Opportunistic infections were treated with antibiotics and fungicides, as well as interferon, which inhibited the growth of KS [22]. We now know that KS is caused by herpes virus-8, [23] so that the interferon may directly inhibit this virus. However, interferon did not have a consistent inhibitory effect on HIV replication. Since HIV is a retrovirus, the obvious pharmaceutical target was an enzyme unique to retroviruses, the reverse transcriptase. Massive screening of oligonucleotides known to affect either DNA or RNA synthesis took place at drug companies. Most of these compounds had been developed as potential anti-cancer agents. In 1986 a drug that appeared to inhibit the HIV reverse transcriptase was rediscovered—this was zidovudine or azidothymidine (AZT), first synthesized in 1964 by Gertrude Elion[2]

[2] Gertrude Elion received the Nobel Prize for her work in 1988. She was responsible for the development of many anti-cancer agents affecting nucleic acid metabolism. She was also one of the only scientists to receive the Nobel Prize who did not have a Ph.D. Elion graduated from Hunter College in New York City with a BS degree, but could not obtain admission to a graduate program, either because she was a woman or because she was Jewish. She later received an MS from New York University. George Hitchings at Burroughs Wellcome, a large pharmaceutical company in drug design, eventually employed her to develop drugs against cancer, and she shared the Nobel Prize with him. I knew her personally from participation in meetings on purine metabolism. She was a very unassuming woman whose husband died of a bacterial infection in 1941, in the days before antibiotics.

at Burroughs Wellcome as an anti-cancer agent, which at that time proved ineffective and was placed in storage. However, when screened on retroviruses, it proved to be an inhibitor of reverse transcriptase, slowing down virus replication. This drug became the standard treatment for individuals with AIDS; it also became the prototype small molecule developed against HIV and other viruses. AZT reduced the amount of circulating virus in the blood and lymph nodes and prevented the collapse and decrease in CD4 lymphocytes, the target cells of the virus. It could be used until either the virus mutated and became resistant, or until the side effects were too severe. Initially it was used only after the effects of full-blown AIDS occurred, but is now recommended early in the infection. Since those early days, other inhibitors of the reverse transcriptase have been proven effective, but it took another 3 years for the next inhibitor to reach the clinic.

Over the next few years, comprehensive studies on antiretroviral drugs resulted in the successful introduction of highly active antiretroviral therapy (HAART) into clinical practice for the treatment of HIV/AIDS. However, there is still a need for new therapeutic approaches, since HAART cannot eradicate HIV-1 and, unfortunately, can be associated with long-term toxicity and the development of drug resistance. Because of the mechanism of HIV replication, it is nearly impossible to "cure" HIV because the virus brings in its own enzyme the reverse transcriptase and is integrated into the chromosome of the host cell. This enzyme is error prone, resulting in mutations that escape the immune system. It is possible that an enzyme (Tre-recombinase) could be developed that removes the virus (or segments of it) from the human chromosome, and such an enzyme is undergoing testing in cells and in mice, but it will take years of research before it can be used in humans; the difficulty will be in targeting it to all infected cells. Initial data show that it is effective in cultured cells, but this is a long way from clinical development.

The current aim of antiretroviral therapy is not to cure the disease but to extend the life of the patient, prevent opportunistic infections and maintain a decent quality of life. The objective is to suppress HIV replication to undetectable levels, i.e., less than 50 copies per ml of blood as measured by PCR. This requires the infected person to take an expensive regimen of drugs (HAART), which may have many side effects. The CDC has issued guidelines for the use of HIV medication: anti-retroviral medication should be administered to pregnant women who are positive for HIV to reduce the risk to the child and transmission of the virus in milk.

Treating people with HIV with anti-retroviral drugs lowers the amount of virus in their body and can dramatically reduce their risk of transmitting HIV to others, underscoring the importance of HIV testing and access to medical care and treatment. In fact, a landmark clinical trial in 2011 indicated that people with HIV who began taking anti-HIV medications early (before their immune systems were significantly weakened) experienced a 96 % reduction in their risk of transmitting HIV to their sexual partners [24].

The most surprising recommendation of the CDC is the use of pre-exposure prophylaxis—taking anti-retroviral drugs before sexual activity in high-risk groups to prevent the infection. The drug tenofovir (Viread), produced by Gilead Biotech,

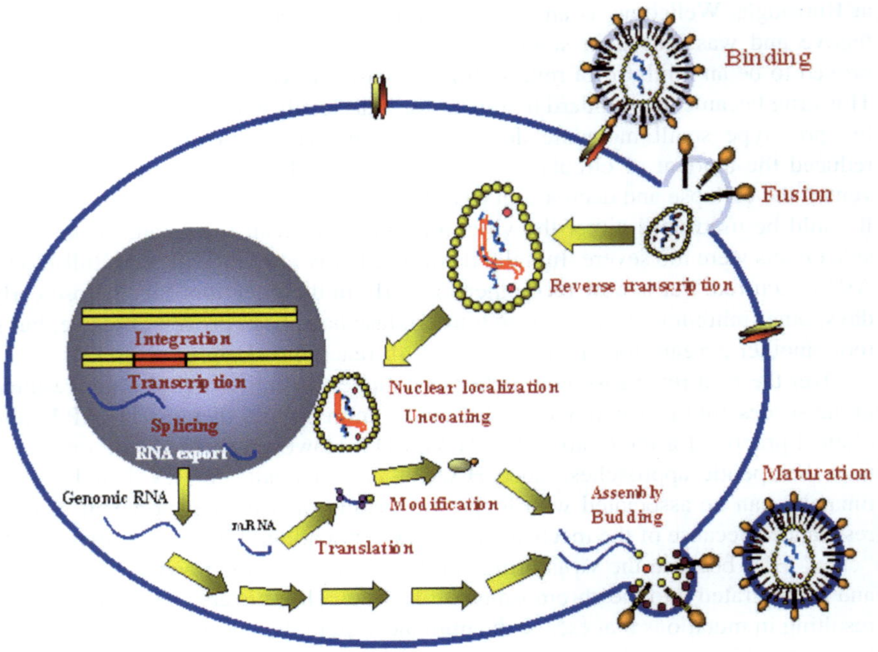

Fig. 15.3 Site of action of anti-retroviral drugs (http://pathmicro.med.sc.edu/ with permission)

is an anti-reverse transcriptase drug that when administered to drug addicts in Thailand reduced infection by 49 % in an at-risk student population compared to a placebo. The principal investigator of the study wrote in the Lancet, "We now know that pre-exposure prophylaxis can be a potentially vital option for HIV prevention in people at very high risk for infection, whether through sexual transmission or injecting drug use" [25]. Similar studies with other anti-retroviral drugs have given similar results.

Anti-retroviral drugs target different segments of the virus, either directly inhibiting viral proteins or targeting essential viral enzymes such as proteases, integrase (the enzyme involved in recombination between the viral DNA and the host chromosome) or blocking attachment and uptake of the virus. These various targets are shown in Fig. 15.3 and Table 15.1. A description of each class can be found in more detail at the reference below the table.

15.9 Vaccine Development

It has been extremely difficult to develop a preventative vaccine against HIV-1. A great deal of effort and research have gone into vaccine development, and various approaches, including using inactivated virus and the cloning of viral segments

Table 15.1 Target of drugs developed against HIV infection

Nucleoside analogs (reverse transcriptase inhibitors)
Non-nucleoside reverse transcriptase inhibitors
Protease inhibitors
Entry or fusion inhibitors
Integrase inhibitors
Disruption of polyprotein formation
Antisense drugs
Assembly inhibitors
Removal of integrated sequences
Cytokines such as IL7, interferons, etc

For more information on the large number of drugs of each class, see http://www.fda.gov/ForConsumers/ByAudience/ForPatientAdvocates/HIVandAIDSActivities/ucm118915.htm or http://www.aidsmeds.com/list.shtml)

into other viral vectors, have been attempted but without much success. This stems from the two major hurdles: the ability of HIV to integrate into the chromosome and be maintained in the chromosome as DNA and not as a free virus particle, as is the case with most other viruses, and the fact that HIV, like other RNA viruses, mutates at a very high rate. Thus within one individual, multiple "strains" or "quasi-species" of viruses may develop, so that viral escape from antibodies induced by the vaccine will occur very quickly. It is true that vaccines have been developed against yellow fever and polio—both RNA viruses—but these viruses replicate in the cytoplasm, do not integrate into the host chromosome, and enter the bloodstream, where they can be neutralized by antibodies. During the natural infection with HIV, antibodies are produced (the basis of the AIDS test), yet they are not sufficient to eliminate the virus, since the virus is "hiding" in immune cells in lymph nodes.

Despite these problems, in 2010 there were 30 clinical trials in progress, using 40 different vaccines, on four continents, and involving 24 countries—yet none of these trials was successful at producing a neutralizing antibody. Neutralizing antibodies are essential to eliminate virus from the blood and lymph nodes. They are mostly of the classical preventative "vaccine" type. Other groups are working on a prophylactic vaccine, which would be given to individuals already infected with HIV, and stimulate the immune system (innate) against the virus. One of the largest trials was of a recombinant adenovirus containing the internal genes of HIV, but in trial this vaccine not only did not work to protect against HIV, it also appeared to increase the frequency of HIV infection. Other vaccine constructs have not enhanced antibody production enough to destroy the virus. A new approach may be necessary for "vaccine" production in which the emphasis is on activation of the innate immune system so that it recognizes infected cells more efficiently [26]. This is difficult, since many immune cells producing cytokines are themselves targets of the virus.

Another problem with vaccine administration is the inability to differentiate the person carrying an infectious virus from the non-carrier who now produces antibody to the virus as a result of the vaccine itself. Thus by receiving the vaccine, the individual will seroconvert—that is mimic the infection—without being infected. Unless infected cells can be eliminated, antibody production by itself will not be sufficient to prevent infections. A novel vaccine was recently tested in rhesus macaques infected with a virulent strain of SIV; the monkeys responded to a vaccine in which SIV genes were expressed in a cytomegalovirus vector. It took some time for the vaccine to have an effect, but no virus was detectable in any of the organs of the cured monkeys [27]. Cytomegalovirus is common in humans, and most humans have antibodies to the virus, thus inhibiting it as a vaccine. Incorporating HIV protein sequences into other herpes virus needs to be attempted. It appears that the presence of the cytomegalovirus caused a strong killer T-cell reaction, and the replication-competent SIV persisted for months in the monkeys before disappearing. The authors conclude that "These data provide compelling evidence for progressive clearance of a pathogenic lentiviral infection, and suggest that some lentiviral reservoirs may be susceptible to the continuous effector memory T-cell-mediated immune surveillance elicited and maintained by cytomegalovirus vectors."

15.10 The Scandals of Contaminated Blood

As mentioned above, it was difficult to convince the American Red Cross and companies such as Bayer, Baxter, and others that contaminated blood might be a source of HIV, particularly in the preparation of factor VIII and IX used in treating hemophiliacs. In the U.S. about 10,000 hemophiliacs became HIV positive as a result of contaminated or untested blood products. In Canada, the Canadian Red Cross delayed implementation of blood treatment (heating) to destroy HIV and hepatitis B, resulting in contaminated blood, used to prepare factors for hemophiliacs. However, in many other countries, contaminated blood was used for blood transfusions and not only in the preparation of factor VIII. In some cases, companies producing these factors for hemophiliacs continued to sell contaminated products in Asia, to avoid dumping the products on hand, as late as 1985. Such products were unacceptable in the U.S. and the factors were produced by an alternate method, and there were strict controls on who could sell whole blood. (See follow-up article at http://www.nytimes.com/2003/05/22/business/2-paths-of-bayer-drug-in-80-s-riskier-one-steered-overseas.html?pagewanted=all&src=pm, also [28, 29].)

15.11 Romania

Physicians in many countries, particularly Eastern Europe, believe that fresh transfusions of blood can alleviate many symptoms of illnesses, not necessarily related to blood diseases themselves. Not only was contaminated blood used in most cases, but needles or other instruments were not sterilized between uses. The most glaring instance of this (or the most published) was the situation of orphans in institutions in Romania, which came to light after the fall of Ceausescu and the communist government. There was a complete lack of hygiene in these institutions, with thousands of children infected with HIV and hepatitis B. These HIV children, known as "Ceausescu's abandoned children," were discovered 20 years ago and many still live in orphanages under dire conditions. Most of them were not HIV-positive from maternal transmission but as a result of the lack of care and the use of contaminated needles for injection. These babies or children received blood as a nutrient with one needle used in a ward of a few hundred children.

Before 1989, birth control and abortions were illegal in Romania for any woman who had fewer than five children, and this led to the abandonment of many children in state institutions under the most appalling conditions. By 1985, many of these children had HIV, but the government refused to recognize that such a disease could exist in Romania, and physicians were too afraid to discuss it [30]. Under the communist regime, one was living in a perfect society where homosexuality and prostitution did not exist. By mid-2002, after the fall of Ceausescu, 12,559 cases of HIV were registered in Romania (9,936 of them in children), of which 2,699 had already died. Most were infected with a clade (group) known as type F, which was unique to that area of the world. Many of these orphans have reached adulthood, and still suffer from the disease or other conditions. The Romanian government has instituted treatment centers, and was one of the first countries in Eastern Europe—after the collapse of communism—to grapple with this problem. However, there are still economic difficulties in housing and treatment of the survivors of this epidemic. Romania is now often held up as a model on how to deal with the AIDS epidemic, and treatment with anti-retrovirus drugs is free. However, there is still a sizeable part of the population that is unaware that they are carriers of the disease [30–32].

15.12 Contaminated Blood in China and Other Asian Countries

Even though HIV was discovered and characterized as a sexually and blood-transmitted disease by the mid-1980s, many countries still continued to use contaminated blood in hospitals, thereby increasing the number of victims, particularly among children. In the Henan province in China, farmers sold their blood for \$5.00 per 400cc, with some of them giving blood as much as 50 times. The blood was pooled and the serum separated; red blood cells from the pool were then transfused back into the sellers; this was thought to prevent blood loss. Thus any contaminated blood, even from one individual, would then reach many. The conditions of collection and transfusion were unsatisfactory. Around 80 % of the inhabitants of Donghu are today HIV-positive, among the highest in the world, [33] and many already have the symptoms of AIDS. However, similar methods are still being used in other provinces in China. The tragedy, in such cases, is not only the spread of AIDS to the total population, but also to newborn children from infected mothers. The Chinese government has tried to suppress this information, and there has been little done in terms of treatment outside of the large cities [34, 35]. Part of the problem was the profit motive, in which government officials ran the blood-collecting stations and sold the serum to pharmaceutical companies for the production of serum by-products. Until 1994, there was no screening of blood supplies for HIV in China.

A similar incident occurred in Kazakhstan in the summer of 2006; [36, 37] One hundred and eighteen children became HIV-positive as a result of blood transfusions ordered by health workers. By 2007, ten of them had died, and the doctors were charged with malpractice. Patients had paid \$20 for blood collection; the physicians received \$10 and the local blood bank the remaining \$10. One child was given as many as 25 unnecessary blood transfusions. On July 2007, 17 of the 21 health care workers were given jail sentences of a few months to 8 years. The disease has spread throughout the country, and it is estimated that there are anywhere from 6,000 to 30,000 carriers of the disease. As in other countries of Eastern Europe and the former Soviet Union, the government has passed legislation to protect victims of the disease and ensure human rights and laws against discrimination.

Nonetheless, despite a decrease in new cases of HIV in the U.S., Western Europe and Africa, the incidence of HIV/AIDS is increasing in Eastern Europe and Central Asia. The latest statistics show that 1.5 million people in Eastern Europe and Central Asia are living with HIV, 3.5 times the number for 2001 (410,000) and more than 10 times than in 1991. In these countries, there is a lack of antiretroviral drugs, and most of the disease occurs in marginalized populations such as drug addicts, and men who have sex with men, sex workers, prisoners and migrants. Tuberculosis and end-stage liver disease caused by viral hepatitis C infection are among the leading causes of death among people living with HIV/AIDS, particularly those who are also addicted to drugs. At 170,000 new HIV infections, the region trailed only sub-Saharan Africa and South and Southeast Asia in terms of new HIV infections in 2011. But the numbers in Eastern Europe and Central Asia are on an upward trend, unlike those in sub-Saharan Africa and South and Southeast Asia, which appear to be stabilizing and declining.

15.13 Origin of HIV

Where did HIV come from? When did it first enter the human population? Is it a "new" virus or an old virus that has resurfaced?

Although there was an extensive search for HIV-1 among primates immediately after the discovery of the virus, a nearly identical virus was not discovered until 1999 [38]. A large number of similar retroviruses have been discovered in different species of primates and in other mammals, such as cats. Such viruses cause immunodeficiency in these animals and have been termed SIV (simian for monkey), and FeIV (cat immunodeficiency). However, a virus nearly identical to HIV-1 was discovered in preserved tissue of a sub-species of chimpanzee, *pan troglodytes troglodytes* that are native to Gabon, Cameroon, and Equatorial Africa in Central West Africa. This virus itself appeared to be a hybrid between two different strains of SIV, SIVcpz and SIVgor. Chimpanzees and gorillas are the only non-human primates to carry a virus closely related to HIV. Phylogenic analysis has indicated that the chimpanzees were infected the virus SIVgor from gorillas, and that a recombinant virus between the two SIVs jumped species into humans to form HIV subgroups infecting humans. It is thought that this SIV jumped species at least four separate times. The species have been hunted for a long time and used as meat, so it was very easy for the virus to be transmitted in blood to humans through handling or eating poorly cooked meat.

HIV-2 is related to an SIV that occurs in the sooty mangabey, a species of monkey, and is thought to have been transmitted to humans in the 1960s in Guinea-Bissau in Western Africa.

Following are the earliest known instances of HIV infection:

A plasma sample taken in 1959 from an adult male who died in the Congo was shown to be HIV positive [39]. A lymph node sample taken in 1960 from an adult female, also from the Democratic Republic of the Congo, was HIV positive [40]. Both isolates were of a different subtype. These examples indicated heterogeneity in the virus, indicating that it may have been present in Kinshasa for a considerable time.

Robert Rayford was an American teenager admitted to Barnes Hospital in St Louis, Missouri, who died a few years later from a "mysterious" disease in 1969. At the time of his death, samples of tissue were frozen. When AIDS appeared in the 1980s, samples from these tissues were tested for the presence of HIV by Western blots, and HIV proteins were detected. Rayford was suspected of being a male prostitute. He did not travel abroad, indicating that HIV was already present in the U.S. as early as the 1960s. Many other cases were probably incorrectly diagnosed and were due to HIV.

HIV-positive samples from a Norwegian sailor who died around 1976. He contracted the disease in Africa. Arvid Noe (his name) began working as a sailor in 1961 at age 15. He is known to have visited Africa at least twice, to have worked as a truck driver in Europe, and is assumed to have frequented prostitutes in the

Cameroons. Beginning in 1966, Noe suffered from joint pain, lymphedema, and lung infections. His condition stabilized with treatment until 1975, when Noe's symptoms worsened, and he developed motor control difficulties, dementia, and died. His wife developed similar symptoms; she and their nine-year-old daughter both died in 1977. About 10 years after their death, doctors in Norway examined tissue samples of all three and found that they were positive for HIV1; the type of HIV was similar to that found in the Cameroons. Noe may have transmitted the virus to prostitutes in Africa and Europe.

Grete Rask was a Danish surgeon who travelled to Zaire in 1972. She was probably exposed to blood from Congolese patients. Returning to Denmark in 1976, she was ill and her symptoms baffled her colleagues. She died in 1977 and in 1984 her tissue was examined for HIV and found to be positive.

Workers returning from the Congo to Haiti probably brought HIV to the Western Hemisphere. The first reported case was in 1966, but of course at that time the symptoms were not recognized as AIDS. A 1998 analysis of the plasma sample from 1959 suggested that HIV-1 was introduced into humans around the 1940s or early 1950s. The virus has evolved since then into 11 different subtypes (clades). Scientists believe that 1959 is near the branch point where the B and D subtypes branched off. The B subtype is the dominant one in the U.S; using sequence and statistical analysis, it is hypothesized that HIV had existed in the U.S. for 12 years before the epidemic was recognized [40]. There appears to be no doubt that the virus originated in the Congo in the 1940s or 1950s or even earlier, was then transmitted to Haiti, possibly in a single case, and then transmitted to the U.S. in the 1960s with a group of Haitian immigrants. From there it eventually made its way into the general population.

15.14 AIDS in 2013

There is no doubt that the AIDs epidemic is under control in most parts of the world, and that there is an increase in the number of people living with AIDS, mostly due to the success of the HAART treatment. Although expensive, this treatment is now available in most countries, although there are a few exceptions, either for economic or political (ideological) reasons. The number of new cases is still high, and there has been an unfortunate tendency, particularly among young adults, to think of AIDS as just another chronic disease that can be kept under control. Most of the new infections arise from intravenous drug use and not using condoms during sexual activity. Large sums of money and time have been spent on educational activities, and this has influenced large segments of the population towards safe sex, in particular in Africa; circumcision (discussed below) is also a factor. Statistically, there has been an obvious decline in both the number of deaths and new cases since the peak of the epidemic, which was around 1997–2000. These statistics in a countrywide fashion can be found at http://apps.who.int/gho/data/node.main.620?lang=en. The decline in each of the categories, new cases,

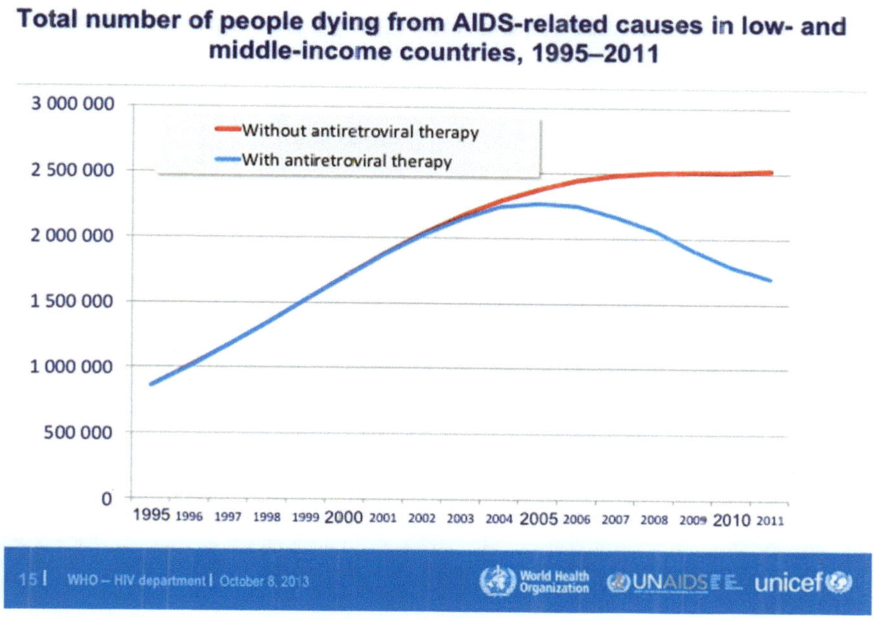

Fig. 15.4 Total number of people dying from AIDS. Note decrease due to anti-retroviral therapy (UNAIDS-UNICEF)

deaths, etc., are presented below, before a discussion of individual countries and continents.

In 2012 there were 35 million people living with HIV worldwide (this is not the same as living with AIDS).

Since the start of the epidemic, there have been about 75 million people infected with HIV.

New HIV infections have fallen by 33 % since 2001. Worldwide, 2.3 million people became newly infected with HIV in 2012, down from 3.4 million in 2001.

New HIV infections among adults and adolescents decreased by 50 % or more in 26 countries between 2001 and 2012.

New HIV infections among children have declined by 52 % since 2001.

Worldwide, 260,000 children became newly infected with HIV in 2012, down from 550,000 in 2001 (Figs. 15.4, 15.5).

15.15 AIDS in Africa

The population of the African continent is approximately one billion. When one considers this population, one has to take into consideration the fact that the continent is made up of various ethnic groups, and various countries, each with its own

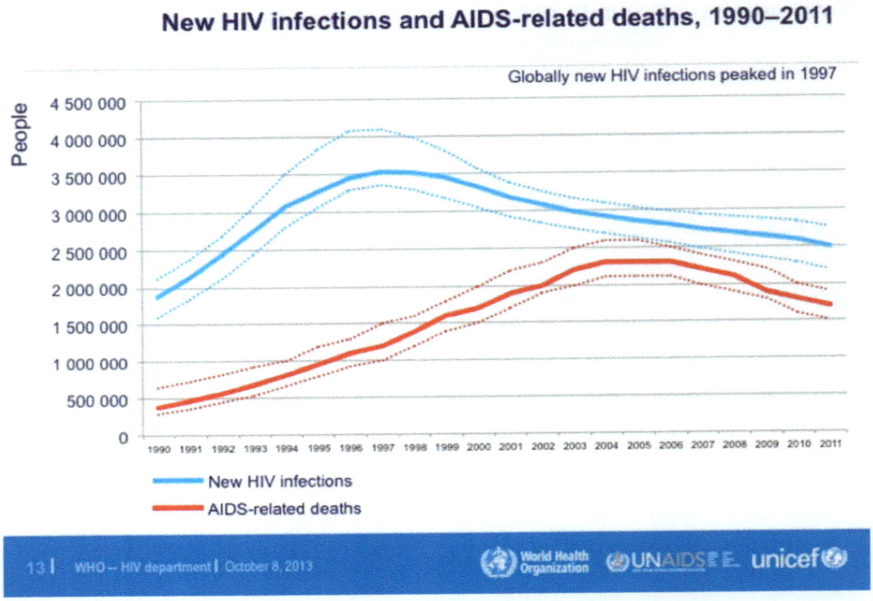

Fig. 15.5 New HIV infections and AIDS- related deaths. Note that peak of new infections was in 1997

culture. The sexual mores of many of the African countries and tribes are different from those of Western Europe and should not be judged in the same way. Polygamy is common, and in many countries of Africa, sex outside the home, or outside the family, is frequent. Rituals for boys to achieve "manhood" may involve sex or mutilation, and similar practices may occur in young girls. In some countries, male circumcision is practiced and that could lead to a lower incidence of AIDS; in other countries it is resisted. Circumcision is associated with different tribes and ethnicities. It is estimated that in 2011 there were 22.9 million adults and children living with HIV in sub-Saharan Africa. In 2010, around 1.2 million people died there from AIDS, and 1.9 million people became infected with HIV. Since the beginning of the epidemic, 14.8 million children have lost one or both parents to HIV/AIDS (see more at: http://www.avert.org/hiv-and-aids-africa.htm#sthash.g6q92aat.dpuf).

In Somalia and Senegal, the HIV prevalence is under 1 % of the adult population, whereas in Namibia, Zambia and Zimbabwe, around 10–15 % of adults are infected with HIV. Southern Africa is the worst impacted by AIDS; in South Africa, the HIV prevalence is 17.8 % and in three other Southern African countries, the national adult HIV incidence rate now exceeds 20 %. These countries are Botswana (24.8 %), Lesotho (23.6 %) and Swaziland (25.9 %).

Western Africa has been less affected by HIV and AIDS than Southern Africa, but some countries are experiencing rising HIV prevalence rates. In Cameroon, HIV occurrence is now estimated at 5.3 % and in Gabon it stands at 5.2 %. In Nigeria, HIV incidence is low (3.6 %), compared to the rest of Africa. However, because of its large population (it is the most populous country in sub-Saharan Africa), this means that around 3.3 million people are living with HIV. Adult HIV prevalence in East Africa exceeds 5 % in Uganda, Kenya and Tanzania.

Overall, rates of new HIV infections in sub-Saharan Africa appear to have peaked in the late 1990s, and HIV prevalence seems to have declined slightly, although it remains at an extremely high level of 5 % among adults.

In 2012, the number of new infections with HIV was in decline due to education and the availability of condoms, and the decline from previous years is quite startling. There was also a decline in the number of AIDS cases in children due to increased availability of anti-retroviral drugs at time of delivery.

The consequences of the AIDS epidemic have been quite disastrous in some countries, such as South Africa. It has left vast numbers of orphans that have to be cared for by the State. Education, agriculture and industrial development have all been disrupted. Each community faces the challenge of treating HIV, using scarce resources for health facilities, and dependency on foreign aid. Large strides have been made in the distribution of anti-retrovirus drugs and in testing for HIV. The large number of cases of HIV in South Africa is due to either ignoring the early stages of the epidemic (a disease affecting only a marginalized segment of society—homosexuals and prostitutes) or due to bad government policy, which relied on AIDS deniers as advisors, and pseudo-science [41]. In 1995 the decision was made by the government (Ministry of Health) to launch a musical based on a popular film, "Sarafina," with an anti-AIDS message. This was to be called "Sarafina II," at a cost of over $1.5 million. The message of the play was not clear, and AIDS advocacy groups were not consulted. The money had come from the European Union and led to a scandal investigated by the public prosecutor. This led to a cancellation of the program and worsening relations between AIDS groups, NGOs and the government.

The second major scandal that involved politics and corruption was the attempt to develop an "African" medication for AIDS. This was the development of Virodene in 1997, which turned out to be an industrial solvent (similar to anti-freeze) that had been tested for the treatment of cancer and had been shown to be useless and toxic. A group at the University of Pretoria administered it to a group of AIDS patients without permission of the equivalent of the local human subject committee. The owners of the company developing Virodene complained to the government that various AIDS action organizations were blocking the development of this drug and harming their patent. The Medicine Controls Council (equivalent to the U.S. FDA) rejected the drug, but pressure was put on various authorities to accept it, and it raised hopes in the AIDS population. It was also rumored that Thabo Mbeke, head of the African National Council, was financially

Table 15.2 Number of orphans due to AIDS in countries of Africa

Nigeria	2,500,000
South Africa	1,900,000
Tanzania	1,300,000
Uganda	1,200,000
Kenya	1,200,000
Zimbabwe	1,000,000
Zambia	690,000
Mozambique	670,000
Malawi	650,000

Number of orphans due to AIDS, alive in 2009 (see more at http://www.avert.org/children-orphaned-hiv-and-aids.htm#sthash.Tu6wNa6b.dpuf)

involved in the development of this drug and held up the use of other anti-retrovirus treatments. In 1999, Mbeke became president and the policy towards AIDS and treatment worsened. Members of the Medicine Control Council who opposed him were replaced.

In 1998, then-Health Minister Dlamini Zuma refused to allow the use of AZT for the treatment of HIV, in particular in pregnant women, on the excuse that it was too costly to test for HIV and too costly to subsidize formula for lactating mothers. since carriers of HIV should not breastfeed their babies. Despite the appointment of a new minister of health, the policy of denial continued. Peter Duesberg, a professor at the University of California Berkeley and the best-known denialist in the U.S., was appointed an advisor. His appointment is blamed for the refusal of the South African government to treat patients with anti-retroviral drugs with the resulting millions of deaths. The conflict between the president and the scientific community continued with a number of public letters and statements. Mbeki inferred that scientists who stuck to the established view that HIV causes AIDS were suppressing freedom of speech or were servants of the pharmaceutical industry. Three South African scientists offered a rebuke in *Nature*, writing that "As long as Mr. Mbeki is being advised by people with no credibility, we as South African scientists feel dangerously marginalized in the search for solutions to HIV/AIDS."

The influence of Mbeki's skepticism about the causes of AIDS continued until his party asked to step down as president in 2008. Things changed for the better in 2009 with the appointment of Jacob Zuma as president. He has encouraged the use of accepted medications and the rate of new HIV incidence has fallen.

The percentage of people with HIV/AIDS in South Africa is about 17.3 %. In Kwa-Zulu Natal province, the percentage is as high as 40 % of 15–49 year olds, an astounding number. The number of orphans in Africa, not only in South Africa, is also astounding (see Table 15.2); the age of orphans, however, is fairly consistent across the countries. Surveys suggest that overall, about 15 % of orphans are 0–4 years old, 35 % are 5–9 years old, and 50 % are 10–14 years old.

15.16 Male Circumcision and AIDS

There have been numerous reports, both scientific studies and observations, that male circumcision offers a significant protective effect against the spread of AIDS and other sexually transmitted diseases [42]. Studies done in sub-Saharan Africa indicated that a majority of men were willing to be circumcised. A randomized, controlled trial of male circumcision in 2005 to examine the effect on HIV incidence in Orange Farm, South Africa, was stopped prematurely due to an observed high protective effect of circumcision. There was a 63 % decrease in heterosexual transmission of HIV compared to that of uncircumcised men. This was a large study, with 3,273 men participating. Similar studies in Kenya and Uganda have given similar results. Uncircumcised men were twice as likely to contract AIDS as a result of heterosexual transfer as circumcised men. Studies of rates of HIV have shown a direct correlation between male circumcision and HIV prevalence in African countries. In general, about 70 % of African men are circumcised, and countries or tribes in which circumcision was not practiced show a significantly higher rate of HIV. The foreskin, which is removed from the penis during circumcision, is very rich in Langerhans cells, a dendritic cell that easily takes up HIV and can thus transport the virus to other sites in the body. The foreskin is also easily damaged (cuts, abrasions) during intercourse. Thus, removing the foreskin removes cells of the immune system that are targets of the virus.

15.17 AIDS in the USA in 2013

By 2012, about 650,000 people had died of AIDS in the U.S., about 68 % of them under the age of 45. The peak of deaths from AIDS in the U.S. was reached in 1994, 15 years after the disease was first identified. The numbers of deaths have decreased since then and have leveled out to around 20,000 per year. AIDS was the sixth leading cause of death among 22–45-year-olds, down from number 1 in 1994 and 1995, due to the introduction of a regimen of anti-HIV drugs known as HAART in 1995. This development was a major innovation in the treatment of HIV/AIDS. Many physicians talk about "the days before HAART and the days after HAART." Immediately after putting a patient on the HAART regimen, the CD4 count increases and within a short time reaches normal levels.

It is estimated that as of 2013 there are about 1.1 million individuals with HIV/AIDS in North America. Although there has been a decrease or leveling off of the incidence of the disease, there has been an increase in recent years among people in the 16–25 age group because of the increase in the use of recreational drugs and an increase in sexual activity without precautions. This is a result of the acceptance of AIDs as just another chronic disease, like diabetes or asthma. The number of new infections has been about 50,000 per year, and has remained constant for the last few years. New cases have decreased among women, but increased among gay and bisexual men. Although the number of people being tested has increased, a sizeable proportion of the population has not been tested, including people who are newly diagnosed with AIDS (about 32 % diagnosed late had not been previously tested). Most cases of AIDS occur in the major cities, although there are pockets in the southern states; the hardest hit are California, Florida, and Texas, with African Americans and Latinos having the highest rates of new infections. In 2010 the rates of infection in African Americans was about eight times higher than in whites, and in Latinos about three times higher. Teens and young adults continue to be at risk, with those under 35 accounting for 56 % of new HIV infections in 2010 (those aged 13–24 accounted for 26 % and those aged 25–34 accounted for 31 %). Most young people are infected sexually. While estimates show that gay men comprise only about 3–4 % of the U.S. population, this group accounts for most new HIV infections (63 % in 2010, with an additional 3 % occurring in gay people with a history of intravenous drug use). Between 2008 and 2010, annual new HIV infections increased 12 % among homosexual men (Fig. 15.6).

The cost of the epidemic is enormous. In 2012 the federal funding to combat AIDS was $27.7 billion—53 % of it was for care, 10 % for research, 3 % for prevention, and 21 % for international assistance. (More information on all of these topics can be found at http://kff.org/hivaids/fact-sheet/the-hivaids-epidemic-in-the-united-states/.)

15.18 AIDS in Asia

From 2005 to 2009 there have been striking increases in the number of people with HIV in East Asia, Eastern Europe, and Central Asia. In Asia, there are now about 5 million people with HIV/AIDS. As a percentage of the population, this is much lower than in sub-Saharan Africa or the U.S. In many Asian countries, the sex trade is flourishing; men pay for sex and then transmit the disease back to their wives. However, in Eastern Europe, most of the infections are related to intravenous drug use. In India we are just beginning to see the epidemic appear, with about 2 million people infected out of a total population of 1 billion. The virus is found predominantly in rural areas, where information on treatment and prevention is scarce, and no medication is available. There are currently (2011) an estimated 780,000 people living with HIV in China, including about 154,000 AIDS patients, and in 2011 around 28,000 people died from AIDS. This may be an underestimate, since for many years the Chinese authorities denied the existence of AIDS in the country. These numbers must be considered in the context of China's extremely large population, which is estimated at around 1.3 billion. Although China's HIV epidemic remains one of low prevalence overall (0.058 %), there are pockets of high

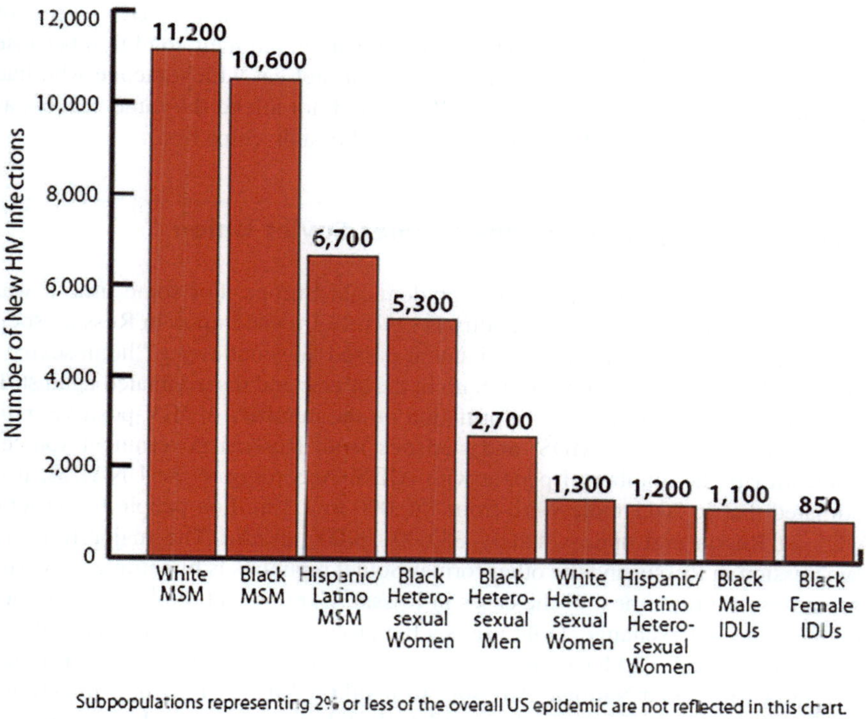

Fig. 15.6 Estimated new HIV infections in the United States, 2010, for the most affected subpopulations. *MSM* men having sex with men (CDC)

infection among specific sub-populations, such as farmers in outlying provinces, and the danger of the epidemic spreading further into the general population persists. This became particularly evident in 2009 when China reported that AIDS had become the country's leading cause of death from infectious diseases for the first time ever, surpassing both tuberculosis and rabies. Since the mid-1990s, the Chinese government has begun to clean up the blood supply; it has succeeded to some extent, although there are still pockets of blood-collecting stations where conditions are still unhygienic. The Chinese government has initiated programs of treatment and education of the public, although most Chinese are unaware of what causes the disease, and do not know how to prevent it. In a recent poll, reported in a Web site run by the journal *Science*, 1 in 6 had never heard of HIV or AIDS. Of those who did, 48 % thought it could be transmitted by mosquitoes and were unaware of sexual transmission [43, 44] (numbers from WHO).

Nevertheless, a study of over 6,000 residents in 2008 found that lack of knowledge and stigma remain a problem. It found that 48 % of respondents thought they could become infected from mosquito bites; 18 % believed they could become infected from a sneezing or coughing person with HIV; 32 % of interviewees thought people with HIV and AIDS deserved their condition because of their drug use or sexual behavior; 48 % would not eat with someone who had HIV; and 30 % thought children with HIV should not attend the same schools as uninfected children (from http://www.avert.org/hiv-aids-china.htm).

15.19 Aids in Russia and the Former Soviet Union

Russia may now has the largest HIV epidemic in Europe. For some time it was ignored or denied, since homosexuality is severely frowned upon in Russian society. In fact, in 2013 the Russian parliament passed laws outlawing "homosexual" propaganda. "Gays" are often beaten up in the streets and discriminated against. It is very difficult to get accurate information on the numbers of HIV-positive individuals or the extent of AIDS, and statistics from Russian government sources underestimate the number of people with AIDS. According to the UNAIDS, it is estimated that in 2011 there were from 780,000 to 1.3 million people living with AIDS in Russia and approximately 38,000–78,000 deaths. The major route of transmission in Russia and the other former Soviet republics is by intravenous drug use and prostitution; most of the cases in Russia, Ukraine and the Baltic countries are due to needle exchange. Drug use is a big problem in Russia, and an estimated 5–8 % of men under 30 have injected drugs. It was not until 2006 that the Russian government formed a National Advisory Council on AIDS. Of the new infections, about 44 % are in women. In total, about 1.6 % of the population of the Ukraine and 1.1 % of the population of Russia are infected, which is the highest percentage outside of Africa.

New infections in Eastern Europe and Central Asia have risen 22 % since 2005 and show no sign of slowing, according to the World Health Organization's European Action Plan for HIV/AIDS 2012–2015. As in "AIDs in Africa," the control of this disease rests to large extent on the attitudes of the people and the government toward sexual behavior, and changes instituted in this behavior. As was initially true in the U.S., it was assumed that HIV infection was not a general phenomenon, but occurred in a small segment of the population (in the U.S. in "gays"), and in Russia of today, it occurs mostly among drug addicts and people at the fringes of society. Thus the Russian government has not seen the necessity of investing large sums of money on treatment or education. In fact, Russia and many of the other former Soviet Republics, incarcerate drug addicts for possession of small amounts of drugs, and such people are considered the "dregs" of society who should be ignored rather than treated. The disease also spreads among the sexual partners of drug addicts, and thus there is a high percentage of women with AIDS. So far, AIDS has not spread into the "general" population (as it did in Africa), and is thus not identified as a major problem. Most treatment centers and funding are derived from non-government medical organizations (NGOs), many of which are frowned upon by the Russian and other governments as interfering in internal affairs.

15.20 The Public Relations Campaign Against AIDS

After the recognition of AIDS as an infectious disease, a vigorous campaign was launched to prevent the spread of the disease. This has included exhortations for complete abstinence of sexual activity to safe sex using condoms. It has taken the form of advertisements on public billboards, reports on radio and television, and even in parts of Africa, of acting troupes simulating the putting on of condoms and singing, "Just say no". In large part, this campaign has been successful. In some countries where AIDS had reached epidemic proportions, such as Uganda, the campaign has had an effect; part of its success arises from the open discussion of this previously "taboo" subject by leaders, and by testing for the presence of HIV in the population. In many parts of Africa, the use of condoms has not been accepted, and therefore testing and education have become the prime focus. One also has to target these messages to the most sexually active sector of the population—the young (teens and young adults), sex workers, and also truck drivers and workers who move from one location to another. In parts of Africa the slogan has been "ABCD": Abstain, Be faithful, Condomize, Disclose. This and other slogans are printed on T-shirts for distribution so that the population is aware of AIDS at all times. A very good presentation of methods for educating a population is discussed in http://www.etu.org.za/toolbox/docs/aids/prevention.html, produced by an organization in South Africa, where AIDS was ignored for too long and not accepted as an infectious disease.

Although AIDS can be considered a "plague"—just like influenza or even the black death—because of the long latent period may take 10 years or more, it has not had the same impact as those other infections in which the onset of disease and death occurred within a few days of each other. One did not see thousands of corpses on the streets of New York, Philadelphia, or London as in other plagues; even in sub-Saharan Africa, one does not see masses of bodies, since death is a slow, painful process and provisions can be made for the impending demise of the infected individual. After the initial shock and panic from the initial outbreak of AIDS in the U.S., there appears to be a readjustment to the disease, and an acceptance that many people will die at a young age, or that in some societies there will be large numbers of orphans. A vigorous anti-HIV campaign, through alterations in sexual behavior, has had limited success in many countries, but the extent of the AIDS epidemic has not hit home in many other places, where it is denied or ignored (e.g., Russia and China). At least in the U.S. by 2012 it appears that the disease is accepted as a chronic condition that can be kept at bay with large numbers of drugs, so AIDS is no longer considered to be a life-threatening disease. This has resulted in the lowering of the guard against the disease and an increase in the number of infected individuals in the last few years.

15.21 Conspiracy Theories and AIDS Denial

Although the evidence that HIV is the causative agent of AIDS is generally accepted, particularly in light of sequencing data and discovery of HIV sequences in samples from pre-epidemic and the tracking of the virus from its beginning, there are still those who are skeptical of these reports and have come up with other theories. There are also many "conspiracy theories" about how HIV evolved, as a result of human activities to deliberately cause an infection or by accident. These are presented below.

15.21.1 Oral Polio Vaccine Theory

According to this theory, HIV was introduced as a contaminant in the oral polio vaccine developed by Hilary Koprowski and administered in the Congo. Koprowski was in competition with Sabin and Salk for development of the first polio vaccine in the 1950s; in fact, his vaccine was the first to be given to children in the U.S., and then administered to populations in Africa. The virus used in these vaccines was produced in chimpanzee cells. This theory is presented in a well-documented book, *The River*, by Hooper [45] a BBC correspondent. However, the examination of cell lines dating back to this time period, and also to the vaccine itself, could not confirm any of these accusations. Moreover, studies of the chimpanzees of this area of the Congo show that the chimpanzees are endemically infected with SIVcpz, which is vastly different from HIV, ruling them out as a source of the virus. Recent calculations using DNA sequences indicate that HIV was first transmitted to man about 30 years before the oral polio vaccine trials and it went unrecognized as a unique disease.

15.21.2 Virus Developed by Humans for Political Purposes

In an interview with Time Magazine (2004), the Nobel Peace Prize laureate Wangari Maathai alleged that AIDS was a biological weapon manufactured by white people to wipe out the black race. This statement in itself is absurd, since the first victims of the disease in the U.S. were predominantly Caucasian males. Maathai, who is Kenyan, received the Nobel Prize for her contribution to sustainable development, democracy and peace. In 2004 she made her accusation against Caucasian scientists or claimed that the virus resulted from a botched experiment, and she did not believe that it originated from primates. However, she did retract her statement in 2006, issuing the following statement: "I have warned people against false beliefs and misinformation such as attributing this disease to a curse from God or believing that sleeping with a virgin cures the infection. These prevalent beliefs in my region have led to an upsurge in rape and violence against children. It is within this context, also complicated by the cultural and religious perspective, that I often speak. I have therefore been shocked by the ongoing debate generated by what I am purported to have said. It is therefore critical for me to state that I neither say nor believe that the virus was developed by white people or white powers in order to destroy the African people. Such views are wicked and destructive."

A number of similar theories involving the U.S. government have been proposed at various times. Jakob Segal, a former biology professor at Humboldt University in East Germany, proposed that HIV was manufactured by Robert Gallo at Fort Detrick, by splicing together two viruses—Visna (a retrovirus—lentivirus), and HTLV-1 (a human leukemia virus). According to this theory, the virus was tested on prisoners who volunteered for these experiments in exchange for early release. He claimed that HIV was predominantly a Visna virus. However, RNA analysis of HIV has shown little or no relationship to the above two viruses. It is thought that Segal might have been a Soviet agent distributing false information, but there is quite a bit of correspondence on the Internet on this theory. (For anyone interested, see http://www.africa.upenn.edu/Urgent_Action/AIDS_Contract.html)

Others have come up with similar claims. Dr. Alan Cantwell says that HIV is a genetically modified organism developed by U.S. scientists, and it was introduced into the human population through hepatitis B experiments performed on gay and bisexual men between 1978 and 1981. He presented his theory in a book titled *Queer Blood*, in which he claims that this was a plot to eliminate gay people from the U.S. Many other conspiracy theories have been proposed, in particular one by Dr. Leonard Horowitz that is based on flimsy evidence. Many of these conspiracy theories can be found on the Internet, although none is supported by scientific evidence.

15.21.3 AIDS Denial

Perhaps the most controversial opinion about the relationship between HIV and AIDs has come from an eminent virologist, Peter Duesberg, who refuses to accept the linkage between the virus and the condition we call AIDS. In his opinion, HIV is a harmless "passenger" virus, and the number of cases of AIDS has been grossly exaggerated, particularly in Africa. To summarize: Duesberg claims that AIDS is the result of indulgence in drugs, both as stimulants and for sexual arousal. He states that there is no linkage between the presence of the virus and AIDS. Millions in Africa will test positive for HIV but not have symptoms of AIDS. Most of the symptoms we do see are part of the normal disease state of the African population. As far as Europe and the Americas are concerned, most AIDS cases are due to recreational drugs. The epidemic does not follow the normal paths of an epidemic, i.e., a bell-shaped curve as the virus spreads, and then a petering out as the population becomes resistant. In fact, the rate of cases of AIDS over the last few years has been constant. The HIV-AIDS hypothesis offers no explanation for why less than 1 % of HIV antibody-positive people develop AIDS and why the mean latency between infection and disease is five years, whereas antiviral immunity is established in a few weeks. One might expect the destruction of T-cells immediately if the virus is going to replicate in such cells. A latent period for pathogenicity that exceeds the latent period for immunity is unambiguous evidence of a co-factor or another causative agent altogether. Finally, the hypothesis cannot support a claim that the virus is even necessary for AIDS, in view of the fact that it is barely present and consistently latent, even in persons with the disease. Since the transmission of AIDS depends on frequent contacts involving the exchange of cells, the case for a viral cause remains open. Moreover, Duesberg claims that HIV does not meet Koch's first postulate, that it has not been possible to identify free virus in all cases of AIDS [46, 47]. Almost all of these statements have now been contradicted as methodologies have improved. In fact, there is an answer to most of Duesberg's arguments; the strongest is that individuals contaminated accidentally in handling AIDs patients all test positive for HIV and most have succumbed to the illness. Likewise, there was a direct correlation between AIDS and HIV in blood products received by hemophiliacs. Moreover, the virus can be detected in most or all AIDS patients and replicates in lymph nodes even if not found in the blood stream. In support of Duesberg's theories, Lawrence Broxmeyer and Alan Cantwell claim that there is an association between acid-fast tuberculosis mycobacterium infection and AIDS. There is a high correlation between the incidence of AIDS and tuberculosis in Third World countries, which might be the missing co-factor preventing the curing of AIDS. Most of these ideas have been discredited, but studies performed on hemophiliacs who are infected with HIV and those who are not, show a high correlation between AIDS and AIDS-related opportunistic infections. The belief in Duesberg's theory has led to the deaths of HIV-positive individuals who refused retrovirus therapy and to high rates of AIDS deaths in South Africa, whose president, Thabo Mbeki, accepted Duesberg's ideas. President Mbeki and his health minister believed that HIV does not cause AIDS, and that there is no such disease. Thus there was a belief that nothing needs to be done. In 2003 the South African health minister advocated a diet of beetroot, garlic, lemon juice and olive oil as protection against AIDS. In 2008 both Mbeki and his health minister were forced to resign, and replaced by people who did believe that HIV and AIDS were connected and should be treated with anti-retroviral drugs. A group from Harvard has estimated that 365,000 people might have been saved from premature death if they had been provided with medication, particularly pregnant women. As a result of this denial, South Africa has one of the highest incidences of HIV.

In a recent paper, Duesberg and his colleagues have challenged the statistics used in the above-mentioned Harvard study, and claim that there has not been any drop in the population growth of South Africa or other African countries, which would have been predicted if the disease had behaved as proposed by epidemiologists [48]. However, Duesberg's arguments do not take into account the unique nature of HIV-1, the lengthy latent period before symptoms occur, and do not discuss the case of hemophiliacs and others who advance to AIDS as a result of accidental infection with HIV. In all fairness, I copy the abstract of Duesberg's last paper on the subject:

"Since the discoveries of a putative AIDS virus in 1984 and of millions of asymptomatic carriers in subsequent years, no general AIDS epidemic has occurred by 2011. In 2008, however, it has been proposed that between 2000 and 2005 the new AIDS virus, now called HIV, had killed 1.8 million South Africans at a steady rate of 300,000 per year and that anti-HIV drugs could have saved 330,000 of those. Here we investigate these claims in view of the paradoxes that HIV would cause a general epidemic in Africa but not in other continents, and a steady rather than a classical bell-shaped epidemic like all other new pathogenic viruses. Surprisingly, we found that South Africa attributed only about 10,000 deaths per year to HIV between 2000 and 2005 and that the South African population had increased by 3 million between 2000 and 2005 at a steady rate of 500,000 per year. This gain was part of a monotonic growth trajectory spanning from 29 million in 1980 to 49 million in 2008. During the same time Uganda increased from 12 to 31 million, and Sub-Saharan Africa as a whole doubled from 400 to 800 million, despite high prevalence of HIV. We deduce from this demographic evidence that HIV is not a new killer virus. Based on a review of the known toxicities of antiretroviral drugs, we like to draw the attention of scientists, who work in basic and clinical medical fields, including embryologists, to the need of rethinking the risk-and-benefit balance of antiretroviral drugs for pregnant women, newborn babies and all others who carry antibodies against HIV".

It is true that the epidemic did not reach the number predicted by epidemiologists in the 1980s, at the outset of the disease. But this could be down to change in sexual behavior, and the introduction of new medications that have allowed HIV survivors to live longer, and testing which has possibly prevented the spread. Admittedly there are unanswered questions, such as is our evidence of the origins of the virus correct, and how much does certain life styles decrease life expectancy and lead to other diseases.

15.22 The Virus

HIV belongs to the genus of viruses known as *lentiviruses* (Fig. 15.7) within the genus of retroviruses. Examples of lentiviruses are:
Simian immunodeficiency virus—SIV
Feline immunodeficiency virus—FIV
Equine infectious anemia virus—EIAV
Visna/Maedi virus of sheep
Caprine (goat) encephalitis leukemia virus

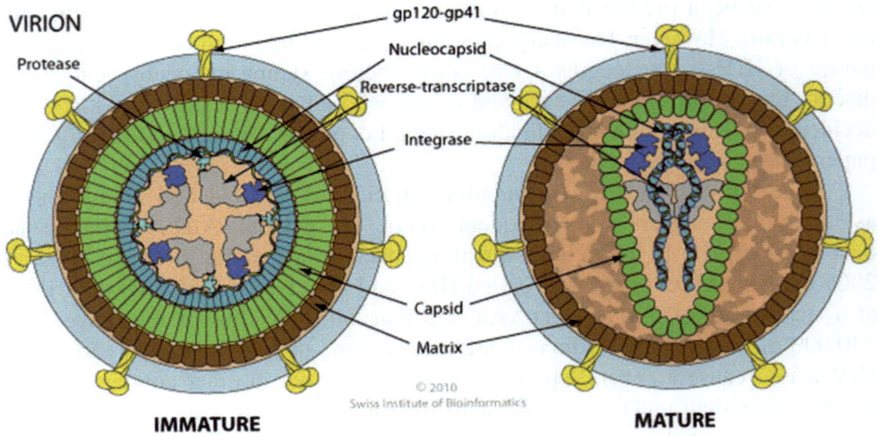

Fig. 15.7 Schematic of HIV-1. Enveloped, spherical to pleomorphic in shape, 80–100 nm in diameter. Mature capsid contains 1,572 capsid proteins

The major characteristic of these viruses is the long period of latency before infection in the animal is manifested.

HIV is taken up by the cell by the binding of the viral surface gp120/gp41 protein to the CD4 protein (receptor) on the cell surface. This causes a conformational change of the viral proteins, bringing it close to CCR5 or CXCR4 (chemokine receptors) on the CD4 cells and thus allowing for fusion with cell membrane and consequent uptake into the cell. Individuals with mutants in this chemokine receptor gene are partially or fully resistant to HIV infection. Drugs have been developed to interfere with virus uptake by binding to CCR5 or CXCR4.

15.23 Pathogenesis

After viral infection occurs, the infected individual may have flu-like symptoms within a few days for a short time. The cells infected are predominantly CD4-helper lymphocytes or other cells with the CD4 receptor. Antibodies to the virus are formed anywhere between 3 and 12 weeks after the initial infection. This is the window period in which an individual does not know whether he/she is infected and would test negative, using ELISA or Western blot. Eventually the virus will destroy the CD4 lymphocytes, leading to immunodeficiency if treatment is not

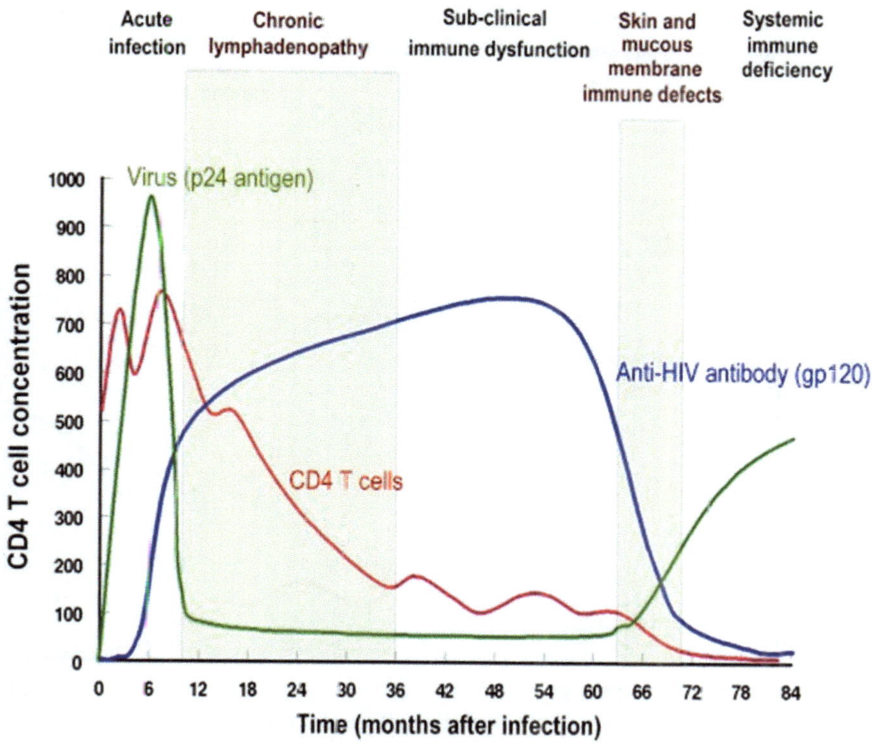

Fig. 15.8 Course of infection with time. Effect of HIV on CD4 cells (http://www.microbiologybook.org/lecture/hiv3.htm)

initiated. However, the latent period may be very long (10 years) or very short (6–8 weeks). As discussed previously testing for HIV measures the presence of antibodies in the blood, or for viral nucleic acid.

Figures 15.8 and 15.9 represent the course of the infection over time. There is a decrease in CD4 cells with the resulting onset of opportunistic infections and death of the infected individual. Later in infection, with the drastic decrease in CD4 cells, CD8 cells will also collapse. Below 200 CD4 cells per ml is considered AIDS and is usually accompanied by opportunistic infections. A large number of different opportunistic infections may occur. Initially there are no real symptoms, or minor flu-like symptoms. This is accompanied by seroconversion

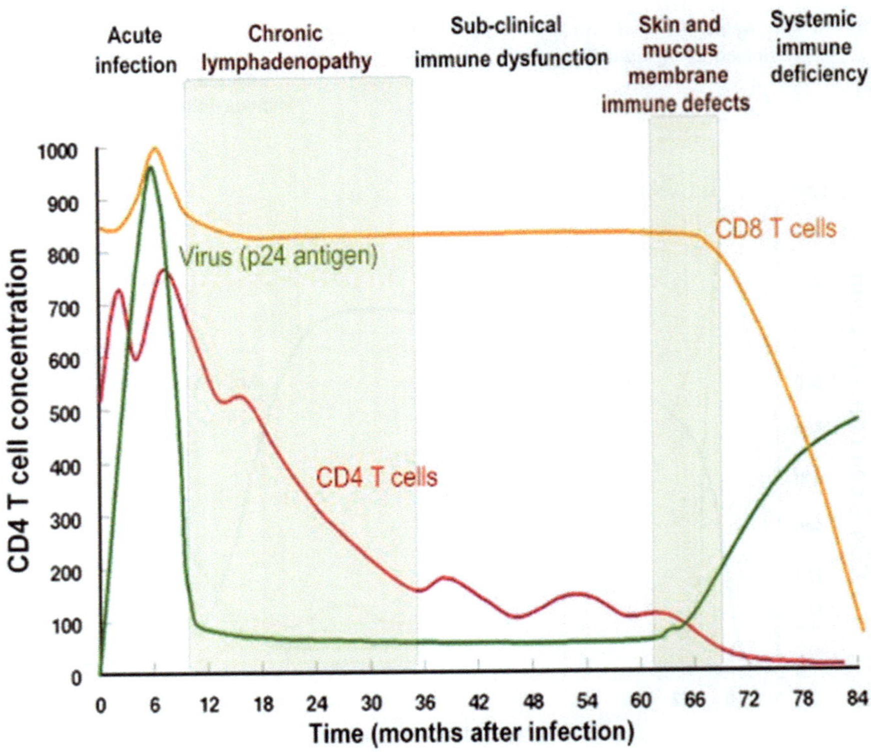

Fig. 15.9 Effect of HIV infection on CD8 cells (http://www.microbiologybook.org/lecture/hiv3.htm)

to antibody positive. Much later, there are changes in oral and skin conditions with various diseases such as thrush and KS, and a generalized infection of the lymph nodes may be obvious. The progression to AIDS can take as long as 10 years or more.

References

1. Grmek MD. (1990). *History of AIDS*. Princeton, NJ: Princeton University Press.
2. Gottlieb, M. S., Schroff, R., Schanker, H. M., Weisman, J. D., Fan, P. T., Wolf, R. A., et al. (1981). *Pneumocystis carinii* pneumonia and mucosal candidiasis in previously healthy homosexual men: evidence of a new acquired cellular immunodeficiency. *New England Journal of Medicine, 305*(24), 1425–1431.
3. Vogt, M., Luthy, R., & Siegenthaler, W. (1982). GRID syndrome. *Deutsche Medizinische Wochenschrift, 107*(41), 1539–1542.
4. Gottlieb, M. S., Groopman, J. E., Weinstein, W. M., Fahey, J. L., & Detels, R. (1983). The acquired immunodeficiency syndrome. *Annals of Internal Medicine, 99*(2), 208–220.

5. Weintrub, P. S., Koerper, M. A., Addiego, J. E, Jr, Drew, W. L., Lennette, E. T., Miner, R., et al. (1983). Immunologic abnormalities in patients with hemophilia A. *Journal of Pediatrics, 103*(5), 692–695.
6. Pitchenik, A. E., Fischl, M. A., Dickinson, G. M., Becker, D. M., Fournier, A. M., O'Connell, M. T., et al. (1983). Opportunistic infections and Kaposi's sarcoma among Haitians: Evidence of a new acquired immunodeficiency state. *Annals of Internal Medicine, 98*(3), 277–284.
7. Vieira, J., Frank, E., Spira, T. J., & Landesman, S. H. (1983). Acquired immune deficiency in Haitians: Opportunistic infections in previously healthy Haitian immigrants. *New England Journal of Medicine, 308*(3), 125–129.
8. Stine, G. J. (2003). *AIDS update 2003* (11th ed.). Englewood Cliffs, NJ: Prentice Hall.
9. Marx, J. L. (1983). Human T-cell leukemia virus linked to AIDS. *Science, 220*(4599), 806–809.
10. Karpas, A. (1983). Unusual virus produced by cultured cells from a patient with AIDS. *Molecular Biology and Medicine, 1*(4), 457–459.
11. Purtilo, D. T., Lipscomb, H., Harada, S., Tatsumi, E., Hinuma, Y., Karpas, A., et al. (1983). Human T cell leukemia virus-1 antibodies not detected in AIDS. *AIDS Research, 1*(4), 237–241.
12. Gallo, R. C., Salahuddin, S. Z., Popovic, M., Shearer, G. M., Kaplan, M., Haynes, B. F., et al. (1984). Frequent detection and isolation of cytopathic retroviruses (HTLV-III) from patients with AIDS and at risk for AIDS. *Science, 224*(4648), 500–503.
13. Sarngadharan, M. G., Popovic, M., Bruch, L., Schupbach, J., & Gallo, R. C. (1984). Antibodies reactive with human T-lymphotropic retroviruses (HTLV-III) in the serum of patients with AIDS. *Science, 224*(4648), 506–508.
14. Zagury, D., Bernard, J., Leibowitch, J., Safai, B., Groopman, J. E., Feldman, M., et al. (1984). HTLV-III in cells cultured from semen of two patients with AIDS. *Science, 226*(4673), 449–451.
15. Popovic, M., Sarngadharan, M. G., Read, E., & Gallo, R. C. (1984). Detection, isolation, and continuous production of cytopathic retroviruses (HTLV-III) from patients with AIDS and pre-AIDS. *Science, 224*(4648), 497–500.
16. Crewdson, J. (2002). *Science fictions*. Boston, New York, London: Little, Brown and Company.
17. Microbicides 2010 (International Conference on Microbicides). "Use of lubricants with anal sex could increase risk of HIV." ScienceDaily. ScienceDaily, 25 May 2010. http://www.sciencedaily.com/releases/2010/05/100525094900.htm.
18. Kingsley, L. A., Detels, R., Kaslow, R., Polk, B. F., Rinaldo, C. R, Jr, Chmiel, J., et al. (1987). Risk factors for seroconversion to human immunodeficiency virus among male homosexuals. Results from the Multicenter AIDS Cohort Study. *Lancet, 1*(8529), 345–349.
19. Winkelstein, W, Jr, Lyman, D. M., Padian, N., Grant, R., Samuel, M , Wiley, J. A., et al. (1987). Sexual practices and risk of infection by the human immunodeficiency virus. The San Francisco Men's Health Study. *JAMA, 257*(3), 321–325.
20. Winkelstein, W, Jr, Samuel, M., Padian, N. S., & Wiley, J. A. (1987). Selected sexual practices of San Francisco heterosexual men and risk of infection by the human immunodeficiency virus. *JAMA, 257*(11), 1470–1471.
21. Stickle DF, Pirruccello SJ, Swindells S, Hinrichs SH. (2002). Discrepant results of 2 screening tests for anti-HIV antibody. *Clinical Infectious Disease* (An Official Publication of the Infectious Diseases Society of America), *35*(6):773–774; author reply 774–775.
22. Krown, S. E. (1986). AIDS and Kaposi's sarcoma: interferons in pathogenesis and treatment. *Interferon, 7*, 185–211.
23. Camparo, P., Gessain, A., & Arborio, M. (1996). Kaposi's sarcoma and human herpesvirus type 8: Recent data. *Medecine Tropicale (Mars), 56*(3), 231–237.

24. Cohen, M. S., Chen, Y. Q., McCauley, M., Gamble, T., Hosseinipour, M. C., Kumarasamy, N., et al. (2011). Prevention of HIV-1 infection with early antiretroviral therapy. *New England Journal of Medicine, 365*(6), 493–505.
25. Choopanya, K., Martin, M., Suntharasamai, P., Sangkum, U., Mock, P. A., Leethochawalit, M., et al. (2013). Antiretroviral prophylaxis for HIV infection in injecting drug users in Bangkok, Thailand (the Bangkok Tenofovir Study): A randomised, double-blind, placebo-controlled phase 3 trial. *Lancet, 381*(9883), 2083–2090.
26. Klein, F., Mouquet, H., Dosenovic, P., Scheid, J. F., Scharf, L., & Nussenzweig, M. C. (2013). Antibodies in HIV-1 vaccine development and therapy. *Science, 341*(6151), 1199–1204.
27. Hansen, S. G., Piatak, M., Jr., Ventura, A. B., Hughes, C. M., Gilbride, R. M., Ford, J. C., et al. (2013). Immune clearance of highly pathogenic SIV infection. *Nature, 502,* 100–104.
28. Tenenbaum, S. A., Morris, C. A., Alexander, S. S., McFerrin, H. E., Garry, R. F., & Leissinger, C. A. (2005). Evidence of HIV exposure and transient seroreactivity in archived HIV-negative severe hemophiliac sera. *Virology Journal, 2,* 65.
29. Dunn, K. (1993). HIV and Canada's hemophiliacs: looking back at a tragedy. *CMAJ: Canadian Medical Association Journal (journal de l'Association medicale canadienne), 148*(4), 609–612.
30. Nedelcu, I. (1992). AIDS in Romania. *American Journal of the Medical Sciences, 304*(3), 188–191.
31. Hersh, B. S., Popovici, F., Apetrei, R. C., Zolotusca, L., Beldescu, N., Calomfirescu, A., et al. (1991). Acquired immunodeficiency syndrome in Romania. *Lancet, 338*(8768), 645–649.
32. Cernescu, C. (1999). Romanian HIV-AIDS epidemic after a decade of evolution. *Romanian Journal of Virology, 50*(1–4), 5–15.
33. Rosenthal, E. (2001). *Silent plague. A special report: Deadly shadows darkens remote Chinese village*. New York: New York Times.
34. Cohen, J. (2004). HIV/AIDS in China. An unsafe practice turned blood donors into victims. *Science, 304*(5676), 1438–1439.
35. Cohen, J. (2004). HIV/AIDS in China. A scoop of international proportions. *Science, 304*(5676), 1439.
36. Ahmad, K. (2007). Kazakhstan health workers stand trial for HIV outbreak. *The Lancet Infectious Diseases, 7*(5), 311.
37. Morris, K. (2006). Transfusion-related HIV outbreak in Kazakhstan children. *The Lancet Infectious Diseases, 6*(11), 689.
38. Gao, F., Bailes, E., Robertson, D. L., Chen, Y., Rodenburg, C. M., Michael, S. F., et al. (1999). Origin of HIV-1 in the chimpanzee *Pan troglodytes troglodytes*. *Nature, 397*(6718), 436–441.
39. Zhu, T., Korber, B. T., Nahmias, A. J., Hooper, E., Sharp, P. M., & Ho, D. D. (1998). An African HIV-1 sequence from 1959 and implications for the origin of the epidemic. *Nature, 391*(6667), 594–597.
40. Worobey, M., Gemmel, M., Teuwen, D. E., Haselkorn, T., Kunstman, K., Bunce, M., et al. (2008). Direct evidence of extensive diversity of HIV-1 in Kinshasa by 1960. *Nature, 455*(7213), 661–664.
41. Nattrass, N. (2004). *The moral economy of AIDS in South Africa*. Cambridge: Cambridge Press.
42. Westercamp, N., & Bailey, R. C. (2007). Acceptability of male circumcision for prevention of HIV/AIDS in sub-Saharan Africa: A review. *AIDS and Behavior, 11*(3), 341–355.
43. Tan, X., Pan, J., Zhou, D., Wang, C., & Xie, C. (2007). HIV/AIDS knowledge, attitudes and behaviors assessment of Chinese students: a questionnaire study. *International Journal of Environmental Research and Public Health, 4*(3), 248–253.
44. Zhang, J., & Jiang, N. (2012). AIDS: China must provide education on HIV. *Nature, 486*(7403), 323.

45. Hooper, E. (2000). *The River: A journey to the source of HIV and AIDS*. Boston: Back Bay Books.
46. Duesberg, P., Koehnlein, C., & Rasnick, D. (2003). The chemical bases of the various AIDS epidemics: recreational drugs, anti-viral chemotherapy and malnutrition. *Journal of Biosciences, 28*(4), 383–412.
47. Duesberg, P. H. (1989). Human immunodeficiency virus and acquired immunodeficiency syndrome: Correlation but not causation. *Proceedings of the National academy of Sciences of the United States of America, 86*(3), 755–764.
48. Duesberg, P. H., Mandrioli, D., McCormack, A., Nicholson, J. M., Rasnick, D., Fiala, C., et al. (2011). AIDS since 1984: No evidence for a new, viral epidemic–not even in Africa. *Italian Journal of Anatomy and Embryology (Archivio italiano di anatomia ed embriologia), 116*(2), 73–92.

Phage Therapy and the Future 16

Abstract

Who really discovered bacteriophage? Ernest Hankin, a British scientist working in India in the nineteenth century, claimed that the lack of bacterial contamination in the Ganges River was due to the presence of anti-bacterial substance. The idea of using phage to counteract bacterial infections was promoted by Felix d'Herelle. In the 1930s there were large, successful clinical trials in the U.S. using phage to counteract carbuncles and other infections. The pharmaceutical company Eli Lilly advertised the sale of bacteriophage to counteract various bacterial infections, mistakenly including herpes. Phage therapy was accepted in Eastern Europe and the Soviet Union from 1930 until the 1950s, with phage being used to treat dysentery among soldiers in World War II. There was a decline in its use with the collapse of the Soviet Union, but phage therapy is actively being investigated as a treatment to prevent food spoilage, and as a veterinary medicine, and the time is ripe to begin phage therapy clinical trials in the West. There is renewed interest in phage therapy because of the high risk of antibiotic-resistant strains of bacteria.

16.1 Early Days of Phage Therapy

Were Frederick Twort and Felix d'Herelle really the first to discover bacteriophage? There are reports from earlier sources indicating a similar discovery, but without the isolation of the phage themselves. Ernest Hankin, a British microbiologist, reported in 1896 that there was some anti-bacterial activity (anti-cholera) in the waters of the Jamuna and Ganges Rivers in India. This substance was filterable and heat-inactivated. His description of the conditions in the Ganges and Jamuna make for interesting reading, with his descriptions of these rivers as being

M. W. Taylor, *Viruses and Man: A History of Interactions*,
DOI: 10.1007/978-3-319-07758-1_16,

cleaner than European or British rivers, despite the fact that people bathe in them, throw corpses into them and defecate nearby [1].

"Crossing the Ganga or the Jamuna that flows through big Indian cities, when one sees thousands of inhabitants washing themselves, their cattle and their clothes in the dirty cloudy waters and one thinks of the half-burnt corpses that often have their final journey in the river, one could be excused for thinking that these waters should be quite dangerous for consumption and that the worship of these sacred rivers by the Hindus proves their ignorance of all idea of health and cleanliness. The European authorities seem to think so, and as regards the spreading of Cholera, they would like to consider that the Ganga is the main cause for transmitting the disease in its country of origin, and the foster father of the microbe".

"However, a simple microscopic examination of the waters of these two rivers reveals a noticeable difference compared to the equally cloudy waters of European rivers. The latter are found to have an abundance of plant and animal matter, a lot of germs and living animal and plant forms. Water from the Ganga and Jamuna, on the other hand, does not contain any trace of organic matter, unless it has been collected from near a bathing ghat downstream from the city. The silt brought by the river is almost exclusively formed of sand or mica. Bacteriological examination shows that microbes are much more scarce than those found in the major rivers of Europe."

Most of the water flowing into these rivers comes from the melting of snow from the Himalayas. The corpses are in many cases burnt, and large turtles or vultures eat those not burnt, very quickly. There is an increase in cholera bacteria in the vicinity of the corpses, but it appears to clear quite quickly. Hankin rigorously tested his "anti-bacterial" material, derived from the water of the river and reported in the Annals of the Pasteur Institute that it had potent anti-bacterial activity, however he could not identify it [1, 2]. We now assume that in fact he was dealing with bacteriophage. Hankin went on to also propose the use of potassium permanganate as a general antiseptic against bacteria in India, adding it to wells. He showed that it had definite anti-cholera activity. While in India a few years ago I was instructed only to eat "purple" salad. This is salad treated with permanganate, primarily served in restaurants and prevents diarrhea in the non-immune visitor. Thus although Hankin's paper was written in 1898, the practice continues until this day. The observations of Hankin are important in considering what came later with the use of phage to "cure" epidemics of cholera.

Whether it was Twort or d'Herelle who first discovered bacteriophage is unimportant in the context of phage therapy, since there is no doubt that d'Herelle championed this zealously, as discussed in Chap. 4. The first actual case of phage therapy was the treatment of a 12-year-old boy suffering from severe dysentery at the Hôpital des Enfants-Malades in Paris in 1919. D'Herelle first consumed a phage preparation that he had prepared, as did some other colleagues, as a test of toxicity (an early version of a phase 1 clinical trial!) before giving it to the patient. The boy recovered within 24 h, and with this success, three other patients were treated with a single phage preparation and recovered. The first published report in a scientific journal was one in which phage were used to treat staph infections of the skin, by injection around the open wound, and again it was reported that the infection cleared up within 24–48 h. There were many anecdotal cases, as well as many studies published in reputable medical journals of phage therapy.

In 1919 d'Herelle used phage to treat individual cases of dysentery that were brought to his attention at the Pasteur Institute, and all the cases were cured after treatment. In order to prove that there was no danger, d'Herelle himself imbibed doses of phage well above those administered in treatment. Other researchers soon were trying phage therapy. Belgian researchers [3] reported the use of staphylococcal phage to treat carbuncles with 1.5–2.0 cc of phage and reported that the boils dried up within 24–48 h. This was done with only six patients and thus could not be considered a clinical trial. In 1922, André Gratia of the Pasteur Institute in Brussels, reported similar results. However, as expected, there were also failures. The number of papers produced during this time in support of d'Herelle's ideas increased exponentially. In 1930 alone, there were 60 papers published on phage therapy (reviewed in [4]).

16.2 U.S. Clinical Trials

A large and exciting clinical trial was performed at the Indiana University School of Medicine and reported in the *American Journal of Medical Sciences* in 1930, [5]. This was a trial using "stock" bacteriophage—that is, a mixture of phage that could lyse different strains of bacillus and staphylococcus. The physicians, headed by Dr. Thurman Rice, treated approximately 300 patients who had carbuncles and open wounds with phage preparation, in most cases applying the phage as a solution on bandages or injecting or placing the phage in the vicinity of the wound. Of 66 children with severe carbuncles, 55 recovered (the lesions disappeared) within a few days. This group applied phage to various types of infections with quite a bit of success; even bedsores were treated with positive results. In untreated patients, or controls in which one side of the face was treated with phage and the other with mercurial ointment, the phage-treated side healed much faster. Ninety percent of all the patients treated responded favorably to phage therapy, even though there were expected complications in many cases from diabetes, syphilis and other conditions; given these results, it is surprising that this method of treatment was not more widespread. It is obvious from the report that the physicians were not quite certain of how the effect was working, but were willing to go ahead with the treatment and there was no report of side effects. This was in the days before drugs had to be approved by the FDA and formal clinical trials were carried out. In general, streptococcus infections were the most difficult to treat and the success rate was much lower.

There is no doubt that there was some success with phage therapy in the case of bubonic plague in India. D'Herelle was also asked to explore the use of phage to treat cholera, and this was carried out in various parts of India. Some of the trials considerably lowered the mortality from cholera, and in some cases recovery seemed to be instantaneous. He stated that the oral administration of Vibrio cholera phage could reduce the mortality rate from 62 to 8 % (quite an achievement) and

Table 10.1 A list of phage therapy products

French company	Eli Lilly
Bacte-coli-phage	Colo-lysate/colo-jel
Bacte-rhino-phage	Entolysate/ento-jel
Bacte-intesti-phage	Neiso-lysate
Bacte-pyo-phage	Stapho-lysate/staphylo-jel
Bacte-staphy-phage	

that the addition of phage to wells acted in a preventative fashion. Other trials were not so successful, and may have been due to different strains of cholera, the wrong phage, or to political upheavals occurring in India at the time. In looking at the data objectively today, as presented by d'Herelle and others, there was a positive effect. It is puzzling that this approach was not pursued further in more controlled clinical trials. One of the criticisms offered against phage therapy was the lack of a control population or placebo in the study. For moral reasons, in dealing with diseases such as bubonic plague and cholera, there was no justification for not treating a population, when there really is no therapeutic alternative, and the disease is fatal in such a short time. Placebos have not been used in many clinical trials in the U.S. in diseases such as hepatitis C and HIV.

D'Herelle, as previously mentioned, set up a laboratory in Paris to produce specific phage for specific illnesses, and a similar production line was started in the U.S. in the 1940s at the Eli Lilly Company in Indianapolis. L'Oreal, the pharmaceutical and cosmetic company, later bought out the French company. The products were either cultures of lysed bacteria or water soluble gels (Table 10.1).

In the 1940s, the Eli Lilly Company (Indianapolis) produced seven phage products for human use, including preparations targeting staphylococci, streptococci, *Escherichia coli*, and other bacterial pathogens. However, with the discovery of antibiotics, these products were phased out in the U.S. In fact, many commercial preparations were quite inactive, and in some cases thought to be lacking quality control, which may be the reason that many trials were unsuccessful. There was a great deal of variations within the commercial products, and the most successful were those prepared "in the laboratory," where the action of the phage could be tested on specific strains of bacteria. In fact, these commercial products were hyped and advertised against herpes and other viral diseases—on which, of course, there would be no effect.

16.3 Phage Therapy Between 1920 and 1940

Between 1920 and 1940 there were hundreds of publications on the treatment of specific diseases with phage therapy. Most of them were reported in non-English publications, and thus not available to American and British physicians or scientists. Many cases were successful, but many were equally unsuccessful, or doubtful.

In fact, some researchers found that they could not repeat their own work. In the case of cholera, there seemed to be more success than failure, since some studies were performed in a large area and in large populations of India the results and methods used have been questioned. In the case of both cholera and dysentery, the infecting organism is confined to the lumen of the intestine, and phage can be taken orally, although it may be necessary to take it with a basic substance to counteract the acidity of the stomach and digestive tract. Many of the papers reported mixed results with dysentery and with typhoid fever; a vaccine for the latter is probably more effective than phage therapy.

A more positive note can be detected in the results of the treatment of boils and carbuncles. Here the success rate is as high as 80 % in many cases, both in the U.S. and in France; even eye sties have been successfully treated. However, as in other studies, other physicians have reported less of an effect, and one is left wondering whether this is due to variations in the phage preparations or to the lack of pre-testing to determine whether the phage were really active and specific for that particular bacterium. One physician had treated 24 cases of neonatal impetigo (a staph infection), and treated the disease with subcutaneous injections, resulting in a cure in 24–48 h in every case, depending on the stage of the sores. Phage were also used in the treatment of peritonitis, and in surgery for ruptured appendicitis. The mortality rate after this type of surgery was reduced by 50 %, and in fact, one of the surgeons of this period (Tsouloukudzi in France) described phage therapy as the greatest discovery in medicine. However, despite the hundreds of papers, one is left with the feeling that there was doubt as to whether this was a practice with potential (these studies have recently been reviewed in [4]). All reviews of the period insist that phage therapy needs more study, and it is puzzling why this attitude persisted in the West (in the U.S., and in France, to a lesser extent), and not in eastern Europe.

To quote a concluding paragraph from a 1933 paper by Janet Stout [6]: "This mysterious agent has very reluctantly yielded up its secrets and no man can say what the future will reveal, but the possibilities are infinite. Even though bacteriophage therapy is still in its infancy, its value even now is firmly established in the treatment of diseases with which we are confronted daily".

There does not appear to have been a comprehensive plan to test this new therapy with a control population in the way that antibiotics were tested a few years later. Could this be because the commercial importance of phage therapy was not realized or that there was no commercial value in phage? Could the individual strains be patented? Most such strains of phage would be in the public domain unless engineered by the company, something that would not happen until the twenty-first century.

16.4 Phage Therapy in the 1960s

Nevertheless, phages continued to be used therapeutically, together with, or instead of antibiotics in eastern Europe and in the former Soviet Union. Phage therapy was part of the standard care in the Soviet Union in the 1960s and 1970s, even when antibiotics were available. Several institutions were actively involved in therapeutic phage research and production, with activities centered at the Eliava Institute of Bacteriophage in Tbilisi, Georgia, and the Hirszfeld Institute of Immunology and Experimental Therapy (HIIET) of the Polish Academy of Sciences in Wroclaw (Breslau), Poland. Hundreds of papers have been published, all extolling the success of phage therapy in the Soviet Union and in Poland and other countries of eastern Europe, and it would take a complete book to describe all these trials. Much of this has been reviewed in Thomas Hausler's *Viruses vs Superbugs* [7] (and in the mini review of Sulakvelidze [2001]) [8]. This paper lists the major human phage therapy studies performed within recent years in Poland and the Former Soviet Union.

The Eliava Institute at one time employed 1,200 people and supplied phage preparations (cocktails of mixed phage) to the Soviet Army and to hospitals all over the Soviet Union. Work at the institute reached its peak in the 1980s, when it produced two tons of phage products twice a week, 80 % of it for the Soviet military—a market it lost, along with its high status under the Soviet Ministry of Health, with the 1991 dissolution of the Soviet Union.

Following the collapse of the Soviet Union, the institute went through hard times and was basically disbanded and abandoned. It has only very recently been revitalized with new trials of phage therapy, both in the West as well as in the East. The Georgian Academy of Sciences, and later the Georgian Ministry of Education and Science, manage the institute. It once more does research, some of it supported by the U.S.

The Ludwig Hirszfeld Institute in Wroclaw, Poland, was founded in 1945, and named for the microbiologist Ludwig Hirszfeld, a physician who had worked on blood groups and anaphylaxis, and was one of the discoverers of the ABO blood groups. He was also involved in testing pregnancy for incompatibility through differences in blood groups (Rh factor). In 1945 he became director of the Institute for Medical Microbiology in Wrocław and dean of the medical faculty. He taught at the institute, now affiliated with the Polish Academy of Sciences and named after him, until his death in 1954 at age 69.

A major study of the effect of phage therapy in Poland was published in 1987 [9]. This was a report on the treatment of 550 cases of septicemia (what used to be called "blood poisoning") patients, the majority of whom were antibiotic-resistant. These patients ranged in age from 1 week to 86 years, and the treatment was performed at various clinics. The bacterial infections included *E. coli*, *Staphylococcus*, *Pseudomonas*, *Klebsiella* and *Salmonella*, and lytic phages were first selected for the appropriate organism from a collection of 250 lytic phages. The patients were either given phage orally (after neutralizing stomach acid), by applying phage-containing dressing on the wounds, or by applying it to the eye, ear or nasal mucosa as needed. If phage resistance developed, other phage were used. The duration of treatment was 1–16 weeks, depending on the response. The success rates were an astounding 92 % overall, and 94 % in the antibiotic-resistant patients. A large number of similar studies were carried out in the Soviet Union.

Most of these did not have placebo controls, since the methodology was accepted as being well tried and successful. An excellent review of all the trials, mostly in the Soviet Union, has recently been published [4].

16.5 Phage Therapy Today

In order to be accepted into medical practice in the U.S., it is essential to carry out pre-clinical trials in animals. For most anti-cancer drugs, preliminary experiments are done on experimental tumors in mice and in rats, although occasionally other animal models are used. One advantage of phage therapy is that it may also have veterinary use, and this may be of commercial importance and allow for start-up companies to test its efficacy before being tried on human subjects. Successful experiments have been performed on mice, calves, piglets and sheep, using a mixture of *coli* phage against *E. coli* pathogenic strains in these animals in the U.K. This particular strain of *E. coli* causes severe diarrhea in calves and other farm animals. In fact, the authors [10] state that the phage treatment was much more effective than treatment with a series of antibiotics used as controls. Only streptomycin gave similar results. In some cases, resistant strains of *E. coli* developed but did not interfere with recovery. Phage were maintained in the environment for some time, in the stalls or bedding, and infected other calves and cured them from a similar infection. Moreover, Smith observed [10], as did d'Herelle 60 years earlier, that some calves appeared to be protected from this strain of *E.coli*. On examination, these calves contained a bacteriophage different from the one used in treatment and kept the bacterial infection under control. These experiments [10] were done with controls, and have been reviewed by others and stimulated a new interest in phage therapy in the West.

It is interesting that the toxin produced by the pathogenic strain of *E. coli* O157:H15 itself appears to be the result of the presence of two lysogenic phage in *E. coli*, yet despite this, the strain could be lysed by other phage isolated from cultures.

16.6 Phage as an Alternative to Antibiotics

One of the major concerns of environmentalists in the U.S. (as well as in the U.K.) is the use of large amounts of antibiotics in animal food that is fed to cattle and chickens. The use of such antibiotics has also led to the development of antibiotic-resistant strains of bacteria. At the same time, there have been outbreaks of salmonella poisoning (sometimes lethal) from the digestion of meat, particularly "rare hamburgers," and even from vegetables that have become contaminated with pathogenic *E. coli* and *salmonella*. The *E. coli* variant is known as 0157:H7 and causes what has been termed "hamburger disease." In some people, especially children between the ages of 1 and 10, hemolytic uremic syndrome (HUS) can result from infection by *E. coli* O157:H7 and can lead to kidney failure and death. A child may become infected after consuming contaminated food or beverages, such as meat, especially undercooked ground beef; unpasteurized juices; contaminated water; or through contact with an infected person. In 2010 there were outbreaks of *E. coli* O157:H7 due to contaminated cheese in some western states of the U.S., shredded romaine lettuce in some central and eastern states, and beef nationally from National Steak and Poultry. The CDC estimates that roughly one in six Americans becomes sick from food-borne diseases each year. Among bacterial pathogens, *salmonella* serotype enteritis accounted for the largest number of outbreaks and outbreak-related cases, and the largest number of deaths, followed by *Listeria monocytogenes*. The availability of a cocktail of phage would be very useful not only in treating these cases, but also in applying it to the possibly contaminated foodstuffs itself [11, 12].

Klebsiella pneumonia is another bacteria that can cause pneumonia in the immunocompromised individual and occurs quite frequently as a hospital-acquired disease. *Klebsiella* is a normal part of the human flora and not normally pathogenic. Recent experiments performed in Russia show that phage are very efficient at curing experimental infections in mice and guinea pigs when administered either nasally or by injection. There was no toxicity in the animals. This was followed in 1992 by a test in humans, in which 109 patients infected with *Klebsiella* were treated with a phage cocktail against different strains of the bacterium. Again, the phage preparations were non-toxic and appeared to be effective.

Another use for phage therapy is in the treatment of burns, which are a large problem both in total numbers of cases and severity, particularly in times of war and after terrorist attacks. Infection remains the leading cause of death among patients who are hospitalized for burns. Treatment with phage both in humans in the Soviet Union or, more recently, at the Tiblisi Institute, has met with remarkable success. "Three woodsmen in the mountains of Georgia stumbled upon a pair of canisters that were, oddly, hot to the touch. Then men lugged the objects back to their campsite to warm themselves on a bitterly cold night. That turned out to be a terrible mistake: the canisters, Soviet relics once used to power remote generators, were intensely radioactive and burned two of the men severely. The victims were rushed to the capital, Tbilisi, where doctors plied them with antibiotics but failed to prevent *Staphylococcus* bacteria from invading the deep wounds. Septic shock seemed just around the corner" [13]. Georgian physicians treated the wounds with bandages impregnated with anti-staph phage and within a short time the wounds healed. The phage products from the Eliava Institute are now being marketed worldwide.

As in all cases of medicine and drug development, it is important that the right phage be picked for the correct target. One has to avoid phage that may undergo lysogeny, or may even introduce toxic genes into the bacteria, increasing its pathogenicity rather than killing the bacteria. A new approach in selecting phage may be necessary and include the sequencing of the DNA to identify any toxic genes, or genes involved in integration. This can now be done routinely and quite cheaply with the advent of fast throughput DNA sequencing. RNA phage may also be used effectively, and here there is less danger of integration into the bacterial chromosome.

As in the case of antibiotics, phage resistance will occur. In fact, this was one of the factors used in the 1950s to discourage further experiments with phage therapy. However, the rate of mutation to resistance to phage is about 100 times lower than that of resistance to antibiotics. Using a cocktail of phage with binding to different receptors on the bacterial host should prevent the development of drug resistance. This has been the approach taken in treating HIV and AIDS, where a cocktail of drugs prevents the selection for drug resistance.

16.7 Conclusions

It is now almost 100 years since the discovery of bacteriophage, and about the same time since d'Herelle proposed using phage to treat bacterial infections. The experience of many physicians from the 1930s to 1940s—before the advent of antibiotics—suggests that this treatment works, once the specific phage for any specific strain of bacterium has been identified. Using genetic engineering technology and selection, it should be possible to engineer phage with broad-spectrum or a narrow-spectrum, depending on the need. There are obviously three major hurdles to development: one is to overcome resistance of the host bacterium; the second to outwit the restriction enzymes of the host if active against the phage and the third to create a government regulatory approval system for clinical trials, since the current FDA regulations do not apply to "living biologicals." An alternative approach is to treat phage in a manner similar to a vaccine, where live agents may be used, e.g., the attenuated polio vaccine.

One subject that has to be addressed is the reason for the skepticism in the West as a result of the mixed results of the 1930s. This appears to have been due to many factors, not least of all the hype by commercial companies on using phage to cure diseases such as herpes and other viruses, the presence of toxins and contaminations in the preparations, and the lack of specificity of the phage towards the disease being treated. Another non-scientific reason may have been the objections to using a method developed in the Soviet Union, although the U.S. was willing to accept the Soviet space program as a competitor. In order to meet U.S. and European standards, it will be necessary to purify the phage, using most recent advanced technology to remove bacterial debris, to select appropriate phage for each strain of bacterium, and to determine efficacy before use. Such phage can be maintained at specific centers, such as the American Type Culture Collection or medical centers and would be available as needed, very much on the model originally developed in Tbilisi, Georgia. It would also be more efficient to use a mixture of phage against different organisms (cocktail), as has been developed at the Eliava Institute. Since commercial biotech companies are interested in patents and profit, it will be necessary to patent modifications made to the phage, as well as combinations with antibiotics, and perhaps even methods of application.

One of the major problems facing the medical establishment, and of course the general population, is the increase in multi-drug-resistant strains of bacteria. These have arisen from the uncontrolled use of antibiotics. It was estimated that in 2005 there were 100,000 serious cases of multidrug-resistant staph infections in the U.S., resulting in some 20,000 deaths. The Centers for Disease Control in Atlanta now estimates that as many as 80,000 hospital patients are infected with antibiotic resistant *S. aureus* every year. Also blood poisoning (sepsis) occurs in the U.S. at the annual rate of 750,000 individuals, resulting in 210,000 deaths. Since the problem of antibiotic resistance is getting worse (many cases of tuberculosis are drug-resistant), alternatives to antibiotics are a must, and phage therapy would seem to offer such an alternative. Because drug companies are loath to do research on new antibiotics, it is likely that they will be even more loath to do research on phage, which may not be patentable. Evidence indicates that in many cases, a combination of phage and antibiotics (in the case of non-antibiotic resistant bugs) is much more effective than either treatment alone. Although there were successful trials carried out in 1959 against staph infections in the U.S., applying phage by different routes, and 80 % of the patients recovered, the staph phage lysate was suspended by the FDA in 1990. The reasons for the withdrawal of license are not clear.

During the last 3 years (2008–2011), phage preparations against secondary infections in cystic fibrosis patients were used. In these patients, mucus accumulates in the lungs, causing difficulty in breathing. The application of phage decreases the bacterial count in the lungs, the patients appeared to be healthier, and there were longer periods between rounds of bacterial infection. Small-scale experiments have recently been carried out in the Burns Center of the Queen Astrid Hospital in Brussels, Belgium. This was a "phase 1" trial, with no adverse effects. Another small-scale trial carried out by a private company, Biocontrol Ltd. in London, on 24 volunteers with chronic ear infections, showed no adverse effects, and recovery from the conditions for some time. Thus a beginning is being made in the west. Biocontrol has recently merged with Targeted Genetics to form a new company, AmpliPhi Biosciences, which has applied for a U.S. patent to cover the use of phage to treat biofilms of a *Pseudomanas aerogenosa* in combination with antibiotics. According to the patent, "Examples of specific treatment regimens covered include the treatment of human and animal infections resulting from a skin burn or skin wound, a lung infection, an eye, ear or urinary infection or an infection associated with a medical device or implant."

In dealing with a treatment such as bacteriophage or even viral oncolysis, discussed later, the rules developed by the FDA for single chemical elements may have to be changed. This was done in initial trials with gene therapy, and a mechanism may exist for approval of such trials by special boards. The evidence that phage therapy works in certain bacterial diseases, such as staph infections, and perhaps in cholera, is quite overwhelming and it is surprising that this methodology is not used much more in the U.S. and Western Europe. With the problem of multidrug resistance in the last few years, we are seeing new biotech companies arising that are developing phage therapeutics. According to the Phage

International Web site, which promotes and advertises phage therapy, there are at least a dozen companies involved in developing phage therapy, including two using the phage collection at the Eliava Institute in Georgia. In the U.S., Intralytex, a company in Maryland, has received grants from the USDA and from the U.S. Army to develop phage therapy, using phage as a probiotic against *shigella* infection among soldiers, and also using phage to prevent food-borne diseases such as *salmonella* and *pseudomonas*. Another company involved is Gangagen, which is producing phage and phage products (lysins) against drug-resistant *Staphylococcus auries*. Even a company as large as the Nestlé food corporation has become involved in testing phage against pathogenic strains of *E. coli* and it is currently recruiting patients in Bangladesh and other developing countries using phage in the treatment of childhood diarrhea.

Thus, there are all the indications that phage therapy will be the alternative to antibiotics in the treatment of some diseases and in treating antibiotic-resistant bacteria.

References

1. Hankin, M. E. (1896). The bactericidal actions of the waters of the Jamuna and Ganges River on cholera microbes. *BrMedJ 1*.
2. Hankin, M. E. (1898). Cholera in Indian villages. *BrMedJ 3*.
3. BRaM, J. (1921). Au sujet del'unite du principe bacteriophage. *Comptes Rendus Acad Sci de Biol, 85*, 1122.
4. Chanishvili, N. (2012). Phage therapy–history from Twort and d'Herelle through soviet experience to current approaches. *Advances in Virus Research, 83*, 3–40.
5. Rice, T. B. (1930). The use of bacteriophage filtrates in the treatment of suppurative conditions: Report on 300 cases. *American Journal Medical Sciences, 179*, 345–360.
6. Stout, B. F. (1933). Bacteriophage therapy. *Texas State Journal Medicine, 29*, 205–209.
7. Hausler, T. (2006). *Viruses versus, superbugs*. Macmillan: Basingstoke. What city?
8. Sulakvelidze, A., Alavidze, Z., & Morris, J. G, Jr. (2001). Bacteriophage therapy. *Antimicrobial Agents and Chemotherapy, 45*(3), 649–659.
9. Weber-Dabrowska, B., Dabrowski, M., & Slopek, S. (1987). Studies on bacteriophage penetration in patients subjected to phage therapy. *Archivum Immunologiae et Therapiae Experimentalis, 35*(5), 563–568.
10. Smith, H. W., & Huggins, M. B. (1982). Successful treatment of experimental Escherichia coli infections in mice using phage: Its general superiority over antibiotics. *Journal of General Microbiology, 128*(2), 307–318.
11. O'Flaherty, S., Ross, R. P., & Coffey, A. (2009). Bacteriophage and their lysins for elimination of infectious bacteria. *FEMS Microbiology Reviews, 33*(4), 801–819.
12. Kutateladze, M., & Adamia, R. (2010). Bacteriophages as potential new therapeutics to replace or supplement antibiotics. *Trends in Biotechnology, 28*(12), 591–595.
13. Stone, R. (2002). Bacteriophage therapy. *Stalin's Forgotten Cure. Science, 298*(5594), 728–731.

Viral Oncolysis or Virotherapy

17

Abstract

The concept of using a virus to "kill" a tumor dates back to the beginning of the 1990's when it was noticed that occasionally a tumor would shrink following a viral infection. This concept was tested from the 1950s until the 1970s in animal models, and in a few human trials. Success was limited and the practice of viral oncolysis, or "virotherapy," was met with skepticism. Renewed interest arose following the development of recombinant DNA technology. Recombinant adenovirus and herpesvirus were explored as anticancer agents. Initial efforts to develop their use was delayed due to the death of a patient. New vectors were produced in the 1990s using various viruses, engineered to bind preferentially to tumor cells. Some of these viruses contain transgenes coding for proteins that stimulate the immune system. Within the last few years, clinical trials using various genetically engineered viruses have been initiated. There have not been any toxic side effects, although long-term data are still not available. Preliminary data from a construct of herpes simplex 1 carrying the gene for GCSF shows a positive response against melanoma. A number of biotech companies are now involved in virotherapy; thus, one can conclude that the future of virotherapy as a major tool for the treatment of cancer looks promising.

17.1 A Short History of Viral Oncolysis

"Viral oncolysis", or "virotherapy" as it is known today, is a concept that developed at the turn of the twentieth century with the observations that occasionally, cancer regressed following a viral infection [1, 2]. This led to a very simple idea: that viruses might be able to target cancer cells, a magic bullet that would destroy the tumor but leave surrounding cells intact. Throughout the 1930s and 1950s, attempts were made to treat cancers in animal models and humans with

M. W. Taylor, *Viruses and Man: A History of Interactions*,
DOI: 10.1007/978-3-319-07758-1_17,

viruses, but with limited or little success [3–5]. A critical review of the progress in the field ended in a very pessimistic tone, [4] leading to skepticism in the scientific community and lack of funding from government agencies.

As an example, in the 1970s I discovered that a picornavirus (bovine enterovirus), when injected directly into transplantable tumors of mice, triggered tumor regression [6, 7]. However, this research never reached the clinic because of skepticism, and, in turn, lack of funding from government agencies. This continued to be the pattern until the late 1990s, when interest in virotherapy was rekindled, concurrent with advances in gene therapy and the development of viral vectors containing transgenes (genes not usually found in that virus or vector). The progression of this interest can be seen in a graph plotting the number of papers containing the words "virotherapy" or "viral oncolysis" in their titles or abstracts, versus time (Fig. 17.1a, b).

The ability to engineer viral vectors completely changed the world of virotherapy: The attachment of the virus to host results in the uptake of the virus through the cell membrane by endocytosis, and the ultimate un-coating of the virus leads to virus gene expression that may result in cell lysis. The first step of this process requires recognition between proteins of the virus coat or their appendages and protein domains (receptors) on the surface membrane of the host cells. Through genetic engineering, the virus genome is altered so that it codes for proteins that bind selectively to receptor domains of the tumor cells in preference to domains of other host cells. In addition, the viral genome can be engineered, so that it codes for proteins that bind to and activate extracellular molecular complexes, leading to additional enhancement of virus-cell interaction. It is possible to introduce genes so that they activate specific components of the immune system in the tumor environment, and unique promoters can be placed in front of viral genes or in front of the transgenes to enhance their expression in the tumor cell environment [8].

Two viruses were initially of heightened interest: adenovirus-5 and herpes simplex virus-1 (HSV-1). Adenoviruses, both in a replicative form and non-replicative form, can incorporate foreign genes into their DNA. HSV-1 can be manipulated so that it targets tumor cells, and because of its large size, incorporates non-viral genes toxic to tumors. A large number of such vectors have been constructed and successfully tested in clinical trials (Table 17.1).

A positive feature that impelled virotherapy forward was the establishment of biotech companies whose goal is to construct and test novel viral vectors—initially for gene therapy and later as anti-cancer agents. Many of the vectors constructed by these companies and in academic laboratories are in phase I and phase II clinical trials. One, talimogene laherparepvec (OncoVEXGM-CSF), is currently in phase III trial of stage-3 and stage-4 melanoma. Phase I examines the toxicity or tolerance to the treatment; phase II tests the therapeutic efficacy in a small number of patients; and phase III is a large-scale test of both therapeutic efficacy and tolerability.

Table 17.1 lists the viruses that are currently being tested as anti-cancer agents in clinical trials.

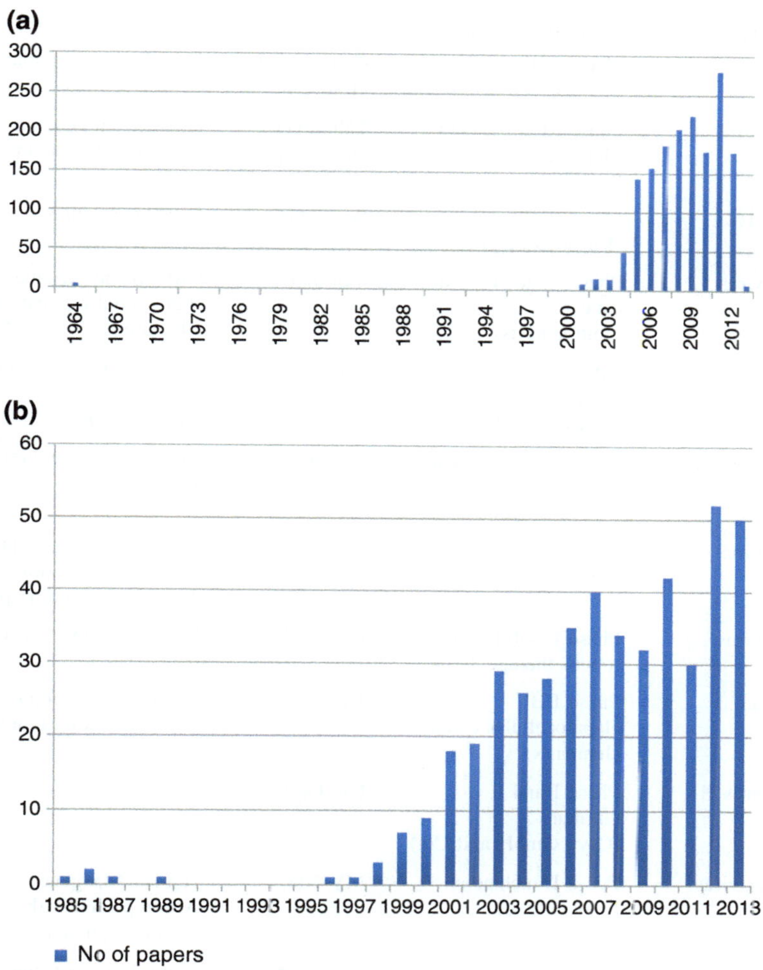

Fig. 17.1 **a** Number of papers with "virotherapy" in title or abstract. **b** Number of papers with "viral oncolysis" in title or abstract (from PubMed)

17.2 Adenovirus

Adenovirus (Ad-5) is a preferred tool in gene therapy research. In humans, adenoviruses cause mild respiratory illness in infants but are harmless in healthy adults. In hamsters, some human adenoviruses are tumorigenic. The viral replication cycle is divided into the two phases, early-gene and late-gene expression. Deletions of the early genes, such as E1 and E2, prevent viral replication, and such deletions are used in the construction of non-replicating adenovirus vectors

Table 17.1 Selective clinical trials ongoing in 2013

Virus	Phase and target	Commercial viral name	Genetic modifications
Herpesvirus	Phase III completed Melanoma (Amgen-BioVex)	Talimogene laherparepvec	GM-CSF in genome Specific deletions in virus
HSV1716	Phase II: Glioma (Virtuu Biologics)	Seprehvir	ICP34.5 deletion
Adenovirus	Phase I and II: bladder cancer (cold genesis)	CG0070	GM-CSF in genome Preferential replication in RB-tumors
Adenovirus	Approved therapeutic in China (Sunway Biotech)	Oncorine (H101)	Preferential replication in tumors E1B deleted
Adenovirus chimera	Phase I and II: metastatic cancer PsiOxus therapeutics	ColoAd1	Chimera of adenovirus 3 and 11
Adenovirus	Phase I: gioblastoma (DNAtrix)	DNX-2401	Conditionally replicating adenovirus encoding An integrin-binding peptide
Adenovirus	Phase I: solid tumors (oncos therapeutics)	CTGT-102	Virus encoding GM-CSF
Vaccinia	Phase IIB Hepatocellular carcinoma (Jennerex)	JX-594	GM-CSF in genome Thymidine kinase deleted
Parvovirus H-1	Phase I and II Glioma (Oryx GmbH and Co.)	ParvOryx	None
Measles virus	Phase I: melanoma (mayo clinic)	MV-NIS	Human NIS gene Accompanied with cyclophosphamide
Poliovirus (sabin strain)	Phase I (Duke University)	PVS-RIPO	Control of ribosomal entry site
Coxsackievirus A21	Phase I: melanoma (Viralytics)	Cavatak	None
VSV	Phase I: hepatocellular carcinoma (Mayo clinic)	VSV-hIFN-β	Contains interferon gene
Toca-511 (replicating retrovirus)	Phase I and II: glioma (Tocagen)	Toca511	Contains cytosine deaminase
Reovirus	Phase III: squamous cell carcinoma (Oncolytics biotech)	Reolysin	None

Company supporting the work in brackets in Column 2. NIS: thyroidal sodium iodide symporter. Table adapted from [8, 9] containing a transgene. Deletion of the E3 gene does not affect viral replication, and is the most common site to insert transgenes. However, deletion of the E3 region does affect the immune response to the virus. Replication-competent viruses are preferred in virotherapy, since theoretically the virus should lyse the tumor cells and destroy them. This is in contrast to vectors for gene therapy where one would prefer a non-replicating virus (see Chap. 18).

Unfortunately, gene therapy experiments with adenovirus were temporarily halted in 1999 because of a tragic accident in an 18-year-old patient, Jesse Gelsinger, who was diagnosed with ornithine carboxylase (OTC) deficiency; an apparent overdose of the adenovirus containing the OTC gene resulted in his death. All human experimentation with adenovirus vectors was stopped for a number of years until safety could be improved. However, researchers continued to engineer the virus.

A series of phase I and II clinical trials were performed with the adenovirus Onyx-015, the first virus specifically engineered to destroy tumor cells defective in the tumor suppressor gene p53 [10, 11]. The virus was not toxic and inhibited tumor growth. Onyx-015 was developed on the hypothesis that adenovirus, containing a mutation in the E1b gene, would only replicate in tumor cells with a mutation in the p53 gene. This is a tumor suppressor gene that inhibits tumor formation by regulating the cell cycle, controlling DNA synthesis and cell replication. p53 protein monitors DNA damage and induces cell death (apoptosis). Deletions or mutations of p53 occur frequently in human tumors, often making them resistant to chemotherapy or radiation therapy. In normal cells, the presence of p53 would prevent the mutated virus from replicating, since the interaction of p53 and E1b is essential for viral replication.

Onyx-015 was extensively tested on tumor cell lines, and efficiently infected tumors containing a mutated or inactivated p53, but not normal cells, or tumors containing wild-type p53. It replicated in p53-deficient cells as rapidly as wild-type virus. Similar results were obtained in human tumors transplanted into nude mice. Since adenovirus in general does not replicate in mice, in these experiments it was impossible to examine the immune response to the virus. There were two problems with this system: the virus had to be injected directly into the tumor, and its effect on metastasis could not be measured. Later experiments indicated that this conditionally defective virus could replicate in the presence of p53, although not to the same extent as in p53-deficient tumors (Fig. 17.2).

Onyx-015 was administered to over 250 cancer patients in approximately 15 clinical trials. There was some limited tumor regression in head and neck cancers; however, the viral treatment was augmented with either chemotherapy or radiation therapy. The virus was not powerful enough to spread to metastatic lesions or completely eliminate the primary tumor. This was a great disappointment, and brought experiments with Onyx-015 to a halt in the U.S. In part, this was due to a lack of funding for further work, and led to the sale of the patent and the virus to Shanghai Sunway Biotech of China.

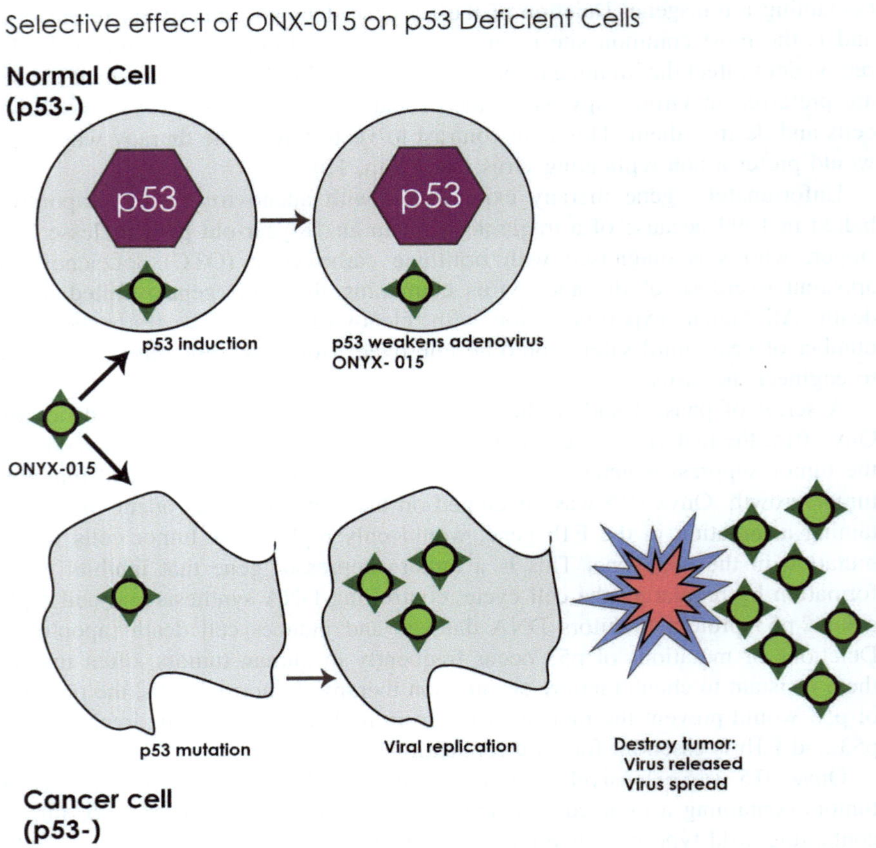

Fig. 17.2 Effect of Onyx-015 on p53-deficient and wild-type (p53+) cells

A modified virus, adenovirus H101—similar to Onyx-015—was approved by the State Food and Drug Administration of China for use in combination with chemotherapy in the treatment of late-stage nasopharyngeal carcinoma and head and neck cancers. H101 was tested without side effects in phase I and phase II trials in China, also in combination with chemotherapy. In a large phase III trial of head and neck cancers with 160 patients at 13 hospitals, there was a 79 % remission rate when the virus was used with chemotherapy against a 40 % rate of remission with chemotherapy alone. This strain of adenovirus (oncorine) was officially approved for clinical use in China in 2005 and is now used in treating many different types of cancer. A second adenovirus containing the p53 gene has also been tested and approved.

17.3 Adenovirus Plus GM-CSF

An important landmark in viral cancer treatment has been the incorporation of "granulocyte macrophage colony stimulating factor" (GM-CSF) as a component of the virus. GM-CSF activates T-cells and dendritic cells against tumor antigens and stimulates dendritic cells to produce interferon-gamma [12]. The GM-CSF gene has been incorporated into the genome of the larger DNA viruses such as adenovirus, HSV-1 and vaccinia. Adenovirus (CG0070) with GM-CSF is currently being tested on bladder cancer in patients whose BCG treatment has not been successful. This is a phase II/III trial, which will not be completed until 2019: the phase I trial showed no adverse effects [13]. Various doses of adenovirus were tested, and toxic levels were not reached. Patients received either single or multiple doses of the virus at levels of 1×10^{12}–1×10^{13} or 3×10^{13} viral particles, and viral replication was detected in all patients. The response rate in patients receiving multiple injections of CG0070 was 63.6 %. In this particular trial, the presence of GM-CSF had a profound effect on the response.

Almost all humans have antibodies to adenovirus-5, or will develop antibodies during treatment, an immune response that may reduce or inhibit the effect of multiple doses of the oncolytic virus. To avoid inhibition by anti-bodies, chimeric adenoviruses have been constructed [14]. Using a chimera of adenovirus 5/3 and a transgene for GM-CSF, a Finnish group [14] demonstrated both an oncolytic effect of the virus and an enhanced immune response against the tumor in several groups of patients. In 115 patients treated with CGTG-102 (Ad5/3-D24-GMCSF), median overall survival was 111 days following single and 277 days after serial treatment in non-randomized comparison. Using different adenovirus or chimeric viruses had no adverse effects.

Despite this initial success, there are still problems with the adenovirus system. These include the slow rate of virus replication; the inability of the virus to spread to other cancer sites (although this is helped by adding immune-stimulants as transgenes to the virus as discussed above); and the possibility that virus is inhibited by the host immune response. There is also the need for the virus receptor to be expressed on the tumor cell surface, which does not happen in all cancers. Viruses have also been engineered so that they recognize novel sites on the tumor. To summarize, many different adenovirus vectors have been produced with modifications of receptor proteins, unique promoters and insertion of genes that stimulate the immune system. The number of modified adenoviruses are too numerous to discuss in such a review, and this field has been reviewed extensively [15].

17.4 Herpes Viral Vectors

Herpes simplex-1 (HSV-1) is associated with cold sores in the mouth (see Chap. 14). A number of features make HSV-1 an attractive vector for virotherapy. These are:

1. Despite its tendency to infect predominantly neural tissue, HSV-1 has a very wide cellular host range, infecting non-dividing as well as dividing cells, and expresses exogenous genes inserted into its DNA with excellent efficiency.
2. Large segments of HSV DNA can be deleted and replaced by genes for drugs or other anti-cancer agents in both replication-competent and replication-defective HSV-1 mutants, allowing for simultaneous delivery of multiple transgenes with different promoters.
3. The HSV-1 genome does not integrate into the cellular DNA, eliminating the possibility of insertional mutagenesis.
4. Recombinant HSV-1 can be readily constructed, and purified stocks of virus containing 10^{10} infectious particles per millilitre can be prepared routinely without contamination of wild-type virus.
5. The large size of the virus facilitates the insertion of multiple genes.
6. Evidence would suggest that the replication of herpesvirus in tumor cells is more efficient than adenovirus vectors.
7. Non-replicating herpesviruses can be used as carriers of anti-cancer drugs or pro-drugs, since enzymes that transform non-toxic compounds to toxic compounds are coded for by the cell or virus.

HSV1716 was first constructed as a non-neurotropic deletion mutant of wild-type HSV-1[16]. This virus was no longer neurotropic but grew in other cell types. It could enter a latent stage but was defective in activation [17]. It was first tested in human neuronal tumors in vitro and in mice [18, 19] and later in brain tumors in mice with an oncolytic effect, destroying only the tumor cells and not the surrounding cells [20]. A large number of laboratory studies had concluded that HSV 1716 (by now called ICP34.5 deletion) could be useful in the treatment of brain tumors. Furthermore, results of animal studies demonstrated that HSV 1716 was oncolytic for a variety of human tumors growing in mice, including mesotheliomas, melanomas, and ovarian cancer. The first phase I trial of HSV1716 was performed in Britain with patients who had a relapsed malignant glioma. Of nine patients treated, four had recovered after 14–24 months [21]. Similar results were found with melanoma [22].

A British biotech company, Biovex, developed a second generation of HSV1716 with the incorporation of the cytokine gene (GM-CSF) granulocyte-macrophage colony-stimulating factor [23]. Thus not only was this virus oncolytic, it also stimulated the immune system to destroy remaining cancer cells. A phase II trial reported a 28 % objective positive response in 2009. Meanwhile, Biovex was bought by the American biotech giant, Amgen. A phase III trial was initiated with advanced melanoma patients. This virus, about which one will hear much in the future, is now known as "T-vec" or "talimogene." The preliminary results of the phase III trial have recently been reported. The data show a 16 % durable response rate—compared to only 2 % for the GM-CSF arm of the study—with a 26 % overall response rate (6 % for GM-CSF). Four hundred patients were recruited for this trial. There is a possibility that this vector will be used in the future to treat many types of cancers, and the study will be completed in 2014. Brain tumors are

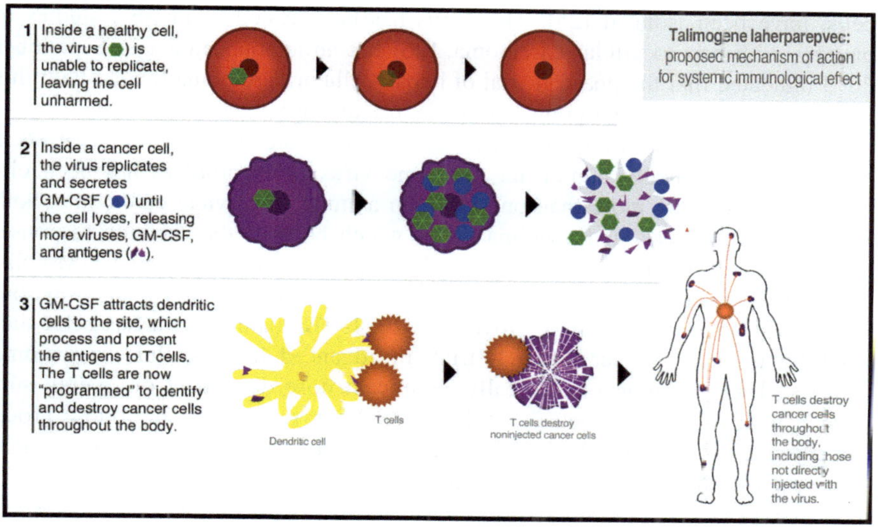

Fig. 17.3 Method of action of oncolytic herpes (from http://en.wikipedia.org/wiki/Talimogene_laherparepvec)

untreatable, so that successful experiments will be a breakthrough in cancer therapy (Fig. 17.3).

The strain of HSV-1 currently used in clinical trials in Japan is HF10, a spontaneous deletion mutant of the virus, which is replication-competent but less pathogenic than wild-type, and it has been used to treat breast and pancreatic cancer [24]. In cases where two nodules of breast cancer were injected at the same time, one with saline and one with the virus, the tumor regressed in response to the virus and not to the saline. Preliminary results have been very favorable. A second-generation construct has been built to incorporate the IL12 cytokine gene, which activates naïve T-cells and turns them into T-helper cells; in addition, it stimulates interferon production and activation of NK (natural killer) cells. This construct has worked very well in a murine glioblastoma model with an enhanced immune response leading to destruction of tumor stem cells. A large phase I trial is currently under way with this vector for the treatment of head and neck cancers and melanoma.

17.5 Vaccinia Virus

If one were searching for an oncolytic virus, vaccinia would be an obvious choice. The virus has been used as a vaccine for years with few side effects and is very lytic in culture. It contains a large amount of DNA, some of which is expendable, and can be replaced by transgenes, or pro-drug genes. As in the case of herpes virus, vaccinia virus thymidine kinase mutants, which only replicated in tumor cells, have been isolated [25]. This virus has been tested in phase I studies in patients with hepatocellular carcinoma, [26] but an announcement in September 2013 indicated that the phase II trial of hepatocellular carcinoma treated with this virus did not meet its end point (survival) in late-stage liver cancer.

As already discussed for adenovirus, one of the problems with viral oncolysis is the need to repeatedly administer the same virus. This problem has been circumvented by using more than one virus. In a study of cervical carcinoma, both Semiliki Forest Virus and vaccinia virus, which boosted the immune response, [27] were used. A clinical trial with JX594, a thymidine kinase negative, GM-CSF-carrying vaccinia virus, has enrolled 129 patients and is currently under way, but results will not be known until November 2014. This same vector has been successfully used in another phase II trial, in advanced hepatocellular carcinoma patients, demonstrating a statistically significant dose dependent on overall survival benefit with 14.1 months' median overall survival for the high-dose group compared to 6.7 months for a low dose group (p value 0.02) [28]. This virus is now being used to treat colon and pancreatic cancer.

17.6 Poliovirus

Another virus that has come to the forefront in oncolytic trials is a modified poliovirus isolated from experiments in which poliovirus was created in a cell-free system [29]. This virus was oncolytic for neuroblastoma cells in mice sensitive to poliovirus. The oncolytic effect appeared to be due to the enhancement of T-cell activity and took place even in the presence of polio antibodies.

A second approach is to make poliovirus less neuropathic by engineering in rhinovirus (common cold virus) sequences. Glioblastoma has a high number of polio receptors, a characteristic of many tumors. Although not yet at phase I clinical testing, physicians at Duke University used the virus in June 2013 and may have saved the life of a young woman with a brain tumor, which had reoccurred after surgery. The 22-year-old patient appears to have recovered and she would be one of the first patients with a brain tumor to be successfully treated with a virus.

Other cancers that have proven to be untreatable by chemotherapy or radiation therapy, such as neuroblastomas and glioblastomas, are now being treated in model systems with a genetically altered recombinant poliovirus. PVSRIPO is a recombinant virus derived from the live attenuated, oral (SABIN) serotype 1 poliovirus (PV) vaccine. It contains a heterologous internal ribosomal entry site (IRES) from human rhinovirus type 2 (HRV2). PVSRIPO recognizes Necl-5, a cancer-related cell adhesion molecule and tumor antigen widely expressed in malignancies such as glioblastoma [30]. (Details on the clinical trial can be found at http://clinicaltrials.gov/show/NCT01491893.) The virus will be injected directly into the tumor, and results are expected in 2015–2016.

17.7 Other Viruses

Measles virus is another virus that has been engineered in the treatment of tumors. It has been claimed to have a very wide host range active against mesothelioma (lung tumors caused by exposure to asbestos), lymphomas, glioblastomas, prostate and many others. Part of the viral therapy effect is due to enhancement of the immune system, with an increase in CD8 T-cells. A phase I trial of the Edmonston (vaccine) strain of measles virus, with or without cyclophosphamide, an anti-cancer compound, in patients with recurrent or refractory multiple myeloma, is under way.

One of the major problems in treating cancers is destroying secondary tumors that have spread as a result of metastasis. Unfortunately, clinical trials with adenovirus suggested that this was not possible. The use of adenovirus as an oncolytic agent may in fact have inhibited this field of research, since it did not spread systemically to other sites. Other viruses have the ability to spread to and destroy metastatic lesions. In particular, this has been true of Newcastle disease virus (NDV), measles, and vaccinia. Mesothelioma is a particularly virulent untreatable tumor caused by exposure to asbestos. NDV (a chicken virus) has been shown to be effective in treatment of this tumor in a mouse model [31].

Oncolytic biotech, a company in Calgary, Canada, has been testing reolysin, a reovirus in combination with cisplatinum and other chemotherapeutic agents on squamous cell carcinoma. Reovirus acts by both direct tumor cell killing and priming of antitumor immunity; reovirus does not replicate in normal cells but does replicate in cancer cells. Normal cells express the ras gene, a known oncogene, which inhibits reovirus replication. In the tumor cells that carry a mutation of ras, the virus can replicate. The major mechanism of reovirus oncolysis of cancer cells has been shown to occur through apoptosis [32, 33]. In addition, the synergistic anti-tumor effects of reovirus in combination with radiation or chemotherapy has been demonstrated for reovirus-resistant and moderately sensitive tumors. Phase III clinical trials are ongoing and early results of the phase III trial are very promising. Of the 105 total patients with evaluable metastatic tumors, 86 % ($n = 50$) of those in the test arm of the study exhibited tumor stabilization or shrinkage, compared with 67 % of patients ($n = 55$) in the control arm. (More information is available at http://clinicaltrials.gov/ct2/show/results/NCT01166542). In most of the clinical trials undertaken to date, an anti-reovirus immune response likely to circumvent anti-tumor efficacy has been seen. Newer methods are being used to deliver reovirus to tumor sites, such as loading the virus into dendritic cells and killer lymphocytes, which helps in activating the immune system, and abrogates the effect of specific anti-reovirus antibodies [34]. Reovirus has also been used as an oncolytic agent in phase I trials against multiple myeloma. It is considered to have a great deal of oncolytic potential.

Parvoviruses are small, single-stranded DNA viruses, among the smallest viruses affecting mammals. Parvovirus H1 is an autonomous replicating virus that has been shown to grow in culture, preferentially in tumor cells. H1 has oncolytic effects on human tumors in mice; these effects are enhanced by the incorporation of cytokine genes into the H1 genome [35–37]. A phase I/II trial has been initiated with the H1 parvovirus in the treatment of glioblastoma; this study has just gotten under way and patients are being recruited.

Other viruses currently undergoing clinical trial include rhabdoviruses, in particular vesicular stomatitis virus, an RNA virus that replicates very quickly, and at least in culture lyses cells within 24 h, giving it an advantage as an oncolytic agent. Early studies, as in the case of all the oncolytic viruses, have proved their usefulness in culture and in tumors in mice. However, VSV is very sensitive to interferons (in fact, this virus is used as an indicator strain in interferon assays) which would inhibit its oncolytic potential, unless the interferon pathway were blocked, but this in turn might inhibit tumor apoptosis. In fact, the first clinical trial with VSV includes a virus carrying an interferon gene, which seems counter-intuitive. The aim of the study reads, "The primary purpose of this study is to evaluate the safety of a viral agent called vesicular stomatitis virus for the use in patients with liver cancer. The study virus has a gene inserted into it which will allow for the production of interferon beta, which is a substance that will have the dual functions of restricting the spread of the virus to the tumor cells and not healthy liver cells and also to have some independent anti-cancer activity. Although the primary goal of this study is to evaluate the safety of delivery of this viral agent to people, patients may benefit clinically by having shrinkage or stabilization of their tumor or reduction in their cancer related symptoms (e.g., pain)" (http://clinicaltrials.gov/show/NCT01628640).

A novel non-lytic replicating retroviral vector (TOCA511) has been derived from mouse murine leukemia virus encoding yeast cytosine deaminase (CD). This enzyme converts the prodrug 5-fluorocytosine (5-FC) to the potent anticancer drug, 5-fluorouracil (5-FU) in an infected tumor [38, 39]. This retrovirus has been extensively tested in animal systems and is now in clinical trial. Three of the ongoing trials involve brain tumors in which the virus is directly injected into the brain. In some cases these are patients who have not responded to chemotherapy; in others, the virus is injected after tumor removal. All will be treated with 5-flurouracil for 6 months after initial treatment with the virus. To quote, "This is a multicenter study evaluating the safety and tolerability of Toca 511, a retroviral replicating vector, injected into the resection cavity of patients with recurrent or progressive Grade III or Grade IV gliomas who have elected to undergo surgical removal of their tumor. Approximately 6–7 weeks after injection of Toca 511, the patient will take an oral course of 5-FC, an antifungal antibiotic. These courses of 5-FC will be repeated during the 6-month study. MRI scans will be performed approximately every 2 months. Three subjects will be evaluated at up to 5 dose levels of Toca 511. The dose of Toca 511 a patient receives will depend upon the number of previous study participants and how well they have tolerated the study drugs." http://clinicaltrials.gov/show/NCT01470794.

17.8 Conclusions

As more trials with oncolytic viruses have occurred, it has become obvious that one has to take into account the extracellular environment of the tumor and the surrounding tissue. Viruses may affect the blood supply to the tumor, either by diminishing it or by increasing the flow. The virus may also enhance the immune response to the tumor or may in fact inhibit its own growth by inducing interferon (which could also inhibit tumor growth). Thus it has become clear that what started off as a simple idea is much more complicated than originally assumed. From these experiments, certain valid principles can be established: namely, one has to use non-pathogenic virus that replicates in the tumor, and not in the healthy tissue of the patient. The virus should be first tested in cell culture to examine host range and tissue specificity, very much as is done with bacteriophage therapy on specific host bacteria. It is apparent from in vitro work that most lytic viruses, that is, viruses that replicate and destroy the cell, could be used for the treatment of tumors. Some are without modification, but others with modification, in order to neutralize serious side effects on the patients. A major problem is the production of antibodies and cytotoxic T-cells to the virus, which requires the use of a succession of different viruses during treatment; the other is to direct the virus to metastatic sites. This might be achieved by intra-venous injection of the virus, but this may lead to a systemic infection, or the virus may be destroyed by a strong immune response.

After a long haul, it finally seems that viral oncolysis has found a place in cancer therapy. The ultimate aim is to eliminate or reduce the need for chemo-therapy. Viral oncolysis may be enhanced with the aid of immunotherapy by adding cytokines either directly with the virus, or as a transgene in the virus. Other cytokines, such as interferon (despite its side effects) and IL-2 or others of the many newly discovered cytokines, could be incorporated into potential oncolytic viruses and their effects studied. Obviously the recent breakthroughs imply that more research is needed.

17.9 Virotherapy and Biotech Companies

It does appear that biotech companies that specifically specialize in this area are carrying out the majority of the studies on viral oncolysis. The success obtained by Shanghai Sunway Biotech with an adenovirus vector, Amgen with a herpes vector; and Oncolytic Biotech with reovirus (Reolysin) are very encouraging. Among other companies now engaged in phase II trials, a Finnish company, Oncos bio-tech, with a new generation of replicating adenovirus vectors, and Cold Genesis of Irvine, California, utilizing an adenovirus vector also with GM-CSF in the treat-ment of BCG-resistant bladder cancer. Also of interest is Jennerex, using a recombinant vaccinia virus, lacking the thymidine kinase gene, and incorporating GM-CSF. This virus is now being used in the treatment of colon and pancreatic cancer. An announcement in September, 2013, of another trial indicated that this phase II trial did not meet its end point (survival) in late-stage liver cancer. Viralytics is running clinical trials to investigate the use of CAVATAK,™ a natural, genetically unaltered Coxsackie virus (part of the common cold family of viruses). Virttu Biotech is a Glasgow-based company using a herpes simplex 1 modified virus in the treatment of mesothelioma; a phase I/II has just been started. PsiOxus is an English/German company using adenoviruses (other than adenovirus 5 used by other companies) in the treatment of colorectal cancer. Again, this is a phase I/II trial.

DNAtrix is developing oncolytic adenoviruses specifically to attack glioblastoma. "There is now compelling evidence that DNX-2401 can (i) replicate in human tumors for a period of weeks to months, (ii) elicit tumor necrosis within weeks of injection, (iii) trigger intratumoral immune cell infiltration and (iv) lead to long term tumor destruction detectable by MRI." Perhaps more importantly, the study agent, DNX-2401 was safe and well tolerated in the clinic with no serious product-related toxicities to date. The phase I trial, which measures tolerability, also indicated a response on the tumor itself. Phase II trials are ongoing, there is no doubt that viral oncolysis has been given a second life. We have to await the results as to whether this treatment shrinks tumors and prolongs life. I am optimistic that virotherapy will become a regular part of cancer treatment.

References

1. Dock, G. (1904). The influence of complicating diseases upon leukemia. *American Journal of the Medical Sciences, 127*, 563–592.
2. De Pace, N. (1912). Sulla scompara di un enome cancro vegetante del collo dell' utero senza cura chirugica. *Ginecologia, 9*, 82–89.
3. Southam, C. M., & Moore, A. E. (1952). Clinical studies of viruses as antineoplastic agents with particular reference to Egypt 101 virus. *Cancer, 5*(5), 1025–1034.
4. Moore, A. E. (1954). Effects of viruses on tumors. *Annual Review of Microbiology, 8*, 393–410.
5. Webb, H. E., & Smith, C. E. (1970). Viruses in the treatment of cancer. *Lancet, 1*(7658), 1206–1208.
6. Taylor, M. W., Cordell, B., Souhrada, M., & Prather, S. (1971). Viruses as an aid to cancer therapy: Regression of solid and ascites tumors in rodents after treatment with bovine enterovirus. *Proceedings of the National Academy of Sciences of the United States of America, 68*(4), 836–840.
7. Sedmak, G. V., Taylor, M. W., Mealey, J, Jr, & Chen, T. T. (1972). Oncolytic effect of bovine enterovirus on mouse and human tumours. *Nature New Biology, 238*(79), 7–9.
8. Miest, T. S., & Cattaneo, R. (2014). New viruses for cancer therapy: Meeting clinical needs. *Nature Reviews Microbiology, 12*(1), 23–34.
9. Sheridan, C. (2013). Amgen announces oncolytic virus shrinks tumors. *Nature Biotechnology, 31*(6), 471–472.
10. Ganly, I., Kirn, D., Eckhardt, G., Rodriguez, G. I., Soutar, D. S., Otto, R., et al. (2000). A phase I study of Onyx-015, an E1B attenuated adenovirus, administered intratumorally to patients with recurrent head and neck cancer. *Clinical Cancer Research, 6*(3), 798–806.

11. Nemunaitis, J., Khuri, F., Ganly, I., Arseneau, J., Posner, M., Vokes, E., et al. (2001). Phase II trial of intratumoral administration of ONYX-015, a replication-selective adenovirus, in patients with refractory head and neck cancer. *Journal of Clinical Oncology, 19*(2), 289–298.
12. Wang, J., Snider, D. P., Hewlett, B. R., Lukacs, N. W., Gauldie, J., Liang, H., et al. (2000). Transgenic expression of granulocyte-macrophage colony-stimulating factor induces the differentiation and activation of a novel dendritic cell population in the lung. *Blood, 95*(7), 2337–2345.
13. Burke, J. M., Lamm, D. L., Meng, M. V., Nemunaitis, J. J., Stephenson, J. J., Arseneau, J. C., et al. (2012). A first in human phase 1 study of CG0070, a GM-CSF expressing oncolytic adenovirus, for the treatment of nonmuscle invasive bladder cancer. *Journal of Urology, 188*(6), 2391–2397.
14. Kanerva, A., Nokisalmi, P., Diaconu, I., Koski, A., Cerullo, V., Liikanen, I., et al. (2013). Antiviral and antitumor T-cell immunity in patients treated with GM-CSF-coding oncolytic adenovirus. *Clinical Cancer Research, 19*(10), 2734–2744.
15. Cerullo, V. K. A., Vaha-Koskela, M., & Hemminki, A. (2012). Oncolutic adenoviruses for Cancer Immunotherapy: Data from Mice. *Hamsters and Humans. Advance Cancer Research, 115*, 266–318.
16. MacLean, A. R., ul-Fareed, M., Robertson, L., Harland, J., & Brown, S. M. (1991). Herpes simplex virus type 1 deletion variants 1714 and 1716 pinpoint neurovirulence-related sequences in Glasgow strain 17 + between immediate early gene 1 and the 'a' sequence. *Journal General Virology, 72*, 631–639.
17. Robertson, L. M., MacLean, A. R , & Brown, S. M. (1992). Peripheral replication and latency reactivation kinetics of the non-neuneo?Rovirulent herpes simplex virus type 1 variant 1716. *Journal of General Virology, 73*(Pt 4), 967–970.
18. Jia, W. W., McDermott, M., Goldie, J., Cynader, M., Tan, J., & Tufaro, F. (1994). Selective destruction of gliomas in immunocompetent rats by thymidine kinase-defective herpes simplex virus type 1. *Journal of the National Cancer Institute, 86*(16), 1209–1215.
19. Kesari, S., Randazzo, B. P., Valyi-Nagy, T., Huang, Q. S., Brown, S. M., MacLean, A. R., et al. (1995). Therapy of experimental human brain tumors using a neuroattenuated herpes simplex virus mutant. *Laboratory Investigation, 73*(5), 636–648.
20. Randazzo, B. P., Kesari, S., Gesser, R. M., Alsop, D., Ford, J. C., Brown, S. M., et al. (1995). Treatment of experimental intracranial murine melanoma with a neuroattenuated herpes simplex virus 1 mutant. *Virology, 211*(1), 94–101.
21. Rampling, R., Cruickshank, G., Papanastassiou, V., Nicoll, J., Hadley, D., Brennan, D., et al. (2000). Toxicity evaluation of replication-competent herpes simplex virus (ICP 34.5 null mutant 1716) in patients with recurrent malignant glioma. *Gene Therapy, 7*(10), 859–866.
22. MacKie, R. M., Stewart, B., & Brown, S. M. (2001). Intralesional injection of herpes simplex virus 1716 in metastatic melanoma. *Lancet, 357*(9255), 525–526.
23. Hu, J. C., Coffin, R. S., Davis, C. J., Graham, N. J., Groves, N., Guest, P. J., et al. (2006). A phase I study of OncoVEXGM-CSF, a second-generation oncolytic herpes simplex virus expressing granulocyte macrophage colony-stimulating factor. *Clinical Cancer Research, 12*(22), 6737–6747.
24. Nakao, A., Takeda, S., Shimoyama, S., Kasuya, H., Kimata, H., Teshigahara, O., et al. (2007). Clinical experiment of mutant herpes simplex virus HF10 therapy for cancer. *Current Cancer Drug Targets, 7*(2), 169–174.
25. Zeh, H. J., & Bartlett, D. L. (2002). Development of a replication-selective, oncolytic poxvirus for the treatment of human cancers. *Cancer Gene Therapy, 9*(12), 1001–1012.
26. Park, B. H., Hwang, T., Liu, T. C., Sze, D. Y., Kim, J. S., Kwon, H. C., et al. (2008) Use of a targeted oncolytic poxvirus, JX-594, in patients with refractory primary or metastatic liver cancer: A phase I trial. *The Lancet Oncology, 9*(6), 533–542.
27. Zhang, Y. Q., Tsai, Y. C., Monie, A., Wu, T. C., & Hung, C. F. (2010). Enhancing the therapeutic effect against ovarian cancer through a combination of viral oncolysis and antigen-specific immunotherapy. *Molecular Therapy, 18*(4), 692–699.

28. Heo, J., Reid, T., Ruo, L., Breitbach, C. J., Rose, S., Bloomston, M., et al. (2013). Randomized dose-finding clinical trial of oncolytic immunotherapeutic vaccinia JX-594 in liver cancer. *Nature Medicine, 19*(3), 329–336.
29. Cello, J., Paul, A. V., & Wimmer, E. (2002). Chemical synthesis of poliovirus cDNA: generation of infectious virus in the absence of natural template. *Science, 297*(5583), 1016–1018.
30. Goetz, C., Dobrikova, E., Shveygert, M., Dobrikov, M., & Gromeier, M. (2011). Oncolytic poliovirus against malignant glioma. *Future Virology, 6*(9), 1045–1058.
31. Sinkovics, J. G., & Horvath, J. C. (2000). Newcastle disease virus (NDV): Brief history of its oncolytic strains. *Journal of Clinical Virology: The Official Publication of the Pan American Society for Clinical Virology, 16*(1), 1–15.
32. Thirukkumaran, C., & Morris, D. G. (2009). Oncolytic viral therapy using reovirus. *Methods Mol Biol, 542*, 607–634.
33. Thirukkumaran, C. M., Shi, Z. Q., Luider, J., Kopciuk, K., Gao, H., Bahlis, N., et al. (2013). Reovirus modulates autophagy during oncolysis of multiple myeloma. *Autophagy, 9*(3), 413–414.
34. Jennings, V. A., Ilett, E. J., Scott, K. J., West, E. J., Vile, R., Pandha, H., et al. (2014). Lymphokine-activated killer and dendritic cell carriage enhances oncolytic reovirus therapy for ovarian cancer by overcoming antibody neutralization in ascites. *International Journal of Cancer, 134*(5), 1091–1101.
35. Chen, Y. Q., de Foresta, F., Hertoghs, J., Avalosse, B. L., Cornelis, J. J., & Rommelaere, J. (1986). Selective killing of simian virus 40-transformed human fibroblasts by parvovirus H-1. *Cancer Research, 46*(7), 3574–3579.
36. Shi, Z. Y., Ma, C. W., Huang, J., Lin, W. M., Dong, R. C., & Luo, Z. Y. (1997). Inhibition of parvovirus H-1 on transplantable human hepatoma and its histological and histobiochemical studies. *Shi Yan Sheng Wu Xue Bao, 30*(3), 247–259.
37. Shadan, F. F., & Villarreal, L. P. (2000). Parvovirus-mediated antineoplastic activity exploits genome instability. *Medical Hypotheses, 55*(1), 1–4.
38. Perez, O. D., Logg, C. R., Hiraoka, K., Diago, O., Burnett, R., Inagaki, A., et al. (2012). Design and selection of Toca 511 for clinical use: modified retroviral replicating vector with improved stability and gene expression. *Molecular Therapy, 20*(9), 1689–1698.
39. Huang, T. T., Hlavaty, J., Ostertag, D., Espinoza, F. L., Martin, B., Petznek, H., et al. (2013). Toca 511 gene transfer and 5-fluorocytosine in combination with temozolomide demonstrates synergistic therapeutic efficacy in a temozolomide-sensitive glioblastoma model. *Cancer Gene Therapy, 20*(10), 544–551.

Gene Therapy and Viral Vectors 18

Abstract

Gene therapy is the administration of a normal gene to replace a defective gene and thus cure or alleviate suffering from genetic diseases. Researchers are currently looking at a number of genetic diseases as targets, such as severe combined immunodeficiency (SCID), cystic fibrosis, and hemophilia, among others. In most cases, the genes are incorporated into a viral vector. These viruses may have a segment of genetic material removed and replaced with a "normal" gene, known as a "transgene." The first clinical trial, which was a failure, was in 1980, when a small number of patients with thalassemia were treated with the human globin DNA. In 1995 the first successful trial was performed in children with a defect in adenosine deaminase (ADA) (SCID-syndrome). Treatment of SICD-X1 children, a condition leading to immuno-deficiency, was also successful, but the retrovirus vector underwent integration into the host chromosome, resulting in a few cases of leukemia and one death. A large number of different viral vectors have been designed for use in gene therapy, including retrovirus, lentivirus, adenovirus and parvovirus. These are each discussed separately. Despite great efforts and a large number of clinical trials, only one vector system—AAV, Glybera—has been approved for gene therapy for the treatment of patients with lipoprotein lipase deficiency, a rare genetic disease.

18.1 An Introduction to Gene Therapy

Inherited diseases account for a substantial number of diseases, including up to 70 % of all admissions to children's hospitals and 10 % of admissions to adult hospitals. Furthermore, many of the most common inherited disorders, including thalassemia, cystic fibrosis, and sickle-cell anemia, and some of the rarer ones,

M. W. Taylor, *Viruses and Man: A History of Interactions*,
DOI: 10.1007/978-3-319-07758-1_18,

such as adenosine deaminase (ADA) deficiency (severe combined immune deficiency) still lack treatments beyond palliative care although there has been limited success in treating these conditions with gene therapy. I began writing this chapter with a great deal of hope that gene therapy using viral vectors had made large strides since the first clinical trials in the 1980s, and was close to clinical realization. By the end of the review of the literature,—and it is a large literature—I concluded that this was still a challenge, with a few successes and many failures along the way. Only now, after 40 years of experimental trials, are there some positive results, but just one vector-gene therapy system has been approved for usage in the U.S. and western Europe. A few viral vectors have been approved in China, but for viral oncolysis, which is different from gene therapy.

Based on studies of bacterial transformation where one gene can be replaced by another, it seemed obvious that once one could excise genes with restriction enzymes, one should be able to incorporate a normal gene into a virus and use that virus as a vehicle to infect a cell or body to replace a defective gene. The replacement could be done ex vivo, that is, in cell culture, and then such cells infused back into the patient (Fig. 18.1), or done by direct inoculation into the blood system or into a targeted organ.

During the last 50 years, many attempts have been made to realize these goals. Experiments performed in cell culture have shown that it is possible to express the transgene, as the introduced gene is called. Experiments performed in mice indicate that a foreign gene can be expressed from a viral vector, but often its activity

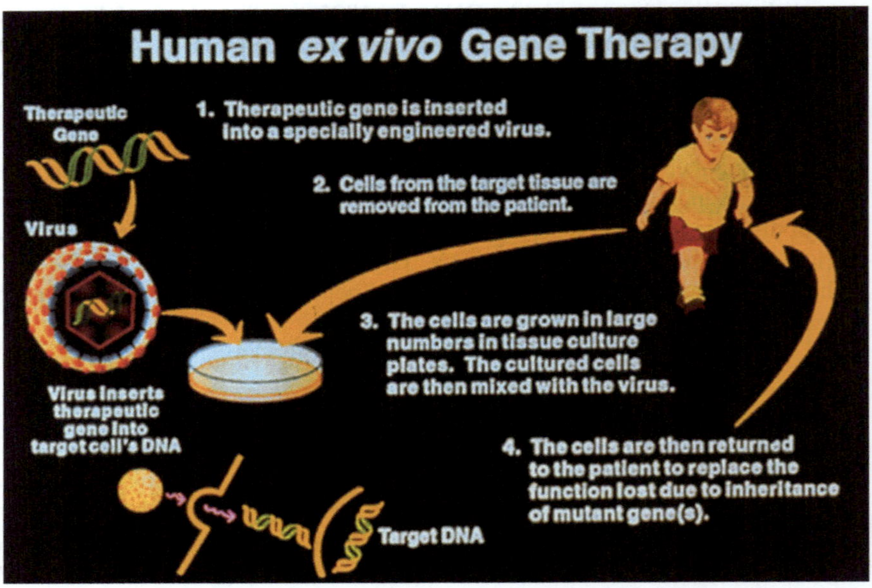

Fig. 18.1 Ex vivo gene therapy (from the NIH .http://history.nih.gov/exhibits/genetics/Sect.4.htm)

is extinguished by the wrong promoter or controlling element, or by the production of antibodies or other components of the immune response. In other cases, the expression of the transgene has been short lived. In human experimentation, safety has to be taken into account. There is danger of an inflammatory reaction to the virus, even if defective, and an immune response, including the production of antibodies, which would prevent reintroducing the same virus vector more than once if necessary, for continuous treatment.

Along the way, there have been disappointments and failures as well as tragic accidents. To date, there have been 1,700 clinical trials in gene therapy, but most of them have used viruses to treat cancer (viral oncolysis) and very few are gene replacement therapy trials or supplying a missing function.

In the gene therapy field, as in other endeavors of science, much has depended on the roles of key individuals and their personalities. The main players in this field were French Anderson, Ted Friedman, Inder Verma and Jim Wilson. French Anderson was a leader in the field, and announced to the world the first "cure" in a child with severe combined immune-deficiency (although there might have been a little hype in the presentation and announcement).

The idea of gene therapy was really born in the 1970s as a result of the advances made in molecular biology and the development of recombinant DNA technology. It was possible to "cut out" a gene and surrounding controlling elements with restriction enzymes (enzymes that spliced out segments of DNA depending on the sequence), joining such segments with other segments, and insert such sequences into a virus without altering its ability to be engulfed by a cell. Of course, although this sounds simple, problems and questions arose even in initial experiments in culture. These included: What controls the expression of the gene inserted into the virus? Is expression enough to provide the missing or defective function? Does the virus need to replicate in the cell or is the expression from a single viral infection enough? When it came to performing gene therapy in an animal model (usually mouse), the major problems multiplied. How does one target the gene to the site of the genetic defect, e.g., liver, lung, neurons, etc.? Can one get the right amount of expression, i.e., not too little and not too much? Do we need a replication-competent virus? Does the patient have antibodies to this virus that would inhibit expression? These are among the multitudinous questions that arose, and many have been answered as a result of actual trials.

There are also ethical issues to be considered. Gene therapy as described in this chapter, and as performed in clinical trials to date, is using the gene (or the cognate protein) as a therapeutic or pharmaceutical agent. This is "somatic" gene therapy and implies replacing a defective gene with a normal copy of the gene, thus providing the normal protein product of that gene. It does not involve embryonic stem cell modification, which would result in subsequent generations bearing the "transduced" gene. One could argue that somatic gene therapy is harmful to mankind, in that if the "cured" individual has children, they will carry the defective gene. To date, there has not been an attempt to alter embryonic stem cells that would give rise to altered progeny—which is a much-debated subject. Not so long ago it was thought that this was not possible, but with the cloning of animals such as sheep and cows, it is conceivable that one would learn how to change the genetic makeup of man and thus subsequent generations. However, who would decide whether such changes should be made? How might it affect other genetic characters of the individual? Modification of a single gene may have many pleiotropic effects.

Historically, there have been two methods of performing gene therapy: one ex vivo and the other in vivo. In ex vivo gene therapy, lymphocytes or other cell types are collected from the patient and cultured for a period to select the desired cell type. These cells are transformed in culture, with a virus carrying the desirable gene, grown for a period of time in culture, and then returned into the patient. This may be accompanied either with or without full bone marrow transplant and radiation treatment. This method has been used successfully in treating children with a number of different immune-deficient diseases (see Fig. 18.1).

A second method is to directly inject into the body (intramuscular or intravenous) a virus carrying the selected gene, or just DNA. A major problem is targeting the DNA or virus to the desired organ, and in preliminary trials gene expression has been weak and only transitory. Many methods have been developed to enhance the efficiency of uptake and to prevent the destruction of the DNA.

18.2 Early Days of Gene Therapy

As discussed in Chap. 3 (History of Cell Culture), the key to early gene therapy experiments was the availability of mutant somatic cells in culture. These were of two kinds: cells isolated from humans or mice carrying a genetic defect, or induced mutations resembling a known genetic defect. A number of model systems were established, most with defects in purine metabolism. Purines are essential for cell growth and animal development. The purines, adenine and guanine, are the building blocks of nucleic acids, and also precursors of key molecules such as ATP and GTP, the energy source in the cells. Many mutants were isolated in the purine salvage pathway, a metabolic pathway that utilizes free purines derived from the breakdown of nucleic acids, and differs from the de novo pathway, where purines are synthesized from simpler precursors such as amino acids and bicarbonate through multiple steps. Three genetic defects were used as a model system: hypoxanthine-guanine phophoribosyl transferase (HGPRT), adenine phosphoribosyl transferase (APRT) and adenosine deaminase (ADA).

HGPRT catalyzes the formation of guanosine monophosphate (GMP), an essential building block in nucleic acid formation. A defect in this gene in humans results in the Lesch-Nyhan syndrome, which may result in severe mental retardation and self-mutilation. A partial defect of this enzyme can cause high levels of uric acid, resulting in gout. This defect only occurs in males since it is X-linked (only one X chromosome is present in males). Cell lines defective in this gene can be isolated following treatment with mutagens and selection in thioguanine. Such mutants are easy to isolate and occur at a high frequency because of one copy of the active gene.

APRT is an enzyme that salvages free adenine to form AMP. APRT deficiency is a rare defect in man, but a lack of this enzyme may lead to kidney stones. Cells defective in APRT activity are selected in di-aminopurine, which is toxic to cells. Such mutants are relatively rare, both in culture and in man, since they are autosomal-recessive. In cells in culture, there are probably two or more copies of the APRT gene. Many cell lines defective in APRT and in HGPRT were isolated and used later for in vitro gene therapy experiments [1, 2].

ADA deficiency results in a severe defect of the immune system in man. A mutation in this gene results in a severe combined immunodeficiency (SCID) phenotype, a complete lacking of humeral immunity. Babies born with this defect were at one time kept in plastic bubbles and known as "bubble babies." This is an autosomal recessive mutation that leads to a defect in the functioning of T-cells, B-cells and NK cells. It became one of the first targets for gene therapy in humans [3].

Thus the availability of mutant cell lines allowed one to test whether a recombinant virus containing a transgene would repair defects in the cell [4]. Although the idea of using viruses to transfer genetic material developed from the research with transducing phage in the early days of molecular biology, it was difficult to envisage how this could be performed with animal viruses. It was known that both DNA and RNA tumor viruses (retroviruses) integrated into the host cell after infection and brought about transformation in culture ("transformation" implies extended growth, and phenotypic alteration of the cell). Until the finding of restriction enzymes and their use to splice out segments of DNA, it was difficult to conceptualize how foreign genes could be incorporated into a virus. Gene therapy experiments were performed using chemical means of introducing DNA into cells and even into mice and later humans, but gene expression was very low.

Martin Cline performed the first human gene therapy experiments in 1980, on patients in Italy and in Israel. He harvested bone marrow cells from individuals with thalassemia (an inherited blood disorder in which hemoglobin synthesis is affected), transfected these cells in vitro with plasmid containing the human globin gene, and transplanted the cells back into the patients after radiation. The Human Subjects Committee at UCLA had not approved the experiment, although it was approved by the hospitals in Naples and Jerusalem. Many scientists thought the experiments premature, and without sufficient background experiments in mice, although as pointed out by Cline, mice do not get thalassemia. The result was that the experiments were condemned as being too hasty, and Cline lost his position at UCLA. The experiments failed, but nonetheless these failed experiments gave a boost to the field of gene therapy [5, 6].

In 1983 a group of scientists gathered at Cold Spring Harbor Laboratories to discuss the future of gene therapy. The leaders in the field, French Anderson, Paul Berg, and Ted Friedmann, organized a meeting that was cautiously upbeat. This was a period when retrovirus vectors were being constructed, and appeared ideal for gene therapy. Genes were cloned into a retrovirus, making it defective, and helper genes or a helper virus could be grown in the same cells as the defective virus, resulting in an infectious defective virus (Fig. 18.3). Retroviruses were

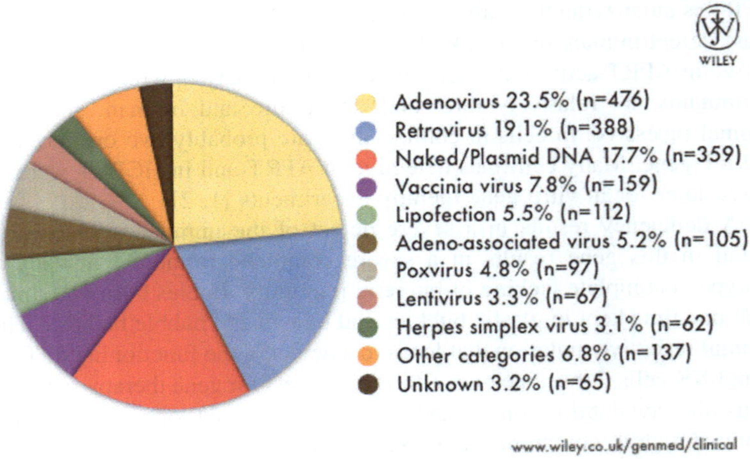

Fig. 18.2 Vectors used in clinical trials for gene therapy up until 2013 (with permission of John Wiley and Sons Ltd)

shown to infect many different cell types and give rise to stable transfected cells. The virus mostly used was mouse leukemia virus, assuming that defective mouse retroviruses would be harmless to humans. Retroviruses carrying the HPRT gene could overcome the defect in HPRT-deficient cells, and retroviruses with ADA could also restore ADA activity in human lymphocytes.

Within a few years it was shown that viruses carrying exogenous genes could transform many different cell types. Other vectors were also created, including adenoviruses and adeno-associated viruses (AAV), herpesviruses and vaccinia vectors … In fact, during the 1980s, ex vivo gene transfer experiments were not only performed for ADA and HGPRT deficiencies but also for coagulation deficiencies for factors XIII and IX, for hypocholeterolemia and many other genetic conditions. My laboratory developed adenovirus vectors and AAV vectors carrying the APRT gene and the interferon gene [7, 8]. The vectors and methods eventually used in gene therapy clinical trials are presented in Fig. 18.2.

18.3 Gene Therapy of ADA Deficiency

Figure 18.3 presents schematically the retrovirus system used to construct gene therapy vectors. The aim was to create a defective non-replicating virus that can be grown in a packaging cell line, but cannot replicate in vivo. Cells containing this defective virus can be selected by including an antibiotic-resistant marker, which would kill non-transformed cells.

ADA deficiency results in the accumulation of a substrate of the enzyme ADA, deoxyadenosine, which is converted into the toxic compound deoxyadenosine triphosphate in T-cells, thus disabling the immune system. Children with ADA

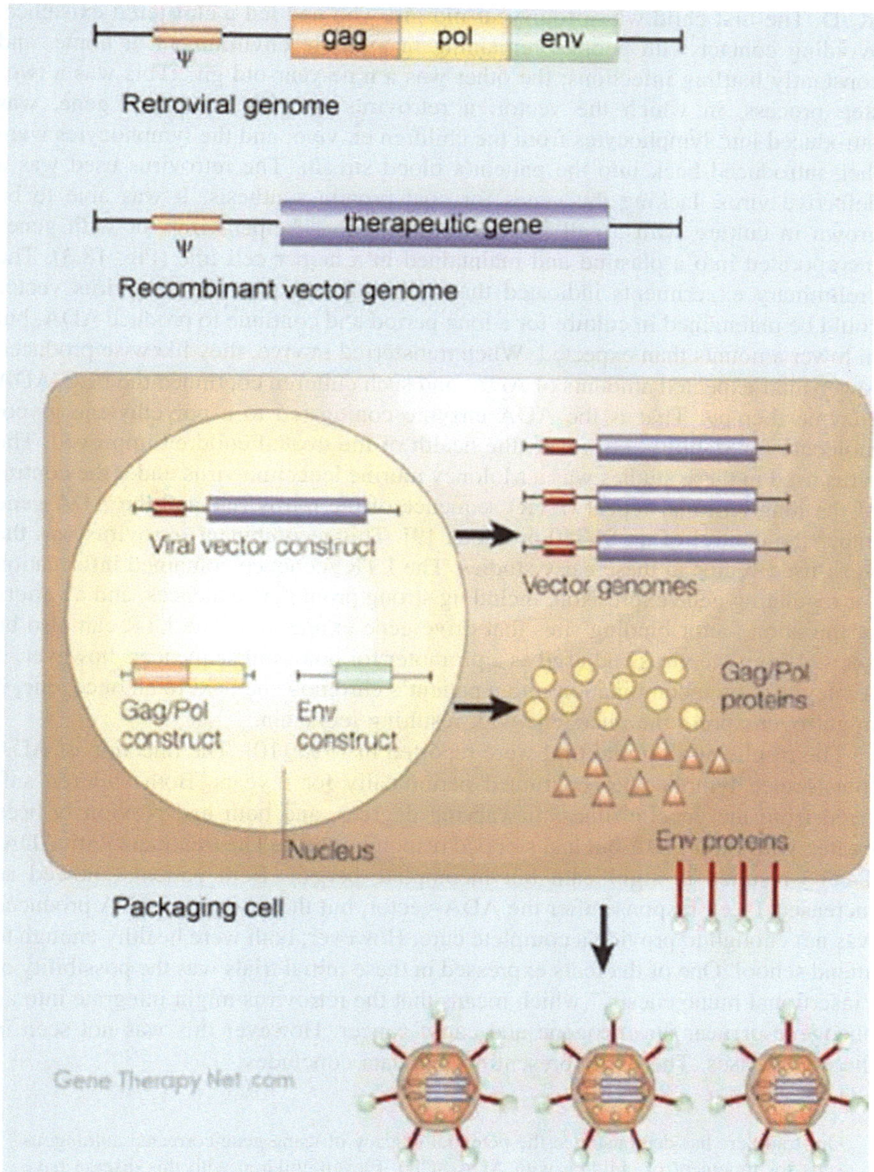

Fig. 18.3 Retrovirus construct in which therapeutic gene replaces components of the retrovirus With permission from Gene therapy.net (http://www.genetherapy.net.com)

deficiency suffer from opportunistic infections. The first approved clinical trial was launched in September 1989. R. Michael Blaese and French Anderson and colleagues at the NIH introduced the gene for ADA into two children suffering from

SCID. The first child was a four-year-old girl who had led a cloistered existence, avoiding contact with people, remaining in a sterile environment at home, and constantly battling infections; the other was a nine-year-old girl. This was a two-step process, in which the vector, a retrovirus carrying the ADA gene, was introduced into lymphocytes from the children ex vivo, and the lymphocytes were then introduced back into the patient's blood stream. The retrovirus used was a defective virus, lacking the genes for coat protein synthesis. It was able to be grown in culture with a cell infected either with a helper virus, or with genes incorporated into a plasmid and maintained in a helper cell line (Fig. 18.3). The preliminary experiments indicated that cells infected with the retrovirus vector could be maintained in culture for a long period and continue to produce ADA, but in lower amounts than expected. When transferred in vivo, they likewise produced lower-than-expected amounts of ADA, and such children continued the PEG-ADA enzyme therapy. That is the ADA enzyme conjugated to a polyethylene glycol molecule for stability. However, the health of the treated children improved. The virus used in these studies was a Moloney murine leukemia virus under the control of the long terminal repeat (LTR) sequence of the retrovirus, and the ADA gene under the control of an SV-40 promoter [9]. This recombinant retrovirus was the most-used vector in these early studies. The LTR sequence contained information for regulating gene expression, including strong promoter sequences. and a variety of initiation factor binding sites that drive gene expression. The LTR can also be excised from the virus and used as a promoter for non-viral sequences; however, if it integrates by recombination into a patient's chromosome next to an oncogene, it might overexpress the oncogene with resulting leukemia.

The results of an initial trial were reported in 1995 [10]. The injection of ADA transformed lymphocytes continued periodically for 2 years. Both children suffered from the genetic illness in varying degrees, and both had previously been treated with PEG-ADA but had stopped responding to it. The treatment with ADA-T-cells resulted in significant but incomplete benefit. Both patients showed an increased T-cell response after the ADA-vector, but the amount of ADA produced was not enough to provide a complete cure. However, both were healthy enough to attend school. One of the fears expressed in these initial trials was the possibility of "insertional mutagenesis," which means that the retrovirus might integrate into an oncogene or near an oncogene and cause cancer. However this was not seen in these two cases. The paper presenting this data concludes,

> Our trial here has demonstrated the potential efficacy of using gene-corrected autologous cells for treatment of children with ADA-SCID. Eleven children with this disease have been enrolled in various gene therapy protocols, each using different strategies and retroviral vector designs and focusing on different target cell populations. The experience gained from these approaches should provide guidance for gene therapy as a treatment for this disorder as well as for a larger array of inherited and acquired diseases.

A major breakthrough occurred in France when 11 children with immunodeficiency–X1 (SCID-X1), a defect in the gene for the interleukin receptor γc gene, received a retrovirus, Moloney murine leukemia virus, containing an exogenous γc cytokine receptor gene. SCID X1 is characterized by faulty T-cell and NK-cell differentiation. It is found in "bubble" boys, is X-linked, and most children afflicted with it die at about age one. After a 10-month follow-up period, γc transgene-expressing T- and NK-cells were detected in two patients. Cell counts and function, including antigen-specific responses, were comparable to those of age-matched controls [11]. Similar experiments were performed in England using a different retrovirus system—gibbon ape leukemia virus pseudotype retrovirus. This vector uses the coat protein of the gibbon ape leukemia virus, which has an extended host range and also enters non-dividing cells.

Although most infants showed dramatic improvement following gene therapy, four of the nine infants that were successfully treated in Paris developed leukemia between 3 and 6 years after the treatment, and one of the infants successfully treated in London also developed a form of leukemia, T-cell acute lymphoblastic leukemia (T-ALL). All of these cases are the result of the retrovirus gene integrating into a specific chromosomal site. In both studies the transgene was integrated into the LMO2 chromosomal gene, a known oncogene, and in one other case into the cyclin D2 gene, essential in cell cycle regulation. Multiple mutations were also found in the children with leukemia; except for one, all responded to chemotherapy.

18.4 Lentivirus as Vectors

In recent years, research has focused on the use of lentivirus vectors, which, like retrovirus vectors, are devoid of viral proteins, free from replication-competent virus, and unlike retrovirus vectors, are able to transduce non-dividing cells [12]. This characteristic is advantageous in many gene therapeutic applications targeting highly differentiated cells. They can carry up to 8 kb of exogenous material and are normally not inactivated by antibodies in the body. Currently, these lentivectors are applied in about 3.3 % of clinical trials (see Fig. 18.2).

Lentiviruses include HIV and Visna virus. They are called lentivirus because it takes a long time for symptoms to manifest themselves in an infection (lenti = slow in Latin). The favorite virus for gene therapy is, surprisingly, HIV. To obtain a lentiviral gene therapy vector, a reporter gene or therapeutic gene is cloned into a vector sequence that is flanked by LTRs and the Psi-sequence of HIV (see above for function of LTRs: Psi sequence is needed for packaging). The LTRs are necessary to integrate the therapeutic gene into the genome of the target cell, just as the LTRs in HIV are necessary to integrate the dsDNA copy of the virus into its host chromosome. The Psi-sequence acts as a signal sequence and is necessary for packaging RNA with the transgene in virions. Viral proteins, which make virus shells, are provided in the packaging cell line, but are not in the genome with the LTRs and Psi-sequences and so are not packaged into virions. Thus, virus particles are produced that are replication-deficient, designed so that

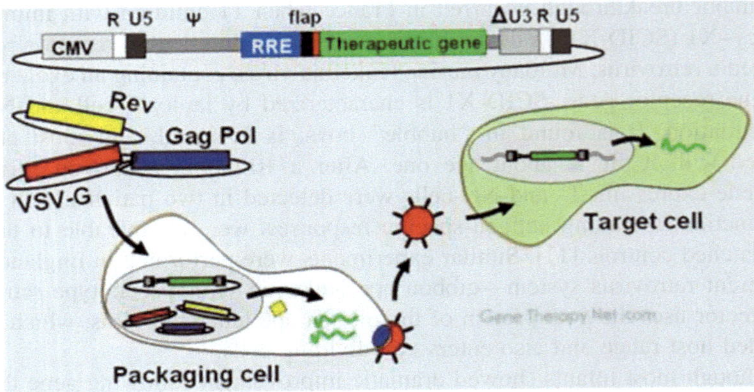

Fig. 18.4 Lentiviruses vector. With permission from gene therapy.net (http://www.genetherapy.net.com)

they are unable to infect their hosts after they deliver their therapeutic content. Newer lentivirus vectors contain self-inactivating LTR sequences that prevent the integration or recombination with the host chromosome. These vectors contain a deletion in the downstream LTR that when transduced into target cells results in the transcriptional inactivation of the upstream LTR and substantially diminishes the risk of vector recombination with the host chromosome and other viruses. A 400-base pair region includes the "tat" region (trans-activator of transcription, an important region in HIV expression and regulation), which is deleted from the LTR (Fig. 18.4). Hybrid lentiviral vectors have also been derived from non-human lentiviruses (for example, simian, equine, feline, caprine and bovine), similar to those used for HIV-derived vectors, using the rationale that they would be more acceptable for clinical application because the parental viruses are not infectious to humans. Lentiviruses are currently being tested in a large number of phase I and phase II trials [13].

Lentiviruses have been used in many experimental systems. For example, a model of mouse hemophilia is corrected by expressing wild-type platelet factor VIII, the gene mutated in human hemophilia. Diabetic mice have been treated with the gene encoding platelet-derived growth factor (PDGF), a therapy being considered for use in humans. The host range of the lentivirus vector is expanded by using the VSV–G protein. This has been used in experimental animals by direct injection of the vector with the transgene, rather than ex vivo use.

18.5 Adenovirus Vectors

Adenovirus was used and tested as a vector for gene therapy in the 1980s. There are a large number of species of human adenoviruses, very few of which appear to be very pathogenic to man. The subtype selected for gene therapy was adenovirus-5,

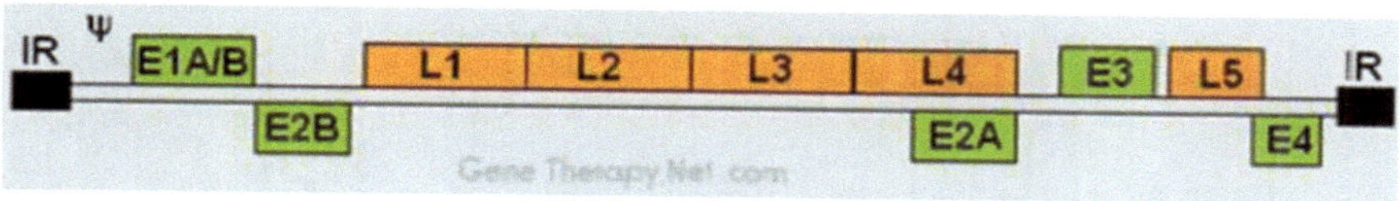

Fig. 18.5 Order of the adenovirus genes: removal of E3 still produces a replication-competent virus; deletion of E1 or E2 produces a defective virus. With permission from gene therapy.net (http://www.genetherapy.net.com)

which causes symptoms in man similar to the common cold. Adenovirus-5 can cause tumors in hamsters, and does not replicate in mice. As discussed in the chapter on viral oncolysis, viral vectors are produced by removing specific adenovirus genes, in particular the E3 gene (involved in immune response) and the E1a-E1b genes. Removal of the E3 gene results in a replicative competent virus, while removal of E1 gives rise to a defective non-replicating virus. Since the adenovirus will not replicate without the E1 gene, packaging cell lines similar to those described above for retrovirus have been constructed. The cell line most used is 293, and contains an integrated E1 sequence (Figs. 18.5 and 18.6). A recombinant adenovirus carrying the rat ornithine transcarbamylase deficiency (OTC) gene under the control of the adenovirus late promoter was constructed and injected into newborn mice. An increase in OTC activity was observed with a decrease in orotic acid in the urine. The transferred gene was active for more than a year following the injection, and this was the first case where it could be shown that a recombinant adenovirus bearing an exogenous gene could restore impaired metabolism [14]. This gene was later used in the human experiments discussed below.

A major paper in the developing field of gene therapy was published in *Science* in 1991 [15], indicating that it was possible to deliver an adenovirus with the human α-1-anti-trypsin gene to the lung of cotton rats. A replication-deficient adenoviral vector (Ad-αl-AT) containing an adenovirus major late promoter and a recombinant human α-1 anti-trypsin gene was used to infect epithelial cells of the rat respiratory tract both in vitro and in vivo. After in vivo intra-tracheal administration of Ad-αrlAT to the rats, human α-l-anti-trypsin messenger RNA was observed in the respiratory epithelium, the human protein was synthesized and secreted by lung tissue, and the human gene product was detected in the epithelial lining fluid for at least 1 week [15].

The main defect in cystic fibrosis is chloride transport. The CF gene—its product being the cystic fibrosis transmembrane conductance regulator (CFTR), and its major mutation, Delta F508, was discovered in 1989. The research presented above in rats stimulated clinical trials with the normal cystic fibrosis transmembrane conductance regulator gene in humans. Initial experiments were successfully performed in vitro on rat lung epithelial cells [16]. Jim Wilson's group in Philadelphia performed similar studies with non-human primates with an E1 deleted, and thus defective adenovirus in baboons. A total of 14 animals were used, receiving escalating doses. Individual animals were killed and lung tissue examined after 4 and 21 days. The expression of the transgene was proportional to

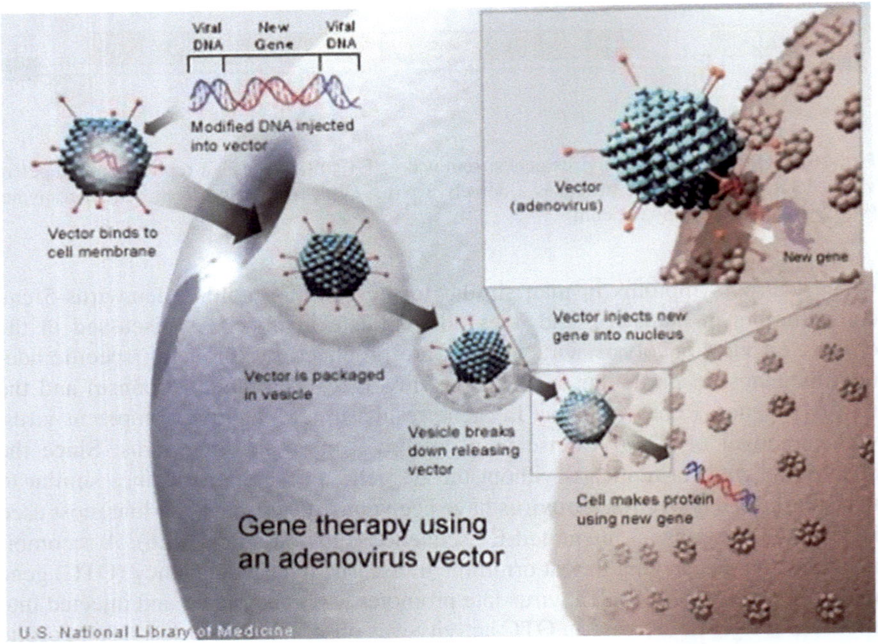

Fig. 18.6 The adenovirus vector system (gene therapy net; U.S. National Library of Medicine)

the dose of virus, and diminished with time. These studies indicated that for gene transfer to work in cystic fibrosis patients, there was a need for multiple doses [17]. A small clinical trial of three patients with CF was performed by nasal administration of the E1-deleted adenovirus carrying the CFTR gene. No adverse effects were found, and there was restoration of CFTR activity [18]. A similar trial was performed in four other patients, and similar results were obtained. After 6 months there did not seem to be adverse effects. Doses as high as 2×10^9 virus were used, but expression of the transgene was only transient, and repeat injections (or nasal administration) of the virus did not enhance expression of the CFTR gene due to antibody production against the virus. One possibility considered was to administer various adenovirus serotypes with the gene.

However, numerous trials using various vectors failed to cure cystic fibrosis. In part, this was due to a lack of knowledge of the complexity of the lung as well as lack of knowledge of the biology of adenovirus. In an article in 2012, Ron Crystal writes,

> We found that it is possible to use viral vectors to transfer the CFTR gene to the airway epithelium and "cure" CF in a limited region of the nasal or airway epithelium, but that the innate and acquired immune defenses of the lung eliminate the foreign gene within a short time, limiting the period of effectiveness of the therapy. Subsequent efforts in the 1990s to make vectors "stealth" failed to circumvent the lung's host defenses. … But is gene therapy for CF feasible in the near future? After all, the gene therapy field has had

> recent remarkable successes in therapy for blindness, immunodeficiency, and hemophilia. As much as I would like to see CF gene therapy added to this list, with the advantage of more than 20 years of hindsight, I think successful gene therapy for CF is still a long way off. The physical and host defense barriers of the lung are simply too great. While I applaud the efforts and admire the creative new approaches, it is doubtful there will be successful clinical trials for CF in the next decade. This does not mean that the McCray and other laboratories toiling away at CF gene therapy should give up, because their work is critical for continued progress toward therapy for CF, but they and the CF community should not expect a gene therapy 'cure' in the near or midterm future [19].

The field of gene therapy was shocked and set back in 1999 by the death of a young man undergoing gene therapy for OTC. Until then, adenovirus had been regarded as a harmless virus. Jesse Gelsinger was an 18-year-old who had a "mild' form of OTC deficiency; his was not the typical inherited form, but occurred after conception since some of his cells did metabolize normally. On September 13, 1999, he was injected with an adenovirus vector carrying the gene to test the safety of the procedure. He died 4 days afterward of multiple organ failure, and it appeared that there was a massive immune response to the virus. The FDA investigated the case and censured the scientists involved: Jesse was not a suitable subject for the trial because of high ammonia levels; the scientists had not reported adverse effects in two other patients; and they had not reported in the consent forms that monkeys had died following the same treatment. The scientist involved, Jim Wilson, refuted the above accusations, indicating that the trials with the monkeys were for a different disease, different vector and different gene. However, Jesse's death brought about a cessation in gene therapy trials for quite some time. The center for gene therapy at the University of Pennsylvania closed in 2002, and Jim Wilson resigned as director. There was talk of financial improprieties, since both Wilson and the university were involved with a company that had patents on the vectors and was supporting the research. This company was later sold for large sums of money. (A full disclosure and details of the Gelsinger case can be found at http://cirge.stanford.edu/May%209%20-%20Gelsinger%20vs.%20UPenn/Gelsinger%20-%20Oxford%20Textbook.pdf.)

Adenovirus vectors continued to be developed. A debate ensued as to whether a vector should be defective, as used up until now with limited results, or non-defective. Perusing the literature of the last 20 years, it seems that adenovirus use has changed from being a tool of gene therapy, in the sense of repairing a genetic defect, to one of viral oncolysis of cancer. There are a large number of clinical trials utilizing adenovirus for this function (see Chap. 17). It would appear that our knowledge of the virus and of human physiology is still lacking and that much further research needs to be done.

Fig. 18.7 Genome of AAV. Transgene replaces the rep and cap genes. With permission from gene therapy.net (http://www.genetherapy.net.com)

18.6 Adeno-Associated Virus

AAV belongs to the family of parvoviruses. This class of viruses is among the smallest of animal viruses and has very limited genetic material, about 5,000 base pairs. This small DNA virus has been studied extensively as a possible gene therapy vector. As the name implies, it is dependent on adenovirus for growth and is thus itself a defective virus. It has a broad host range and can be grown as an integrated virus or lytic virus. There is a large number of serotypes, all non-pathogenic to humans. High-copy numbers of the virus can be obtained by superinfecting the cells with adenovirus [20]; however, its major limitation is the small packaging capacity. The viral genome consists of only two genes (Fig. 18.7), each producing many polypeptides. "Rep" is required for viral genome replication and "cap" encoding structural proteins. These two genes are rounded by viral repeat sequences that are 145 nucleotides in length. Vectors are created by incorporating the desired therapeutic gene between the two ITR sequences, having the rep and cap gene on a separate plasmid, and adding adenovirus as a third component (Fig. 18.8).

AAV has a propensity to integrate into human chromosome 19, but this is lost when the rep gene is removed. Thus, vectors are not incorporated into the chromosome. Since the virus used in transduction does not have any viral sequences (other than the ITR), it is non-toxic and non-pathogenic. The vector can be delivered in vivo to multiple tissues and is under test for many different disease situations. The virus used in most of the initial studies was AAV2. Important advances have been made in altering the target range of AAV, by single alterations of amino acid in the coat protein [21].

Several phase I clinical trials for gene therapy of inherited defects using first-generation AAV2 vectors have been completed or are currently in progress. For instance, AAV2 vectors have been evaluated for gene transfer within the liver in the treatment of hemophilia B; in the lung for treatment of cystic fibrosis; in the brain for treatment of Parkinson's disease; within the joints for patients with rheumatoid arthritis; and in the eye for treatment of Leber's congenital amaurosis and age-related macular degeneration. Leber's congenital amaurosis is an eye disorder that primarily affects the retina, which is the specialized tissue at the back of the eye that detects light and color. People with this disorder typically have severe visual impairment beginning in infancy. The visual impairment tends to be stable, although it may worsen very slowly over time and is the most common cause of blindness in children. Perhaps the most striking example of successful

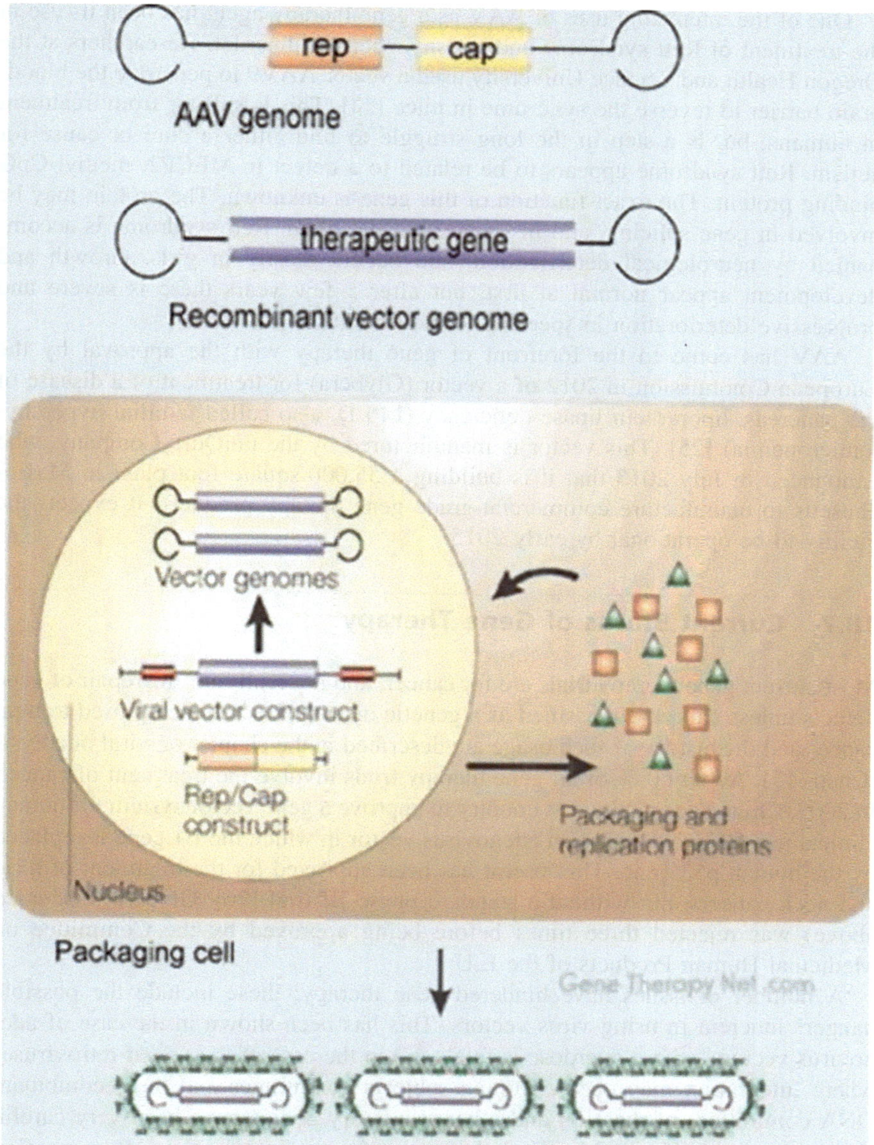

Fig. 18.8 The AAV system. With permission from gene therapy.net (http://www.genetherapy.net.com)

gene therapy in the clinic is the phase I trial of Leber's congenital amaurosis. Persistent improvement in the vision of affected patients has been reported in over 40 patients treated with AAV2 vectors delivering a corrective version of the RPE65 gene [22, 23].

One of the interesting uses of AAV as a gene therapy agent has been its use in the treatment of Rett syndrome and autism spectrum disorder. Researchers at the Oregon Health and Science University used a vector AAV9 to penetrate the blood-brain barrier to reverse the syndrome in mice [24]. This is still far from treatment in humans, but is a step in the long struggle to find either a cure or cause for autism. Rett syndrome appears to be related to a defect in MECP2, methyl CpG binding protein. The exact function of this gene is unknown. The protein may be involved in gene splicing, and in synapsis in the brain. Rett syndrome is accompanied by neurological deterioration, and occurs mostly in girls. Growth and development appear normal at first, but after a few years there is severe and progressive deterioration in speech and movement.

AAV has come to the forefront of gene therapy with the approval by the European Commission in 2012 of a vector (Glybera) for treatment of a disease of the pancreas, lipoprotein lipase deficiency (LPLD, also called familial hyperchylomicronemia) [25]. This vector is manufactured by the uniQure Company, who announced in July 2013 that it is building a 55,000-square foot plant in Massachusetts to manufacture commercial-grade gene therapy products. It expects the facility to be operational by early 2015.

18.7 Current Status of Gene Therapy

Most current gene therapy trials are for cancer, and not really for the repair of gene defects unless cancer is classified as a genetic disease. Vectors being used to treat cancer and the history of such usage are described in the chapter on viral oncolysis (Chap. 17). About 60 % of all gene therapy trials involve the treatment of cancer. In 2003, China became the first country to approve a gene vector system for human clinical use. "Gendicine" is an adenovirus vector in which the E1 gene is replaced by the human p53 gene. This vector has been approved for the treatment of head and neck cancers, but without a standard phase III trial [26]. Glybera (discussed above) was rejected three times before being approved by the Committee on Medicinal Human Products of the E.U.

A number of issues have hindered gene therapy; these include the possible dangers inherent in using virus vectors. This has been shown in the case of adenovirus vectors, where overdose is fatal, and in the case of the use of retroviruses where integration may occur. Human subjects committees and the recombinant DNA committees of the NIH and other regulatory bodies have been very careful about allowing clinical trials. The other aspect is a technical one: gene expression is not optimum in most cases. One can argue that more time is needed to test promoters; perhaps one should use natural promoters rather than viral promoters for the transgene. We lack knowledge on gene regulation gene regulation, and a major problem is that antibodies and the immune system may limit the number of times one can deliver the transgene in a virus vector, which appears to be the case with adenovirus vectors. Although there eventually will be gene therapy for the medical treatment of specific diseases such as cystic fibrosis, hemophilia, etc., we may still have to wait a number of years. (A list of vectors discussed in this chapter (and others) with their characteristics can be found at http://www.genetherapynet.com/viral-vectors.html.)

References

1. Taylor, M. W., Tokito, M., & Gupta, K. C. (1979). Lack of enhanced purine biosynthesis in HGPRT- and Lesch-Nyhan cells. *Human Heredity, 29*(3), 187–192.
2. Taylor, M. W., Tokito, M. K., Gupta, K. C., & Pipkorn, J. (1978). Regulation of de novo purine biosynthesis in normal and 8-azaguanine-resistant Chinese hamster cells. *Biochimica et Biophysica Acta, 517*(1), 1–13.
3. Verma, I. M. (1990). Gene therapy. *Scientific American, 263*(5), 68–72, 81–64.
4. Friedmann, T. (1992). A brief history of gene therapy. *Nature Genetics, 2*(2), 93–98
5. Sun, M. (1981). Cline loses two NIH grants. *Science, 214*(4526), 1220.
6. Wade, N. (1981). Gene therapy caught in more entanglements. *Science, 212*(4490), 24–25.
7. Wang, Q., & Taylor, M. W. (1993). Correction of a deletion mutant by gene targeting with an adenovirus vector. *Molecular and Cellular Biology, 13*(2), 918–927.
8. Konan, V., Sahota, A., Graham, F. L., & Taylor, M. W. (1991). Transduction of the CHOaprt gene into mouse L cells using an adeno-5/APRT recombinant virus. *Somatic Cell and Molecular Genetics, 17*(4), 359–368.
9. Culver, K., Cornetta, K., Morgan, R., Morecki, S., Aebersold, P., Kasid, A., et al. (1991). Lymphocytes as cellular vehicles for gene therapy in mouse and man. *Proceedings of the National Academy of Sciences of the United States of America, 88*(8), 3155–3159.
10. Blaese, R. M., Culver, K. W., Miller, A. D., Carter, C. S., Fleisher, T., Clerici, M., et al. (1995). T lymphocyte-directed gene therapy for ADA-SCID: Initial trial results after 4 years. *Science, 270*(5235), 475–480.
11. Cavazzana-Calvo, M., Hacein-Bey, S., de Saint Basile, G., Gross, F., Yvon, E., Nusbaum, P., et al. (2000). Gene therapy of human severe combined immunodeficiency (SCID)-X1 disease. *Science, 288*(5466), 669–672.
12. Kay, M. A., Glorioso, J. C., & Naldini, L. (2001). Viral vectors for gene therapy: The art of turning infectious agents into vehicles of therapeutics. *Nature Medicine, 7*(1), 33–40.
13. Escors, D., & Breckpot, K. (2010). Lentiviral vectors in gene therapy: Their current status and future potential. *Archivum immunologiae et therapiae experimentalis, 58*(2), 107–119.
14. Stratford-Perricaudet, L. D., Levrero, M., Chasse, J. F., Perricaudet, M., & Briand, P. (1990). Evaluation of the transfer and expression in mice of an enzyme-encoding gene using a human adenovirus vector. *Human Gene Therapy, 1*(3), 241–256.
15. Rosenfeld, M. A., Siegfried, W., Yoshimura, K., Yoneyama, K., Fukayama, M., Stier, L. E., et al. (1991). Adenovirus-mediated transfer of a recombinant alpha 1-antitrypsin gene to the lung epithelium in vivo. *Science, 252*(5004), 431–434.
16. Rosenfeld, M. A., Yoshimura, K., Trapnell, B. C., Yoneyama, K., Rosenthal, E. R., Dalemans, W., et al. (1992). In vivo transfer of the human cystic fibrosis transmembrane conductance regulator gene to the airway epithelium. *Cell, 68*(1), 143–155.
17. Engelhardt, J. F., Simon, R. H., Yang, Y., Zepeda, M., Weber-Pendleton, S., Doranz, B., et al. (1993). Adenovirus-mediated transfer of the CFTR gene to lung of nonhuman primates: Biological efficacy study. *Human Gene Therapy, 4*(6), 759–769.
18. Zabner, J., Couture, L. A., Gregory, R. J., Graham, S. M., Smith, A. E., & Welsh, M. J. (1993). Adenovirus-mediated gene transfer transiently corrects the chloride transport defect in nasal epithelia of patients with cystic fibrosis. *Cell, 75*(2), 207–216.
19. Crystal, R. G. (2012). The challenge of using gene- or cell-based therapies to treat lung disease. *Molecular Therapy, 20*(6), 1077–1078.

20. Hermonat, P. L., & Muzyczka, N. (1984). Use of adeno-associated virus as a mammalian DNA cloning vector: Transduction of neomycin resistance into mammalian tissue culture cells. *Proceedings of the National Academy of Sciences USA, 81*(20), 6466–6470.
21. Asokan, A., Schaffer, D. V., & Samulski, R. J. (2012). The AAV vector toolkit: Poised at the clinical crossroads. *Molecular Therapy, 20*(4), 699–708.
22. Maclaren, R. E., Groppe, M., Barnard, A. R., Cottriall, C. L., Tolmachova, T., Seymour, L., Clark, K. R., During, M. J., Cremers, F. P., Black, G. C., et al. (2014). Retinal gene therapy in patients with choroideremia: Initial findings from a phase 1/2 clinical trial. *The Lancet, 383*, 1129–1137.
23. McClements, M. E., & MacLaren, R. E. (2013). Gene therapy for retinal disease. *Translational Research: The Journal of Laboratory and Clinical Medicine, 161*(4), 241–254.
24. Garg, S. K., Lioy, D. T., Cheval, H., McGann, J. C., Bissonnette, J. M., Murtha, M. J., et al. (2013). Systemic delivery of MeCP2 rescues behavioral and cellular deficits in female mouse models of Rett syndrome. *The Journal of Neuroscience: The Official Journal of the Society for Neuroscience, 33*(34), 13612–13620.
25. Salmon, F., Grosios, K., & Petry, H. (2014). Safety profile of recombinant adeno-associated viral vectors: Focus on alipogene tiparvovec (Glybera((R))). *Expert Review of Clinical Pharmacology, 7*(1), 53–65.
26. Peng, Z., Yu, Q., Bao, L. (2008). The application of gene therapy in China. *IDrugs: The Investigational Drugs Journal, 11*(5), 346–350.

Vaccines Against Viral Infections

19

Abstract

Vaccines are one of the miracles of modern medicine. Without vaccines, the population—particularly children—would be troubled with multitudinous infectious diseases such as diphtheria, scarlet fever, whooping cough, and measles, just to name a few. Vaccines are available against both viral and bacterial infections; they have saved millions of lives and continue to do so. The World Health Organization hopes to have vaccinated all children under a year old against most infectious diseases by 2020. The history of the public's acceptance of vaccines has been a stormy one. Anti-vaccination movements have been active since the creation or development of the smallpox vaccine in the eighteenth century and in government-mandated vaccination until the present. An example of this is the fabricated information spread about the relationship between measles vaccine and autism. A successful polio vaccine was developed in the 1950s thanks to the research of three groups, led by Koprowski, Salk and Sabin, although each of these vaccines has its advantages and disadvantages. An oral vaccine has the advantage of ease of administration and a herd effect. Polio vaccine workers have become targets of extremists in Nigeria and Pakistan. Recently, recombinant DNA technology, was used to develop new vaccines in order to avoid the side effects, since live virus is not involved. The future may see the production of vaccines in edible plants, allowing for cheap production and ease in worldwide distribution.

19.1 Introduction

In 2011, it is estimated that 6.9 million children under five died, mostly from diseases that are preventable by vaccination. This number is down from 12.4 million who died in 1990; thus, in 11 years, over 5 million lives were saved as a

M. W. Taylor, *Viruses and Man: A History of Interactions*,
DOI: 10.1007/978-3-319-07758-1_19,

result of the development of new vaccines, and a World Health Organization campaign to vaccinate children in Africa, India and other less-developed countries, through the "Global Vaccine Action Plan" (GVAP). The most common causes of death among young children are acute respiratory infections, diarrhea from contaminated sources, measles, malaria, HIV and malnutrition, and most of these deaths could have been prevented by vaccination. Table 19.1 lists the vaccines that are administered to children (and adults) worldwide, when they were developed or licensed, and the number of cases and coverage where available.

The World Health Organization launched a program, "Decades of Vaccines—Global Action Plan 2011–2020" to promote more equitable access to vaccines throughout the world. One of the aims of this program is the complete eradication of poliovirus. However, this program has already run into trouble, since inhabitants of the tribal areas of Pakistan, and certain groups in Nigeria, see the vaccination as a CIA plot, and this has hindered the complete elimination of polio. Polio workers carrying out the vaccination program, were shot and killed in Pakistan by Al Qaeda and the Taliban. This results from the participation of a physician involved in the polio vaccine campaign helping the US military in physicians to collect information from the Osama bin Laden household which aided in identifying the Al Qaeda leader.

When the WHO launched the expanded program on immunization in 1974, less than 5 % of the world's children under age one were immunized against killers such as polio, measles, and whooping cough; by 2020, the aim is to reach 80–90 % of these children with available vaccines. It is currently estimated that 83 % of the world's children under age one receive vaccines against most killer diseases, and new vaccines against hepatitis A and B are now administered (see Table 19.1). However, one-fifth of the world's children—about 22.4 million infants—are not immunized against these diseases. An estimated 1.5 million children died from vaccine-preventable diseases in 2011.

It is obvious that vaccination has had a profound effect on the incidence of disease in the U.S., and attempts to vaccinate the rest of the world against many of these diseases are under way. Smallpox has been eradicated and polio is near eradication. Tables 19.2 and 19.3 illustrate the effect those vaccines have had on health in the U.S.

There has been much discussion recently on whether there is the need to vaccinate children against so many infectious diseases. In particular, the necessity of vaccinating against measles has been questioned, with the troubling menace of measles outbreaks occurring in various parts of the world, specifically in the U.K. Our guard against a number of known childhood diseases has been lowered because of misinformation on the Internet, or even in scientific publications relating vaccines to autism and other pathologies. Neglecting polio vaccination has resulted in epidemics in various parts of the Middle East and Pakistan, and in India in 2013. Just this year, emergency vaccinations against polio had to be given in Israel and Syria.

The history of vaccine development, but more so the history of the public's acceptance of vaccines, has been a stormy one. Anti-vaccination movements have been active since the introduction of the smallpox vaccine and

Table 19.1 Available vaccines and date of introduction

Disease	Year of vaccine development or license	Number of cases (2012)	Percentage vaccine coverage (%)	Comments
Diphtheria	1920s	4,425	83	Caused by corynebacterium diphtheria
Measles	1963 Edmonston B strain, attenuated	227,245	84	Caused by paramyxovirus; highly lethal (50 %)
Mumps	First strain developed in 1949. Today at least 10 strains of vaccine	Endemic; numbers not available	59	Predominantly childhood disease
Polio	1950–1956	220	99	Attenuated vaccines developed by Sabin and Koprowski, "killed" vaccine by Salk
Yellow fever	1936, developed by Theiler	200,000 or more	37	Increase in incidence despite vaccine
Smallpox	1796 (vaccinia)	0	No longer applied	Eradicated in 1977
Hepatitis A	1993–1996	1.4 million/year	Worldwide endemic	
Hepatitis B	1981	240 million chronic infections	Not covered	1st recombinant vaccine. 600,000 deaths per year
Papilloma	2006–2007	529,000 new cases of cervical cancer in 2012	Not covered	274,000 deaths due to cervical cancer
Rabies	1885 (Pasteur attenuated vaccine), 1967 killed vaccine: 1984 recombinant vaccine:	55,000 die per year, mostly in Africa and Asia	15 million receive vaccine per year	
Varicella (Chickenpox)	1988 in Japan and Korea, 1995 in USA	6,800 deaths	Not known	Mild disease
Zoster (Shingles)	2006	N/A	N/A	For people over 60 years of age

(continued)

Table 19.1 (continued)

Disease	Year of vaccine development or license	Number of cases (2012)	Percentage vaccine coverage (%)	Comments
Influenza	1945, first given to army personnel in USA	N/A	N/A	Vaccine changes annually. Tripartite, containing three different viruses
Rubella	1969	100,000	Most countries do not vaccinate	Given as part of MMR vaccine
Rotavirus	2006	537,000 deaths per year	No coverage	Causes severe diarrhea in children

Table 19.2 The impact of vaccination on the total incidence of viral disease comparing the average incidence in 1930–1950 with the year 2010 (USA)

Disease	Annual morbidity in twentieth century	Reported cases in 2010	Decrease (%)
Smallpox	29,005	0	100
Polio (paralytic)	16,316	0	100
Measles	530,217	61	>99
Mumps	162,344	2,528	98
Rubella	47,745	6	>99
Congenital rubella syndrome	152	0	100

Data from table derived from http://www.cdc.gov/vaccines/pubs/pinkbook/downloads/appendices/G/impact-of-vaccines.pdf

Table 19.3 Comparison of pre-vaccine era with current morbidity estimates

Disease	Pre-vaccine era annual	2008 estimates	Decrease (%)
Hepatitis A	117,333	11,049	91
Hepatitis B	66,232	11,269	83
Rotavirus (hospitalizations)	62,500	7,500	88
Varicella	4,085,120	449,363	89

Data from table derived from http://www.cdc.gov/vaccines/pubs/pinkbook/downloads/appendices/G/impact-of-vaccines.pdf

government-mandated vaccination. In the U.S. and U.K., opponents to vaccination have often been religious. Arguments against variolation, and later vaccination, go back to the eighteenth century, even before Jenner and his discovery of cowpox as protection against smallpox. Preachers pronounced diseases as punishment for sins, and as such should not be treated. A national poll conducted between 2005 and 2010 showed that 60 % of Americans did not want to get flu shots since they worried about side effects and over-vaccination of the population. Recent opposition stems from a report in *The Lancet* in 1998, linking the measles mumps, and rubella vaccine to the onset of autism. This paper was later withdrawn following the exposure of fabricated data, but the measles vaccine program in the U.K., and to a lesser degree in the U.S., was severely damaged.

A discussion of the history and development of individual vaccines is presented below.

19.2 Smallpox

A history of the smallpox vaccine is presented in Chaps. 1 and 9. To summarize: Dr. Edward Jenner (1749–1823), a practicing physician in Berkeley, England (Gloucestershire), had noted, as had others, that milkmaids appeared to be immune to smallpox. Instead, their hands often had blisters picked up from cows who had developed pocks on their teats, later shown to be caused by the cowpox virus (later called "vaccinia"). Jenner, who had a small laboratory in the shadows of Berkeley Castle and its torture chambers in Northern England, inoculated a number of children, including his 11-month-old son first with cowpox, and then reported protection against smallpox. Over 100,000 people had been vaccinated worldwide by the year 1800. Most physicians accepted Jenner's vaccine, although there was some resentment against Jenner by other physicians who did not appreciate that the "cure" for smallpox was developed by a country physician. Others may have resented the loss of revenue from treating smallpox [1]. In 1802, the British Parliament discussed Jenner's vaccine, upheld its usefulness and his claim that he was the inventor of the method. Despite its success, and its use worldwide, there continued to be opposition to the vaccine.

Not with standing this initial success, the production of the smallpox vaccine was plagued with problems. Not least was the method of growing the cowpox virus in amounts large enough to supply the demand. Initially it had to be isolated from pocks on the cow's udder. The virus was unstable over long periods and could not be shipped long distances. It could be kept in dry storage for a few months but could not tolerate heat or exposure to the sun. The supply could not keep up with the demand, since one had to search for cows with cowpox, which was not all that common.

A unique method of transferring the virus was "arm-to-arm"—that is, to transfer the material from one individual after the formation of the scab from the vaccination site to another individual. Although hard to believe, a girl who had been recently vaccinated was sent to an orphanage in St. Petersburg, Russia, and served as a source of the smallpox vaccine for all the children more than a week old. The orphanage continued this practice from one child to another for 92 years (1801–93) [2]

Then there was the Balmis Expedition, the goal of which was to bring the vaccine to the New World and the American Indian population. The King of Spain's daughter had been stricken with smallpox in 1798. In 1803, Charles IV ordered his personal physician, de Balmis, to deliver the vaccine to the inhabitants of North and South America. To maintain the vaccine during its long trip across the Atlantic, de Balmis recruited 22 young boys from orphanages in Spain, ages 3–9, who had never had smallpox. During the trip, de Balmis sequentially vaccinated the boys in a living chain and this was continued until the ship reached Venezuela. More than 100,000 people were thus vaccinated in Latin America. The "Royal Smallpox Expedition," as it was called, then recruited 22 more orphans and continued the vaccination program in the Philippines, Macao and Canton [2].

At a medical congress in France in 1864, a better method of growing the virus was reported from Italy and spread worldwide after the presentation. Cowpox could be grown in large quantities by making a series of cuts in the flank of a calf, inoculating the virus into the animals' skin, and harvesting the material from the pocks 5 days later. This was still a crude way of preparing the vaccine, since it was often contaminated with hair, bacteria, and other debris.

Around 1800, a Dr. J. Clinch, a medical missionary in Newfoundland, began using Jenner's vaccine, and by the end of 1801 he had vaccinated approximately 700 people. This was probably the first vaccination of a community in North America. By 1821 compulsory vaccination of infants was in place in Bavaria, Denmark, Norway, Bohemia, Russia and Sweden. The British government passed the vaccination act in 1840, after a smallpox epidemic from 1837 until 1840, made variolation illegal and supplied smallpox vaccine (cowpox) free of charge. In 1853 the Vaccination Act was passed, requiring all children to receive smallpox vaccines by the age of 4 months; later on, the law was changed to allow parents who objected to vaccination to appeal. This was known as the conscientious objection law, hence the term "conscientious objector." In 1905 the U.S. Supreme Court ruled that individual states could impose mandatory vaccination on children.

Despite this, or perhaps because of compulsory vaccination, there were many who continued to oppose vaccination. One of the problems was that Jenner erred when he proclaimed that immunity would last a lifetime. In fact, the immune effect lapsed after 5–10 years and required re-inoculation. Others still objected on religious grounds, saying that it was interference in the "work of the Lord" and there was quite a high level of accidents, in which vaccine was not directly taken from pockmarks of the udder, and was contaminated with other viruses or bacteria. In particular, cases of syphilis were reported both in the "arm-to-arm" method and sometimes from the vaccine itself.

There was quite an active anti-vaccine campaign (see illustrations below of pamphlets and posters from the Library of Medicine Website http://www.nlm.nih.gov/copyright.html) (Fig. 19.1).

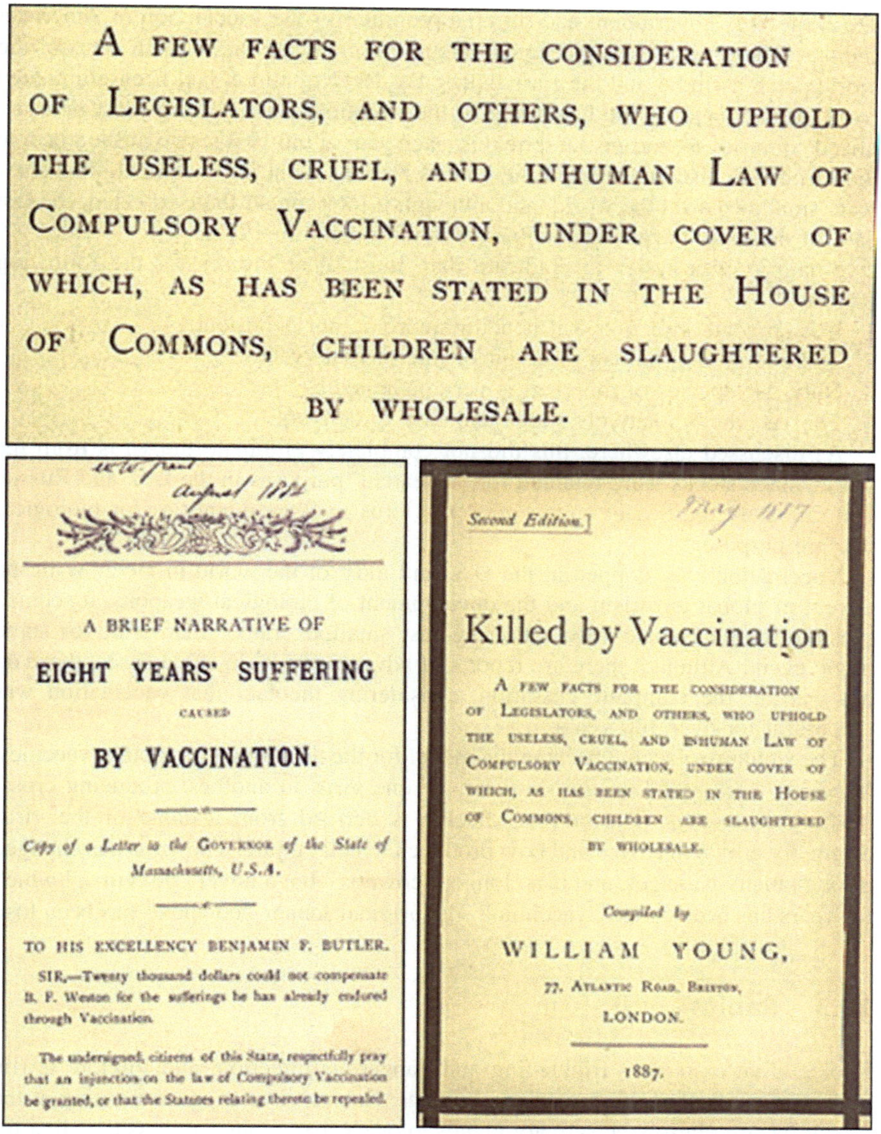

Fig. 19.1 Posters urging the rejection of compulsory vaccination

Smallpox was declared eradicated from the world in 1980, after an intensive struggle to eradicate the virus. In fact, eradication began in 1803 when the Spanish crown attempted to vaccinate the population of the Philippines. In 1813 the U.S. Congress passed the vaccination act to insure safe vaccines for the U.S. public. By

1832, the U.S. government had set up a program for the vaccination of American Indians. There was also a piecemeal attempt to eradicate the virus in areas of the world, such as India and the East Indies. By 1897 smallpox had been eliminated from the U.S. and by the beginning of the twentieth century from most industrialized countries of Europe. Despite this, each year of the 1950s, two million people died from the disease. In 1959 the United Nations launched a campaign to eradicate smallpox from the world, and although it took longer than expected, the last case of naturally occurring smallpox was in Somalia in 1977.

Smallpox was easier to eradicate than most other viruses for the following reasons:

1. It is an acute infection that is neither chronic nor persistent.
2. There is no non-human reservoir or intermediate vector.
3. Only one species of the virus is very pathogenic.
4. The vaccine is relatively stable and easy to deliver.

As discussed elsewhere, this did not completely eliminate the virus from the world, since stocks were retained for "research" purposes in the U.S. and Russia. Unfortunately the Soviet Union grew the virus in large quantities for biological warfare purposes.

Vaccination was stopped in the U.S. and most of the world in 1972. With the advent of global terrorism and the development of biological weapons, it seemed sensible to reintroduce vaccination against smallpox rather than wait for some tragic event. Although there are reports of adverse effects in some cases, these do not warrant the complete cessation, considering the fact that vaccination was successful for over 100 years.

The smallpox vaccine became the model for the development of other vaccines. Vaccines were based on the similarity of one virus to another, producing cross-reacting antibodies. The current vaccine is derived from a clone of the virus originally grown in calves and now produced in cell culture. The virus has changed through many passages, and is no longer "cowpox" but a novel "poxvirus," which for years has been called "vaccinia." The original Jenner "cowpox" has been lost.

19.3 Rabies

Rabies is an extremely frightening and dangerous infection, and folklore is rife with stories of rabid dogs attacking humans. Rabies virus triggers foaming at the mouth by stimulating the salivary glands to produce copious amounts of saliva for the efficient transmission of the virus. Pasteur and Roux developed a rabies vaccine in 1885. The virus was attenuated by passing the most virulent strain isolated from rabid dogs through rabbits' brains, harvested from the nerve tissue, and dried for 5–10 days, resulting in an attenuated virus. The vaccine was delivered to patients in the form of 10–14 intramuscular injections over a period of 2 weeks to anyone suspected of having been in contact with a rabid animal.

The development of this vaccine is mired in controversy since it has been claimed that it was not adequately tested before being used on a young nine-year-old boy, Joseph Meister. Emil Roux, a young colleague of Louis Pasteur's, was studying for his MD degree and using the development of the rabies vaccine as a research topic. Pasteur was an experimental scientist working on the vaccine, but was not allowed by French law to treat patients. Joseph Meister had been mauled by a rabid dog and was brought to Pasteur for help. Pasteur consulted with two physicians, and since there seemed little hope of a cure since rabies is fatal, asked one of the physicians to administer the untested vaccine. Using it on Meister has come into some criticism and controversy in recent years [3] with the release of Pasteur's notebooks. Questions are being asked whether the vaccine was adequately tested before being used in the Meister case, and doubts have been expressed as to whether the boy had been bitten by a rabid dog. Both Pasteur and Roux worked on the vaccine, but Roux's name is not on the paper reporting the treatment of Meister. Some [3] claim that Pasteur lied in his paper, stating that he had treated 40 dogs with the vaccine, when in fact he had just initiated treatment with 20 dogs, and that development of the vaccine was not far enough along to warrant use in humans; there is also speculation that Roux might have opposed the use of the vaccine.

There is obviously a moral question here: What does one do when there is a possibly fatal case or illness? Wait until experimentation is complete? This is an enigma facing many physicians today when faced with new, but untested treatment for cancer or AIDS. Pasteur's work with the rabies vaccine met with much skepticism in the U.S. Doubts were expressed on the efficacy of the vaccine, and whether the vaccine itself led to the deaths of victims, who might not have been exposed to a rabid dog but bitten by a healthy one.

In 1967 a vaccine was developed from attenuated rabies virus grown in cell culture. This vaccine was extensively tested, and four or five subsequent injections resulted in complete protection. Further improvement to the vaccine was made in 1971 by growing the rabies virus in human embryonic cells. Although Hilary Koprowski's team inoculated themselves with this new vaccine, the medical profession did not like the idea of using a "live" rabies virus. Koprowski also cloned the rabies virus glycoprotein into vaccinia and this vaccine is used to control the disease in wild animals [4]. In 1976 an inactive rabies vaccine, called human diploid cell vaccine (HDCV) was licensed in Europe and in 1980 in the U.S. One of its major advantages was that it was administered in five injections rather than the 14–21 injections of the older vaccine, and this cell culture inactive virus vaccine is now the one of choice.

In the 1950s while tending sheep I found an injured animal. Since I handled it, it was suggested I get rabies shots, in case a rabid jackal had bitten the animal. This was in the days before the modern vaccine, and I was given a series of 12 shots in the abdomen as a precaution (I assume that this must have been the Pasteur vaccine). After the shots the animal tested negative!

The number of deaths from rabies in the United States had declined from approximately 100 at the turn of the last century (1900) to one or two per year, and those few cases are due to exposure to wildlife (bats). In the rest of the world, the main source of rabies is still the rabid dog. More than 55,000 people die of rabies each year in Asia and Africa; 40 % of those bitten by suspect rabid animals are children less than 15 years old (WHO report).

19.4 Yellow Fever

When the Panama Canal opened in 1912, there was an urgent need for a vaccine against yellow fever. Although transmitted by mosquitoes, the agent responsible was still unknown. The Rockefeller Institute in New York sent one of its scientists, Hideyo Noguchi, to Guayaquil, Ecuador, a center for epidemic outbreaks, to investigate the possibilities of producing a vaccine. While in Ecuador, Noguchi claimed to have discovered the agent causing yellow fever, a spirochete bacterium, *Leptospira icteroides* [5]. He went on to develop a vaccine from this organism and used it extensively in South America and Africa. Noguchi published many papers reporting success with his vaccine, in a total of 7,964 cases. However, his results were not reproducible, filters retained the agent he used, indicating that it was not a viral disease, the statistics were faulty, and eventually the Rockefeller Institute dropped the project. This material is also presented in Chap. 10.

In 1925 the Rockefeller Institute launched another scientific expedition, this time to West Africa to determine whether African yellow fever was identical to South American yellow fever. The aim was to characterize the causative agent, including further search for *Leptospira icteroides* (the organism studied by Noguchi), and to study its epidemiology. The expedition was led by Major Henry Beeuwkes, a Johns Hopkins-trained bacteriologist retired from the army, and Adrian Stokes, a London-based professor of pathology and expert on leptospirosis; Noguchi and others joined him.

In June 1927, blood from a 28-year-old African yellow fever victim named Asibi, was injected into a rhesus monkey. The monkey became ill, had symptoms of yellow fever and became, thus, the first case of an animal model for yellow fever. A short while later, Stokes himself died of yellow fever. Noguchi pursued his ideas that the disease was caused by *Leptospira icteroides* and inoculated over one thousand monkeys with blood from yellow fever victims, but could not find any *leptospira* in them. Noguchi also contracted yellow fever and died in May 1928, as did a third investigator, who performed the autopsy on him [6]. However, now that an animal model had been discovered, the infecting agent could be removed to the laboratory for further study. Soon after, it was shown that serum from immune humans, who had recovered from the infection, protected monkeys against infection, and immune serum from South American patients protected against the African "virus" (thus suggesting that one vaccine would offer protection globally). This strain of yellow fever was called "Asibi."

A French group in Dakar also collected blood from another patient, and passaged the "virus" in rhesus monkeys. This gave rise to a strain of virus known as the French strain. The virus could be frozen and work now pursued in the laboratory.

Max Theiler, a South African working at the Rockefeller Institute, had been involved in research on *Leptospira icteroides*, the organism proposed by Noguchi as the causative agent of yellow fever. In 1927 he isolated a virus that appeared to be the causative agent of the disease. Being aware of the previous literature on the rabies virus, and on recent work on growing herpes virus in the brain of mice, Theiler inoculated the yellow fever virus into mouse brains and found that it reproduced. When injected into rhesus monkeys, the passaged virus was less pathogenic systemically, but more neurotropic. This research was extremely dangerous. The Rockefeller Institute announced in 1931 that there had been 32 cases of yellow fever in eight laboratories, with five fatalities. The French strain, passaged a hundred more times in mouse brains, was administered to protect laboratory and field workers in 1931. The virus was combined with immune serum from yellow fever-infected lab workers to prevent encephalitis. This vaccine was successful and was used up until World War II. Meanwhile, the French scientists developed a different vaccine, using the mouse brain passaged virus, coating it with oil or egg yolk, freeze-drying the mixture, suspending it in saline, and applying it with the smallpox vaccine by scarification. By 1945, sixteen million Africans had been vaccinated with this combination and with few side effects.

The method originally developed by Theiler required large amounts of human serum; since there was fear of the neurotropic effects, it was thus of limited use. By the 1930s, cell culture had been introduced, and Theiler passaged the virus in chick embryos lacking nervous tissue more than 100 times. This virus lost its neurotropism in mice, due to random mutations. This strain, called 17D, had no side effects, and became the vaccine of choice. By 1939, 1.3 million people had been vaccinated in South America with the Theiler vaccine, without any side effects. However, in 1941, 119 cases of encephalitis were reported, all of the infected having received the vaccine from a single lot; this virus had apparently mutated during sub-culture. Thus there arose the "seed lot" system, where original cultures of the virus were stored for further use.

World War II saw a huge demand for the vaccine for the troops engaged in North Africa. Between January 1941 and April 1942, 7 million doses were distributed. One batch of vaccine did have a serious side effect—there were 26,772 cases of jaundice that occurred about 60–120 days after the injection. The vaccine contained a small amount of human serum for stabilization, and some batches were contaminated with hepatitis B. It was thus decided to switch to a serum-free vaccine and to give the vaccine only to those posted to a yellow fever area [6]. After further improvements, the 17D strain was adapted as the sole yellow fever vaccine, and it is still used today.

19.5 Poliovirus

Tremendous efforts went into developing a polio vaccine in the 1940s and 1950s (this is discussed in Chaps. 3 and 12). A major advance was the discovery that both poliovirus and mumps virus could grow in mouse lung cells. Three major research groups were involved in the development of a polio vaccine: those of Jonas Salk, Albert Sabin and Hilary Koprowski. Unfortunately, there was bitter competition and rivalry between the three groups. The first vaccine licensed in the U.S. was Salk's inactivated polio vaccine (1955), in which the three strains of poliovirus were inactivated by treatment with formalin. The development of this vaccine was a great achievement, and Jonas Salk became a national hero. A second attenuated vaccine containing all three strains of polio was tested and licensed in 1962. Two other groups had independently produced an attenuated vaccine—one headed by Albert Sabin, the other by Hilary Koprowski. A committee of the National Institutes of Health decided on the Sabin vaccine, which consisted of the three mutated strains of polio, extensively passaged in cell culture and mice. Hilary Koprowski had developed the first attenuated poliovirus in 1950 and tested it on himself and on a group of 20 children in New York, with no side effects. Mass vaccination trials of the Koprowski oral polio vaccine took place in the Belgian Congo, Poland, Croatia, and Switzerland. By 1960, approximately 13 million individuals had been vaccinated with the Koprowski polio vaccine and over 11 million with the Sabin vaccine [7].

The Salk vaccine is based on three wild, virulent reference strains, Mahoney (type 1 poliovirus), MEF-1 (type 2 poliovirus), and Saukett (type 3 poliovirus), grown in monkey kidney cell culture (Vero cell line), and the virus inactivated with formalin. Initially, the Salk vaccine was the vaccine of choice in the U.S. In 1955, in the largest trial ever held, more than 400,000 children were vaccinated with the Salk vaccine, which was 90 % effective against type II and type III polio, and 60–70 % effective against type I. Large industrial facilities were built for the production of the vaccine, which was exported to many countries. In 1955, in an event known as the "Cutter" (Cutter was the name of one of the pharmaceutical companies that produced the polio vaccine), some 120,000 doses of the Salk vaccine were created, containing a live poliovirus instead of an inactive one. The administration of this vaccine caused 40,000 cases of polio, with 53 cases of paralysis and 5 deaths. The disease spread, creating a polio epidemic that lead to a further 113 cases of paralytic polio and another 5 deaths. It has been called "one of the worst pharmaceutical disasters in U.S. history;" [8] it of course affected the public's attitude towards the vaccine. In order for a vaccine to be successful, the public has to have confidence in it. In fact, the "Cutter Incident" led to the use of an alternative vaccine (Sabin instead of Salk) for some time.

Many health advocates believed that the Sabin vaccine was a better vaccine, since as a "live" vaccine, it provides "herd immunity"—that is, virus is shed by those being vaccinated—and it can spread throughout the family and provide protection to the surrounding population. It was also assumed that the immunity from the inactivated vaccine would not be long lasting, whereas the Sabin vaccine

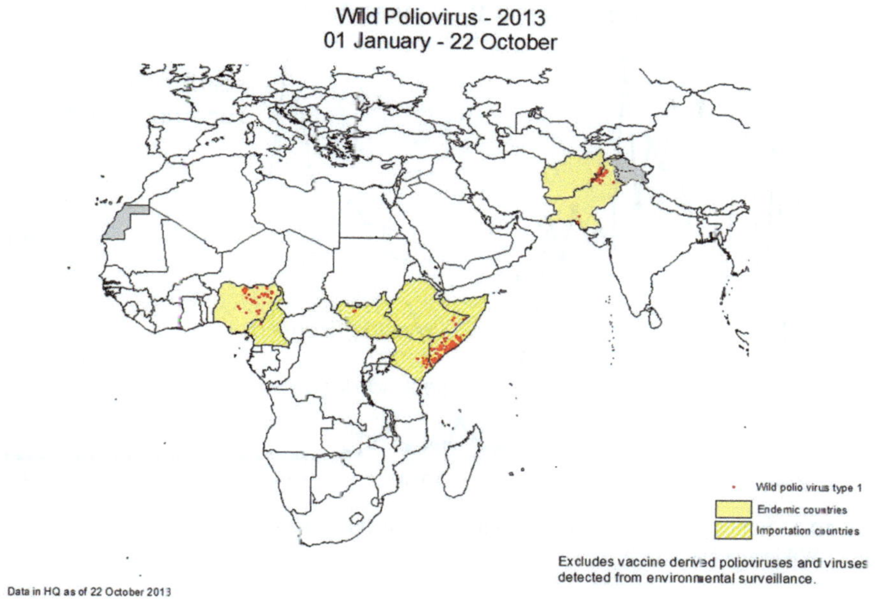

Fig. 19.2 Map of polio endemic areas as of 2013

virus multiplied in the gut, and would confer a stronger immune response. By 1968 the Sabin vaccine replaced the Salk vaccine. Since it was given orally rather than by injection, it was easier for children to take and for parents to administer. However there was a low, but steady number of individuals receiving the vaccine who developed paralytic polio, indicating that the attenuated virus had undergone mutation, or reverted to wild-type. By 2000 the US CDC recommended the Salk, inactivated vaccine. In the WHO campaign to eradicate polio, the oral vaccine is preferred, because of herd immunity. A map of the areas of the globe with endemic polio is shown in Fig. 19.2 (a newer map would include Syria).

19.6 Measles

In the 1960s, vaccines were developed for mumps, rubella (German measles) and measles, and in 1971 the combination of all three viruses (MMR vaccine) was approved for use in the U.S. All three viruses are derived from attenuated strains by passage through chicken eggs or cell culture, and were developed by Maurice Hilleman of Merck Pharmaceuticals; in fact, Hilleman was responsible for the development of 36 vaccines. Of the 14 vaccines routinely recommended in current vaccine schedules, he developed eight: for measles, mumps, hepatitis A, hepatitis B, chickenpox, meningitis, pneumonia and *Haemophilus influenzae* bacteria (Wikipedia).

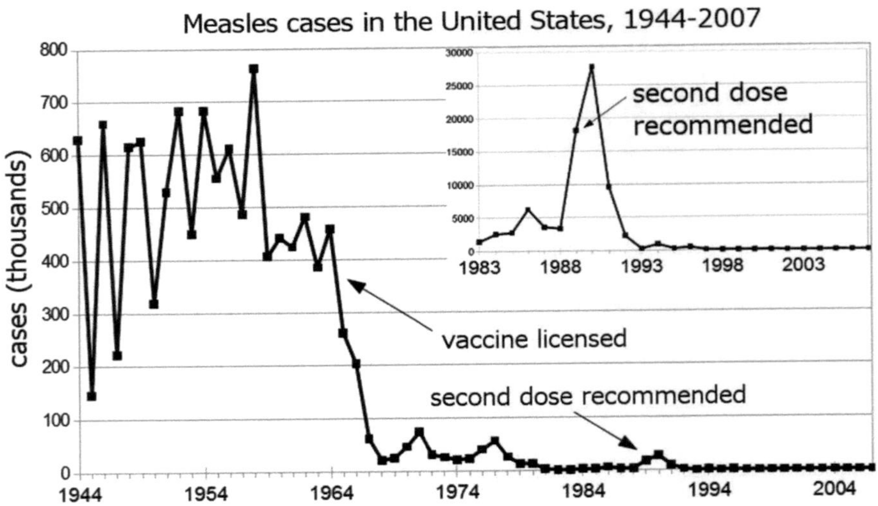

Fig. 19.3 Number of cases of measles in the USA, and decline after introduction of the vaccine (from Wikipedia and CDC)

Measles is a very dangerous childhood disease, leading in some cases to death, blindness, or hearing difficulties. Since the introduction of the vaccine, the number of cases in the U.S. has fallen drastically (Fig. 19.3).

19.6.1 Measles Vaccine and Autism

In 1998 [9] a paper appeared in *The Lancet*, the influential British medical journal claiming that the increase in cases of autism in children was linked directly to the MMR vaccine. This was based on a study headed by Dr. Andrew Wakefield, a British surgeon and specialist in diseases of the gastric tract. In particular, he claimed to have found a correlation between irritable bowel syndrome, autism and the vaccine. His conclusions were based on a study of 12 children. Most of the data presented in the paper was anecdotal, on information from the parents as to when the autistic symptoms first appeared. In most cases this was said to have been within a few weeks of the MMR inoculation. Dr. Wakefield published a follow-up series of papers [10–12] claiming a direct relationship between the vaccine and autism (and irritable bowel syndrome or Crohn's disease). The papers were peer-reviewed, many appearing in *The Lancet* or other influential peer-reviewed journals. The major argument presented was that (1) the vaccine overloaded the immune system with viruses (three types), leading to an abnormal immune response; (2) the injection of measles virus at such a young age altered the normal immune response, resulting in aberrations of the immune system; and (3) that lack of natural exposure to these viruses could lead to irritable bowel syndrome and

Crohn's disease, both autoimmune diseases (although also known to be inherited). These reports caused an uproar in the U.K., resulting in many parents refusing to vaccinate their children; the effects were such that the level of MMR vaccination dropped from 93 to 73 % in the U.K., and to 50 % in London.

Within a few years, large studies were performed to confirm or refute Wakefield's data. One of the largest was a Danish study with 537,303 children in the cohort—children who had received MMR and children who had not received it [13]. No correlation between the MMR vaccine and autism was detected. Other studies quickly followed, showing similar results, although the Wakefield group continued publishing the opposite conclusions. It was not until 2010 that 10 of 12 authors in the first paper published in *The Lancet* [14] retracted their authorship, claiming that they did not agree with the conclusions of the paper. Similar retractions were made in other journals, but by then a great deal of damage had been done. During this period, Wakefield stood by his results and claimed that there was a conspiracy against him. What is astounding is that none of the co-authors on the original paper realized that the data was wrong (or not statistically correct) and that the retraction took so long. An investigation by the General Medical Council of Great Britain found that many of the co-authors could not identify which child was being discussed in the paper. Their names, with their agreement, had been added to the authorship of the papers for the flimsiest of reasons; unfortunately, this is a common occurrence in science today.

The Wakefield case came to prominence due to the investigative reporting of Brian Deer of the Times of London. Deer published his first investigation into Wakefield's papers in 2004. He uncovered the possibility of research fraud, unethical treatment of children, and Wakefield's conflict of interest through his involvement in a lawsuit against manufacturers of the MMR vaccine. Deer published a summary of his work in the *British Medical Journal* in 2011 [15, 16]. He interviewed each of the parents of the 12 (probably 13) children described in the Wakefield paper, and found that the time of appearance of autism did not match that reported in the 1998 paper, nor were the medical records reported correctly. This is shown in Table 19.4.

The summary of the BMJ paper [9] is presented below:

"The *Lancet* paper was a case series of 12 child patients; it reported a proposed 'new syndrome' of enterocolitis and regressive autism and associated this with MMR as an "apparent precipitating event." But in fact:"

"Three of nine children reported with regressive autism did not have autism diagnosed at all. Only one child clearly had regressive autism."

"Despite the paper claiming that all 12 children were "previously normal," five had documented pre-existing developmental concerns."

"Some children were reported to have experienced first behavioral symptoms within days of MMR, but the records documented these as starting some months after vaccination."

"In nine cases, unremarkable colonic histopathology results—noting no or minimal fluctuations in inflammatory cell populations—were changed after a medical school 'research review' to 'non-specific colitis'."

Table 19.4 Comparison of three features of the 12 children in the *Lancet* paper with features apparent in the NHS records, including those from the Royal Free Hospital

Child	Regressive autism		Non-specific colitis		First symptoms days after MMR		All three features	
	Lancet	Records	Lancet	Records†	Lancet	Records‡	Lancet	Records
1	Yes	?	Yes	Yes	Yes	No	Yes	No
2	Yes	Yes	Yes	Yes	Yes	No	Yes	No
3	Yes	?	Yes	No	Yes	?	Yes	No
4	Yes	?	Yes	No	Yes	No	Yes	No
5	Yes	?	Yes	No	No	No	No	No
6	Yes	No	Yes	Yes	Yes	?	Yes	No
7	Yes	No	No	No	Yes	No	No	No
8	No	No	Yes	No	Yes	No	No	No
9	No	No	Yes	No	No	No	No	No
10	No	No	Yes	No	No	No	No	No
11	Yes	?	Yes	No	Yes	No	Yes	No
12	Yes	No	Yes	No	No	No	No	No
Total	9/12	?6/12	11/12	3/12	8/12	?2/12	6/12	0/12

"The parents of eight children were reported as blaming MMR, but 11 families made this allegation at the hospital. The exclusion of three allegations—all giving times to the onset of problems in months—helped to create the appearance of a 14 day temporal link.

"Patients were recruited through anti-MMR campaigners, and the study was commissioned and funded for planned litigation."

As a scientist, it is difficult to discern a motive for this falsification of data. Was it done out of greed? This does seem to be the case. All the shenanigans—creating companies to show a linkage between MMR and autism and Crohn's disease even before the children described in the *Lancet* paper were hospitalized—are presented in http://briandeer.com/mmr/lancet-summary.htm. Companies were created in England, Ireland and in the U.S., patents were applied for, and money accepted from charitable and government agencies on the basis of false data. The hospital and institutions employing Wakefield were only too pleased to take their cuts.

"Moreover," Deer reported, "a lawyer planning to sue vaccine manufacturers on behalf of parents who thought MMR caused their children's problems retained Wakefield as an expert witness 2 years earlier. The parents cited in the *Lancet* article came to Wakefield's clinic in response to an advertising campaign led by the lawyer's group, called Jabs, and not for routine screening." From the London Times: "ANDREW WAKEFIELD, the former surgeon whose campaign linking the MMR vaccine with autism, caused a collapse in immunization rates, was paid more than £400,000 by lawyers trying to prove that the vaccine was unsafe.

The payments, unearthed by The Sunday Times, were part of £3.4 m distributed from the legal aid fund to doctors and scientists who had been recruited to support a now-failed lawsuit against vaccine manufacturers.

"Critics this weekend voiced amazement at the sums, which they said created a clear conflict of interest and were the 'financial engine' behind a worldwide alarm over the triple measles, mumps and rubella shot.

"According to the figure, released under the Freedom of Information Act, Wakefield was paid £435,643 in fees, plus £3,910 in expenses. Wakefield's work for the lawyers began 2 years before he published his now notorious report in The Lancet medical journal in February 1998, proposing a link between the vaccine and autism."

It is interesting that the Wakefield 1998 paper has been cited more than any other retracted paper, over 700 times [10]. The vaccination rate in the U.K. has not completely recovered".

19.7 Influenza

In the U,S,. the annual morbidity from influenza is estimated at between 25–50 million cases with 225,000 hospitalizations. The WHO, extrapolating from these numbers, estimates that there are 1 billion cases of influenza annually, with 3–5 million severe cases and 300,000–500,000 deaths. Most of the deaths in the U.S. are in the over-65 age group [11]. Chapter 11 presents the biological aspects of the influenza virus. The major antigenic components of the virus are the hemagglutinin protein and the neuraminidase protein, the major surface proteins of the virus. As discussed in Chap. 11, the virus undergoes random mutation at a high rate, resulting in antigenic drift. Reassortment and even mixing in the case of two viruses of different origins results in antigenic shift, making it difficult to predict the makeup of the virus causing the seasonal outbreak of the flu. The result of the antigenic changes may mean that a population has no immunity to the virus, and infection can result in a pandemic, as happened in 1918. The vaccine has to be "predictive" from year to year.

There are three types of influenza vaccine currently in circulation. One, containing inactivated subunits of the virus, is given intramuscularly or intradermally; the second is a live vaccine (LIAV live attenuated influenza vaccine) and is given intranasally. Both vaccines are trivalent—that is, have three species of influenza virus in the vaccine. A live attenuated cold adapted influenza virus can be administered as a nasal vaccine. FluMist contains three cold-adapted H1N1, H3N2, and B influenza. These are recombinant viruses containing a backbone of a virus grown in the cold, with the incorporation of H and N genes from different sources, and also influenza B. Mean shedding of the virus is for 7–8 days and longer in children, thus having a herd effect. There have been no cases of illness from this vaccine in the USA.

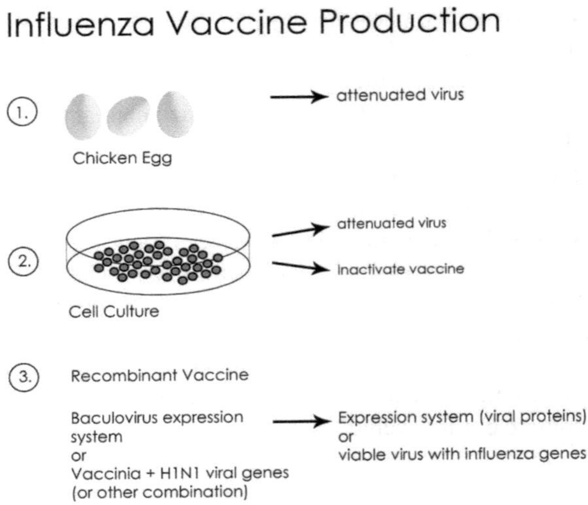

Fig. 19.4 Various methods of producing influenza vaccines

The CDC introduced a new vaccine this year. The quadrivalent flu vaccine protects against two influenza A viruses and two influenza B viruses. A standard dose of quadrivalent flu vaccine, given as a nasal spray, is approved for healthy people aged 2–49; each year the vaccine changes according to predictions of the type of influenza likely to be prevalent in a particular year. For 2013, the FDA chose A/California/7/2009 (H1N1): A/Victoria/361/2011(H3N2) and B/Wisconsin/1/2010—two strains of influenza A and one strain of influenza B. People over 65 should get a new high dose inactivated virus vaccine annually. The strains for the next year's vaccine are chosen in February for use in the following season, and vaccine production begins with strains carrying the "drifted" HA and NA genes of the last season combined with reference strains that grow well in chick eggs. New methods of producing vaccines, in which reverse genetics is used to copy genes, which are inserted into plasmids, and then cloned back into a reference strains of influenza have been developed. New technologies are required to produce sufficient influenza vaccines, and some are already being tried in various companies. These include cell-based rather than egg-based live attenuated and inactive viral vaccines as well as recombinant protein vaccines, where the genes for HA and NA are cloned into baculovirus vectors and administered as purified proteins. Influenza virus gene can also be cloned into other viral vectors such as vaccinia virus or adenovirus and administered as either a replication-competent virus or defective virus (Fig. 19.4). However, this creates problems discussed in the chapter on gene therapy, where antibodies may inactivate such a virus. The development of an influenza vaccine is to some extent a "guessing game," since one cannot predict the appearance of a novel strain of virus.

19.8 Rational Approach to Designing Vaccines

Vaccines fall into many categories. The majority of those discussed above are live virus vaccines in which the virus has been attenuated either by multiple passages through other animals or through multiple passages through cells in culture. During passage, the virus undergoes mutation (particularly if it is an RNA virus) at a very high rate, and the virus emerging is often attenuated. It should be possible to rationally design mutations by in vitro mutagenesis. Influenza vaccines are made by adapting the virus to grow at lower temperatures and could be made by performing re-assortment in the laboratory to make attenuated strains (although recombination in the laboratory might also produce viruses that are more pathogenic). Another approach is to make recombinant vaccines using viral genes produced in bacteria, yeast or baculovirus systems.

The vaccine for hepatitis B is a recombinant vaccine produced in yeast. In Europe a second vaccine is produced in Chinese hamster ovary cells. These vaccines are very highly immunogenic and have few side effects. Large-scale studies have been performed in at-risk patients, and antibody titers have been high. The WHO—with the goal of eradicating HBV—has recommended a universal vaccination program. In Taiwan and Italy, mass vaccination of infants has not only decreased the incidence of HBV infection and the prevalence of chronic carriers, but also the incidence of hepatocellular carcinoma.

Yeast and an insect virus, baculovirus, have also been used to produce the human papilloma virus vaccine. Two vaccines have been developed; both are highly immunogenic. Gardasil (Merck and Co.) is a quadrivalent vaccine against HPV type 6, 12, 16 and 18, while Cervarix (Glaxo Smith Kline Biologicals) is against HPV types 16 and 18. Both vaccines give excellent protection. Gardasil also demonstrated strong protection against genital warts and anal cancers [12].

There has been some controversy surrounding the recommendations for papilloma vaccination. The vaccines are recommended for girls and women aged 9–46 who have not been exposed to HPV. The vaccine has also been approved for men and boys, with the CDC recommending it for boys as young as 9. There has been fear expressed in the public and newspapers that giving the vaccine to teenagers will encourage promiscuity.

Other alternatives proposed for the development of vaccines, include the use of purified proteins, viral proteins expressed by prokaryotes, the use of synthetic peptides and even DNA. Proteins from a variety of viruses have been cloned and expressed in prokaryotes and cells in culture. Most have proven effective in mice but have not been approved by the appropriate regulatory bodies. DNA vaccines, coding for antigens from cancer or specific bacterium such as anthrax are delivered as plasmids are delivered as plasmids or in harmless strains of *E. coli*. The major problem with DNA vaccines has been a weak antigen response. The main method of delivery is intramuscular injection or application to the skin with a gene gun [13]. However, a large number of clinical trials with DNA vaccines are under way.

19.9 Vaccines from Plants

One of these days, it may be possible to eat lettuce and be vaccinated at the same time, or receive antibodies against a bacterial or virus infection. It sounds like science fiction, yet experimental trials of producing viral antigens are under way, both in the research laboratory and in biotech companies, by introducing animal viral genes into the plant genome, and into chloroplasts. Chloroplasts are small organelles of prokaryotic origin that are responsible for photosynthesis in green plants, analogous to mitochondria in higher organisms. Plant-derived vaccines (or other biopharmaceuticals) are cheap to produce (compared to cell culture) and to store, easy to scale up for mass production, and safer than those produced from other sources, since there is less likelihood of contamination by bacteria or even other viruses. Plant systems are much more economical than other industrial systems for producing vaccines and would be ideal for Third World countries. For example, it is estimated that to produce 100 million doses of influenza vaccine by a conventional pharmaceutical company, such as Sanofi in Pennsylvania, would cost $150 million per year. If Novartis were to produce 300 million doses, they would charge $100 billion. Using tobacco plants, the cost would be $15 million per year to produce one billion doses [14]. The technology is already available for the purification, if necessary, of the products, although the ultimate aim might be to digest the vaccine (oral vaccine), despite the fact that precautions need to be taken that the vaccine is not itself digested by stomach enzymes (this obviously does not apply to tobacco plants). To date, the yield of products in experimental systems such as tobacco or lettuce or grains has not been as great as expected; however, the yield can be greatly enhanced by cloning the desired viral genes into chloroplasts, which multiply in large numbers in the plant cell, thus enriching the product. An HSV-2 vaccine has been produced in soybeans using a cauliflower mosaic virus promoter [15]. This antibody was effective in preventing HSV-2 infection in a mouse model and was indistinguishable from a monoclonal antibody produced in cell culture. A whole series of monoclonal antibodies have been produced in a variety of plants using plant virus vectors [16]. The hepatitis B envelope protein has been produced in potatoes, lettuce, and lupines, and was immunogenic when digested orally. The Norwalk virus capsid protein was also produced in potatoes and tobacco plants and was immunogenic. The protein produced resembled empty viral capsids (Table 19.5).

So far, the amount of foreign or viral proteins produced in the plants has not been sufficient for mass vaccination. This has led to further experimentation to increase the yield by using various promoter sequences, and adding polyadenylation sites (as in mammalian tissue) to increase the yield. It is estimated that the yield should be around 1 % of the total weight to make it of practical use. The yield can be increased by using chloroplasts. There can be upto 10,000 copies of chloroplasts per cell and result in the accumulation of 47 % of the total protein being the recombinant protein. The desired genes are incorporated into the chloroplast by recombination. Thus the cloned gene is surrounded on both sides by

Table 19.5 Shows those experimental results

Source of the protein	Protein or peptide expressed	Plant expression system	Immunogenicity
Hepatitis B virus	Envelope surface protein	Tobacco	Virus-like particles, immunogenic on injection
Hepatitis B Virus	Envelope surface protein	Potato	Immunogenic when administered orally
Hepatitis B virus	Envelope surface protein	Lupine	Immunogenic when administered orally
Hepatitis B virus	Envelope surface protein	Lettuce	Immunogenic when administered orally
Norwalk virus	Capsid proteins	Tobacco	Intact virus and immunogenic. Oral vaccine
Norwalk virus	Capsid proteins	Potato	Virus like particles, immunogenic, oral
Rabies virus	Glycoprotein	Tomato	Intact high yield protein
Human Cytomegalovirus	Glycoprotein B	Tobacco	Immunologically related protein
Rabbit hemorrhagic disease virus	VP60	Potato	Immunogenic
Foot and mouth disease virus	VP1	Aridopsis	Immunogenic when administered by injection
Foot and mouth disease virus	VP1	Alfalfa	Immunogenic when given orally
Coronavirus (pigs)	Glycoprotein S	Aridopsis	Immunogenic when injected
Coronavirus (pigs)	Glycoprotein S	Tobacco	Immunogenic when administered by injection
Coronavirus (pigs)	Glycoproteins S	Maize	Protective when administered orally

Table 19.4. Adapted from [16]

chloroplast sequences. The method of transformation is quite different from that of animal cells. Four- to six-week-old leaves are selected for transformation; DNA is coated on the surface of microparticles of gold or tungsten and then shot onto the leaf surface with great force, using a helium-driven particle delivery device (gene gun).

Regeneration. Leaves are then cut into small leaf discs after 48 h of incubation in the dark and put on selection media. Primary shoots generally arise within 4–6 weeks. Initially, a few copies of plastid genome are transformed and leaf cells that are called heteroplastomic. Homoplastomy, a state where all copies of chloroplast genome are transformed, is achieved by performing a few rounds of regeneration under selection. The putative transgenic shoots are then further characterized for transgene integration and protein expression [17].

HIV gag protein and dengue virus proteins have been produced in chloroplasts. However, the product has not yet reached human clinical trials, although there has been a large investment by U.S. government agencies. It may take a few more years until we see clinical trials as production techniques are patented. However, this may be the way of the future, producing cheap vaccines for most infectious diseases. This method is now known in the "trade" as "molecular farming" and it may be the answer to vaccine shortage; a number of biotech companies are now pursuing this course. To quote from an announcement by Fraunhofer center for molecular biotechnology (http://www.fraunhofer.de/en/press/research-news/2013/june/automated-plant-factory-for-the-prodution-of-vaccines.htm) regarding one such plant, "The plants grow in trays with hydroponic cultures of mineral wool as opposed to soil, in specially designed growth modules. Light, water, and nutrients are precisely dosed and distributed. Specially developed robots bring the plants from station to station to carry out the various steps—from inserting the tiny seeds and vacuum infiltration, to harvesting and extraction.

"The plants grow for four weeks before the vector is introduced by means of vacuum infiltration." The process goes as follows: "A robot picks up a tray with plants, turns it upside down, and submerges the tobacco plants headfirst into water." This water holds the vector (biological carrier) containing the genetic information that tells the plants which protein they should produce. Then a vacuum is applied by drawing the air from the water and the plants. As soon as we switch off the vacuum, the plants suck in the water together with the vector. This takes just a few seconds, and then the plants are put back in the growth module to grow further. In about a week they have produced the proteins. Once harvested, the leaves are cut into small pieces and homogenized in fully automated processes. This produces a liquid from which the proteins are extracted. The end product is a clear liquid.

"The pilot facility is capable of producing up to 300 kg of biomass a month, which roughly corresponds to 2.5 million units of vaccine."

Is this the future of vaccine production?

References

1. Kotar, S. L., & Gesser, J. E. (2013). Smallpox: a History. Jefferosn, North Carolina: McFarland and Company.
2. Tucker, J. B. (2001). *Scourge*. New York: Atlantic Monthly Press.
3. Geison, G. L. (1990). Pasteur, roux, and rabies: scientific versus clinical mentalities. *Journal of the History of Medicine and Allied Sciences, 45*(3), 341–365.
4. Plotkin, S. A. (2013). In memoriam: Hilary Koprowski, 1916-2013. *Journal of Virology, 87*(15), 8270–8271.
5. Noguchi, H. (1920). Leptospira Icteroides and yellow fever. *Proceedings of the National Academy of Sciences of the United States of America, 6*(3), 110–111.
6. Frierson, J. G. (2010). The yellow fever vaccine: a history. *Yale J Biol Med, 83*(2), 77–85.
7. Koprowski, H. (2006). First decade (1950-1960) of studies and trials with the polio vaccine. *Biologicals: Journal of the International Association of Biological Standardization, 34*(2), 81–86.

8. Offit, P. A. (2005). The Cutter incident, 50 years later. *New England Journal of Medicine, 352*(14), 1411–1412.
9. Deer, B. (2011). How the case against the MMR vaccine was fixed. *BMJ, 342*, c5347.
10. Fang, F. C., Steen, R. G., & Casadevall, A. (2012). Misconduct accounts for the majority of retracted scientific publications. *Proceedings of the National Academy of Sciences of the United States of America, 109*(42), 17028–17033.
11. Lambert, L. C., & Fauci, A. S. (2010). Influenza vaccines for the future. *New England Journal of Medicine, 363*(21), 2036–2044.
12. Schiller, J. T., Castellsague, X., & Garland, S. M. (2012). A review of clinical trials of human papillomavirus prophylactic vaccines. *Vaccine, 30*(Suppl 5), F123–F138.
13. Smahel, M., Polakova, I., Pokorna, D., Ludvikova, V., Duskova, M., & Vlasak, J. (2008). Enhancement of T cell-mediated and humoral immunity of beta-glucuronidase-based DNA vaccines against HPV16 E7 oncoprotein. *International Journal of Oncology, 33*(1), 93–101.
14. Penney, C. A., Thomas, D. R., Deen, S. S., & Walmsley, A. M. (2011). Plant-made vaccines in support of the millennium development goals. *Plant Cell Reports, 30*(5), 789–798.
15. Zeitlin, L., Olmsted, S. S., Moench, T. R., Co, M. S., Martinell, B. J., Paradkar, V. M , et al. (1998). A humanized monoclonal antibody produced in transgenic plants for immuno protection of the vagina against genital herpes. *Nature Biotechnology, 16*(13), 1361–1364.
16. Daniell, H., Streatfield, S. J., & Wycoff, K. (2001). Medical molecular farming: production of antibodies, biopharmaceuticals and edible vaccines in plants. *Trends in Plant Science, 6*(5), 219–226.
17. ZM, Ahmad N. (2013). Green factories: plastids for the production of foreign proteins at high levels. *Gene Therapy and Molecular Biology, 15*, 14–29.

Emerging Viruses

20

Abstract

Emerging viruses are viruses that appear suddenly in the human population. These are viruses to which man has no history of exposure and thus no or limited immunity; they are not new evolutionary creations, but are viruses than man meets due to environmental changes, such as deforestation, entering into new habitats, or viruses that are transmitted from one species of animal to humans. Most of these viruses are terrifying, and cause hemorrhagic fever, a complete destruction of the circulation system; they include Lassa fever, Nipah virus, Ebola, HIV, Severe acute respiratory syndrome (SARS), and, recently, Middle East respiratory syndrome (MERS), which is the latest in a series of "new" respiratory viruses infecting man. It is possible that unknown emerging viruses are the cause of death, often listed as "death due to an unknown cause," as in the retrospective cases of HIV. Emerging viruses might also include poliovirus and influenza, since their introduction into the human population is (was) often sudden and due to changes in the environment or due to contact with other animal species. For examples, polio was a result of changes in sanitation in the countries of North America and Western Europe, and influenza is constantly jumping from aquatic birds to man and other animal species where genomic reassortment occurs.

20.1 Nipah and Hendra Viruses

An interesting constituent of epidemiology is the detective work to track down the origins of novel viruses. An example of such investigative work is the exploration into the origin of Nipah virus, a paramyxovirus, which caused an epidemic illness in Malaysia in 1998. A mild disease occurred among pigs that eventually spread as a severe, fatal disease to man and dogs. The first outbreak was noted in 1998 in the

M. W. Taylor, *Viruses and Man: A History of Interactions*,
DOI: 10.1007/978-3-319-07758-1_20,

Fig. 20.1 Grey headed flying fox (fruit bat)

village of Ampang, where there were a number of cases of fatal encephalitis among farmers, and respiratory infections and encephalitis among pigs. The disease spread to other parts of Malaysia with the movement of pigs for commercial reasons. The outbreak was arrested by a mass cull, which began on March 20, 1999 and ended on April 26 of that year, with a ban on the movement of pigs [1, 2]. Malaysia had 2.4 million pigs and 1,800 pig farms in January 1999; by July 21, 1.1 million pigs and 956 farms had been destroyed and 48 more pig farms had closed. In 1998 and 1999, there was a total of 265 confirmed human cases of Nipah virus and 105 human deaths. Infection with Nipah virus was associated with encephalitis (inflammation of the brain), characterized by fever, drowsiness, and more serious central nervous system symptoms, such as coma, seizures, and inability to breathe. The virus also spread to cats, dogs, and even to ponies. There have been no further outbreaks in Malaysia since then.

Other outbreaks of Nipah virus have occurred in humans in Bangladesh in 2004, with 75 % fatalities, [3] and in India. At the time of this outbreak, there was no evidence of human-to-human transmission. The outbreak was not associated with pigs, but with the consumption of date palm sap that had been contaminated with fruit-bat droppings. A more severe outbreak occurred in 2012 in Bangladesh where there was evidence of human-to-human transmission, and transmission through fomites or the handling of the dead bodies [4]. From 2001 to 2012 there were 280 cases of Nipah virus infections in humans, with 211 deaths—a mortality rate of 75 %.

Where did this virus come from and how did it arise? In some cases these outbreaks have been associated with land clearing for agricultural purposes. The affected area of Malaysia had been cleared of jungle and planted with fruit trees. These fruit trees were attractive to a particular species of bats known as "flying foxes (Fig. 20.1)". They in turn sprinkled the areas with their droppings into areas foraged by pigs; the droppings contained the virus, which was infectious to pigs and other mammals. Although the bats contained viral antigens and antibodies, they were healthy. Thus there was a chain of events, triggered by land clearing, that resulted in a novel virus infection. In fact, it has been hypothesized that the

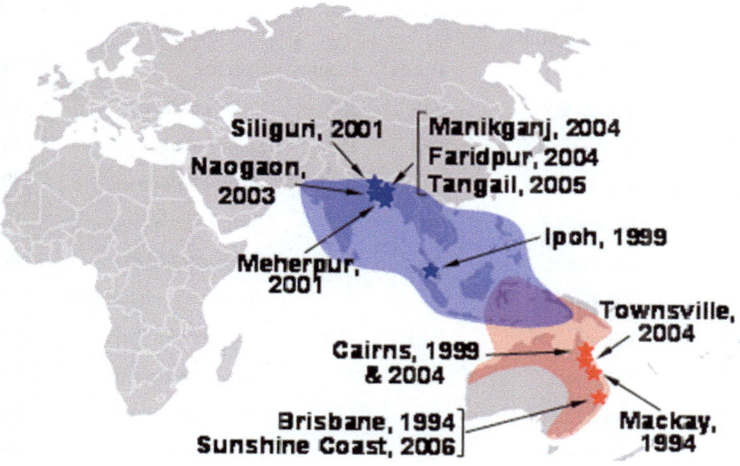

Fig. 20.2 Geographic distribution of Henipavirus outbreaks and fruit bats (*pteropodidae*)

clearing of virgin forests, as is being done in the Amazon, may also result in new infections in man as wildlife attempts to find new habitats.

A similar related virus, Hendra virus (Hendra is a suburb of Brisbane, Australia), affected horses, pigs, and livestock in Australia. There have been a number of occurrences confined to horses, and to three humans working with horses. Two of the three humans infected had respiratory illness with severe flu-like symptoms and died. (For more information see http://www.cdc.gov/ncidod/dvrd/spb/mnpages/dispages/nipah.htm.)

Both of these viruses are paramyxoviruses belonging to a new genus called henipavirus. The Hendra virus was first isolated in 1996, and the Nipah virus in 1999. Figure 20.2 shows the range of spread of *Pteropus* (flying fruit bats) and the sites of outbreaks of Nipah virus and Hendra virus. The National Institute of Allergy and Infectious Diseases (NIAID) has developed an effective vaccine against both viruses.

20.2 Ebola Virus

Ebola outbreaks occur with ferocity and suddenness, and with high mortality; they may originate from bats, and the virus spreads easily to a susceptible human population. Ebola is the most lethal human viral infection known, First identified in 1976 in Zaire and the Sudan, it causes hemorrhagic fever (internal bleeding) with a mortality rate of 88 %. The first recorded outbreak killed 280 out of 318 cases in Zaire, and 151 out of 284 cases in the Sudan. The disease spread very rapidly among

Table 20.1 Cases of Ebola Hemorrhagic fever in Africa, 1976–2012. http://www.cdc.gov/vhf/ebola/resources/distribution-map.html

Country	Town	Cases	Deaths	Species	Year
Dem. Rep. of Congo	Yambuku	318	280	*Zaire ebolavirus*	1976
South Sudan	Nzara	284	151	*Sudan ebolavirus*	1976
Dem. Rep. of Congo	Tandala	1	1	*Zaire ebolavirus*	1977
South Sudan	Nzara	34	22	*Sudan ebolavirus*	1979
Gabon	Mekouka	52	31	*Zaire ebolavirus*	1994
Ivory Coast	Tai Forest	1	0	*Taï Forest ebolavirus*	1994
Dem. Rep. of Congo	Kikwit	315	250	*Zaire ebolavirus*	1995
Gabon	Mayibout	37	21	*Zaire ebolavirus*	1996
Gabon	Booue	60	45	*Zaire ebolavirus*	1996
South Africa	Johannesburg	2	1	*Zaire ebolavirus*	1996
Uganda	Gulu	425	224	*Zaire ebolavirus*	2000
Gabon	Libreville	65	53	*Zaire ebolavirus*	2001
Republic of Congo	Not specified	57	43	*Zaire ebolavirus*	2001
Republic of Congo	Mbomo	143	128	*Zaire ebolavirus*	2002
Republic of Congo	Mbomo	35	29	*Zaire ebolavirus*	2003
South Sudan	Yambio	17	7	*Zaire ebolavirus*	2004
Dem. Rep. of Congo	Luebo	264	187	*Zaire ebolavirus*	2007
Uganda	Bundibugyo	149	37	*Bundibugyo ebolavirus*	2007
Dem. Rep. of Congo	Luebo	32	15	*Zaire ebolavirus*	2008
Uganda	Luwero District	1	1	*Sudan ebolavirus*	2011
Uganda	Kibaale District	11*	4*	*Sudan ebolavirus*	2012
Dem. Rep. of Congo	Isiro Health Zone	36*	13*	*Bundibugyo ebolavirus*	2012
Uganda	Luwero District	6*	3*	*Sudan ebolavirus*	2012

* Numbers reflect laboratory confirmed cases only

hospital personnel treating the sick, and among the general population through contact with ill persons, handling the dead, and reuse of needles. Because there is a danger of infection in preparing the dead for burial, there arose the practice of burning the bodies and the huts where the victims had lived; therefore, strict precautions are necessary for handling Ebola infections. A subtype of Ebola, known as Ebola Reston, occurred among monkeys imported from the Philippines in the state of Virginia and later in Texas, in 1990 and 1992. Luckily this strain was non-pathogenic for humans although some handlers did develop antibodies to the virus [5, 6]. During the 1990s there were fatal outbreaks of Ebola in the Congo; in one case, 16 people became sick from eating a chimpanzee found dead in the jungle. Other cases have involved an index case among hunters and among people associated with hunting

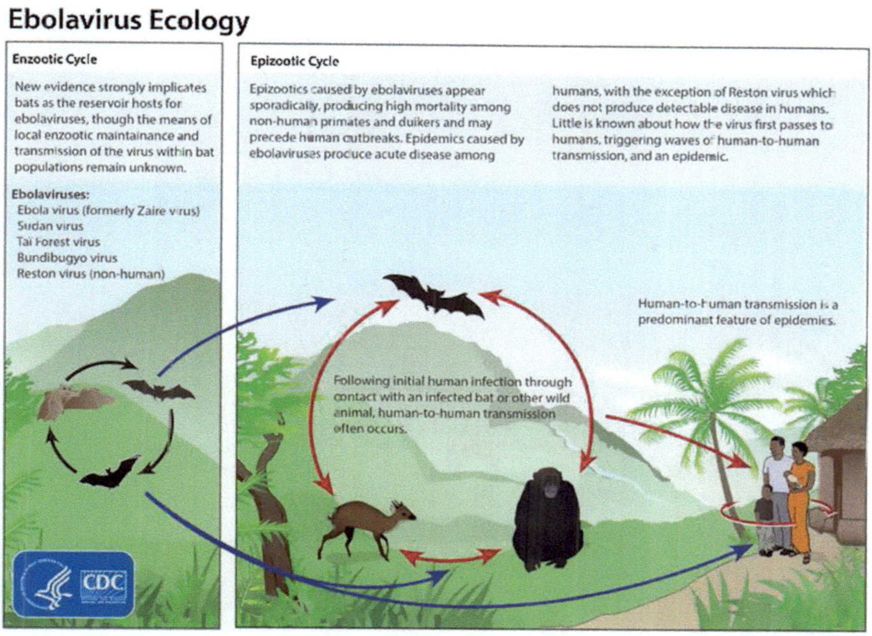

Fig. 20.3 The ecology of Ebola virus (CDC). This graphic shows the life cycle of the ebolavirus. Bats are strongly implicated as both reservoirs and hosts for the ebolavirus. Of the five identified ebolavirus subtypes, four are capable of human-to-human transmission. Initial infections in humans result from contact with an infected bat or other wild animal. Strict isolation of infected patients is essential to reduce onward ebolavirus transmission

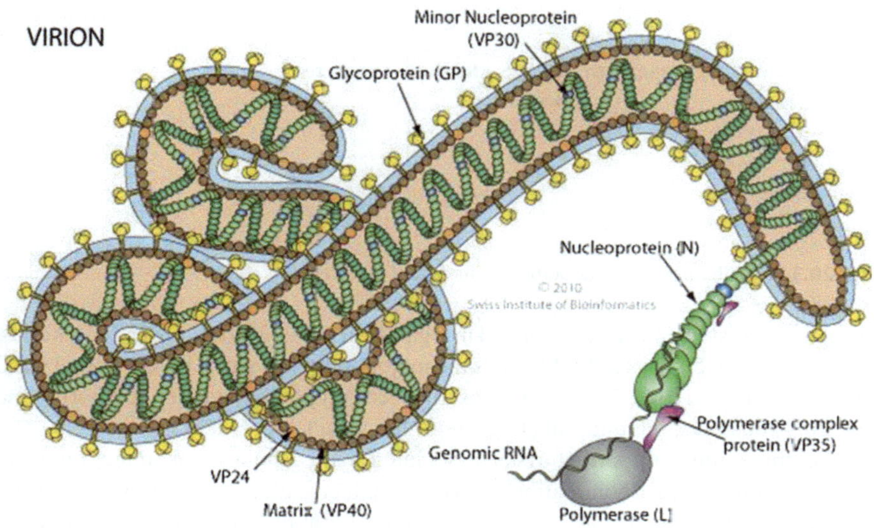

Fig. 20.4 Schematic of Ebola virus (ViralZone, SIB Swiss Institute of Bioinformatics)

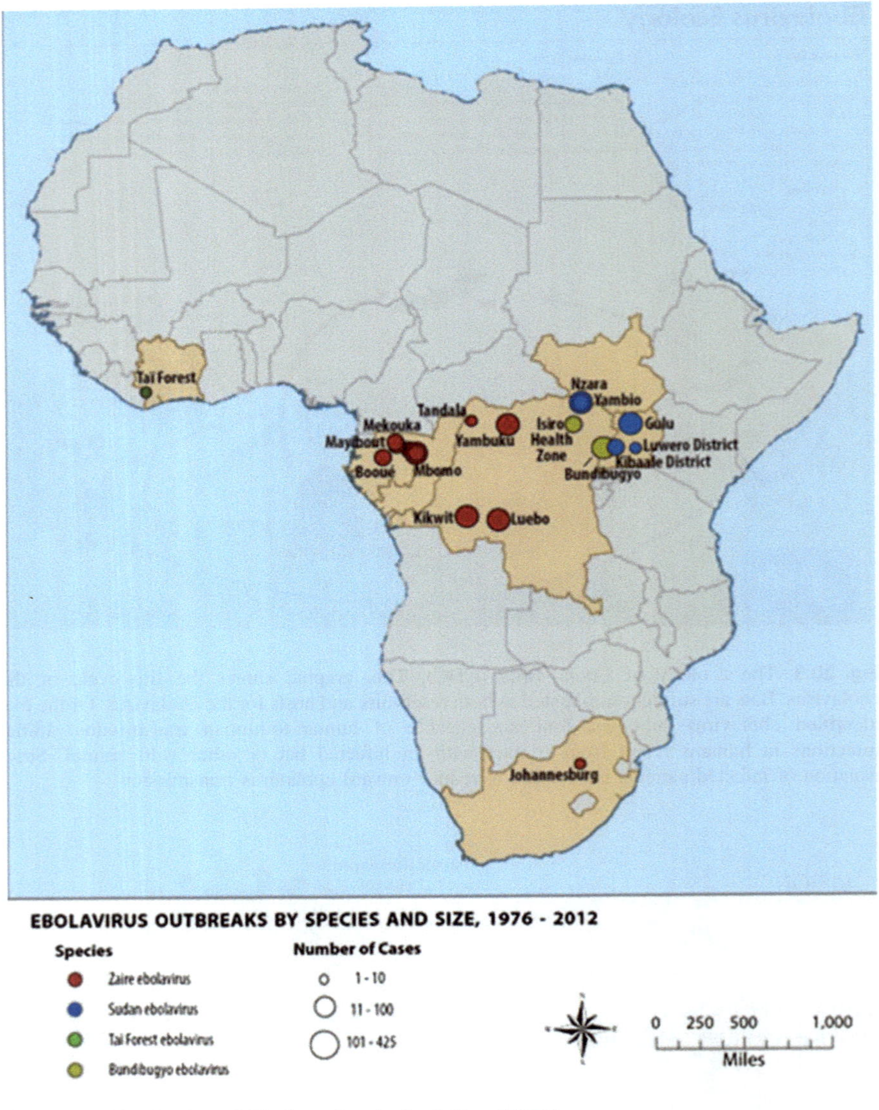

Fig. 20.5 Distribution of Ebola outbreaks in Africa (WHO)

primates. Sizeable outbreaks occurred in Uganda in 2000–2001, with 425 cases and a mortality rate of 53 %. Table 20.1 lists the major outbreaks since 1976 (source: CDC Ebola resources) (Fig. 20.3 illustrates the ecology as known, of the virus and Fig. 20.4 illustrates the unusual structure of the virus).

The three most important risks associated with Ebola virus infection were attending funerals of Ebola patients, having contact with the sick in one's family, and providing medical care to Ebola patients without using adequate personal protective measures [7, 8]. At the time of this writing (2014), there are outbreaks of Ebola in the Congo and Uganda. The disease also infects and is fatal for non-human primates such as monkeys, gorillas and chimpanzees. According to the CDC and WHO, the zoonose origin of Ebola is still unknown, although bats are suspect in this case and also in a similar disease, Marburg hemorrhagic fever. Figure 20.5 presents the distribution of Ebola on the African continent.

Ebola virus belongs to a family of viruses known as "filoviruses." This is a long, 970 nm filamentous negative-stranded RNA virus (Fig. 20.2). The viral proteins inhibit interferon activity. The viral RNA codes for 7–8 proteins.

20.3 SARS

SARS first appeared in 2002. Between November 2002 and July 2003 there were approximately 800 cases in southern China, with a death toll of 10 %. The major "hot spot" was Hong Kong, with 9 % fatality, but within a few months SARS spread worldwide, carried by unsuspecting travelers. The range and the speed of worldwide transmission are very well documented. (A day-to-day account of the epidemic is presented both at http://www.cdc.gov/about/history/sars/timeline.htm and http://en.wikipedia.org/wiki/Progress_of_the_SARS_outbreak.)

In the days before air transport, it may have been confined to this area of southern China.

The infectious virus was identified as a coronavirus in 2003 (Fig. 20.6).

SARS typically begins with flu-like signs and symptoms—fever, chills, muscle aches and occasionally diarrhea. After about a week, signs and symptoms include fever of 100.4 °F (38 °C) or higher, dry cough, and shortness of breath.

The SARS epidemic (or pandemic) showed how quickly infection can spread in a highly mobile and interconnected world. The SARS epidemic also showed how international cooperation among health care experts can effectively contain the

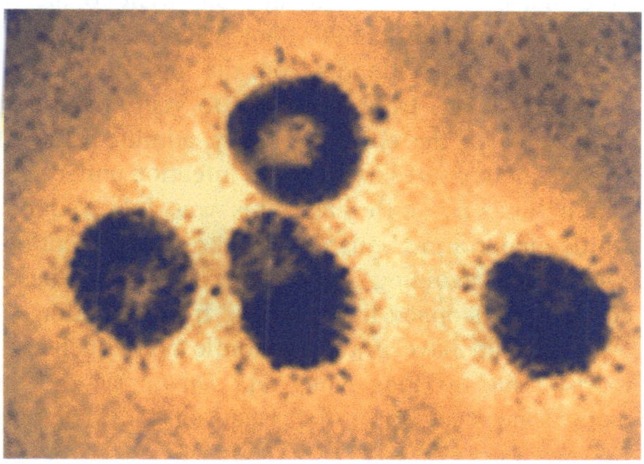

Fig. 20.6 Electron micrograph image of SARS (http://www.cdc.gov/niosh/topics/SARS)

spread of the disease. Since 2004, known instances of SARS transmission have fallen to zero worldwide.

The virus spread from southern China to Singapore, Taiwan, the U.S. and Canada (Ontario). To date, there is no vaccine against SARS; however, the passive transfer of antibodies inhibits the virus. If an outbreak is suspected, patients should be quarantined, and a strict regimen of sterility enforced. Most important is for health care workers to avoid contact with the patients, to wear surgical masks, and to regularly wash their hands.

The detective work involved in tracking the SARS epidemic is fascinating in its extensiveness and detail. Every case appears to have been followed and documented (see references to CDC and Wikipedia above).

The first case of SARS appeared to have been a farmer from the Guangdong area of China (that borders Hong Kong). The Chinese authorities reported the outbreak to the WHO, with 305 cases (including 105 health care workers) and five deaths. This may have been an underestimate of the severity of the outbreak [9]. The first clue of the outbreak appears to be on November 27, 2002, when Canada's Global Public Health International Network (GPHIN), an electronic warning system that is part of the World Health Organization's Global Outbreak and Alert Response Network (GOARN), picked up reports of a "flu outbreak" in China through Internet media monitoring and analysis, and the data was forwarded to the WHO. Most news media in China did not report the extent of the outbreak, in keeping with the government's policy of not reporting bad news. This hindered the treatment and the recognition that this was a major new epidemic illness. The illness spread extremely quickly, primarily among health care workers and physicians attending to the sick. What is rather amazing is the speed with which the virus was sequenced, recognized, and brought under control by quarantine and international cooperation. Scientists are still puzzled as to how 64-year-old Dr. Liu Jianlun could have transferred—during his brief stay—the virus to 16 other occupants of the same floor of his hotel. These guests subsequently travelled to other destinations, including Toronto, Vietnam, and Singapore, thus spreading the virus. In less than 4 months, some 4,000 cases and 550 deaths from SARS outside of China can be traced to this one visit to the Metropol Hotel. (The hotel is considering turning the 9th floor, where Dr. Jianlun stayed, into a museum.) [10].

It took about 5 months from the initial outbreak for the Chinese authorities to allow physicians from the WHO to visit Guodong province, and to admit to the seriousness of the epidemic. Once admitted, the Chinese authorities carried out Draconian measures to bring the infection under control, even firing the mayor and health minister in Beijing. This was followed by mass quarantine of hospital wards, under armed guards, and threat of execution of anyone knowingly having SARS and avoiding quarantine. However, these measures and international cooperation stopped the epidemic from spreading.

SARS did have grave economic consequences. It substantially damaged the tourist trade in Hong Kong and Taiwan as well as in Toronto. Taiwan in part blamed the recession in that island on the SARS outbreak (Fig. 20.7).

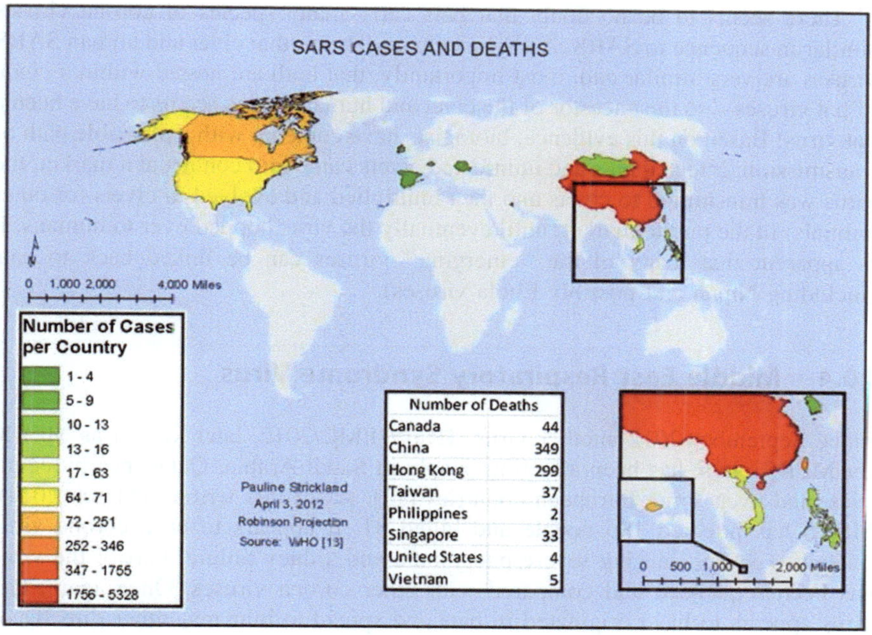

Fig. 20.7 Distribution of SARS cases and deaths. http://en.wikipedia.org/wiki/File:Sars_Cases_and_Deaths.pdf

The identification of the virus occurred rapidly following the initial outbreak of the infection. The isolate (initially unknown whether viral or not, but at one time suspected of being *Chlamydia*) [9] was grown in culture and tested against sera from recovering individuals and uninfected controls. The virus was inhibited by their sera, thus allowing for a diagnostic test to be developed. However, at this early stage it was impossible to identify the virus. The virus was later identified as a previously unknown corona virus. Corona viruses normally cause symptoms of a cold, and take about 10 days' incubation to manifest. In this case, the long incubation period helped the spread of the virus.

Where did the virus come from? Initial testing indicated that the virus came from civets (a member of the cat family), a food delicacy in China and found in the open markets. Further analysis of civets in the wild did not find traces of the virus. Thus the civets in the market must have also been "dead-end" victims of SARS. In 2005 two teams of researchers reported the presence of similar corona viruses in Chinese horseshoe bats. These are probably the culprits, and the virus was probably passed on through an intermediate host in the market. Using DNA sequencing techniques (although this is an RNA virus), scientists have been able to work out the relationship between the virus discovered in humans, bats, and civets [11, 12].

There seems to be no doubt that bats carry many species of corona viruses similar in sequence to SARS. A phylogenic tree shows that civet and human SARS viruses are very similar and, most importantly, that both are nested within a *clade* of bat viruses—so the ancestor of the civet and human strains seems to have been a bat virus! Based on this evidence, biologists have come up with a plausible path of transmission: infected bats and uninfected civets came into contact at a market, the virus was transmitted to civets and then multiplied and evolved in civets (or other animals) in the public market, until eventually the virus hopped over to humans. It is apparent that many of the "emerging" viruses can be linked back to bats (including Nipah and possibly Ebola viruses).

20.4 Middle East Respiratory Syndrome Virus

Since September 2012 another virus (HCoV-EMC/2012, later known as MERS, for MERS virus, has been affecting people in Saudi Arabia, Qatar, nearby countries, and even some European countries [13]. As of this writing (March 2014), MERS has infected 186 people and killed 81. It appears to be a corona virus related to SARS, causing severe pneumonia and kidney failure. Part of the virus has been sequenced and compared with other corona viruses. Once again, this virus appears to have originated in bats and spread to humans, either directly or through an intermediate host. Bats in Saudi Arabia contain many different corona viruses. It was possible to isolate a virus with nucleotide sequences identical to that in an infected patient in one case [14]. The virus does appear in clusters and there is human-to-human transmission. The fatality rate is high—over 30 %. It has been traced to an Egyptian tomb bat, with the sequence of the viral RNA identical to that of the index case.

Thus, this virus appears to be genetically related to the SARS virus. Recent evidence suggests that dromedary camels found in southern Oman may have also been infected in the past with the MERS virus, since antibodies to components of the virus have been found in the sera of such animals, but there is no evidence that this is the source of the current epidemic.

20.5 HantaVirus

Hantavirus causes a severe respiratory infection and is transmitted by a variety of different rodents, mostly mice. The name comes from the Hantan River in South Korea. The virus caused an outbreak of hemorrhagic fever among American and Korean troops during the Korean War of 1951–1953, with more than 3,000 troops being sick with kidney failure and internal bleeding; the mortality rate was 10 %. The virus was not isolated until 1976 and found in a striped field mouse [15] and urban rats.

This disease was first recognized in the U.S. (Four Corners Disease) in 1993. An unexplained pulmonary (lung) illness occurred in an area shared by Arizona, New Mexico, Colorado and Utah. Quite a number of previously healthy young

Fig. 20.8 Cotton rat. The cotton rat, *Sigmodon hispidus*, is a Hantavirus carrier that becomes a threat when it enters human habitation in rural and suburban areas

adults died from an acute respiratory infection; the outbreak was related to an increase in crop yield with a concomitant increase in the deer mouse population. The virus was isolated from a deer mouse captured close to the home of one of the victims, and identified as a new virus, first called "Muerto Canyon virus," but later changed to "Sin Nombre virus" (SNV). Why did this outbreak occur in 1993? Previous years had seen periods of drought, and not enough food to maintain a large deer mouse population. In 1993 there was plenty of rain and snow, and thus the plant yield was greater than usual. It is estimated that the deer mouse population grew tenfold during this period, and thus had more contact with humans [16]. Since the initial outbreak of Hantavirus, other strains of the virus have been identified that have not been associated with the Four Corners states [17]. This is probably not a new virus; the Navajo Indians of the region appeared to be familiar with a respiratory infection that was associated with increases in the mouse population. Certain strains of Hantavirus can be transmitted between humans.

Hantaviruses have now been identified in large areas of the U.S., all carried by different rodents. [These rodents are described in http://www.cdc.gov/hantavirus/rodents/white-footed-mouse.html (Fig. 20.8).]

Since November 2012 there have been 10 confirmed cases of Hantavirus infection in Yosemite National Park, with three fatalities. From the CDC website: "NPS public health officials believe that 9 of the 10 people with confirmed hantavirus infection were exposed to the virus while staying at the Signature Tent Cabins in Curry Village in Yosemite National Park. The other park visitor with Hantavirus infection was probably exposed to the virus while hiking or staying at the High Sierra Camps, located about 15 miles from Curry Village. Some of the infected cabins have now been destroyed." (An interesting article on the Yosemite outbreak can be found at http://www.outsideonline.com/adventure-travel/north-america/united-states/national-parks/Death-at-Yosemite-The-Story-Behind-Last-Summers-Hantavirus-Outbreak.html.)

The Hantavirus belongs to the family of *Bunyaviridae*, a single-stranded RNA virus with a segmented genome. The virus encodes four to six proteins.

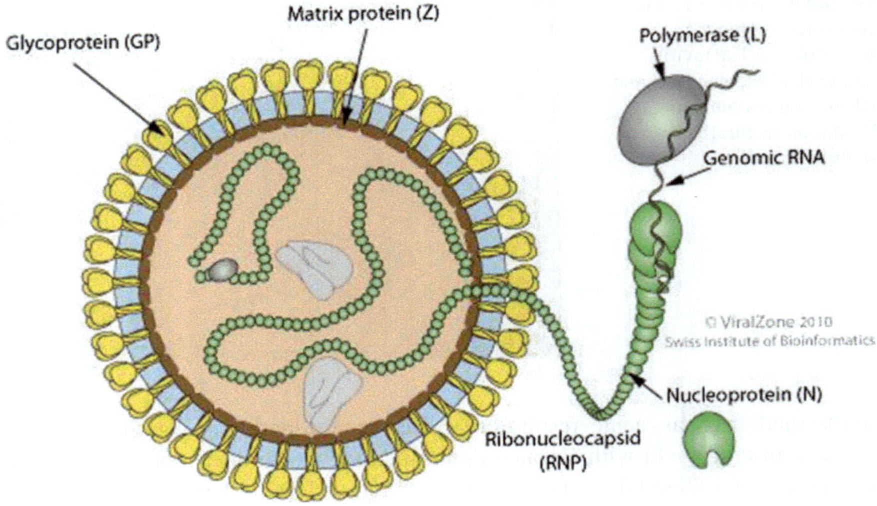

Fig. 20.9 Schematic of an arenavirus. Note that there are two separate strands of RNA (ViralZone, SIB Swiss Institute of Bioinformatics)

20.6 Lassa Fever

There are many other types of hemorrhagic fever viruses, not as lethal as Ebola, but still very dangerous. These include the arena viruses, a large class of viruses that include Lassa fever, and other hemorrhagic fever viruses named for their countries of origin, such as Bolivian, Venezuelan, and Brazilian hemorrhagic fevers; these viruses are associated with rats and mice (Fig. 20.9). They do not cause obvious illness in the rodent population and the virus is shed in the urine or droppings of the rodents. The virus can be inhaled as an aerosol, or in some cases transferred by person-to-person contact, as discussed above for Ebola. Lassa virus was first identified in 1969 when two missionary nurses died in Nigeria, West Africa. The virus causes significant morbidity and mortality. It is endemic in western Africa and the estimates are of between 100,000 and 300,000 infections per year. The mortality rate varies between 5–25 %, depending on the outbreak. Many more may be infected without symptoms. The reservoir, or host, of Lassa virus is a rodent known as the "multimammate rat" of the genus *Mastomys*. This is a small rat that lives among humans in large areas of West, Central and East Africa. The virus is spread in the droppings and urine of the rats, and may be picked up in dust. The rats are often consumed as food, and if infected, transmission to humans may occur during handling. Human-to-human transmission may also occur through body fluids and sexual transmission, since the virus can be found in semen. Eradicating the rats, storing food in rodent-proof containers, trapping and removing the animals and avoiding them for food, can control the

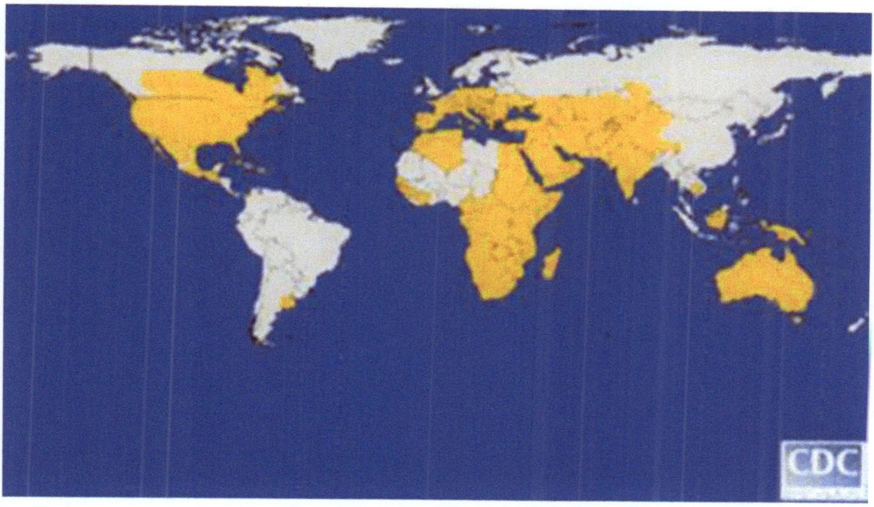

Fig. 20.10 Global distribution of West Nile virus (CDC)

virus. However, the population of rats is so great that annihilation of the population is very difficult.

Lassa fever belongs to the genus of viruses known as "arenavirus." Five arenaviruses are known to cause human illness: Lassa virus, Junin virus, Machupo virus, Guanarito virus and Sabia virus. The structure of the nucleic acid is rather unusual, in that it is segmented, each piece of opposite polarity. Both segments are ambisense. The viral RNA-dependent RNA polymerase (L) binds to a promoter on each encapsidated segment, and transcribes a messenger RNA. Transcription is terminated by a strong hairpin sequence at the end of each gene. mRNAs are capped, probably by L protein during synthesis. An ambisense genome is one in which both nucleic acid strands encode for proteins (Fig. 20.9).

20.7 West Nile Virus

Mosquitoes or other insects transmit many "emerging" viruses. Recently there have been outbreaks in the U.S. of West Nile virus, a virus carried by mosquitoes and transmitted to birds and often to humans. West Nile virus was first isolated from a feverish woman in the West Nile District of Uganda in 1937. The ecology was characterized in Egypt in the 1950s. The virus became recognized as a cause of severe human meningitis or encephalitis (inflammation of the spinal cord and brain) in elderly patients during an outbreak in Israel in 1957. An equine form of the disease was first noted in Egypt and France in the early 1960s. WNV first appeared in North America in 1999, with encephalitis reported in humans and horses. Since then there have been a large number of cases of WNV in birds and in

Table 20.2 Number of incidents of West Nile virus by state (USA) from 1999–2012

West Nile virus disease cases reported to CDC by state, 1999–2012 State	1999	2000	2001	2002	2003	2004	2005	2006	2007	2008	2009	2010	2011	2012	Total
Alabama	0	0	2	49	37	16	10	8	24	18	0	3	5	62	234
Alaska	0	0	0	0	0	0	0	0	0	0	0	0	0	0	0
Arizona	0	0	0	0	13	391	113	150	97	114	20	167	69	133	1,267
Arkansas	0	0	0	43	25	28	28	29	20	9	6	7	1	64	260
California	0	0	0	1	3	779	880	278	380	445	112	111	158	479	3,626
Colorado	0	0	0	14	2,947	291	106	345	576	71	103	81	7	131	4,672
Connecticut	0	1	6	17	17	1	6	9	4	8	0	11	9	21	110
Delaware	0	0	0	1	17	0	2	0	1	1	0	0	1	9	32
Dist. of Columbia	0	0	0	34	3	2	5	2	0	8	2	6	15	10	87
Florida	0	0	12	28	94	41	21	3	3	3	3	12	24	73	317
Georgia	0	0	6	44	50	21	20	8	50	8	4	13	22	99	345
Hawaii	0	0	0	0	0	0	0	0	0	0	0	0	0	0	0
Idaho	0	0	0	0	1	3	13	996	132	39	38	1	3	17	1,243
Illinois	0	0	0	884	54	60	252	215	101	20	5	61	34	290	1,976
Indiana	0	0	0	293	47	13	23	80	24	4	4	13	9	77	587
Iowa	0	0	0	54	147	23	37	37	30	6	5	9	9	31	388
Kansas	0	0	0	22	91	43	25	30	40	31	13	19	4	56	374
Kentucky	0	0	0	75	14	7	5	6	4	3	3	3	5	23	148

(continued)

Table 20.2 (continued)

West Nile virus disease cases reported to CDC by state, 1999–2012 State	1999	2000	2001	2002	2003	2004	2005	2006	2007	2008	2009	2010	2011	2012	Total
Louisiana	0	0	1	329	124	109	171	180	40	49	21	27	10	335	1,396
Maine	0	0	0	0	0	0	0	0	0	0	0	0	0	1	1
Maryland	0	0	6	36	73	16	5	11	10	14	1	23	19	47	261
Massachusetts	0	0	3	23	17	0	6	3	6	1	0	7	6	33	105
Michigan	0	0	0	614	19	16	62	55	17	17	1	29	34	202	1,066
Minnesota	0	0	0	48	148	34	45	65	101	10	4	8	2	70	535
Mississippi	0	0	0	192	87	51	70	183	136	65	53	8	52	247	1,144
Missouri	0	0	0	168	64	36	30	62	77	15	5	3	10	20	490
Montana	0	0	0	2	222	6	25	34	202	5	5	0	1	6	508
Nebraska	0	0	0	152	1,942	53	188	264	163	47	52	39	29	193	3,122
Nevada	0	0	0	0	2	44	31	124	12	16	12	2	16	9	268
New Hampshire	0	0	0	0	3	0	0	0	0	0	0	1	0	1	5
New Jersey	0	6	12	24	34	1	6	5	1	10	3	30	7	48	187
New Mexico	0	0	0	0	209	88	33	8	60	8	8	25	4	47	490
New York	62	14	15	82	71	10	38	21	22	46	7	128	44	107	670
North Carolina	0	0	0	2	24	3	4	1	8	3	0	0	2	7	54
North Dakota	0	0	0	17	617	20	86	137	369	37	1	9	4	89	1,386
Ohio	0	0	0	441	108	12	61	48	23	15	2	5	21	121	857

(continued)

Table 20.2 (continued)

West Nile virus disease cases reported to CDC by state, 1999–2012 State	1999	2000	2001	2002	2003	2004	2005	2006	2007	2008	2009	2010	2011	2012	Total
Oklahoma	0	0	0	21	79	22	31	48	107	9	10	1	1	191	520
Oregon	0	0	0	0	0	3	7	69	26	16	11	0	0	11	143
Pennsylvania	0	0	3	62	237	15	25	9	10	14	0	28	6	60	469
Puerto Rico	0	0	0	0	0	0	0	0	0	0	0	0	0	1	1
Rhode Island	0	0	0	1	7	0	1	0	1	1	0	0	1	4	16
South Carolina	0	0	0	1	6	2	5	1	5	1	3	1	0	29	54
South Dakota	0	0	0	37	1,039	51	229	113	208	39	21	20	2	203	1,962
Tennessee	0	0	0	56	26	14	18	22	11	19	9	4	18	33	230
Texas	0	0	0	202	720	176	195	354	260	64	115	89	27	1,868	4,070
Utah	0	0	0	0	1	11	52	158	70	26	2	2	3	5	330
Vermont	0	0	0	1	3	0	0	0	0	0	0	0	1	3	8
Virginia	0	0	0	29	26	5	1	5	5	1	5	5	9	30	121
Washington	0	0	0	0	0	0	0	3	0	3	38	2	0	4	50
West Virginia	0	0	0	3	2	0	0	1	0	1	0	0	2	10	19
Wisconsin	0	0	0	52	17	12	17	21	13	8	1	2	3	57	203
Wyoming	0	0	0	2	375	10	12	65	181	8	12	6	3	7	681
Total	62	21	66	4,156	9,862	2,539	3,000	4,269	3,630	1,356	720	1,021	712	5,674	37,088

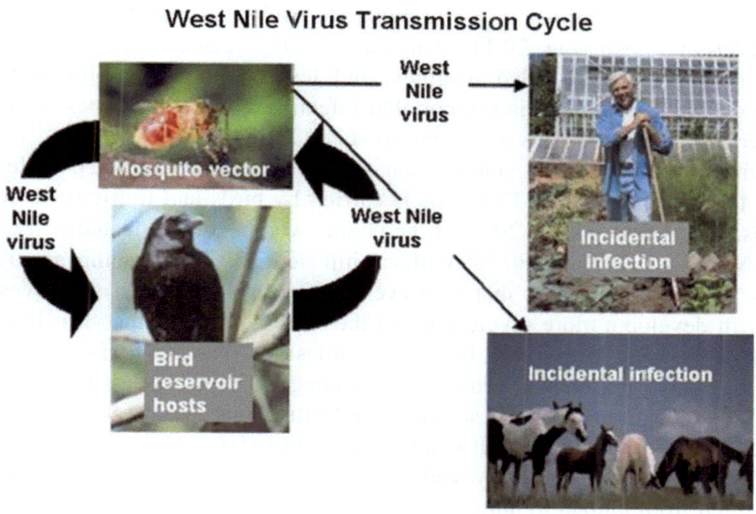

Fig. 20.11 West Nile transmission cycle. Note that man and horse are "dead-end" hosts (http://www.cdc.gov/westnile/index.html)

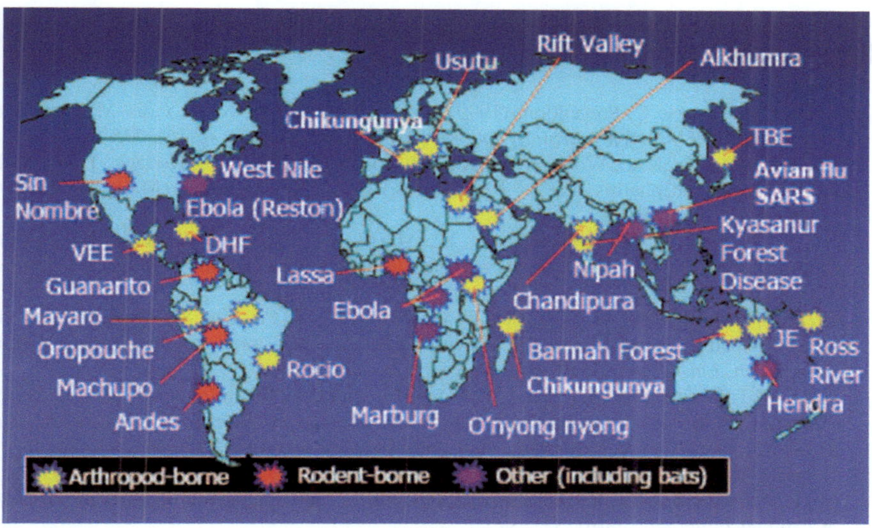

Fig. 20.12 Map of emerging viruses showing the animal reservoir, or the vector (American Medical Veterinary Association)

humans, and it is now considered one of the most widespread of the "emerging viruses." The virus has now spread globally with new strains being identified in 2012 (Fig. 20.10).

The virus is transmitted in a bird-mosquito cycle, with humans being considered dead- end hosts (Fig. 20.11). The major birds infected include the American crow, blue jays, black-billed magpies, and less commonly, robins and house sparrows. West Nile virus is a reportable infection and the CDC has tracked the number of cases in humans according to state (Table 20.2).

Numbers fluctuate from year to year after a large increase in 2002. In 2012, Texas had the most cases. The virus is carried by birds and transmitted by mosquitoes. It is estimated that 20 % of the people who become infected will develop West Nile fever, leaving 80 % of those infected not experiencing any type of illness. It is also estimated that 1 in every 150 persons infected with West Nile virus will develop a more severe form of the disease. In 2012 the virus killed 286 people in the U.S., with Texas being the hardest hit. Those who survive encephalitis may have permanent neurological damage. The virus has been found in almost all rodents in the U.S., so when controlling this disease it is important to control mosquito populations. It is also important to wear protective clothing and to spray clothing and the body with mosquito repellent. West Nile virus is not transmitted from human to human (see Fig. 20.5); it is a *flavivirus*, the same class as yellow fever and hepatitis C. Its mode of replication is similar to that of those two viruses. Figure 20.12 is a summary figure of the known emerging viruses as of 2012.

References

1. Centers for Disease C, Prevention. (1999). Update: Outbreak of Nipah virus–Malaysia and Singapore, 1999. *MMWR Morbidity and Mortality Weekly Report, 48*(16), 335–337.
2. Chua, K. B., Goh, K. J., Wong, K. T., Kamarulzaman, A., Tan, P. S., Ksiazek, T. G., et al. (1999). Fatal encephalitis due to Nipah virus among pig-farmers in Malaysia. *Lancet, 354*(9186), 1257–1259.
3. Hsu, V. P., Hossain, M. J., Parashar, U. D., Ali, M. M., Ksiazek, T. G., Kuzmin, I., et al. (2004). Nipah virus encephalitis reemergence Bangladesh. *Emerging Infectious Diseases, 10*(12), 2082–2087.
4. Sazzad, H. M., Hossain, M. J., Gurley, E. S., Ameen, K. M., Parveen, S., Islam, M. S., et al. (2013). Nipah virus infection outbreak with nosocomial and corpse-to-human transmission Bangladesh. *Emerging Infectious Diseases, 19*(2), 210–217.
5. Geisbert, T. W., & Jahrling, P. B. (1990). Use of immunoelectron microscopy to show Ebola virus during the 1989 United States epizootic. *Journal of Clinical Pathology, 43*(10), 813–816.
6. Geisbert, T. W., Rhoderick, J. B., & Jahrling, P. B. (1991). Rapid identification of Ebola virus and related filoviruses in fluid specimens using indirect immunoelectron microscopy. *Journal of Clinical Pathology, 44*(6), 521–522.
7. Crowcroft, N. S., Morgan, D., & Brown, D. (2002). Viral haemorrhagic fevers in Europe–effective control requires a co-ordinated response. *Euro surveillance: bulletin Europeen sur les maladies transmissibles = European Communicable disease bulletin, 7*(3), 31–32.
8. Francesconi, P., Yoti, Z., Declich, S., Onek, P. A., Fabiani, M., Olango, J., et al. (2003). Ebola hemorrhagic fever transmission and risk factors of contacts Uganda. *Emerging Infectious Diseases, 9*(11), 1430–1437.
9. Cyranoski, D. (2003). China joins investigation of mystery pneumonia. *Nature, 422*(459), 456–459.

10. Fleck, F. (2003). How SARS changed the world in less than six month. *Bulletin of World Heath Organization, 81*(8), 625–626.
11. Lau, S. K., Woo, P. C., Li, K. S., Huang, Y., Tsoi, H. W., Wong, B. H., et al. (2005). Severe acute respiratory syndrome coronavirus-like virus in Chinese horseshoe bats. *Proceedings of the National Academy of Sciences of the United States of America, 102*(39), 14040–14045.
12. Li, W., Shi, Z., Yu, M., Ren, W., Smith, C., Epstein, J. H., et al. (2005). Bats are natural reservoirs of SARS-like coronaviruses. *Science, 310*(5748), 676–679.
13. de Groot, R. J., Baker, S. C., Baric, R. S., Brown, C. S., Drosten, C., Enjuanes, L., et al. (2013). Middle East respiratory syndrome coronavirus (MERS-CoV): Announcement of the Coronavirus Study Group. *Journal of Virology, 87*(14), 7790–7792.
14. Memish, Z. A., Mishra, N., Olival, K. J., Fagbo, S. F., Kapoor, V., Epstein, J. H., et al. (2013). Middle East respiratory syndrome coronavirus in bats, Saudi Arabia. *Emerging Infectious Diseases, 19*(11), 1819–1823.
15. Lee, H. W., Baek, L. J., & Johnson, K. M. (1982). Isolation of Hantan virus, the etiologic agent of Korean hemorrhagic fever, from wild urban rats. *Journal of Infectious Diseases, 146*(5), 638–644.
16. Jenison, S., Hjelle, B., Simpson, S., Hallin, G., Feddersen, R., & Koster, F. (1995). Hantavirus pulmonary syndrome: Clinical, diagnostic, and virologic aspects. *Seminars in Respiratory Infections, 10*(4), 259–269.
17. Morzunov, S. P., Feldmann, H., Spiropoulou, C. F., Semenova, V. A., Rollin, P. E., Ksiazek, T. G., et al. (1995). A newly recognized virus associated with a fatal case of hantavirus pulmonary syndrome in Louisiana. *Journal of Virology, 69*(3), 1980–1983.

Manufacturing Viruses

21

Abstract

The manufacturing of viruses in the test tube begins with the observations that viral capsid proteins undergo self-assembly when ionic and pH conditions are suitable. Components of *Escherichia coli* phage, such as T4 and λ, assemble spontaneously to form whole phage particles, and components of icosahedral viruses in the presence or absence of nucleic acid form spherical structures. Viral RNA was copied by the reverse transcriptase into cDNA and expressed on a plasmid with the formation of intact virus. By 2002, poliovirus was synthesized from full-length cDNA. With the progress in technology for the synthesis of long stretches of DNA and improvements in sequencing, it was a natural next step to construct a virus in a cell-free system using commercially available DNA. Poliovirus and $Q\beta$ bacteriophage were synthesized in vitro; similar techniques were used to sequence and reconstitute the 1918 flu pandemic virus, and to modify the host range of avian influenza virus. These latter experiments were criticized as being dangerous to humans and to the environment, and new regulations were introduced to control such experiments.

21.1 Development of In Vitro Techniques for Virus Assembly

In 2002, a controversial paper was published in *Science* [1] announcing the full in vitro synthesis of the poliovirus in a cell-free extract, devoid of ribosomes, mitochondria and nuclei. This was performed using cDNA (complementary DNA) manufactured synthetically from the known sequence of the viral RNA. Some condemned the paper as irresponsible, as a recipe for terrorists, as a stunt, or an ego trip, while others hailed it as a major scientific breakthrough.

M. W. Taylor, *Viruses and Man: A History of Interactions*,
DOI: 10.1007/978-3-319-07758-1_21,

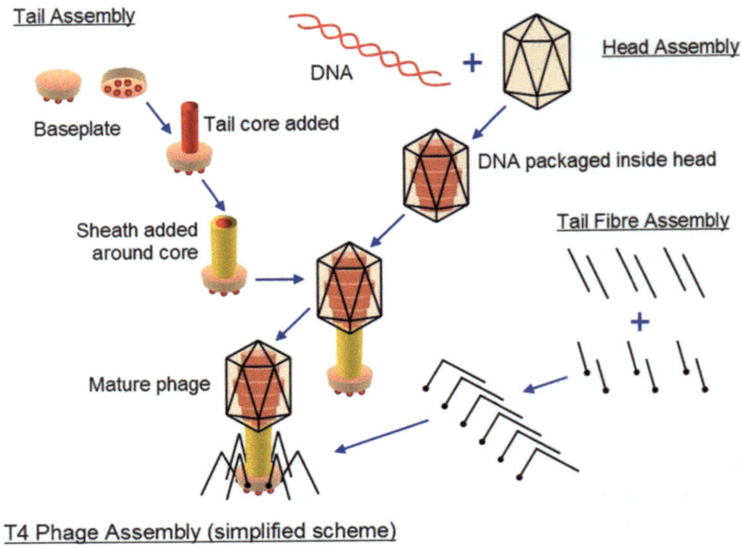

Fig. 21.1 Phage assembly (with permission, www.Cronodon.com)

This scientific advance is part of a natural progression in the field of virology, starting with the research in the 1960s of creating phage T4 in vitro from components that underwent self-assembly [2]. When phage components—such as phage heads, tails, and tail fibers—were mixed in a cell-free extract, whole phage particles were made. This was done by using temperature-sensitive mutants of phage blocked in the assembly of one component, and mixing them with extracts from phage blocked in a second component. The original experiments were performed with phage T4 (Fig. 21.1) and phage λ [3]. T4 assembly to the intact phage does require scaffolding proteins, and the packaging of the DNA into the phage head does need an energy source, ATP. However, the head and tail assembly occurs spontaneously, without other interactions.

Small icosahedral viral capsid proteins, such as those of picornavirus, undergo self-assembly when the concentration reaches a critical mass, and the pH and salt concentrations are optimum. During picornavirus replication (poliovirus, mengovirus, or bovine enterovirus), subviral particles containing subunits of the capsid proteins can be found and isolated. Such subviral particles will self-assemble and form empty capsids with icosahedral structures identical to whole virus particles as viewed by electron microscopy [4]. Similar studies of self-assembly of capsid proteins and nucleic acid were performed with cowpea chlorotic mottle virus and brome mosaic virus. In both cases, viable infectious virus was created in vitro by mixing capsid protein and nucleic acid [5, 6].

It can be argued that in vitro self-assembly of viruses was first shown in the elegant experiments of Heinz Fraenkel-Conrat with tobacco mosaic virus and Holmes' ribgrass virus. Viral nucleic acid was mixed with the coat proteins of the respective viruses and the nucleic acid was packaged in vitro. However, the purpose of these experiments was to prove that RNA was the genetic material and that TMV RNA could be packaged in Holmes' ribgrass coat protein and reverse [7].

A major advance occurred when Taniguchi et al. [8] showed that it was possible to make a full-length DNA copy of a phage RNA using reverse transcriptase (the enzyme found in retroviruses). Phage Qβ is a phage of RNA with a length of approximately 4,500 nucleotides, about half that of poliovirus RNA. The RNA was copied into positive and negative strands and cloned into a plasmid (a small circular piece of DNA that replicates in a bacterial system) and transformed (inserted) into *Escherichia coli* HB101. The resulting transfection produced full-length and active Qβ virus. Although the nucleotides were not chemically synthesized but rather through enzymatic reactions, this was the in vitro synthesis of a viral genome using reverse genetics. The concepts and techniques developed in the Weissmann laboratory could be adapted to the synthesis of any RNA virus into DNA, with the incorporation of alternative nucleotides to create mutants. Both positive- and negative-stranded RNA can be copied into cDNA. In the case of a positive-stranded RNA virus, the cDNA is transfected directly into the host cell to produce virus. In the case of negative-stranded RNA, the cDNA is cloned into a plasmid under the control of a cellular promoter (regulatory sequence), which in turn is added to the host cell to produce virus. Similar studies were initiated with the human pathogen, poliovirus [9, 10], with double-stranded cDNA copies of polio being cloned into a commonly used *E. coli* plasmid (pBR322). The full-length virus (representing 7,440 nucleotides) was ligated together from three segments that were generated, using reverse transcriptase. When this cDNA clone was transfected into HeLa cells, infectious virus was produced. The virus produced by this method was identical to wild-type virus.

Thus it is possible to use in vitro mutagenesis to construct site-specific mutants and pose essential questions on viral replication and viral virulence. Utilizing reverse genetics, full-length infectious cDNA has been made from a large number of different viruses, including rabies, respiratory syncytial virus, measles, Ebola, Bunyavirus, and even segmented viruses such as reovirus and influenza A. This research is summarized in [11]. These constructs can be used to develop novel vaccines and to study virus replication and assembly in detail. However, it is important to emphasize that all were made with naturally occurring RNA or DNA.

21.2 DNA Synthesis

In parallel with these studies, there were major developments in the techniques of synthesizing DNA in vitro. DNA synthesis was initially expensive, only short fragments could be synthesized, and it was time consuming. The first DNA sequence of biological importance synthesized in vitro was the DNA sequence coding for the alanine tRNA of yeast and later the tyrosine suppressor tRNA from *E. coli*. The first was a 75-base pair sequence, and the second a 207-base

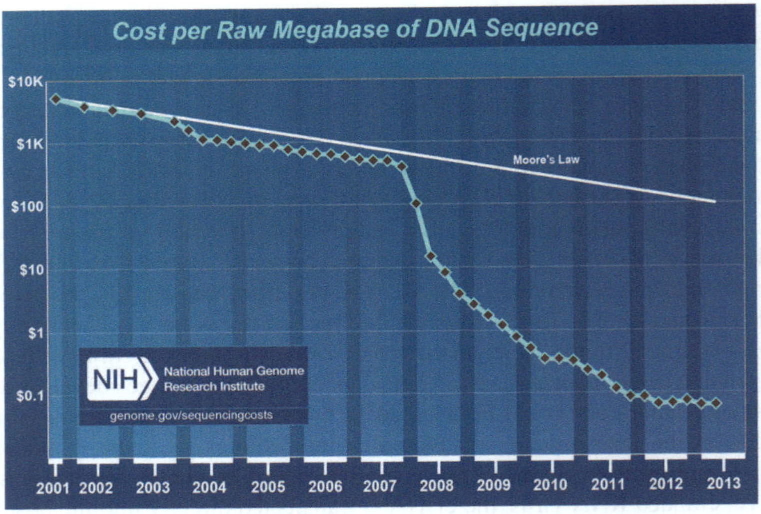

Fig. 21.2 Cost of synthesizing DNA over the years

pair [12, 13]. Their synthesis was hailed as major biological (biochemical) breakthroughs. When I was a graduate student, I remember the excitement generated by the work of Har Gobind Khorana in announcing their synthesis; Khorana received the Nobel Prize in Medicine in 1968 together with Marshall Nirenberg and Robert Holley for work showing how triplet nucleotides formed the genetic code. He also developed the methodology for the synthesis of what at that time was considered long oligonucleotide sequences [14]. Since then, the speed and cost of synthesizing DNA sequences have decreased substantially, and commercial companies now perform DNA (oligonucleotide) synthesis.

In the year 2000, it cost $12 per base pair to synthesize oligonucleotides under special arrangement, so that it cost approximately $85,000–$100,000 to synthesize the genome of polio virus [1]. Today it costs about $5,000. Figures 21.1 and 21.2 demonstrate the decline in the costs of synthesizing DNA and of sequencing DNA. In Fig. 21.1, the cost of generating DNA sequences using Sanger-based chemistries and capillary-based instruments ("first-generation" sequencing platforms) shows a slow decline up to 2007. Beginning in January 2008, the data represent the costs of generating DNA sequence using "second-generation" (or "next-generation") sequencing platforms. The change in instruments represents the rapid evolution of DNA sequencing technologies that occurred in recent years. To exemplify the nature of the reductions in DNA sequencing costs, each graph also shows hypothetical data reflecting Moore's Law, which describes a long-term trend in the computer hardware industry that involves the doubling of "computing power" every 2 years. Technological improvements that keep up with Moore's Law are widely regarded to be doing exceedingly well, making it useful for comparison. Figure 21.2 represents the decline in the cost of sequencing over the years; this also fell drastically with the introduction of newer sequencing methods and technologies. The data in Table 21.1 makes it even clearer. In September 2001 it cost $5,292 to sequence a megabase (1 million base pairs) DNA; as of April 2013, the cost was 6¢ (Fig. 21.3).

With the cost of creating a viral genome under $10,000, we have entered a "brave new world" of virology. This is obviously a bonanza for viral research as well as a danger that viral genetic material may easily fall into the wrong hands. It allows virologists to develop custom-made viruses—that is, alterations in sequences to explore pathology and relationships between gene and structure as well as to easily construct attenuated viruses for vaccines.

The original method, known as the Sanger method, was the method of choice until a few years ago. In 2005, a new method of sequencing was begun; it was called "cyclic array sequencing." This dramatically reduced the cost and speed of sequencing. The cost of sequencing a whole human genome dropped from the millions of dollars to approximately $10,000, so that both the cost of obtaining the full sequence and the manufacture of the DNA is considerably less.

21.3 Viruses Produced In Vitro Without a Natural Template (Until 2013)

Obviously, construction of a viral genome requires knowledge of the sequence of that genome. How many viral genomes are there? At the time of writing this chapter, the NIH Genome Repository contains the sequences of 3,570 complete viral genomes, including animal viruses, bacteriophage, plant viruses, and satellite viruses. These range in size from the mimivirus with 1,181,404 base pairs to $Q\beta$ with 4,215 base pairs.

The first virus synthesized from commercially available synthetic DNA outside a living cell was poliovirus. To correct a misconception, the virus is formed from synthetic DNA as described. The DNA was added to a cell-free extract (that is, an extract derived from cells, but without nuclei or mitochondria), but that contains ribosomes, which are required for protein synthesis. This was a step different from previous work performed by the same group, in which cDNA was made from a viral template and transfected into cells to produce replicating wild-type virus. In one, the virus is made in a test tube; in the other, the virus is made in a cell. The virus produced in this artificial cell-free system appeared to be similar to wild-type virus except for subtle changes made to distinguish it from possible contaminants.

The synthesis of poliovirus did not require living cells. Subsequent to its chemical synthesis, the cDNA was transcribed in vitro into infectious viral RNA that, in turn, yielded infectious polio upon incubation in an extract of non-infected HeLa cells (Fig. 21.4). For the chemist, therefore, poliovirus is nothing more than a chemical. When the virus enters a cell, however, it has a program for survival. It will subvert cellular compartments and turn them into viral factories, in which it will proliferate subject to the evolutionary laws—heredity, genetic variation, selection toward

Table 21.1 Above data from Wetterstrand KA

Date	Cost per Mb	Cost per genome
Sep-01	$5,292.39	$95,263,072
Mar-02	$3,898.64	$70,175,437
Sep-02	$3,413.80	$61,448,422
Mar-03	$2,986.20	$53,751,684
Oct-03	$2,230.98	$40,157,554
Jan-04	$1,598.91	$28,780,376
Apr-04	$1,135.70	$20,442,576
Jul-04	$1,107.46	$19,934,346
Oct-04	$1,028.85	$18,519,312
Jan-05	$974.16	$17,534,970
Apr-05	$897.76	$16,159,699
Jul-05	$898.90	$16,180,224
Oct-05	$766.73	$13,801,124
Jan-06	$699.20	$12,585,659
Apr-06	$651.81	$11,732,535
Jul-06	$636.41	$11,455,315
Oct-06	$581.92	$10,474,556
Jan-07	$522.71	$9,408,739
Apr-07	$502.61	$9,047,003
Jul-07	$495.96	$8,927,342
Oct-07	$397.09	$7,147,571
Jan-08	$102.13	$3,063,820
Apr-08	$15.03	$1,352,982
Jul-08	$8.36	$752,080
Oct-08	$3.81	$342,502
Jan-09	$2.59	$232,735
Apr-09	$1.72	$154,714
Jul-09	$1.20	$108,065
Oct-09	$0.78	$70,333
Jan-10	$0.52	$46,774
Apr-10	$0.35	$31,512
Jul-10	$0.35	$31,125
Oct-10	$0.32	$29,092

(continued)

Table 21.1 (continued)

Date	Cost per Mb	Cost per genome
Jan-11	$0.23	$20,963
Apr-11	$0.19	$16,712
Jul-11	$0.12	$10,497
Oct-11	$0.09	$7,743
Jan-12	$0.09	$7,666
Apr-12	$0.07	$5,901
Jul-12	$0.07	$5,985
Oct-12	$0.07	$6,618
Jan-13	$0.06	$5,671
Apr-13	$0.06	$5,826

DNA sequencing costs: data from the NHGRI Genome Sequencing Program (GSP) Available at www.genome.gv/sequencingcosts. (Accessed 12 September 2013)

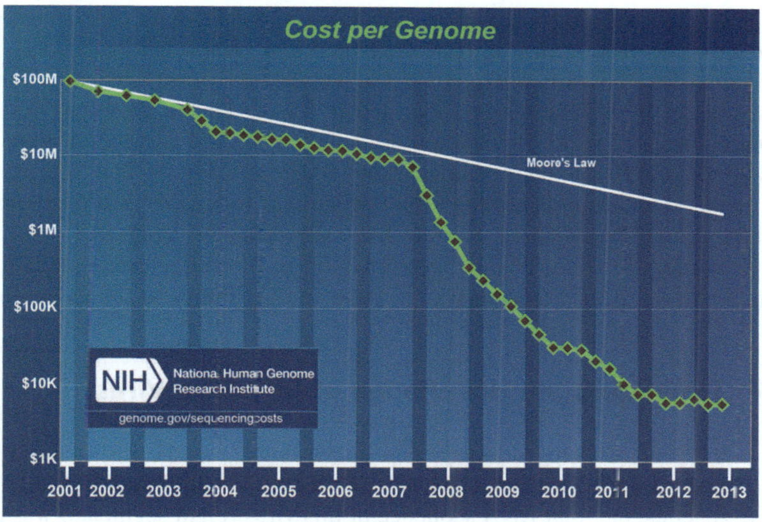

Fig. 21.3 Cost of sequencing genomic DNA over the years

fitness, evolution into different species and so on. That is, poliovirus obeys the same rules that apply to living entities. One could even argue that poliovirus has sex in the infected cell, as it readily recombines with sibling progeny or with related viruses should they co-infect the same cell. This fascinating dual nature of viruses as nonliving and living entities that is, an existence as chemicals with a life cycle—has been largely ignored in response to the chemical and biochemical synthesis of poliovirus, which was published in 2002 [11].

However, when this manufactured poliovirus was injected into the brains of mice containing the polio virus receptor, it was found to be greatly attenuated. When tested on human neuroblastoma cells, the virus had a temperature-sensitive phenotype. Specific changes had been made in sequences considered non-essential in the manufactured sequence to act as proof that this was the original reconstructed virus and not a wild-type contaminant. These changes were for identification purposes only. This attenuation was found to be due to a UA-to-GG exchange at nucleotides 102/103, mapping to a region between the cloverleaf-like structure and the internal ribosome entry site (IRES) in the 5′-nontranslated region of the viral genome, not previously considered important. Thus, this simple switch altered the characteristics of the virus, and led in fact to a virus with oncolytic potential, a fortuitous finding [15].

21.4 In Vitro Synthesis of Bacteriophage ϕX174

The first bacteriophage to be produced "artificially" was the bacteriophage ϕX174 [16]. This virus consists of a single-stranded circular DNA of 5,386 nucleotides and its in vitro synthesis was universally hailed as a great scientific success. It was produced in the laboratory of Craig Venter, who had previously announced the first full sequence of the human genome, and it was to be the prototype for larger genomes. Designing the oligonucleotides for the ϕX174 virus only took about eight hours, and the actual synthesis about four days. Figure 21.5 illustrates the steps required to manufacture the virus, which was tested for infectivity and was identical to wild-type. There was no discussion after the announcement of this virus construct of the possible dangers inherent in "manufacturing" such a virus. The main thrust of the research was to show the feasibility of the technology.

21.5 Expanding the Host Range of Influenza Virus

One of the most exciting developments in virology was the resurrection of the 1918 pandemic influenza virus. This is discussed in detail in Chap. 11. Using reverse genetics, the total RNA sequence of this virus (eight segments) was copied into cDNA and used to infect human cells, thereby recreating the deadly virus. Segments of this virus were mixed with segments from other influenza viruses and the virulence tested to pinpoint the modifications responsible for virulence. This work was criticized, since the total sequence of each gene was published, and the virus could thus be manufactured by anyone with skills in molecular biology. Since the 1918 strain itself was recreated, even with the safeguards of a high security laboratory, it is feasible that the virus could accidently escape from the laboratory and infect the human population.

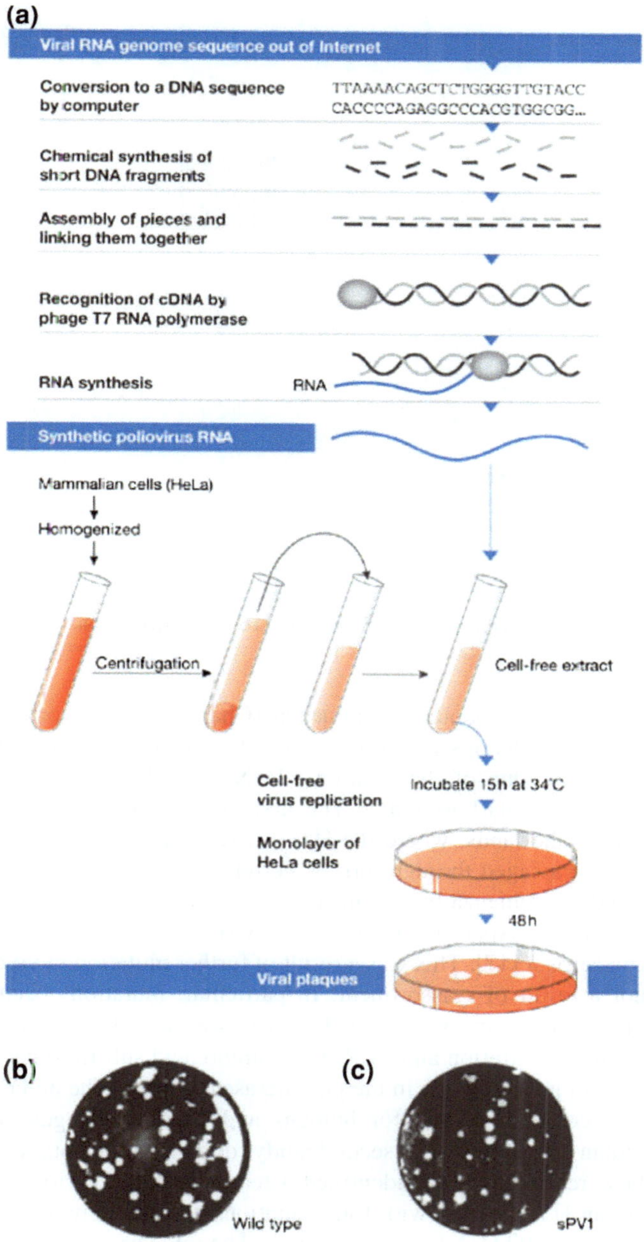

Fig. 21.4 Construction of polio virus in vitro (with permission of E. Wimmer)

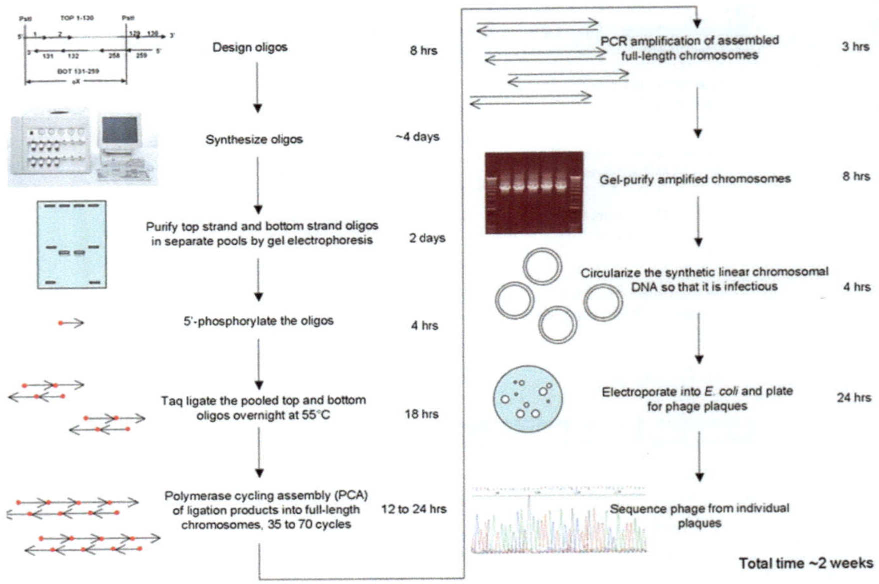

Fig. 21.5 Method used to construct ϕX174 in vitro [16] with permission

As discussed in the chapter on influenza (Chap. 11), the world is currently experiencing two pandemics of influenza infecting poultry, with sporadic transmission to humans. One of these strains, H5N1, has been manipulated in the laboratory to increase its host range. The first experiment of this type was performed in The Netherlands, where the H5N1 virus was mutated by site-directed mutagenesis and passaged through ferrets. Ferrets are the optimum model system and are infected by influenza in a manner similar to the way humans are infected. As a result of these experiments, strains of virus were isolated that could be transmitted by aerosol [17]. This was a result of further mutation in vivo (antigenic drift) and not a result of reassortment. In particular, mutations occurred in the polymerase gene, an unexpected result. The virus was not lethal for the ferrets, but could be detected by titration and PCR. Four amino acid substitutions occurred in the hemagglutinin gene and one in the polymerase complex. The authors point out that this virus could be a risk for humans and acquire changes allowing for human-to-human transmission. A second study, done simultaneously, had similar findings. These researchers [18] identified a reassorted H5N1 virus—comprising H5 HA (from an H5N1 virus) with four mutations and the remaining seven-gene segments from the 2009 H1N1 pandemic virus. The virus produced was capable of droplet transmission in a ferret model.

Both groups argued that such studies would define the changes that might occur to make this a pandemic virus for humans, and could be useful in studying and developing treatment in such a pandemic. As expected, this finding has been severely criticized. If such viruses escaped from the laboratory, they could be dangerous, and even though the virus was not lethal to the ferrets, it could be dangerous in the hands of terrorists or as a biological weapon. Many are questioning whether the research was necessary, and both papers were held up to scrutiny by the National Safety Board of the U.S. and Holland, where some of the work was performed. I personally am not convinced that these experiments were essential. It is true that mutations do occur in nature, as does reshuffling of the influenza genomic segments, but those performed in the laboratory were directed mutations, whereas in nature such events are random, and may not be selected in nature. The publication of these papers led to a storm of condemnations resulting in a moratorium on such research until security matters could be discussed and defined. The NIH issued guidelines under what was termed "dual use research of concern," now known as "DURC."

DURC was defined as "life sciences research that, based on current understanding, can be reasonably anticipated to provide knowledge, information, products, or technologies that could be directly misapplied to pose a significant threat with broad potential consequences to public health and safety, agricultural crops and other plants, animals, the environment, materiel, or national security" (DURC policy, here referred to as "dual use" policy). It includes six categories of potentially problematic scientific experiments:

1. Enhancing the harmful consequences of an agent or toxin
2. Disrupting immunity or the effectiveness of an immunization against an agent or toxin without clinical or agricultural justification
3. Conferring to an agent or toxin resistance to clinically or agriculturally useful prophylactic or therapeutic interventions against the agent or toxin, or facilitating its ability to evade detection methodologies
4. Increasing the stability or transmissibility of or the ability to disseminate an agent or toxin
5. Altering the host range or tropism of an agent or toxin
6. Enhancing the susceptibility of a host population to an agent or toxin

Moreover, the government has declared a list of pathogens that fall into this category:

"… review will focus on research that involves one or more of the agents or toxins listed below, which pose the greatest risk of deliberate misuse with most significant potential for mass casualties or devastating effects to the economy, critical infrastructure, or public confidence, and produces, aims to produce, or is reasonably anticipated to produce one or more of the effects listed above [my wording] Viruses are listed in bold."

1. Agents and toxins:
   (a) **Avian influenzavirus (highly pathogenic)**
   (b) *Bacillus anthracis*

(c) *Botulinum neurotoxin*
(d) *Burkholderia mallei*
(e) *Burkholderia pseudomallei*
(f) **Ebola virus**
(g) **Foot-and-mouth disease virus**
(h) *Francisella tularensis*
(i) **Marburg virus**
(j) **Reconstructed 1918 Influenza virus**
(k) **Rinderpest virus**
(l) Toxin-producing strains of *Clostridium botulinum*
(m) **Variola major virus**
(n) **Variola minor virus**
(o) *Yersinia pestis*

(The full copy of the report can be found at: http://oba.od.nih.gov/oba/biosecurity/PDF/United_States_Government_Policy_for_Oversight_of_DURC_FINAL_version_032812.pdf).

Scientists working on influenza imposed a moratorium on research leading to alterations in the host range of avian influenza. This moratorium continued for one year, but research has been renewed, taking into account the new regulations imposed by the U.S. government on such research. Thus we are entering a "new world" in virology, in which old viruses can be restored to life, or new ones created. It brings to mind the film "Jurassic Park," where the DNA from dinosaur eggs is cloned and "old" species are brought back to life. only to attack mankind. We may be facing a similar situation.

References

1. Cello, J., Paul, A. V., & Wimmer, E. (2002). Chemical synthesis of poliovirus cDNA: generation of infectious virus in the absence of natural template. *Science, 297*(5583), 1016–1018.
2. Edgar, R. S., & Lielausis, I. (1968). Some steps in the assembly of bacteriophage T4. *Journal of Molecular Biology, 32*(2), 263–276.
3. Weigle, J. (1966). Assembly of phage lambda in vitro. *Proceedings of the National Academy of Sciences of the United States of America, 55*(6), 1462–1466.
4. Putnak, J. R., & Phillips, B. A. (1981). Picornaviral structure and assembly. *Microbiological Reviews, 45*(2), 287–315.
5. Bancroft, J. B., & Hiebert, E. (1967). Formation of an infectious nucleoprotein from protein and nucleic acid isolated from a small spherical virus. *Virology, 32*(2), 354–356.
6. Bancroft, J. B., Hills, G. J., & Markham, R. (1967). A study of the self-assembly process in a small spherical virus. Formation of organized structures from protein subunits in vitro. *Virology, 31*(2), 354–379.
7. Fraenkel-Conrat, H., & Singer, B. (1957). Virus reconstitution. II. Combination of protein and nucleic acid from different strains. *Biochimica et Biophysica Acta, 24*(3), 540–548.
8. Taniguchi, T., Palmieri, M., & Weissmann, C. (1978). QB DNA-containing hybrid plasmids giving rise to QB phage formation in the bacterial host. *Nature, 274*(5668), 223–228.
9. Racaniello, V. R., & Baltimore, D. (1981). Cloned poliovirus complementary DNA is infectious in mammalian cells. *Science, 214*(4523), 916–919.

10. Racaniello, V. R., & Baltimore, D. (1981). Molecular cloning of poliovirus cDNA and determination of the complete nucleotide sequence of the viral genome. *Proc Natl Acad Sci USA, 78*(8), 4887–4891.
11. Wimmer, E., Mueller, S., Tumpey, T. M., & Taubenberger, J. K. (2009). Synthetic viruses: a new opportunity to understand and prevent viral disease. *Nature Biotechnology, 27*(12), 1163–1172.
12. Agarwal, K. L., Buchi, H., Caruthers, M. H., Gupta, N., Khorana, H. G., Kleppe, K., et al. (1970). Total synthesis of the gene for an alanine transfer ribonucleic acid from yeast. *Nature, 227*(5253), 27–34.
13. Caruthers, M. H., Kleppe, R., Kleppe, K., & Khorana, H. G. (1976). Total synthesis of the structural gene for the precursor of a tyrosine suppressor transfer RNA from Escherichia coli. 10. Enzymatic joining of chemically synthesized segments to form the DNA duplex corresponding to the nucleotide sequence 86-126. *Journal of Biological Chemistry, 251*(3), 658–666.
14. Khorana, H. G. (1979). Total synthesis of a gene. *Science, 203*(4381), 614–625.
15. De Jesus, N., Franco, D., Paul, A., Wimmer, E., & Cello, J. (2005). Mutation of a single conserved nucleotide between the cloverleaf and internal ribosome entry site attenuates poliovirus neurovirulence. *Journal of Virology, 79*(22), 14235–14243.
16. Smith, H. O., Hutchison, C. A, 3rd, Pfannkoch, C., & Venter, J. C. (2003). Generating a synthetic genome by whole genome assembly: phiX174 bacteriophage from synthetic oligonucleotides. *Proceedings of the National Academy of Sciences of the United States of America, 100*(26), 15440–15445.
17. Herfst, S., Schrauwen, E. J., Linster, M., Chutinimitkul, S., de Wit, E., Munster, V. J., et al. (2012). Airborne transmission of influenza A/H5N1 virus between ferrets. *Science, 336*(6088), 1534–1541.
18. Chutinimitkul, S., van Riel, D., Munster, V. J., van den Brand, J. M., Rimmelzwaan, G. F., Kuiken, T., et al. (2010). In vitro assessment of attachment pattern and replication efficiency of H5N1 influenza A viruses with altered receptor specificity. *Journal of Virology, 84*(13), 6825–6833.

Viruses as Biological Weapons 22

Abstract

Viruses as weapons of mass destruction have been in the forefront of the news following the use of chemical weapons in the Syrian conflict. Historically, smallpox may have been used in the past as a weapon, both by throwing corpses into the battlefield, or supplying American Indians with contaminated blankets. During World War II, the Japanese established a special unit, 731, to test bacterial agents as weapons in China, using humans as experimental animals. In 1972 and entered into force in 1975, 158 countries, including the U.S. and the Soviet Union, signed an agreement known as the "Biological and Toxins Weapons Convention," in which countries agreed to stop the development of offensive biological and chemical weapons. Unfortunately, there was no clause in the treaty for verification. In 1998, Ken Alibek, a defector from the Soviet Union, published a book, *Biohazard,* that unmasked wide segments of the Soviet and then Russian biological war program [1], which included the growth of smallpox and other viruses as offensive agents. Although the U.S had a biological warfare program, following the treaty, it shut down facilities for the production of such weapons. Today, despite the treaty, it is known that many countries are still working on and stockpiling biological weapons.

22.1 Introduction

During the last few years, the Department of Homeland Security has warned of the possibilities and dangers of biological (germ) terrorist attacks. These warnings increased after the September 11, 2001, attack on the World Trade Center and the Pentagon, and during the war in Iraq. It was assumed that the government of Sadam Hussein had stockpiles of biological and chemical weapons, coined as "weapons of mass destruction." Most of these were destroyed following UN inspection, and new ones were not discovered after the U.S. invasion of Iraq. Syria

M. W. Taylor, *Viruses and Man: A History of Interactions,*
DOI: 10.1007/978-3-319-07758-1_22,

also has stockpiles of chemical weapons that were used in the current civil war in that country in 2013; hopefully, these are currently being destroyed (2014). There are fears that terrorists can manufacture biological weapons from bacteria or viruses and use them in crowded cities, where the rate of spread and infection would be extremely high (see table below for aerosol distribution).

The most prominent weapon considered has been anthrax, not a virus but a deadly bacterium, engineered to be extremely pathogenic both to man and livestock. Unfortunately, deadly human viruses have also been adapted for use as weapons, although the use of bacteria or virus to destroy one's enemy is not new (see history below). There are a few documented cases where infectious diseases may have been deliberately introduced into a population. In the early days of the Cold War, both the U.S. and the Soviet Union clandestinely supported the production of bacteria and viruses for offensive purposes, although the U.S. apparently stopped this production in the early 1970s, whereas the Soviet Union continued with this development. Other countries, including Syria, Iran, China, and North Korea may have developed biological weapons.

22.2 History of Biological Warfare

Since the beginning of recorded history there have been many attempts to destroy the enemy by using infected bodies or natural toxins. There were deliberate efforts to poison wells and food supplies. The Assyrians, Spartans, Carthaginians and Persians were all reported to have used poisons or diseased bodies in their battles, although the evidence is unsubstantiated. It is rather difficult to prove, since there is no archaeological evidence, and one depends on historians of the time who might have their own biases. There is a report of poisonous snakes in ceramic vessels being thrown into ships during a battle between the Carthaginians under Hannibal and the Kingdom of Pergamon under King Eumenes II in 184 B.C. The sailors of Eumenes fled after realizing that this was not a joke.

During the Middle Ages and the period of the Black Death in Europe, decaying bodies were catapulted into enemy strongholds with the hope of spreading disease. The Tartars used this approach in the siege of the city of Caffa (modern-day Feodosia) in the Crimea. This event is described in detail by the fourteenth-century Genoese chronicler Gabriele de Mussi. In fact, he suggested that this was the origin of the Black Death that swept Europe in the Middle Ages. He wrote,

> "Oh God! See how the heathen Tartar races, pouring together from all sides, suddenly invested the city of Caffa and besieged the trapped Christians there for almost three years. There, hemmed in by an immense army, they could hardly draw breath, although food could be shipped in, which offered them some hope. But behold, the whole army was affected by a disease which overran the Tartars and killed thousands upon thousands every day. It was as though arrows were raining down from heaven to strike and crush the Tartars' arrogance (Table 22.1). All medical advice and attention was useless; the Tartars died as soon as the signs of disease appeared on their bodies: swellings in the armpit or groin caused by coagulating humours, followed by a putrid fever.

> The dying Tartars, stunned and stupefied by the immensity of the disaster brought about by the disease, and realizing that they had no hope of escape, lost interest in the siege. But they ordered corpses to be placed in catapults and lobbed into the city in the hope that the intolerable stench would kill everyone inside. What seemed like mountains of dead were thrown into the city, and the Christians could not hide or flee or escape from them, although they dumped as many of the bodies as they could in the sea. And soon the rotting corpses tainted the air and poisoned the water supply, and the stench was so overwhelming that hardly one in several thousand was in a position to flee the remains of the Tartar army. Moreover one infected man could carry the poison to others, and infect people and places with the disease by look alone. No one knew, or could discover, a means of defense.
>
> Thus almost everyone who had been in the East, or in the regions to the south and north, fell victim to sudden death after contracting this pestilential disease, as if struck by a lethal arrow which raised a tumor on their bodies. The scale of the mortality and the form which it took persuaded those who lived, weeping and lamenting, through the bitter events of 1346–1348—the Chinese, Indians, Persians, Medes, Kurds, Armenians, Cilicians, Georgians, Mesopotamians, Nubians, Ethiopians, Turks, Egyptians, Arabs, Saracens and Greeks (for almost all the East has been affected)—that the last judgement had come" [2, 3].

This may have been the first authenticated example of biological warfare.

The exception, of course, to bacterial agents was the smallpox virus. Lord Jeffrey Amherst was the commanding general of British forces in North America during the final battles of the French and Indian War (1754–1763), winning victories against the French to acquire Canada for England. Pontiac, an Ottawa chief who had sided with the French, led an uprising against the British after the French surrender in Canada. Indians were angered by Amherst's refusal to continue the French practice of providing supplies in exchange for Indian friendship and assistance and by an imperious British attitude toward them and their land. Colonel Henry Bouquet, an aide to General Amherst, suggested in the postscript of a letter to General Amherst on July 13, 1763, the distribution of blankets to "inoculate the Indians" with smallpox; Amherst, on July 16, 1763, approved this proposal, stating "to try every other method that can serve to Extirpate this Execrable Race." (Letters containing this information can be found at the United States Library of Congress, and the archives are on microfilm.) These letters between Amherst and Bouquet and others indicate an intense hatred for the Indians by Amherst, and the suggestions many times of eradicating the Indians by using smallpox. This did not extend to tactics against the French, whom he admired (http://www.nativeweb.org/pages/legal/amherst/34_40_305_fn.jpeg).

Following is a copy of the correspondence between Colonel Bouquet and General Amherst:

> P.S. I will try to inocculate the Indians by means of Blankets that may fall in their hands, taking care however not to get the disease myself. As it is pity to oppose good men against them, I wish we could make use of the Spaniard's Method, and hunt them with English Dogs. Supported by Rangers, and some Light Horse, who would I think effectively extirpate or remove that Vermine.

On July 16 Amherst replied, also in a postscript:

> P.S. You will Do well to try to Innoculate the Indians by means of Blanketts, as well as to try Every other method that can serve to Extirpate this Execrable Race. I should be very glad your Scheme for Hunting them Down by Dogs could take Effect, but England is at too great a Distance to think of that at present.

On July 26 Bouquet wrote back:

I received yesterday your Excellency's letters of 16th with their Inclosures. The signal for Indian Messengers, and all your directions will be observed.

There was an outbreak of smallpox at Fort Pitts and we assume that the orders were carried out, since an epidemic of smallpox did occur among the Indian tribes the following spring (or months). However, we have no idea whether in fact these two events are connected.

22.3 Twentieth-Century Horrors

During the twentieth century the Japanese used various infectious species of bacteria (cholera, anthrax, pestis) in the war against China [4]. A unique unit ("Unit 731") developed new weapons for germ warfare, and performed human experimentation on a large scale. According to the 2002 International Symposium on the Crimes of Bacteriological Warfare, the number of people killed in germ warfare and human experiment programs by the Japanese Army was around 580,000. According to other sources, the use of biological weapons researched in Unit 731's bio-weapons and chemical weapons programs resulted in possibly as many as 200,000 deaths of military personnel and civilians in China. The cruelty involved in these experiments in indescribable and ranks among the worst atrocities in their extent and methods outside of the Holocaust. (For more details, see http://en.wikipedia.org/wiki/Unit_731. The cruelty described in this article is hard to believe.)

Biological weapons are easy and cheap to produce. The U,S. and other major nations have developed their own weapons of mass destruction, particularly atomic weapons. However, these have a distinct disadvantage in that in order to deliver them, one needs sophisticated bombs or missiles. Biological weapons can be as simple as an infected person ("suicide bomber"), or a vial containing infectious material. It is even possible to have small bombs that on explosion will deliver anthrax or smallpox or other viruses into the air which can either be inhaled, picked up on clothing, or by bodily contact. Chemical weapons such as poisonous gas, may have the same instant effect, but do not spread beyond the area in which they are used. Smallpox or influenza in the trenches is much more effective in the long run than mustard gas. However, there is the danger in a war situation that it might get back to the troops using this to start out with. This may have happened with the use of the tularemia bacteria by the Russians during World War II at Stalingrad [1]. However, some doubt has been cast on this by the U.S. Office of Chemical and Biological Weapons (CBW).

The ease of travel around the globe makes it easy to transport biological weapons. They are small and easy to hide in luggage or even on the body (infected individuals). The urbanization of the world with cities of millions of teeming inhabitants are ideal targets for the terrorist or even the nation wanting to inflict maximum casualties. Viruses can be grown in small laboratories by anyone with limited knowledge of cell culture, and it may be possible to purchase virus stocks through a third party, ostensibly for research purposes. This apparently was the approach taken by the Soviet Union in the 1970s–1980s. Until very recently it was possible to purchase infective agents for research from the American Type Culture Collection; they could be purchased by commercial laboratories and then transferred to the military for development. Today, after the attack on the World Trade Center and Pentagon on September 11, 2001, this is more difficult. However, a country or group wishing to develop viral weapons can establish a molecular biology laboratory and copy genetic information to produce viruses such as poliovirus or other small viruses in vitro (see discussion on creating viruses in the test tube Chap. 21). Following the collapse of the Soviet Union, large numbers of scientists working on the Russian germ war establishment found themselves unemployed. These scientists were targeted by countries such as Iraq and Iran, countries with the financial ability to hire such scientists to continue the work they were doing in the Soviet Union. This is similar to events after World War II, when German scientists were hired by the U.K. and the U.S. to continue some research into nuclear weapons and rocket technology, or, in the case of the Japanese, scientists involved in biological warfare who were not prosecuted after World War II. In fact, the Russians used Japanese blueprints from Unit 731 to establish their clandestine program, even down to using the building plans for the establishment of new centers. One thing that terrorists or rogue states have going for them is an increased public awareness of events through the Internet and other media. This in turn could induce mass panic and enhance the effects of such weapons—for example, by crowding through panic, or alternatively by fleeing and thus transmitting the disease further afield.

In 1972, today 160 countries—including the U.S. and the Soviet Union—signed an agreement known as the "Biological and Toxins Weapons Convention" in which they agreed to stop the development of offensive biological and chemical weapons. This treaty was supposed to go into effect in 1975, but, there were no details on the means of verifying that countries lived up to the agreement. This was missing from the pact and led to several nations clandestinely developing such weapons [5].

Historically, the Geneva Convention of 1925, following World War I, in which the Germans used poisonous gas, banned the use of such weapons. It banned the use of such chemical weapons and also expanded it to biological weapons. During World War II, the U.S. initiated a program to develop biological weapons, which led to the production of anthrax bombs; this was to be used in retaliation for Japanese or Germans' use of such weapons. They were never used, and instead the war against the Japanese ended with the dropping of two atomic bombs on Japan, which was more destructive or just as destructive as the biological weapons. The U.S. in fact continued its program on biological weapons development at Fort Dietrich until 1969 when President Nixon ordered the cessation of this program and the destruction of existing stockpiles.

22.4 Weapons Development in the Soviet Union

Despite being a signatory to the pact, the Soviet Union continued with its program of developing biological weapons on a massive scale. Defectors from the Soviet Union have provided information on a program that employed as many as 50,000 workers in the development of antibiotic-resistant strains of *Yersinia pestis*, development of *tons* of smallpox virus, and two senior Russian microbiologists, Dr. Vladimir Pasechnik and Dr. Ken Alibek (Alibekov), defected to the U.K. and U.S. and gave information on the Soviet program. The former served as director of the Institute of Ultra Pure Biochemical Preparations in Leningrad, and the latter was First Deputy Chief of the Biopreparat System, a large network of biological weapons-oriented facilities (Pasechnik's institute being among them). Moreover, in 1998, Alibek published *Biohazard,* a book that exposed wide segments of the Soviet and then-Russian biological war program [1]. Orders were given by the President of Russia, Leonid Brezhnev, to develop various systems for military use. Factories for the mass production of anthrax, *Yersinia*, and smallpox in large amounts (enough to wipe out humanity many times over) were set up in many locations, and the organisms placed into warheads of ballistic missiles for use against the U.S. Alibek describes in frightening detail the construction and running of such facilities, as well as the hazards of working in them. This work was undertaken because of fear that the U.S. would eventually try to destroy the Soviet Union, and the program continued and expanded even in the time of Gorbachev and *perestroika*.

After a hiatus of many years because of the influence on genetics by Trofim Lysenko, who believed environmental factors to be more important than genetic ones, in the 1970s Russian scientists turned their attention to molecular biology as it was being developed in the West. This meant that such scientists learned how to construct recombinant molecules and to splice and reattach different genes. These techniques were used in developing new strains of organisms, much more virulent than the original "wild-type" strain, and in many cases resistant to antibiotics (in the case of bacteria), and virus strains resistant to commonly used vaccines. At it height, Biopreparat, as this division of the military wing of the Soviet Union was called, employed over 30,000 people, plus about 15,000 scientists, and 10,000 more in the development of agricultural toxins that could wipe out plants or livestock. The whole massive program was carried out under complete secrecy and was unknown to anyone outside the Russian military and Communist leadership. Many of these facilities have been dismantled but it is believed that some are still functioning in secret.

Although smallpox was officially eliminated as a plague on earth, and only two vials were retained for research purposes, the Soviet Union took advantage of this situation and developed even more pathogenic strains, collecting them from an outbreak in India. Although the initiative to eliminate smallpox worldwide came in 1958 from Viktor Zdhanov, a Russian virologist, he headed the biological warfare program for many years, beginning in 1973. Was his intention to eliminate the disease, but at the same time develop smallpox as a weapon that could then be used on a non-immune population? The Soviet scientists learned how to produce these organisms on an industrial scale. Although they introduced stringent conditions of sterility and safety, occasional accidents did happen, and organisms escaped into the environment. One case was the escape of anthrax spores in Sverdlovsk, resulting in the deaths of a great number of people (the exact number is not known), since this was kept secret by the KGB, who blamed the outbreak on contaminated meat sold on the black market. In the late 1980s, Biopreparat began work on even more deadly viruses, such as Marburg and Ebola. When a scientist contracted Marburg by accident and died, the virus that was removed from his body at the autopsy was shown to be an even more pathogenic strain and was put into production. Enough virus was produced to kill everyone on earth. These viruses were put into small bombs that could easily be dropped on populated areas or sprayed from planes. Following the collapse of the Soviet Union, American virologists were allowed to visit the "secret" factories, and even participated in dismantling some of the huge facilities for growing tons of dangerous organisms. However, there is still suspicion that the Russian military has not disclosed the location of all its factories and may still be working on these abhorrent strains of bacteria and virus, perhaps even with funds obtained from the U.S.

Alibek's book makes it clear that most Russian scientists working in this area believed that the U.S. was pursuing biological weapons with the aim of destroying the Soviet Union. It was only after visiting the U.S., and seeing for themselves that such facilities did not exist, or if they did, were not in use, that it became clear that the U.S. did not have a program similar to the one in the Soviet Union. Soviet scientists were brainwashed by their government, and also received the kinds of perks for working in this area of virology that they could never have received working for civilian outfits. The arsenal of weapons developed by the Soviet Union, and possibly other countries include bacteria such as anthrax, tularemia, glanders (a bacterial disease of horses, but also lethal to humans) and typhus; and viral diseases such as smallpox, Venezuela equine encephalitis, dengue fever, Lassa fever, Marburg, Ebola, Bolivian hemorrhagic fever and other hemorrhage-causing agents.

There is no evidence that the U.S. continued it biological warfare program after President Nixon ordered its cessation; they did have such a program during World War II and for a short time afterwards, but compared to the Soviet program, it was a small one and was mostly devoted to the development of anthrax and tularemia as weapons. The American program was similar to the Soviet program but disbanded by President Nixon. The parallels between these programs suggest that Soviet spies transferred information to the Soviet military. It is known that many other countries are involved in developing biological weapons; these include Iran, China, India, and possibly North Korea. Material is readily available, and it is very cheap to develop such weapons. The knowledge of how to reconstruct antibiotic-resistant bacteria or more pathogenic viruses can be found in most molecular biology journals, textbooks or on the Internet.

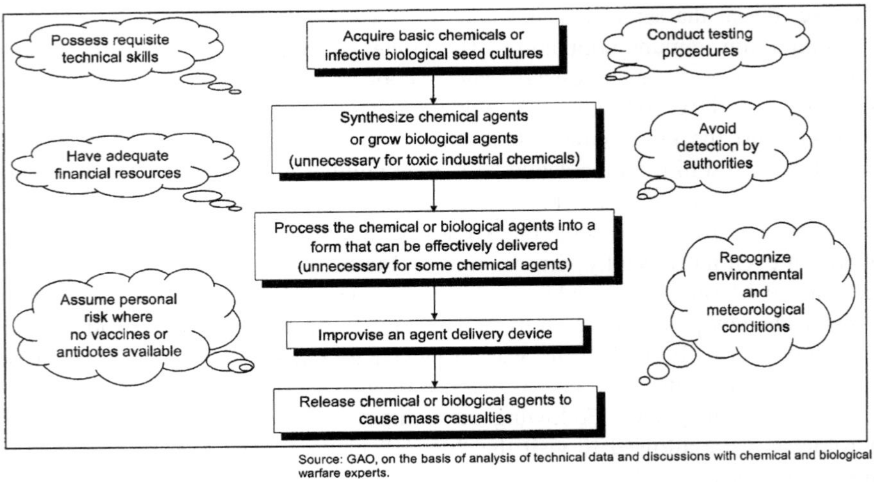

Fig. 22.1 Stages and obstacles for chemical and biological terrorism. *Sources* GAO, on the basis of analysis of technical data and discussions with chemical and biological warfare experts

To summarize: In order to produce biological weapons, one needs to develop pathogenic strains of an organism that can overcome natural and induced immunity, This was done with anthrax and smallpox. For weapons-grade material, one has to develop the means to grow large quantities in large vats under ideal conditions (the Russians have produced large amounts of material). The material has to be dried and maintained in an active state, which is difficult with most viruses, or, in the case of anthrax, maintained as spores so that it can be spread as a fine aerosol (1–10 microns). Of course these agents have to be handled with extreme care, and under circumstances where escape is impossible. This requires facilities with many layers of walls and doors, as well as separate air-handling systems. These conditions are shown in cartoon form in Fig. 22.1.

The American government and military, until the late 1990s and turn of this century, did not believe that biological weapons were a danger or that they would work in the battlefield, which may be one of the reasons for the U.S. abandoning its program in the 1970s. It was thought that atomic weapons were a much better defense against attack. However, with the rise of terrorism, this attitude has changed, and it is realized that even unsophisticated terrorists can manufacture and set off biological weapons easily in crowded malls or subway systems. One of the major problems facing any potential target is the insidiousness of the attack, since initially few people may fall ill, and it may not be recognized as the result of a deliberate attack. For example, most physicians in the West have never seen a case of smallpox, so initially it may be misdiagnosed and transmitted to thousands of others before it is realized that it is the result of an attack. A similar scenario is true of most of the other agents. It thus behooves all countries facing this danger to stockpile vaccines and supplies of antibiotics in case of such an event. However,

Table 22.1 Effect of the release of 50 kg of material upwind of a city (population) of 500,000 along a 2 km line

Agent	Up wind reach	Number dead	Number incapacitated
Rift Valley fever	1 km	400	35,000
Tick-borne encephalitis	1 km	9.500	35,000
Typhus	5	19,000	85,000
Brucellosis	10	500	125,000
Q fever	>20	150	125,000
Tuleremia	>20	30,000	125,000
Anthrax	>20	95,000	125,000

Adapted from [6] (http://www.globalsecurity.org/wmd/world/japan/bw.htm)

Notes Table 22.1: Brucellosis is a bacterial disease usually transmitted by animals and found in uncooked meat; can be treated with antibiotics

Q fever is also a bacterial infection, *Coxiella`burnetii,* found also among animals. Is severe in humans (see http://www.cdc.gov/qfever/)

even then, it may take time to realize that this is a biological attack, and in many instances vaccines and antibiotics will arrive too late to suppress the agent in the initial population. Table 22.1 estimates the number of deaths and number incapacitated if there were an attack in an urban setting. The only way to stop such an attack is vigilance and tight control over countries and inspection of their facilities, as was carried out in the case of Iraq in the 1990s.

Agents listed by the CDC as possible biological weapons:

1. Smallpox
2. Anthrax
3. *Yersinia pestis* (Black Death plague)
4. *Francisella tularensis* (tularemia)
5. Filo viruses (Ebola)
6. Arena viruses (Lassa Fever and other hemorrhagic viruses)

References

1. Alibek, K., & Handelman, S. (1999). *Biohazard.* New York: Random House.
2. Wheelis, M. (2002). Biological warfare at the 1346 siege of Caffa. *Emerging Infectious Diseases, 8*(9), 971–975.
3. Derbes, V. J. (1966). De Mussis and the great plague of 1348. A forgotten episode of bacteriological warfare. *JAMA, 196*(1), 59–62.
4. Barenblatt, D. (2004). *A plague upon humanity.* New York: Harper Collins.
5. Kadlec, R. P., Zelicoff, A. P., & Vrtis, A. M. (1997). Biological weapons control. Prospects and implications for the future. *JAMA, 278*(5), 351–356.
6. Christopher, G. W., Cieslak, T. J., Pavlin, J. A., & Eitzen, E. M, Jr. (1997). Biological warfare. A historical perspective. *JAMA, 278*(5), 412–417.

Index

M. W. Taylor, *Viruses and Man: A History of Interactions*,
DOI: 10.1007/978-3-319-07758-1,

MIX
Papier aus verantwortungsvollen Quellen
Paper from responsible sources
FSC® C105338

If you have any concerns about our products,
you can contact us on
ProductSafety@springernature.com

In case Publisher is established outside the EU,
the EU authorized representative is:
Springer Nature Customer Service Center GmbH
Europaplatz 3, 69115 Heidelberg, Germany

Printed by Libri Plureos GmbH
in Hamburg, Germany